AF581630

12th EASN International Conference on Innovation in Aviation & Space for Opening New Horizons

12th EASN International Conference on Innovation in Aviation & Space for Opening New Horizons

Editors

Spiros Pantelakis
Andreas Strohmayer
Jordi Pons-Prats

MDPI • Basel • Beijing • Wuhan • Barcelona • Belgrade • Manchester • Tokyo • Cluj • Tianjin

Editors
Spiros Pantelakis
University of Patras
Patras, Greece

Andreas Strohmayer
University of Stuttgart
Stuttgart, Germany

Jordi Pons-Prats
Universitat Politècnica de Catalunya
Castelldefels, Spain

Editorial Office
MDPI
St. Alban-Anlage 66
4052 Basel, Switzerland

This is a reprint of articles from the Special Issue published online in the open access journal *Aerospace* (ISSN 2226-4310) (available at: https://www.mdpi.com/journal/aerospace/special_issues/12th_EASN).

For citation purposes, cite each article independently as indicated on the article page online and as indicated below:

LastName, A.A.; LastName, B.B.; LastName, C.C. Article Title. *Journal Name* **Year**, *Volume Number*, Page Range.

ISBN 978-3-0365-8370-9 (Hbk)
ISBN 978-3-0365-8371-6 (PDF)

Contents

About the Editors

Spiros Pantelakis

Spiros Pantelakis is Prof. Emeritus, University of Patras. He has been Director of the Laboratory of Technology and Strength of Materials (2007–2020), Chairman of the Department of Mechanical Engineering and Aeronautics (2009–2013), and Vice President of the Board of Executives of the Research Committee of the University of Patras (2010–2013). He is a founding member of the European Aeronautics Science Network (EASN) Association since its establishment in 2008 and served as its Chairman up to 2019. In 2019, he was awarded the title of Honorary Chairman of the EASN Association. He has been Representative of the European aeronautics' academia in the Plenary of ACARE and Chairman of ACARE's Working Group on Human Resources. He has been a member of the Clean Sky Academy since its establishment in 2015. He participated in the authors' team of the Clean Aviation Partnership Program and co-chaired the group, dealing with the exploration and maturation of technologies. He has been Member of the Board of Directors of Hellenic Aerospace Industry (2015–2020) and is a former Chairman of the Board of Directors of the Greek Metallurgical Society (2008–2011). He has about 45 years' experience in the field of materials and structures and has been involved in more than 100 international aeronautics-related research projects. He is the author of several scientific books, chapters in international textbooks, and more than 250 scientific publications in peer-reviewed international journals and conference proceedings, and he has been the supervisor of more than 20 PhD Theses. He is chairman of about 30 International Scientific Conferences and organizer of several decades of International Workshops.

Andreas Strohmayer

Andreas Strohmayer is a professor of aircraft design at the Institute of Aircraft Design, University of Stuttgart, with a research focus on manned (hybrid) electric flight ("Icaré 2" and "e-Genius") and scaled UAS flight testing. He studied Aeronautical Engineering at TU Munich and graduated in 2001 under Dr. Ing in the field of conceptual aircraft design. From 2002 to 2008, he was director of Grob Aerospace in Mindelheim, Germany, responsible for the design, production, and support of the Grob fleet of all-composite aircraft: specifically, the development of a four-seat aerobatic turboprop, a seven-seat turboprop, and the SPn business jet. From 2009 to 2013, he was program director for a 19-seater commuter aircraft project at Sky Aircraft in Metz, France, and then VP Programs at SST Flugtechnik in Memmingen, Germany, setting up an EASA-approved design organization holding the TC for a six-seater all-composite turbo-prop aircraft. He then left the industry to join the University of Stuttgart in 2015, teaching aircraft design and promoting electric flight. In 2016, he became a member of the Board of Directors of the European Aeronautics Science Network (EASN); since 2019, he has been an EASN Chairman. In this position, he also represents academia in the ACARE General Assembly.

Jordi Pons-Prats

Dr. Jordi Pons-Prats is a Mechanical Engineer, PhD in Aerospace Science and Technology by Polytechnical University of Catalonia (UPC, BarcelonaTech). He is lecturer at UPC since 2020. Before he spent 12 years at CIMNE (International Center for Numerical Methods in Engineering), a research center of UPC. From 2008 he is the responsible of the aerospace engineering group of CIMNE. He also had previous experience as project manager at industrial companies. His research career focused on the field of optimization, robust design and stochastic analysis, while with the integration into UPC it moved to Air Transport, Air Operations and Urban Air Mobility. He has published 21 scientific papers in JCR journals, among other publications like 10 book chapters and 1 conference proceedings. He has more than 250 citations and a h-index of 7 (Source: Web of Knowledge). He is the coordinator of the Iberia East Pilot Center of ERCOFTAC (www.ercoftac.org), and the Spanish National Contact Point and member of the Board of Directors of EASN (www.easn.net).

Preface to "12th EASN International Conference on Innovation in Aviation & Space for Opening New Horizons"

This Special Issue features selected papers presented at the 12th EASN International Conference on "Innovation in Aviation & Space to the Satisfaction of the European Citizens" (https://easnconference.eu/2022/home), which took place successfully from October 18th to 21st, 2022, in Barcelona. The conference included 8 keynote lectures delivered by distinguished personalities from the European Aviation & Space Community, along with 395 technical presentations spread across 74 virtual sessions. Notably, 85 Aviation and Space projects showcased their latest research findings and future trends in their respective technological domains during the conference. The 12th EASN International Conference garnered considerable interest and active engagement, attracting a remarkable turnout of over 470 participants representing 39 countries from around the world.

Out of the numerous submissions received during the 12th EASN Virtual Conference, a total of 20 papers have been selected for publication in this Special Issue.

Focusing on the theme of sustainable aviation and infrastructure, in [1], promising technologies, such as a wingtip propeller and electric green taxiing, are discussed, and their potential impacts on the future of aviation are highlighted. In [2], The development and implementation of the GreAT (Greener Air Traffic Operations) concept within the DLR realtime simulation environment ATMOS is realised. In [3], the results of an energy demand analysis for a future regional airport over three different time horizons are presented.

Focusing on safety and performance analysis, in [4], a methodology has been developed in order to optimize the structure of the parachute harness with the purpose of lightening it. [5] aims to establish capacity models that consider the neurophysiological variables recorded during the development of ad hoc real-time simulation exercises designed for the CRITERIA project. In [6], the impact of nanosecond pulse discharge NSPD on enhancing the flame propagation of CH4/H2/air mixture under ambient temperature and pressure is investigated.

With regard to aviation technologies and innovations, [7] presents the extension of an approach controller support system for diversions around severe weather areas. Within [8], a procedure is developed which enables an efficient design of interface-modified CFRP under impact loads. The aim of [9] is to simultaneously improve the duct's pressure loss and flow distortion under three flight conditions: nominal cruise, low-altitude climbing, and high-altitude cruise.

Focusing on aircraft design and materials, [10] describes the definition and application of the technology evaluation stage of the Advanced Morphological Approach AMA design process based on expert workshops. In [11], different mixtures of both milled and unmilled Cp-Ti grade 2 powder were utilized using the PM method, aiming to synthesize samples with high mechanical properties comparable to those of high-strength alloys. Within [12], the authors focus on the results of the preliminary activities performed within the CleanSky 2/Astib research program, dedicated to the definition of the iron bird of a new regional-transport aircraft. Within [13], a volume coefficient-based vertical tail plane sizing is compared to handbook methods and the possibility to reduce the necessary vertical stabilizer size is assessed with regard to the position of the engine integration and their interconnection.

With regard to advanced technologies and applications, [14] develops a methodology in which the final result is a machine learning model that allows predicting capacity regulations. In [15], a set of three different metaheuristic search algorithms have been considered for failure detection and identification purposes. [16] describes the multidisciplinary design of an aerodynamic drag device used to allow a passive re-entry, i.e., avoiding retropropulsion, of a reusable launch vehicles' first stage. In [17], the robustness of a hybrid MCDM tool to support the selection of sustainable materials in aviation has been assessed. In [18], a rapid and robust sizing methodology is presented, along with a comparative study on the impact of energy source configurations, on the autonomy of a multirotor aerial vehicle. The authors of [19] deal with developing a self-healing resin designed for aeronautical and aerospace applications. Finally, [20] demonstrates the competitive properties of a manufactured part, produced using laser powder bed fusion (LPBF), using a specific build orientation.

The editors of this Special Issue extend their appreciation to the authors for their excellent contributions and for greatly assisting in managing this Special Issue. Additionally, the editors would like to express their gratitude to Ms. Linghua Ding and the Aerospace editorial team for their valuable professional support.

References

[1] Eissele, J.; Lafer, S.; Mejía Burbano, C.; Schließus, J.; Wiedmann, T.; Mangold, J.; Strohmayer, A. Hydrogen-Powered Aviation—Design of a Hybrid-Electric Regional Aircraft for Entry into Service in 2040. *Aerospace* **2023**, *10*, 277. https://doi.org/10.3390/aerospace10030277

[2] Ahrenhold, N.; Stasicka, I.; Abdellaoui, R.; Mühlhausen, T.; Temme, M.-M. Enabling Green Approaches by FMS-AMAN Coordination. *Aerospace* **2023**, *10*, 278. https://doi.org/10.3390/aerospace10030278

[3] Meindl, M.; de Ruiter, C.; Marciello, V.; Stasio, M.D.; Hilpert, F.; Ruocco, M.; Nicolosi, F.; Thonemann, N.; Saavedra-Rubio, K.; Locqueville, L.; et al. Decarbonised Future Regional Airport Infrastructure. *Aerospace* **2023**, *10*, 283. https://doi.org/10.3390/aerospace10030283

[4] Grim, R.; Popela, R.; Jebáček, I.; Horák, M.; Šplíchal, J. Determination of the Parachute Harness Critical Load Based on Load Distribution into Individual Straps with Respect of the Skydiver's Body Position. *Aerospace* **2023**, *10*, 83. https://doi.org/10.3390/aerospace10010083

[5] Zamarreño Suárez, M.; Arnaldo Valdés, R.M.; Pérez Moreno, F.; Delgado-Aguilera Jurado, R.; López de Frutos, P.M.; Gómez Comendador, V.F. Methodology for Determining the Event-Based Taskload of an Air Traffic Controller Using Real-Time Simulations. *Aerospace* **2023**, *10*, 97. https://doi.org/10.3390/aerospace10020097

[6] Mehdi, G.; De Giorgi, M.G.; Bonuso, S.; Shah, Z.A.; Cinieri, G.; Ficarella, A. Comparative Analysis of Flame Propagation and Flammability Limits of CH4/H2/Air Mixture with or without Nanosecond Plasma Discharges. *Aerospace* **2023**, *10*, 224. https://doi.org/10.3390/aerospace10030224

[7] Temme, M.-M.; Gluchshenko, O.; Nöhren, L.; Kleinert, M.; Ohneiser, O.; Muth, K.; Ehr, H.; Groß, N.; Temme, A.; Lagasio, M.; et al. Innovative Integration of Severe Weather Forecasts into an Extended Arrival Manager. *Aerospace* **2023**, *10*, 210. https://doi.org/10.3390/aerospace10030210

[8] Kuhtz, M.; Richter, J.; Wiegand, J.; Langkamp, A.; Hornig, A.; Gude, M. Concepts for Increased Energy Dissipation in CFRP Composites Subjected to Impact Loading Conditions by Optimising Interlaminar Properties. *Aerospace* **2023**, *10*, 248. https://doi.org/10.3390/aerospace10030248

[9] Drężek, P.S.; Kubacki, S.; Żółtak, J. Kriging-Based Framework Applied to a Multi-Point, Multi-Objective Engine Air-Intake Duct Aerodynamic Optimization Problem. *Aerospace* **2023**, *10*, 266. https://doi.org/10.3390/aerospace10030266

[10] Todorov, V.T.; Rakov, D.; Bardenhagen, A. Structured Expert Judgment Elicitation in Conceptual Aircraft Design. *Aerospace* **2023**, *10*, 287. https://doi.org/10.3390/aerospace10030287

[11] Miko, T.; Petho, D.; Gergely, G.; Markatos, D.; Gacsi, Z. A Novel Process to Produce Ti Parts from Powder Metallurgy with Advanced Properties for Aeronautical Applications. *Aerospace* **2023**, *10*, 332. https://doi.org/10.3390/aerospace10040332
[12] Bertolino, A.C.; De Martin, A.; Jacazio, G.; Sorli, M. Design and Preliminary Performance Assessment of a PHM System for Electromechanical Flight Control Actuators. *Aerospace* **2023**, *10*, 335. https://doi.org/10.3390/aerospace10040335
[13] Albrecht, A.; Bender, A.; Strathoff, P.; Zumegen, C.; Stumpf, E.; Strohmayer, A. Influence of Electric Wing Tip Propulsion on the Sizing of the Vertical Stabilizer and Rudder in Preliminary Aircraft Design. *Aerospace* **2023**, *10*, 395. https://doi.org/10.3390/aerospace10050395
[14] Pérez Moreno, F.; Gómez Comendador, V.F.; Delgado-Aguilera Jurado, R.; Zamarreño Suárez, M.; Arnaldo Valdés, R.M. Prediction of Capacity Regulations in Airspace Based on Timing and Air Traffic Situation. *Aerospace* **2023**, *10*, 291. https://doi.org/10.3390/aerospace10030291
[15] Baldo, L.; Querques, I.; Dalla Vedova, M.D.L.; Maggiore, P. A Model-Based Prognostic Framework for Electromechanical Actuators Based on Metaheuristic Algorithms. *Aerospace* **2023**, *10*, 293. https://doi.org/10.3390/aerospace10030293
[16] Piacquadio, S.; Pridöhl, D.; Henkel, N.; Bergström, R.; Zamprotta, A.; Dafnis, A.; Schröder, K.-U. Comprehensive Comparison of Different Integrated Thermal Protection Systems with Ablative Materials for Load-Bearing Components of Reusable Launch Vehicles. *Aerospace* **2023**, *10*, 319. https://doi.org/10.3390/aerospace10030319
[17] Markatos, D.N.; Malefaki, S.; Pantelakis, S.G. Sensitivity Analysis of a Hybrid MCDM Model for Sustainability Assessment—An Example from the Aviation Industry. *Aerospace* **2023**, *10*, 385. https://doi.org/10.3390/aerospace10040385
[18] Chahba, S.; Sehab, R.; Morel, C.; Krebs, G.; Akrad, A. Fast Sizing Methodology and Assessment of Energy Storage Configuration on the Flight Time of a Multirotor Aerial Vehicle. *Aerospace* **2023**, *10*, 425. https://doi.org/10.3390/aerospace10050425
[19] Vertuccio, L.; Calabrese, E.; Raimondo, M.; Catauro, M.; Sorrentino, A.; Naddeo, C.; Longo, R.; Guadagno, L. Effect of Temperature on the Functionalization Process of Structural Self-Healing Epoxy Resin. *Aerospace* **2023**, *10*, 476. https://doi.org/10.3390/aerospace10050476
[20] Kovacs, S.E.; Miko, T.; Troiani, E.; Markatos, D.; Petho, D.; Gergely, G.; Varga, L.; Gacsi, Z. Additive Manufacturing of 17-4PH Alloy: Tailoring the Printing Orientation for Enhanced Aerospace Application Performance. *Aerospace* **2023**, *10*, 619. https://doi.org/10.3390/aerospace10070619

Spiros Pantelakis, Andreas Strohmayer, and Jordi Pons-Prats
Editors

Article

Hydrogen-Powered Aviation—Design of a Hybrid-Electric Regional Aircraft for Entry into Service in 2040

Jona Eissele [1,†], Stefan Lafer [1,*,†], Cristian Mejía Burbano [1,†], Julian Schließus [1,†], Tristan Wiedmann [1,†], Jonas Mangold [2] and Andreas Strohmayer [2]

1 University of Stuttgart, Pfaffenwaldring, 70569 Stuttgart, Germany
2 Institute of Aircraft Design, University of Stuttgart, Pfaffenwaldring 31, 70569 Stuttgart, Germany
* Correspondence: stefan@lafer.biz
† These authors contributed equally to this work.

Abstract: Over the past few years, the rapid growth of air traffic and the associated increase in emissions have created a need for sustainable aviation. Motivated by these challenges, this paper explores how a 50-passenger regional aircraft can be hybridized to fly with the lowest possible emissions in 2040. In particular, the use of liquid hydrogen in this aircraft is an innovative power source that promises to reduce CO_2 and NO_x emissions to zero. Combined with a fuel-cell system, the energy obtained from the liquid hydrogen can be used efficiently. To realize a feasible concept in the near future considering the aspects of performance and security, the system must be hybridized. In terms of maximized aircraft sustainability, this paper analyses the flight phases and ground phases, resulting in an aircraft design with a significant reduction in operating costs. Promising technologies, such as a wingtip propeller and electric green taxiing, are discussed in this paper, and their potential impacts on the future of aviation are highlighted. In essence, the hybridization of regional aircraft is promising and feasible by 2040; however, more research is needed in the areas of fuel-cell technology, thermal management and hydrogen production and storage.

Keywords: FutPrInt50; sustainable aviation; zero emissions; hybrid-electric propulsion; liquid hydrogen; fuel-cell technology; electric green taxiing system; wingtip propulsion; superconducting materials

Citation: Eissele, J.; Lafer, S.; Mejía Burbano, C.; Schließus, J.; Wiedmann, T.; Mangold, J.; Strohmayer, A. Hydrogen-Powered Aviation—Design of a Hybrid-Electric Regional Aircraft for Entry into Service in 2040. *Aerospace* **2023**, *10*, 277. https://doi.org/10.3390/aerospace10030277

Academic Editor: Dimitri Mavris

Received: 2 February 2023
Revised: 1 March 2023
Accepted: 6 March 2023
Published: 11 March 2023

1. Introduction

Over the past 40 years, air traffic has grown 10-fold, and air cargo has grown 14-fold [1]. Despite numerous crises in the 21st century, air traffic has been growing continuously [2]. Therefore, the climate responsibility of the aviation sector is even more important than ever before. The overall goal of halving the carbon emissions by the year 2050 relative to 2000 was set by the aviation industry [3]. In order to meet the European Union (EU) Green Deal main objectives [4], an overall willingness from all parties to take new paths and design bold concepts must be present.

For reaching the goals, the aircraft technology roadmap to 2050 engages in several contributing opportunities [3]. In addition to focusing on the improvement of known technologies, the roadmap foresees the increased use of new-generation technologies from 2030 onward. The EU-funded project FutPrInt50 focuses on the possible technologies for commercial hybrid-electric aircraft for the years 2035–2040 [5]. As part of the project and contributions to the FutPrInt50 Aircraft Design Challenge [5], the authors of this paper designed the hybrid regional aircraft HAIQU (Hydrogen Aircraft desIgned for Quick commUting).

This aircraft design provides capacity for 50 passengers and combines the advantages of battery and fuel-cell technology in a hybrid combination. Therefore, the motivations for HAIQU are to achieve zero emissions during the whole flight mission and to bring regional aviation one step closer to a modern environment as well as to encourage future aircraft designs. Using liquid hydrogen to approach net zero emissions is based on the production method of liquid hydrogen [6]. Depending on the method of production, hydrogen is labeled with different colors depending on the method of production [6,7].

Indeed, green hydrogen is the only climate-neutral type [7,8]. However, the amount of renewable energy required to produce green hydrogen in large quantities is enormous [9]. To ensure the efficiency and effectiveness of the aircraft design, various methods are utilized during the design process, which are outlined in Section 2. The trade-off studies and their results are summarized in Section 3. Finally, in Section 4, the advantages and disadvantages of the aircraft design are discussed to provide a comprehensive understanding of the design's strengths and limitations.

2. Materials and Methods

In the following sections, the classical aircraft design methods build the base of the familiar design process. Moreover, the unconventional design methods for the hybrid-propulsion technologies are described in detail. As this aircraft design is part of the FutPrInt50 Aircraft Design Challenge, the guidelines [5] that were predefined by the challenge committee had to be strictly followed in order to ensure comparability between the different design proposals. Hereby, the main requirements were defined in the top level aircraft requirements (TLAR) [5] and elaborated by Eisenhut et al. [10], and these are listed in Table 1. The aim of the design process is to achieve the TLAR under the lowest possible direct operating costs (DOC).

Table 1. Top level aircraft requirements [5].

TLAR	Value
Number of passengers	50
Passenger weight	106 kg per passenger (incl. luggage) = 5300 kg
Design range	800 km
Design cruise speed	≤Ma 0.48
Maximum payload	5800 kg
Reserve fuel policy	185 km + 30 min holding
Rate of climb (MTOM, SL and ISA)	≥1850 ft/min
Time of climb to FL 170	≥13 min
Maximum operating altitude	7620 m (25,000 ft)
Take-off field length	≥1000 m
Landing field length	≥1000 m
Benchmark for DOC	Design payload with 400 km mission

Using a reference aircraft facilitates the design process and improves accuracy [11]. Therefore, a market analysis with the goal of finding a proper aircraft was conducted. The regional aviation sector led to aircraft such as the Dash 8 and ATR 42. For the following design process, the concept refers to only one reference aircraft that fit the given TLAR the best. In this case, the ATR 42-600 [12] formed the most promising foundation and was, thus, selected as the reference aircraft.

A basic comparability of the ATR 42 and ATR 72 can be generally assumed under certain conditions, since the ATR 72 is merely an extended ATR 42 with an increased maximum seat capacity from 48 to 78 seats [13]. In addition, the wingspan and engine power are increased from the ATR 42 to the ATR 72 [13]. In the typical cases, the ATR 42-600 was used for reference values. However, if no reference values for the ATR 42-600 were available, relative values from the ATR 72 were used, which are still legitimate due to the scaled relationship between the two aircraft.

2.1. Design Iteration Code

The design process is described in Figure 1. In a first step, assumptions about the propulsion system, the wing configuration, the fuselage shape and the empennage were applied in the pre-design (red box). With these assumptions, preliminary values were defined for the range, passenger and luggage weight, runway take-off length, cruise and landing speed and wingspan. Having these pre-design values, as well as the initial values (blue box) set, the sizing process (green box) began with the weight estimation according to Roskam [14].

In the next step, the sizing diagram was used to identify the design point and, thus, the power and wing area. This calculated wing area was used to size the wing, resulting in the required aerodynamic lift coefficients $C_{A,Land}$ and $C_{A,To}$. Further, the lift coefficient was used to obtain the drag-polar and, thus, the lift-to-drag ratio. This ratio was, in turn, used to derive the weight of the aircraft components and, thus, the center of gravity. With the calculated maximum takeoff mass, the payload-range diagram was created, which then led to the required fuel mass. Bringing the iteration to an end and moving on to the analysis (yellow box), a CAD model was adjusted, and the three-side view was drawn, bringing the mechanical concept to a final freeze.

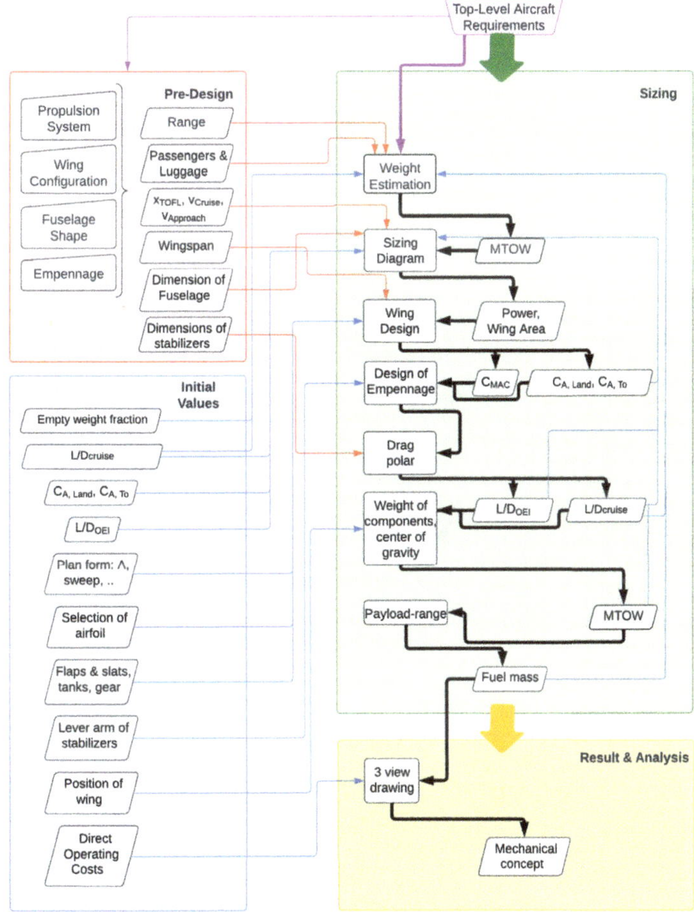

Figure 1. Schematic representation of the sizing code with the pre-design in red, the initial values in blue, the sizing process in green and the results and analysis in yellow.

Range Calculation for Hybrid-Electric Aircraft

Conventional fuel burning aircraft can be sized using the well-known Breguet equation, while battery electric aircraft can be sized using the modified Breguet Equation [15]. It is possible to simplify the mission segments as "equivalent stationary horizontal flight phases" [16]; however, for this application, a more detailed incremental time step approach was chosen. This allows the split between the energy sources to vary from one moment to the next. With this approach, the normal flightpath of an aircraft can be simulated in small time increments, calculating the required energy for each time step, splitting that energy between battery and fuel and deriving the resulting mass change from it. This allows for a very refined energy management strategy between the primary and secondary energy source.

The input parameters for the incremental calculation are the basic aircraft characteristics regarding mass, aerodynamics, efficiencies and specific energy. The output is fuel and energy used per flight segment. This incremental approach will also be used in Section 3.1 to create the payload-range diagram for the aircraft.

2.2. Design Process of the Aircraft

To feed the design iteration code with information about new technologies, they must be thoroughly investigated. For each new component of the aircraft, a literature study was conducted to derive equations and reference values to apply to the design. The reference aircraft was used to calibrate the masses, empennage and sizing diagram.

2.2.1. Configuration Selection

Through a preliminary evaluation process in compliance with the given TLAR [5], some design possibilities were excluded. The remaining design possibilities are shown in Figure 2. In this assessment, the design possibilities were evaluated by performance, environment and economics as well as operational aspects. Since the focus of this work lies in the basic design of the aircraft, no detailed analyses of the individual structures and aerodynamics were performed. Nevertheless, this work is based on scientific evidence by using estimation formulas for the design according to reliable sources [11,14,17,18]. The aim of the configuration selection is to identify a performing concept and to place the performance in relation to the reference aircraft. In the following sections, the reasoning behind every decision is explained.

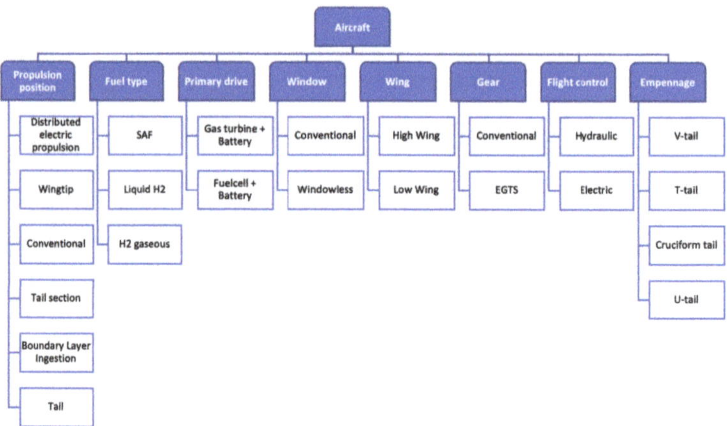

Figure 2. All investigated design possibilities listed by their category.

2.2.2. Propeller and Wing Tip Propulsion

In the regional aircraft sector, the most prominent propulsion method is the propeller, while jets are more often used for longer ranges than the given TLAR for this design. The propeller has the advantage of possible higher efficiencies at low speed [19] combined with lower complexity and easy conversion to an electric drivetrain when compared to jet engines [20,21]. With a small hybridization ratio, i.e., the ratio between electric power and total propulsive power, it is possible to use motor–generator units within a jet turbine.

With power electrification beyond 50%, the power has to be provided by an electric motor driving a propeller. Further, the electrification enables an almost free positioning of the propellers and motors. The current trends are the use of wingtip propulsion (WTP), boundary layer ingestion (BLI) and distributed electric propulsion (DEP) [22,23]. Research regarding BLI showed that the positive effects increased with speed, making the use at relatively low speeds of the aircraft less suitable. DEP showed promising results in an isolated setting; however, further analysis on aircraft level showed no effect or even negative effects for this application [24].

The advantage of using WTP is the reduction of induced drag [25]. Considering the WTP concept individually, it promises an improvement of up to 50% less induced drag during climb and 25% less during cruise [26] regarding the wing and up to 15% less total drag on the aircraft level [27]. Secondary effects are the decrease of the wingtip vortex intensity through the WTP [26], which leads to a significant noise reduction [24] as well as to a reduced wing bending moment, relieving the wing by mounting a mass with a long lever arm on the wing tip.

Concerning the aircraft, the reduction in induced drag is derived as an average of the values of Sinnige et al. [28]. The reduction in drag is measured for a single engine per wing configuration. To account for a possible twin engine per wing architecture, the drag decrease value is assumed to be halved for that setup. In that case, the inside propeller is spinning in the same direction as the wing tip propeller, which could also help in the vortex reduction; however, this requires further investigations. In Table 2, the chosen reference drag decrease values from Sinnige et al. [28] for the single engine case are listed, as well as the assumed correction factor to account for a possible twin engine setup. These reference values are used to reduce the induced drag at the wing level for the calculation within the design code.

Table 2. Drag reduction through wing tip propellers [27].

Reference Drag Reduction—High Power	Reference Drag Reduction—Medium Power	Correction Factor for Four Propellers
35%	10%	−50%

2.2.3. Wing Design

For the wing, the maximum lift coefficient $C_{L,max}$ was calculated using Raymer [18] sizing methods. For the wing area S, the wingspan b and the taper ratio λ, the initial values were defined before the execution of the sizing iterations. In the following design process, S was obtained through the iterating mass estimation, while b and λ were chosen values. With the lift coefficients of the selected airfoils at different positions from the reference aircraft, the maximum lift coefficient was obtained. In contrast to current generation aircraft, the wing is not used to store fuel and is, therefore, not subjected to passive stress relief.

The advantage of the WTP, in this case, is the mass of the motor, gearbox, DC/AC converter and propeller acting as a load reducing bending moment with a long lever arm. This can result in an increase of the local wing bending moment along the wingspan. The additional stress is compensated for by applying a factor to the wing mass since the maximum absolute bending moment occurs at MTOM. This factor is calculated as the ratio between the wing bending moment for an aircraft wing in a theoretical "standard" configuration (inboard engine and fuel in wing; usual estimations applicable) and the selected configuration (WTP and dry wing). The additional wing root bending moment can be reduced by increasing the wingspan and, therefore, increasing the lever arm of the WTP mass. This further improves the aircraft performance by reducing the weight.

2.2.4. Empennage

According to Raymer [18], the empennage should typically be considered to be conventional. Nevertheless, unconventional tail configurations may have advantages in special use cases. The criteria related to structural connections, dimensions, mass and flow interactions must be considered when deciding on the type of empennage [18].

Depending on the chosen propulsion concept, the yawing moment must be considered when designing the vertical tail plane (VTP). If the propulsion concept is not different from the reference aircraft, the empennage is dimensioned with a volume coefficient similar to the one of the reference aircraft. Since the configuration includes wingtip propellers, the volume coefficient of the VTP has to be increased to provide a sufficient yawing moment during wingtip engine failure. The increase is calibrated using the reference aircraft. First, the volume coefficient of the reference aircraft is determined. The ratio between the yawing moment created by the engine in the one engine inoperative (OEI) case and the volume coefficient of the VTP of the reference aircraft is calculated. This ratio is used to increase the conventionally calculated volume coefficient to account for the unconventional architecture.

2.2.5. Fuselage

New propulsion concepts do not necessarily require a new fuselage design. A comparison of aircraft with similar passenger capacities in terms of the number of seats can be useful as demonstrated in the chart provided by [29]. Based on the integration of the fuel tanks and further system technologies, the dimensions of the fuselage could vary. A reasonable fuselage fineness ratio should be between 5.6 and 10 for regional aircraft, according to Roskam [17].

There are multiple approaches to maximizing cabin space; however, it is essential to weigh the balance between passenger comfort and space utilization in the design process [30]. With a market analysis, the current state of the art in cabin layout can be analyzed as well as new concepts, such as a windowless fuselage [31]. Nevertheless, the urgency for a regional aircraft design preferring passenger comfort over the use of space is not relevant. The major change to the fuselage is the integration of the fuel tanks, which is covered in more detail in Section 2.3.3.

2.2.6. Landing Gear and Electric Green Taxiing System

There are two main options for the integration of the main landing gear in reference to other high-wing aircraft [12,32]. The first option is the integration in a belly fairing—similar to the reference aircraft. The second option is the integration in the cowling of the turboprop engine as on the De Havilland Dash 8–400 [32]. The integration in the cowling is challenging due to the propulsion choice of the relatively compact electric motor instead of a gas turbine. To reduce energy use and noise on the ground, we investigated alternative forms of taxiing.

At larger European airports, an aircraft spends 10–30% of its block time taxiing [33]. The electric green taxiing system (EGTS) promises to be a great solution to increase efficiency during this phase. Electric motors have a relatively low noise output, and, locally, they emit no emissions [34]. The technology readiness level (TRL) of onboard electric taxiing systems is between TRL 6 to 7 [35]. With regenerative breaking during the taxiing phase, 15% of the energy could be recovered [36], and the lifetime of the brake system would be increased [37,38]. The utilized dimensions for an electric taxiing system were calculated with the formulas from Heinrich et al. [39].

2.2.7. Mass Calculation and Center of Gravity

The individual component masses were calculated by semi-empirical equations, which were derived by Torenbeek [11]. To start the iterative calculations, first, assumptions were made with the existing data of the reference aircraft [40]. The propulsion concept eliminates the conventional turboprops; therefore, the mass of the electric motors, including cabling, is calculated by linear scaling using the required power [41]. Electric motors are scalable and adjustable within the required power [42].

The mass of the fuel cell and batteries are calculated according to the methods in Sections 2.3.1 and 2.3.4. In case the reference aircraft differs in the composite vs. total structure volume ratio, further mass reduction factors can be introduced [43]. One method for calculating mass reduction factors is to reference existing data, such as the 20% structural mass savings achieved by Boeing through the use of 50% composite materials [44]. However, there is inherent uncertainty in this extrapolation, and thus conservative values as calculated by Kolb-Geßmann [45] were adopted and are presented in Table 3.

Table 3. Component mass reduction factor and composite amount.

Component	Mass Reduction Factor	Composite Amount in %
Wing	0.73	90
Fuselage	0.85	50
Empennage	0.715	95
Cowling	0.85	50

The crew is considered to consist of two pilots and one cabin crew member, which were conservatively assumed at 85 kg each [46]. According to Roskam [14], the crew is an element of the operating mass empty (OME). Furthermore, the center of gravity (CoG) of the components can be determined with the methods of Torenbeek [11]. Considering the aircraft at OME with the added max fuel mass, the maximum static margin can be determined.

2.3. *Powertrain*

New propulsion technologies require new calculation approaches. By scaling the confirmed values of previous work, the component architecture can be calculated as laid out in the following sections.

2.3.1. Fuel-Cell System

The sizing of the fuel-cell system was conducted based on the evaluation of the feasibility of dual use of liquid hydrogen in regional aircraft [47]. The focus of the research of Hartmann et al. [47] is an ATR 72, and thus the results of their baseline scenario are scaled down using the fuel-cell polarization curve for the baseline case in [47]. The ATR 72 is the larger version of the reference aircraft; however, since the powertrain is only influenced by the required power for the aircraft and not other characteristics, e.g., the wingspan and fuselage length, it is viable to scale the powertrain from the ATR 72 to a smaller aircraft.

The linearized section is assumed to be applicable from a current density of 0.2 up to 1.4 A/cm^2. Three different cases were investigated, and the resulting sizing of the fuel cell had to work for all of them. The first scenario is the maximum power during take-off. For this case, the highest current density is expected. The upper limit of 1.4 A/cm^2 results in the smallest size of the fuel cell, while it is desirable to achieve a lower current density since that correlates linearly with higher efficiency. The second case is cruise flight. The segment is the longest of the whole mission; therefore, a high efficiency in this segment results in a high efficiency for the total flight. The third case is the descent phase.

Since the powertrain components utilize superconducting technologies, it is necessary to provide a minimum flow of hydrogen to keep the components within their operating window. This hydrogen will not be vented but instead used to power the aircraft during descent, with the excess power being used to charge the battery. In this case, the current density should not be lower than 0.2 A/cm^2. The minimum flow is calculated in three steps:

1. Calculate the mass flow for the ATR 72 during cruise.
2. Divide the mass flow by the ATR 72 cruise power.
3. Calculate the mass flow of the aircraft by multiplying the result of step 2 with the aircraft's cruise power.

Another limiting boundary condition for the fuel cell sizing is the maximum de-charge rate of the battery during take-off. For durability reasons, this is set as 2C. The mass of the fuel cell is calculated using the stack mass density given for the baseline scenario in [47] using the calculated fuel cell area for the aircraft. The mass of the compressor and the humidifier are linearly scaled from the results for the ATR 72 using the fuel cell power. To account for structures, etc., we introduced a Balance of Plant (BoP) factor. This was set as 20% by Hartmann et al. [47] and was used for the aircraft to add to the total mass. The reference values for the calculations are listed in Table 4.

Table 4. Reference data for the fuel-cell system calculation [47].

Reference Power	Stack Mass Density	Reference Compressor Mass	Reference Humidifier Mass	BoP-Factor
4100 kW	1.65 kg/m²	200 kg	300 kg	0.2

2.3.2. Thermal Management System

The thermal management system (TMS) consists of two different types of cooling loops. One loop uses the liquid hydrogen to absorb the heat from components, while the other uses a conventional cooling cycle [47]. Since the difference between the temperature of the components and the surrounding air is small compared to other appliances, the airflow in the cooling inlet has to be higher to achieve sufficient cooling. To estimate the amount of thermal energy that must be removed from the system, the energy distribution between power used at the propellers and generated heat was taken from Hartmann et al. [47] for the cruise flight segment.

The TMS load is defined as the ratio between heat energy displaced by the TMS and the electric energy provided by the fuel cell. To calculate the TMS load for the aircraft, first, the difference in fuel cell efficiency compared to the reference in Hartmann et al. [47] was added/subtracted to the reference TMS load. These values are listed in Table 5. The second step is multiplying the resulting number with the required propulsive power to result in the heat power distributed to the TMS for the aircraft. Conventional cooling systems consist of an air intake, a radiator and a nozzle to expel the air efficiently. Most applications have an additional fan fitted to keep the performance of the radiator high even if the airspeed is low, for example during the taxi phase.

The additional drag from the radiator can be reduced with the Meredith effect [48]. Kellermann et al. [49] concluded in their studies that the drag increased by 0.7% due to the radiator. Since the study was conducted for a 30% hybridized 180 seat aircraft, the drag can be scaled with the reference heat load and a reference MTOM. To calculate the additional drag for the aircraft, an increase in heat load and a decrease in MTOM result in a higher drag penalty. The reference values from Kellermann et al. [49] are listed in Table 6.

Table 5. Selection basis for thermal management system load [47].

Reference Efficiency of Fuel Cell	Reference TMS Load
52.9%	107.9%

Table 6. Reference values for the thermal management system drag by [49].

Reference Heat Load	Reference Drag Increase	Reference MTOM
192.6 kW	0.7%	80.000 kg

Chapman et al. [50] conducted simulations on the transient behavior of fluid-based TMS. Their findings concluded that designing the TMS for the maximum system power as a steady state can result in oversized cooling systems. Their simple model was tested for a much smaller load and fluid mass compared to this aircraft; however, the results showed that maximum power for less than 2 min leads to a much smaller TMS size. Since the takeoff phase takes less than a minute for regional aircraft, the cruise power was used to size the TMS. The TMS fluid capacity should be sized to allow this simplification in follow up projects.

2.3.3. Fuel Tanks

Liquid hydrogen is stored at temperatures below 20 K for multiple hours within the aircraft's tanks, resulting in the need for an insulated fuel tank. Compared to gaseous hydrogen, which is used in the automotive industry, the tank pressure is low, thereby, resulting in lower strength requirements. The wall thickness and resulting empty tank mass were derived from Silberhorn et al. [51] for their tank concept in the rear of the aircraft and are listed in Table 7. Their fuel tank mass was scaled cubic with the block energy. This was chosen because the tank volume, which is directly proportional to the stored energy, scales with the power of 3 with the radius. The resulting mass from the cubic scaling was further reduced since the mission range in the TLAR is only 14% of the aircraft used by Silberhorn et al. [51], resulting in an estimated reduction of wall thickness of 60%.

Table 7. Reference values for liquid hydrogen fuel tanks by [51].

Reference Block Energy	Reference Tank Mass	Reference Wall Thickness
533 GJ	1651 kg	70 mm

2.3.4. Battery

The automotive and aviation industries are currently dominated by lithium-ion batteries; therefore, to establish the configuration of the battery, we considered six important factors: the Specific Energy, Specific Power, Cost, Life Span, Safety and Performance. Lithium cobalt oxide ($LiCoO_2$)—LCO; lithium nickel manganese cobalt oxide ($LiNiMnCoO_2$)—NMC; and lithium iron phosphate ($LiFePO_4$)—LFP batteries are the three most promising concepts for future batteries [52].

Figure 3 shows a comparison between the main variables investigated for the selection of the battery. Figure 3a presents a comparison between the specific power and the specific energy of the three selected battery types. Those figures, in combination with the requirements for the aircraft powertrain, are used to decide on the battery chemistry. High power needs to be achieved, and high specific energy use is attenuated since the batteries will be used mainly at takeoff and not during the flight. For this reason, a LFP battery that meets those characteristics was chosen.

Figure 3a indicates specific powers up to 2000 W/kg with 140 Wh/kg for LFP batteries. Figure 3b represents a radar chart that relates the aforementioned variables where LFP batteries stand out with respect to their life span, safety and specific power. LFP batteries have an ideal cycle life of 1000–2000 with a charge rate between 1 and C and discharge rate of between 1 and C and 3C. This means that the battery can provide between one and three times its capacity in energy output per hour [53]. However, it is important to note that the discharge rate of LFP batteries is generally higher than the charge rate; however, they maintain better performance during their life cycle compared to the other batteries mentioned [54].

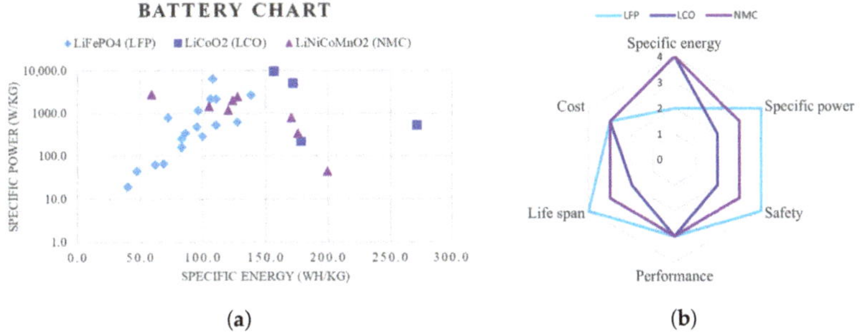

Figure 3. (**a**) Specific power vs. specific energy of Li-ion batteries distinguished by cell chemistry [55]. (**b**) Radar chart of LFP, LCO and NMC battery comparison [56].

2.3.5. Superconducting Motors and Power Electronics

As mentioned in Section 2.3.2, the hydrogen is used to cool the powertrain components. This enables the use of superconducting materials for the motors, inverters and cables. The efficiencies and masses for the cables and inverters were estimated using the values from Hartmann et al. [47]. The superconducting motor mass can be estimated with the equation of Lukaczyk et al. [42], while the efficiency was estimated at 98% [47]. Power densities for high-temperature superconducting (HTS) motors producing 1 MW above 13 kW/kg are feasible according to Yoon et al. [57] with their market readiness expected in the early 2030s [58]. The total power of the aircraft is expected above the reference aircraft's power at about 4000 kW. The selected power density for the motor is 15 kW/kg, resulting in a slightly heavier motor than the estimation equation according to Komiya et al. [41] would predict.

2.4. *Impacts on the Turnaround Process*

One of the key elements of a competitive aircraft is the turnaround process. Compared to the current generation of regional aircraft, there are three major changes:

1. Liquid hydrogen is used instead of jet fuel.
2. The refueling vehicle is connected to the back of the aircraft instead of the wings.
3. The battery may have to be recharged.

Mangold et al. [59] investigated possible solutions to refuel hydrogen aircraft. They concluded that the safety level can be as high as it is today for kerosene. To calculate the total time for the turnaround process, fixed values were used for segments where no fuel is transferred into the aircraft, and a flow rate was used to calculate the time for the actual refueling. For this aircraft, a single fuel line case with purging was assumed with the variables taken from Mangold et al. [59] as listed in Table 8.

Table 8. Calculation basis for refueling with liquid hydrogen by [59].

Sum of Fixed Process Times	Fuel Flow Rate
9 min	20 kg/s

The other areas that we considered for the turnaround process were the passengers, baggage and battery recharging. For all of those segments, the equipment positioning and removal time was taken from Mangold et al. [59]. For the passengers, the boarding and de-boarding rates for a Type I door were taken from Airbus [60]. The cleaning rate can be estimated as 3 seats/min according to Fuchte [61] with an estimation of two cleaners for this aircraft. The baggage is loaded at two separate locations in one compartment each as bulk cargo. The Boeing 737 utilizes the same cargo storage strategy allowing for their loading and unloading rates [62].

Regarding the battery, the before-mentioned maximum rate of 2C for charging and discharging is applied for the energy transfer required on the ground. However, the strategy for the aircraft is to not have the charging time exceed the longest segment in the turnaround process, i.e., the passengers. This is managed by recharging the battery in flight to a level where the battery can reach the desired state of charge during the turnaround in the specified time frame. The selected values used to calculate the turnaround process are listed in Table 9.

Table 9. Turnaround calculation basis for the passengers, baggage and battery.

Boarding Rate	De-Boarding Rate	Cleaning Rate	Loading Rate	Unloading Rate	Charging Rate
12 PAX/min	18 PAX/min	6 Seats/min	10 Bags/min	15 Bags/min	2C

2.5. Cost Calculation

The costs are subject to different variables, such as depreciation, insurance, maintenance, fuel consumption, flight crew, cabin crew, landing fees and passenger services. These variables are grouped into direct operating costs (DOC) and indirect operating costs (IOC) [63]. The IOC depend on the relation between the airline and the client and, thus, is difficult to estimate. For this reason, it was not calculated in the present study. This study focuses on the evaluation of aircraft design in economic terms through the cost per available seat kilometer (CASK) and the DOC. A number of methodologies exist to estimate the DOC with high reliability, e.g., ATA, NASA and AEA1989 [64].

ATA is based on industrial statistics from the United States [65], NASA uses an estimation methodology considering interest [66] and AEA is a comparison methodology used in Europe [67]. In this paper, studies were conducted based on the AEA1989 model. The comparison with the reference aircraft was performed considering a depreciation calculation based on the AEA Method 1989 over 14 years. For the year 2035, consequently, a reduction in hydrogen prices between 1.5 €/kg and 5.5 €/kg is projected [68] as shown in Table 10, while an increase of 10% for fossil fuels is estimated compared to the current price [69].

Table 10. Price of hydrogen according to S&P Global [68].

Hydrogen Type	Price in €/kg
Green	5.5
Blue	2.3
Gray	1.5

On the other hand, Rethink Energy's model establishes, however, that the cost of green H_2 will fall to slightly over $1/kg in 2035 as led by nations such as Brazil or Chile, generating the possibility of reaching even lower prices [70,71]. Consequently, achieving this price for green hydrogen reduces the level of emissions to zero, something that does not occur with blue hydrogen with 10% CO_2 emissions or the total release of gray hydrogen obtained only by steam methane reforming (SMR) or gasification [68].

In contrast, electricity prices will remain constant given the continent's transition to renewables. The implementation of these new environmentally friendly technologies affects the fee costs in the DOC. For the AEA method, it was necessary to add charges related to pollution and noise [72]. Those charges are considered for this study, although they are currently not applied at every airport.

3. Results

The following sections present detailed results from the previous chapter's analytical and empirical analysis. Trade-off studies, which are explained in the next section, provide the basis of the final design choice.

3.1. Payload Range Diagram and Design Point

As hydrogen was chosen as the main energy carrier, the shape of the payload range diagram is significantly different from typical diagrams. The high gravimetric energy density of the fuel leads to a very shallow middle section, making it impossible to place the TLAR mission on this section without breaking the maximum payload requirement. Therefore, the chosen design point is the required maximum payload of 5800 kg for a 400 km mission as shown in Figure 4 with the blue line and dot. The TLAR mission can be achieved as shown by the red dot.

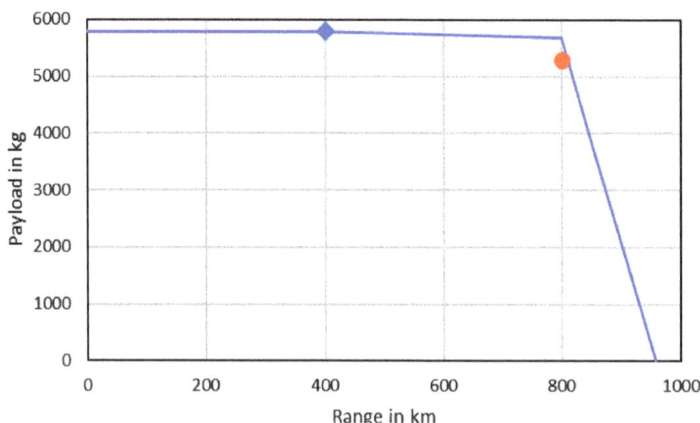

Figure 4. Payload-range diagram with design point (blue) and TLAR mission (red).

3.2. Trade-Off Wingspan

Several trade-off studies are performed with the design iteration code. More wingspan usually leads to better aerodynamic performance; therefore, the calculations were performed for wingspans starting from the ATR 42 at 24.6 m and reaching to the end of Airport Category C at 36 m. The grading criteria are the energy requirements for the 400 km mission since this is a major part of the grading criteria for the whole aircraft, i.e., the DOC. Even though the best aircraft performance is at a span of 36 m, the overhanging WTP tips reduce the available wingspan to about 32 m. To account for structural integrity, the maximum taper ratio was set to 0.25, which limits the wingspan to 28.5 m as can be seen in Figure 5. Therefore, the wingspan of HAIQU was selected to be 28.5 m.

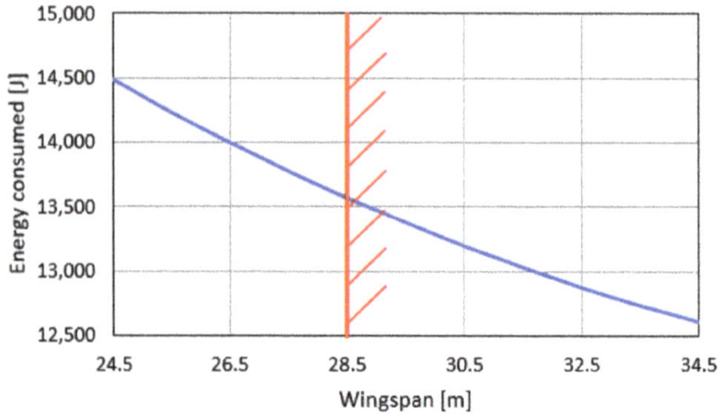

Figure 5. Energy requirements for a given wingspan.

3.3. Result of the Propulsion System Sizing

During the aircraft design process,we observed that using only wingtip propellers would lead to immense yawing moments in the case of a single engine failure regarding the electric motor, thus, leading to an oversizing of the VTP. To mitigate this problem, two further propellers were placed in a conventional location on the wing close to the fuselage. To optimize the power split between the outer and inner propellers, a tradeoff study was conducted with the result being that a slightly more powerful WTP would be the optimum solution regarding the boundary conditions.

However, the design iteration code does not include effects regarding disc loading, blown wing area and similar; hence, the results of Nathaniel [27] were used to define the power split to achieve a higher level of fidelity. They investigated the power distribution of an ATR 42–500 with WTP and conventional propellers. A power split of 50–50 between the WTP and the inbound propellers was identified to be the optimum as the split reduces the disc loading and, therefore, reduces the total required power by 5% compared to a configuration with one propulsor per wing. While the induced drag reduction decreases when using more than one engine per wing, the effects of the disc loading dominate in comparison, thereby, leading to better performance with four propellers.

3.4. Power Split between Battery and Fuel Cell

The battery concept of HAIQU is to keep the mass as low as possible and to use it as emergency backup while also increasing the lifetime of the battery. With the calculated H_2 mass flow for the descent, the minimum fuel cell power setting is 1050 kW, of which 420 kW are not used by the propulsion system or climate control. This overproduction is directly fed into the battery to recharge it at 2C to further reduce the turnaround time. This maximum charging power during descent, in turn, limits the fuel cell takeoff power to 3400 kW. The remaining 440 kW for the takeoff power are provided by the battery system.

With the given power density, the resulting battery mass would be 300 kg; however, the self-set requirement to have a large enough battery to fly at a constant altitude for 10 min (a single traffic pattern in case of fuel cell failure for example) demands a storage capacity of 206 kWh resulting in a battery mass of 1950 kg. The fuel cell has a resulting mass of 1020 kg with the compressor and humidifier weighing 410 kg. The cable, propeller, inverter and gearbox masses are listed as sums in chapter Section 3.5.2. The results are listed in Table 11 for a better overview.

Table 11. Result of powertrain sizing regarding the power and mass.

Fuel Cell Power	Battery Power	PEMFC System Mass	Battery System Mass
3400 kW	440 kW	1020 kg	1950 kg

To increase safety through redundancy, two identical fuel cells in the back of the cabin and two identical batteries in the front of the cabin are used. The energy from both systems is transformed using a DC/DC converter each to achieve the best transport voltage for the superconducting cables. Those cables run from the DC/DC converters to the DC/AC converters located at the motors in the wing as can be seen in Figure 6.

The motors need to provide a power of 900 kW each, resulting in a mass of 60 kg with the gearbox adding another 35 kg each. They are driving a six-bladed propeller with a diameter of 4 m running at low speed for lower noise emissions. Regarding the TMS, the hydrogen is pumped from the tanks through the HTS cables and in a separate line to the motors to supply the right temperature to both components. Both flows recombine at the DC/AC converters and merge back in the fuselage to cool the DC/DC converters before the gaseous hydrogen is heated to 85 °C using the fuel-cell system. This order is derived from Hartmann et al. [47] and shown in Figure 7 but with a few modifications to reduce the fuel line length.

For the cruise flight, a fuel consumption of 2.2 kg/min is achieved at a fuel cell efficiency of 55%. This results in fuel consumption for the whole 400 km flight mission of 172 kg. The electric green taxiing system complies with the requirements, and HAIQU fulfills the Flightpath 2050 goal of emission-free aircraft movements during the taxiing phase [73], as the electric motors produce relatively low noise, and they emit no emissions locally [34].

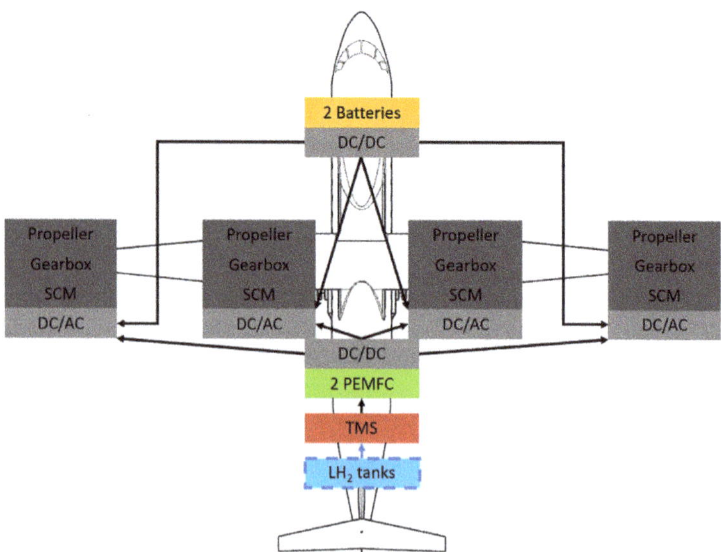

Figure 6. Powertrain component arrangement within HAIQU.

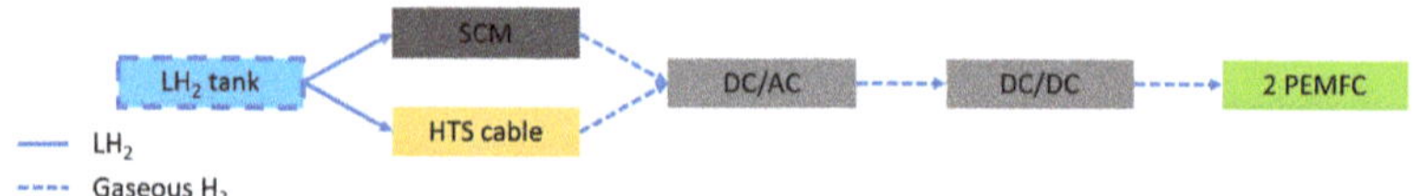

Figure 7. Hydrogen flow from the fuel tanks to the fuel-cell system; full line representing liquid hydrogen and dotted line representing gaseous hydrogen.

3.5. Resulting Aircraft Design—HAIQU

All the components mentioned in the previous chapter are part of the final design of HAIQU. In this section, the full aircraft will be shown and its performance characteristics described. In Figure 8, the three-side view of the final aircraft design is shown, and Figure 9 shows a rendering of the operating plane [74].

3.5.1. Fuselage

In order to keep the main focus of HAIQU on the propulsion technology, the cabin should not present any noticeable change for the passengers and operators. This results in a seamless integration in today's airport infrastructure for all non-fuel-related ground segments.

As shown in Figure 10, the layout consists of a four abreast seat arrangement per row with a single row of two seats on one side at the back. Boarding is conducted in the rear while the baggage is loaded in the front on the left side and on the back on the right side. The rest of the cabin layout is conventional.

The fuel tanks showcased in Figure 11 behind the cabin weigh a total of 300 kg. Concluding the design choices, the fineness ratio for the fuselage is 8.6. The cabin cross section is split into two parts, the upper, pressurized section is reserved for passengers and baggage, and the lower, unpressurized section is used for powertrain components, the TMS and the landing gear. This is shown in Figure 12.

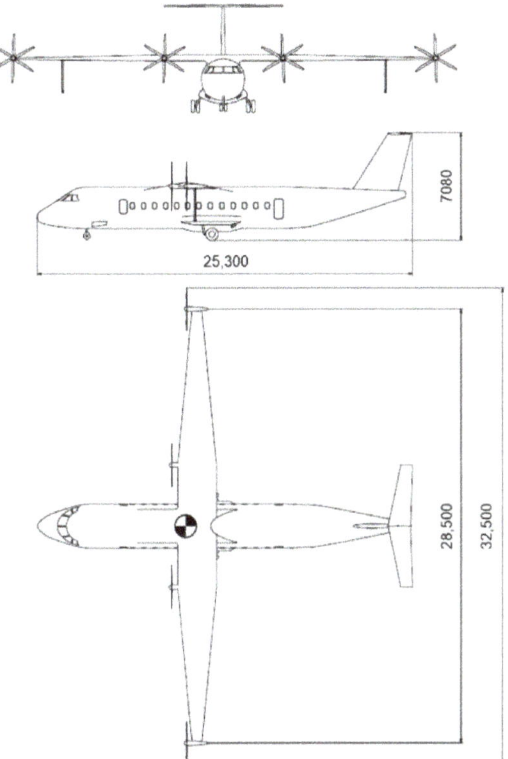

Figure 8. Three-side view drawing of HAIQU.

Figure 9. 3D rendering of the HAIQU aircraft.

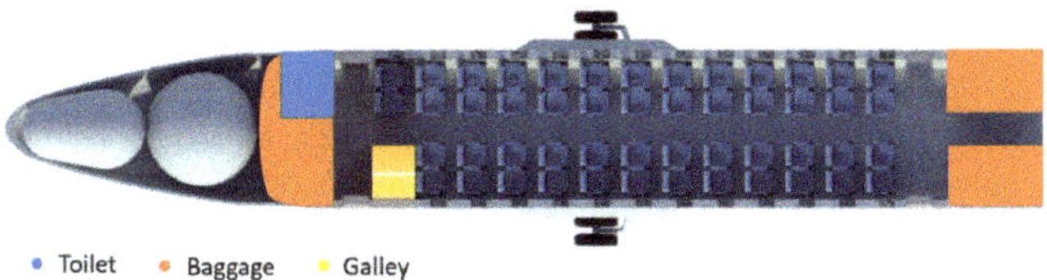

Figure 10. Cabin layout of HAIQU; twelve rows with four abreast, one row with two seats, space for baggage in the front and back and fuel tanks located behind the rear bulkhead.

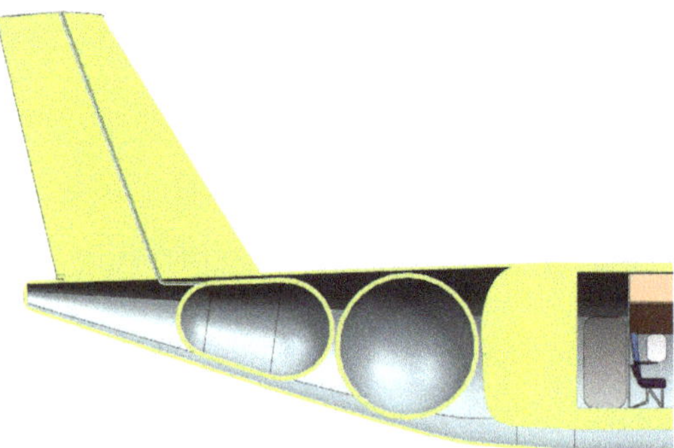

Figure 11. Cross-sectional view of the rear fuselage section showcasing the fuel tank arrangement; venting line routed through the vertical stabilizer.

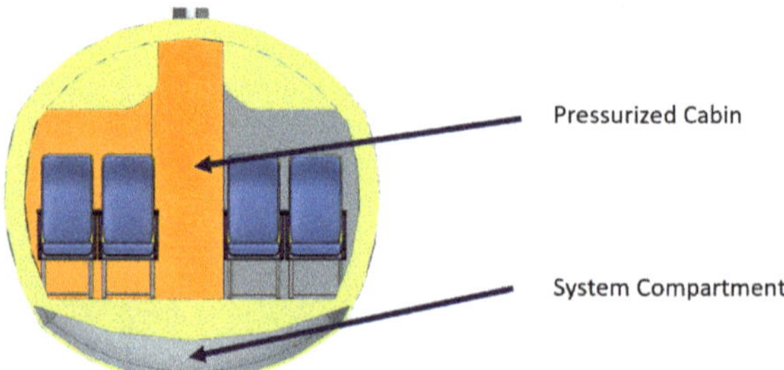

Figure 12. Cross-sectional view of the cabin section, showcasing the pressurized cabin and the unpressurized system compartment.

3.5.2. Masses and Center of Gravity

Compared to the reference aircraft, HAIQU has a higher MTOM due to the additional weight of the system masses of the hybrid powertrain. The resulting individual masses are listed in Table 12. Regarding lightweight construction, HAIQU has a composite vs. total structure volume ratio of 54% leading to four-times higher [43] usage of composite material in comparison with the reference aircraft. In case of the typical 400 km mission, the aircraft masses are listed in Table 13.

Table 12. Summary of the component masses.

Component	Resulting Mass in kg
Wing	900
Fuselage	1950
Empennage	320
Landing Gear	620
Control Mechanism	205
Electric Motors	255
EGTS	120
Flight Systems	2590
Fuel Cell System	2780
Fuel Tank	300
Battery System	1950
Crew and Equipment	1050
Miscellaneous	150
Gearboxes	140
Cowling	170
Total OME	**13,500**

Considering the tank position in the tail, the center of gravity (CoG) shift during the refueling process is important. The CoG is located at 23.8% mean aerodynamic chord (MAC) at the operating mass empty (OME) as shown in Figure 8. Considering the ferry case as the most critical loading case, the CoG is located at 33% MAC.

Table 13. Aircraft masses.

Designation	Mass in kg
Operating Mass Empty	13,500
Fuel + Reserve for 400 km Mission	300
(Max. Fuel + Reserve)	(420)
Max. Payload	5800
Max. Takeoff Mass	19,600

3.5.3. Wing and Empennage

For the wing design, the classical configuration of high wing and T-tail was chosen. The shape of the main wing resulting from the design code is a two-shape design with a rectangular shape from the centerline towards the inboard motor and a trapezoid shape from there until the outboard motor as shown in Figure 8. The wing features single segment Fowler flaps [43] that stretch for 60% of the wingspan with a depth of 17%. As the wing profile, the proven parameters of the reference aircraft were selected leading to a NACA 43018 at the root and a NACA 43013 at the tip [75]. The sizing resulted in an increased wingspan of 28.5 m and an aspect ratio of 15.

Retractable struts on the lower side of the wing are necessary for takeoff and landing in strong side winds to protect the WTPs. During the flight they are retracted and fully enclosed and covered inside the wing, so that no drag is added. In case of an emergency, the struts brake before the wing would be overstressed. The T-shaped empennage was chosen to reduce the wing-and propeller-wake interaction with the tail plane. Furthermore, the volume coefficient of the vertical tail area was set to a larger value than the reference aircraft to provide enough yawing moment during wingtip engine failure.

3.5.4. Cost Results

Given the high fluctuation of fuel costs due to geopolitical variables, the highest price was taken. This affects the projection as at the beginning of this project the price of the LH_2 was three-times lower than the stipulated price. CO_2, NO_x and Noise costs were charged to the reference aircraft based on the methodology of Johanning and Scholz [72].

This new methodology includes the additional cost applied in the form of tax for pollution and noise at airports from 2000. The results for the benchmark mission are shown in Figure 13. These prices are reflected mainly in the reference aircraft. Pollution and noise charges represent 0.2% and 0.02%, respectively, of the total DOC.

The costs were divided into their respective categories, and the unit EUR/Seat-100 km represents the costs in Euros per seat per 100 km. These results indicate a total cost saving of approximately 19% on the total DOC only taking into consideration the prices for green hydrogen. On the design route, this implies a saving of 2 million euros per year per aircraft. However, hydrogen prices can have a significant impact on the cost of the mission depending on the type of hydrogen to be used, i.e., blue or gray hydrogen as can be seen in Figure 13. For example, using blue hydrogen produces a 25% savings in DOC and approximately 28% in the case of using gray hydrogen, actions that nevertheless would not nullify polluting emissions due to the way in which these fuels are obtained.

Additionally, although the variable that most influences operating costs in a positive way is fuel, in this study, a negative performance was observed in other variables, such as depreciation, interest, crew, maintenance and taxes, where there is a slight increase in prices.

3.6. Turnaround Process

As discussed in Section 2.4, the passengers are the critical path for the turnaround. Both refueling and baggage take less time, while the battery recharging takes the exact same time as the passengers. Specifically for the refueling, a total time of 9 min was achieved, while a maximum time of 21 s was required for actual hydrogen flow. Table 14 lists the total times for all considered turnaround segments.

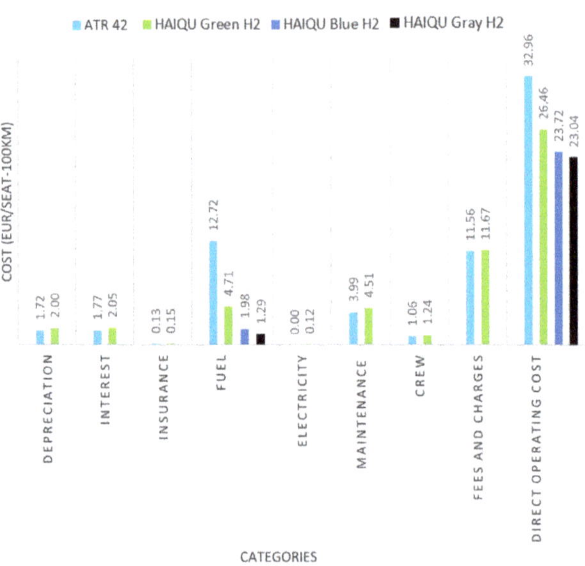

Figure 13. Cost for design mission for HAIQU based on 1500 flights per year.

Table 14. Turnaround process times.

Baggage	Hydrogen	Passengers	Battery
12.5 min	9 min	15 min	15 min

3.7. Performance Data

Evaluating the payload-range diagram in Figure 4, it is clear that the range TLAR are proven to be fulfilled. The aircraft can fly both the design mission of 800 km with a payload of 5300 kg as well as a maximum payload mission of 400 km. The rest of the TLAR are confirmed in Table 15. Now, the performance characteristics of HAIQU can be compared with the reference aircraft. It must be kept in mind that other criteria than the DOC specified for HAIQU may were used in the design of the reference aircraft. In Table 15, the performance characteristics of the reference aircraft [12] are displayed.

HAIQU exceeds the in the TLAR required climb rate at sea level, improving the take-off performance about 30% compared to the reference aircraft. The glide ratio is not ground breaking, which can be attributed to the short take-off distance requirement and the resulting wing design trade-off. At this point, it should be stated again that HAIQU is designed with the aim of fulfilling the specified TLAR of Table 1. This is why the target missions for HAIQU and the reference aircraft may differ and cannot always be compared directly. With a fuel flow of 132 kg/h during cruise, the lost mass in flight is much lower than for similar kerosene aircraft, such as the reference aircraft.

However, it has to be said that this flow of pure hydrogen is combined with oxygen from the air, which results in a production of several tons of water, which is released in fluid form at low temperatures. The total required energy amounts to 2360 kWh for the DOC benchmark mission, which is less than 1/3 of the required energy of the reference aircraft, leading to the significant cost reduction discussed in Section 3.5.4.

Table 15. Performance characteristics.

Aircraft Performance	Achieved Characteristics	Verification	Characteristics Reference Aircraft [12]
Take-off distance at MTOM, ISA, SL	980 m	Calculation	1.107 m
Landing distance at MTOM, ISA, SL	775 m	Calculation	966 m
Maximum rate of climb at MTOM, ISA, SL	2400 ft/min	Calculation	1851 ft/min [76]
Cruise speed, Mach, Altitude	298 kt/0.48 Ma @ FL170	Defined	289 kt @ FL240
L/D in cruise	12.9	Calculation	14.8 [75]
LH_2 consumption in cruise	2.2 kg/min	Calculation	10.3 kg jet fuel/min
Total energy required for DOC mission	2360 kWh	Calculation	7263 kWh [1]

[1] Assuming an energy density of Jet A-1 at 42.8 MJ/kg [77] and an interpolation of the fuel mass for the DOC mission.

4. Discussion

The decrease in DOC shows a great deal of promise for a more environmentally friendly hydrogen-powered and more economical new aircraft. Recent press releases by Embraer [78] and Airbus [79] indicated that the near future of aviation is moving jointly towards fuel-cell powered aircraft for the regional market. Big players, such as Airbus, being part of the movement, aim at aircraft larger than 50 seats.

However, there are many challenges ahead that need to be overcome before the first hydrogen-powered regional aircraft can enter service. Regarding the fuel, there is currently no infrastructure at airports for storing hydrogen and refueling aircraft. There is also no sufficient production for blue or green hydrogen as of now, including the means of distribution from production facilities to consumers. Regarding the technical side, many technologies are still evolving and will need multiple years to achieve market readiness. Right now, no 1 MW superconducting electric motors for the aviation industry exist in the market, and fuel cells are still being tested.

The true impact of the thermal management system on aircraft performance has to be evaluated. From a legal point of view, procedures and criteria for the certification, production, maintenance and operation of hydrogen aircraft have to be developed. In order to be able to make more precise statements concerning the actual effects of these new technologies utilized in the proposed aircraft, more detailed studies using CFD and FEM may be considered for future investigations.

Author Contributions: Conceptualization, methodology, software, formal analysis, validation, writing—original draft preparation, J.E., S.L., C.M.B., J.S. and T.W.; writing—review and editing, J.E., S.L., C.M.B., J.S., T.W., J.M. and A.S.; supervision, A.S. and J.M.; funding acquisition, S.L. All authors have read and agreed to the published version of the manuscript.

Funding: This research project has received funding from the European Union's Horizon 2020 research and innovation programme under grant agreement No 875551. The student team also thanks Stuttgart Airport for the financial support for both the trip to present at the EASN Conference in Barcelona as well as for the trip to Amsterdam to meet with Embraer.

Data Availability Statement: The data presented in this study are available on request from the corresponding author. The data are not publicly available due to being in German language.

Acknowledgments: The findings presented here were studied, acquired and prepared by the student teams independently. The teams were able to obtain support from the expertise of the FUTPRINT50 Consortium. However, the statements made herein do not necessarily have the consent or agreement of the FUTPRINT50 Consortium. These represent the opinion and investigations of the author(s).

Conflicts of Interest: The authors declare no conflict of interest.

Abbreviations

The following abbreviations are used in this manuscript:

BLI	Boundary Layer Ingestion
BoP	Balance of Plant
C-Rate	Charge Rate
CASK	Cost per Available Seat Kilometer
CFD	Computational Fluid Dynamics
CFRP	Carbon Fiber-Reinforced Polymers
CoG	Center of Gravity
CO_2	Carbon Dioxide
DC/AC	Direct Current to Alternating Current converter
DC/DC	Direct Current to Direct Current converter
DEP	Distributed Electric Propulsion
DOC	Direct Operating Costs
EGTS	Electric Green Taxiing System
FEM	Finite Element Method
FL	Flight Level
FutPrInt50	Future propulsion and integration: towards a hybrid-electric 50-seat regional aircraft
H_2	Hydrogen
HAIQU	Hydrogen Aircraft designed for Quick commuting
HTS	High Temperature Superconducting
ISA	International Standard Atmosphere
L/D	Lift-to-Drag Ratio
LFP	Lithium Iron Phosphate
LCO	Lithium Cobalt Oxide
LH_2	Liquid Hydrogen
MAC	Mean Aerodynamic Chord
Ma	Mach number
MTOM	Maximum Take-Off Mass
MTOW	Maximum Take-Off Weight
NACA	National Advisory Committee for Aeronautics
NMC	Lithium Nickel Manganese Cobalt Oxide
NO_x	Nitrogen Oxides
OEI	One Engine Inoperative
OME	Operating Mass Empty
PAX	Passengers
PEMFC	Proton Exchange Membrane Fuel Cell
SCM	Superconducting Motor
SL	Sea Level
SMR	Steam Methane Reforming
TLAR	Top Level Aircraft Requirements
TMS	Thermal Management System
TOFL	Take-Off Field Length
TRL	Technology Readiness Level
VTP	Vertical Tail Plane
WTP	Wing Tip Propulsion

References

1. Bisignani, G. *Vision 2050;* Technical Report; IATA: Montreal, QC, Canada, 2011.
2. ATAG. WAYPOINT 2050. Available online: https://aviationbenefits.org/media/167417/w2050_v2021_27sept_full.pdf (accessed on 15 November 2022).
3. IATA. *Aircraft Technology Roadmap to 2050;* Technical Report; IATA: Geneva, Switzerland, 2020.
4. European Commission and Directorate—General for Research and Innovation. *Fly the Green Deal: Europe's Vision for Sustainable Aviation;* Publications Office of the European Union: Brussels, Belgium, 2022. [CrossRef]

5. FutPrInt50. Second FutPrInt Academy. Available online: https://futprint50.eu/2nd-futprint-academy (accessed on 1 April 2022).
6. Ustolin, F.; Campari, A.; Taccani, R. An Extensive Review of Liquid Hydrogen in Transportation with Focus on the Maritime Sector. *J. Mar. Sci. Eng.* **2022**, *10*, 1222. [CrossRef]
7. Marchant, N. World Economic Forum. Available online: https://www.weforum.org/agenda/2021/07/clean-energy-green-hydrogen/ (accessed on 16 December 2022).
8. Yusaf, T.; Fernandes, L.; Abu Talib, A.R.; Altarazi, Y.S.M.; Alrefae, W.; Kadirgama, K.; Ramasamy, D.; Jayasuriya, A.; Brown, G.; Mamat, R.; et al. Sustainable Aviation—Hydrogen Is the Future. *Sustainability* **2022**, *14*, 548. [CrossRef]
9. Leight, C. Recharge. Available online: https://www.rechargenews.com/transition/a-wake-up-call-on-green-hydrogen-the-amount-of-wind-and-solar-needed-is-immense/2-1-776481 (accessed on 25 August 2022).
10. Eisenhut, D.; Moebs, N.; Windels, E.; Bergmann, D.; Geiß, I.; Reis, R.; Strohmayer, A. Aircraft requirements for sustainable regional aviation. *Aerospace* **2021**, *8*, 61. [CrossRef]
11. Torenbeek, E. *Advanced Aircraft Design*; Wiley: Hoboken, NJ, USA, 2013. [CrossRef]
12. ATR-Aircraft. ATR42-600. Available online: https://www.atr-aircraft.com/wp-content/uploads/2022/06/ATR_Fiche42-600-3.pdf (accessed on 7 July 2022).
13. ATR-Aircraft. ATR Aircraft Family. Available online: https://www.atr-aircraft.com/our-aircraft/aircraft-family/ (accessed on 24 February 2023).
14. Roskam, J. *Airplane Design: Part I, Preliminary Sizing of Airplanes*; Roskam Aviation and Engineering Corp.: Toronto, ON, Canada, 1985.
15. Staack, I.; Sobron, A.; Krus, P. The potential of full-electric aircraft for civil transportation: From the Breguet range equation to operational aspects. *CEAS Aeronaut. J.* **2021**, *12*, 803–819. [CrossRef]
16. Geiß, I.; Voit-Nitschmann, R. Sizing of the energy storage system of hybrid-electric aircraft. *CAES Aeronaut. J.* **2017**, *231*, 53–65. [CrossRef]
17. Roskam, J. *Airplane Design: Part II, Preliminary Configuration Design and Integration of the Propulsion System*; Roskam Aviation and Engineering Corp.: Toronto OT, Canada, 1985.
18. Raymer, D. *Aircraft Design: A Conceptual Approach*, 5th ed.; AIAA: Washington, DC, USA, 2012. [CrossRef]
19. McCormick, B.W. *Aerodynamics, Aeronautics, and Flight Mechanics*; Wiley: Hoboken, NJ, USA, 1994.
20. Hale, F.J. Aircraft Performance and Design. In *Encyclopedia of Physical Science and Technology (Third Edition)*; Academic Press: San Diego, CA, USA, 2003; Volume 3, pp. 365–397.
21. Rendón, M.A.; Sánchez R., C.D.; Gallo M., J.; Anzai, A. H. Aircraft Hybrid-Electric Propulsion: Development Trends, Challenges and Opportunities. *J. Control. Autom. Electr. Syst.* **2021**, *32*, 1244–1268. [CrossRef]
22. FutPrInt50. e-Workshop on "Aircraft Design Specs & Mission". Available online: https://futprint50.eu/news/e-workshop-aircraft-design-specs-mission (accessed on 12 December 2022).
23. Sustainable Skies. DLR's Novel Configurations. Available online: https://sustainableskies.org/dlrs-novel-configurations/ (accessed on 12 December 2022).
24. de Vries, R.; Brown, M.; Vos, R. Preliminary Sizing Method for Hybrid-Electric Distributed-Propulsion Aircraft. *J. Aircr.* **2019**, *56*, 2172–2188. [CrossRef]
25. Minervino, M.; Andreutti, G.; Russo, L.; Tognaccini, R. Drag Reduction by Wingtip-Mounted Propellers in Distributed Propulsion Configurations. *Fluids* **2022**, *7*, 212. [CrossRef]
26. Pfeifle, O.; Notter, S.; Fichter, W.; Bergmann, D.P.; Denzel, J.; Strohmayer, A. Verifying the Effect of Wingtip Propellers on Drag Through In-Flight Measurements. *J. Aircr.* **2022**, *59*, 474–483. [CrossRef]
27. Blaesser, N.J. *Propeller-Wing Integration on the Parallel Electric-Gas Architecture with Synergistic Utilization Scheme (PEGASUS) Aircraft*; American Institute of Aeronautics and Astronautics: Washington, DC, USA, 2019.
28. Sinnige, T.; van Arnhem, N.; Stokkermans, T.C.A.; Eitelberg, G.; Veldhuis, L.L.M. Wingtip-Mounted Propellers: Aerodynamic Analysis of Interaction Effects and Comparison with Conventional Layout. *J. Aircr.* **2019**, *56*, 295–312. [CrossRef]
29. Jenkinson, L.; Simpkin, P.; Rhodes, D. *Civil Jet Aircraft Design*; American Institute of Aeronautics and Astronautics: Washington, DC, USA, 1999. [CrossRef]
30. Wang, L.; Fan, H.; Chu, J.; Chen, D.; Yu, S. Effect of Personal Space Invasion on Passenger Comfort and Comfort Design of an Aircraft Cabin. *Math. Probl. Eng.* **2021**, 1–15. [CrossRef]
31. Bagassi, S.; Lucchi, F.; Persiani, F. Aerospace—Europe. Available online: https://aerospace-europe.eu/media/books/CEAS2015_211.pdf (accessed on 15 November 2022).
32. Bombardier. Smartcockpit. Available online: https://www.smartcockpit.com/docs/Q400-Landing_Gear_1.pdf (accessed on 12 August 2022).
33. Deonandan, I.; Balakrishnan, H. Evaluation of Strategies for Reducing Taxi-Out Emissions at Airports. In Proceedings of the 10th AIAA Aviation Technology, Integration, and Operations (ATIO) Conference, Fort Worth, TX, USA, 13–15 September 2010. [CrossRef]
34. Re, F. Viability and State of the Art of Environmentally Friendly Aircraft Taxiing Systems. In Proceedings of the 2012 Electrical Systems for Aircraft, Railway and Ship Propulsion, Bologna, Italy, 16–18 October 2012. [CrossRef]

35. Lukic, M.; Giangrande, P.; Hebala, A.; Nuzzo, S.; Galea, M. Review, Challenges, and Future Developments of Electric Taxiing Systems. *IEEE Trans. Transp. Electrif.* **2019**, *5*, 1441–1457. [CrossRef]
36. Heinrich, M.T.E.; Kelch, F.; Magne, P.; Emadi, A. Regenerative Braking Capability Analysis of an Electric Taxiing System for a Single Aisle Midsize Aircraft. *IEEE Trans. Transp. Electrif.* **2015**, *1*, 298–307. [CrossRef]
37. Cheaito, H.; Allard, B.; Clerc, G. Proof of concept of 35 kW electrical taxiing system in more electrical aircraft for energy saving. *Int. J. Electr. Power Energy Syst.* **2021** , *130*, 106882. [CrossRef]
38. Lukic, M.; Hebala, A.; Giangrande, P.; Klumpner, C.; Nuzzo, S.; Chen, G.; Gerada, C.; Eastwick, C.; Galea, M. State of the Art of Electric Taxiing Systems. In Proceedings of the 2018 IEEE International Conference on Electrical Systems for Aircraft, Railway, Ship Propulsion and Road Vehicles & International Transportation Electrification Conference (ESARS-ITEC), Nottingham, UK, 7–9 November 2018. [CrossRef]
39. Heinrich, M.T.E.; Kelch, F.; Magne, P.; Emadi, A. Investigation of regenerative braking on the energy consumption of an electric taxiing system for a single aisle midsize aircraft. In Proceedings of the IECON 2014—40th Annual Conference of the IEEE Industrial Electronics Society, Dallas, TX, USA, 29 October–1 November 2014. [CrossRef]
40. IHS Jane's. *Jane's All the World's Aircraft 2010–2011*; IHS Jane's: Coulsdon, UK, 2011.
41. Komiya, M.; Aikawa, T.; Sasa, H.; Miura, S.; Iwakuma, M.; Yoshida, T.; Sasayama, T.; Tomioka, A.; Konno, M.; Izumi, T. Design Study of 10 MW REBCO Fully Superconducting Synchronous Generator for Electric Aircraft. *IEEE Trans. Appl. Supercond.* **2019**, *29*, 1–6. [CrossRef]
42. Lukaczyk, T.W.; Wendorff, A.D.; Colonno, M.; Economon, T.D.; Alonso, J.J.; Orra, T.H.; Ilario, C. SUAVE: An Open-Source Environment for Multi-Fidelity Conceptual Vehicle Design. In Proceedings of the 16th AIAA/ISSMO Multidisciplinary Analysis and Optimization Conference, Dallas, TX, USA, 22–26 June 2015. [CrossRef]
43. Von Croy, A. Airwork. Available online: https://airwork.biz/wp-content/uploads/2009/03/atr_e1.pdf (accessed on 2 June 2022).
44. Hale, J. Boeing 787 from the Ground Up. Available online: https://www.boeing.com/commercial/aeromagazine/articles/qtr_4_06/AERO_Q406_article4.pdf (accessed on 14 July 2022).
45. Kolb-Geßmann, S. Entwurf Eines 50-Sitzigen Regionalflugzeugs für Einen EIS in 2035/2040. Master's Thesis, Institute of Aircraft Design, University of Stuttgart, Stuttgart, Germany, 2021.
46. European Union Aviation Safety Agency. Easy Access Rules for Air Operations (Regulation (EU) No 965/2012). Available online: https://www.easa.europa.eu/document-library/easy-access-rules/easy-access-rules-air-operations (accessed on 23 February 2023).
47. Hartmann, C.; Noland, J.K.; Nilssen, R.; Mellerud, R. Dual Use of Liquid Hydrogen in a Next-Generation PEMFC-Powered Regional Aircraft With Superconducting Propulsion. *IEEE Trans. Transp. Electrif.* **2022**, *8*, 4760–4778. [CrossRef]
48. Piancastelli, L.; Frizziero, L.; Donnici, G. The Meredith Ramjet: An Efficient Way to Recover the Heat Wasted in Piston Engine Cooling. *ARPN* **2015**, *10*, 10–12.
49. Kellermann, H.; Lüdemann, M.; Pohl, M.; Hornung, M. Design and Optimization of Ram Air–Based Thermal Management Systems for Hybrid-Electric Aircraft. *Aerospace* **2020**, *8*, 3. [CrossRef]
50. Chapman, J.W.; Schnulo, S.L.; Nitzsche, M.P. Development of a Thermal Management System for Electrified Aircraft. In Proceedings of the AIAA Scitech 2020 Forum, Orlando, FL, USA, 6–10 January 2020. [CrossRef]
51. Silberhorn, D.; Atanasov, G.; Walther, J.N.; Zill, T. *Assessment of Hydrogen Fuel Tank Integration at Aircraft Level*; Technical Report; DLR: Cologne, Germany, 2019.
52. Saldaña, G.; Martín, J.I.S.; Zamora, I.; Asensio, F.J.; Oñederra, O. Analysis of the Current Electric Battery Models for Electric Vehicle Simulation. *Energies* **2019**, *12*, 2750. [CrossRef]
53. GWL. How to Operate a Lifepo4 Battery. Available online: https://files.gwl.eu/inc/_doc/LFP_Guide_ENG.pdf (accessed on 12 January 2023).
54. Karimov, V. New Tests Prove: LFP Lithium Batteries Live Longer than NMC. Available online: https://www.onecharge.biz/es/sin-categorizar/lfp-lithium-batteries-live-longer-than-nmc/ (accessed on 21 August 2022).
55. Stenzel, P.; Baumann, M.; Fleer, J.; Zimmermann, B.; Weil, M. Database development and evaluation for techno-economic assessments of electrochemical energy storage systems. In Proceedings of the IEEE International Energy Conference (ENERGYCON), Cavtat, Croatia, 13–16 May 2014. [CrossRef]
56. Batteryuniversity. Available online: https://batteryuniversity.com/article/bu-205-types-of-lithium-ion (accessed on 16 August 2022).
57. Yoon, A.; Yi, X.; Martin, J.; Chen, Y.; Haran, K. A high-speed, high-frequency, air-core PM machine for aircraft application. In Proceedings of the 2016 IEEE Power and Energy Conference at Illinois (PECI), Urbana, IL, USA, 19–20 February 2016; pp. 1–4 . [CrossRef]
58. Luongo, C.A.; Masson, P.J.; Nam, T.; Mavris, D.; Kim, H.D.; Brown, G.V.; Waters, M.; Hall, D. Next Generation More-Electric Aircraft: A Potential Application for HTS Superconductors. *IEEE Trans. Appl. Supercond.* **2009**, *19*, 1055–1068. [CrossRef]
59. Mangold, J.; Silberhorn, D.; Moebs, N.; Dzikus, N.; Hoelzen, J.; Zill, T.; Strohmayer, A. Refueling of LH2 Aircraft—Assessment of Turnaround Procedures and Aircraft Design Implication. *Energies* **2022**, *15*, 2475. [CrossRef]
60. Airbus. Aircraft Characteristics—Airport and Maintenance Planning. Available online: https://www.airbus.com/sites/g/files/jlcbta136/files/2021-11/Airbus-Commercial-Aircraft-AC-A320.pdf (accessed on 22 January 2023).

61. Fuchte, J.C. Enhancement of Aircraft Cabin Design Guidelines with Special Consideration of Aircraft Turnaround and Short Range Operations. Available online: https://elib.dlr.de/89599/1/Fuchte%20FB-2014-17%20Version%20Druck.pdf (accessed on 22 January 2023).
62. Boeing. 737 Airplane Characteristics for Airport Planning. Available online: https://archive.aoe.vt.edu/mason/Mason_f/B737.pdf (accessed on 22 January 2023).
63. Gomez, F. Design Evaluation/DOC. Available online: https://www.fzt.haw-hamburg.de/pers/Scholz/HOOU/AircraftDesign_14_DOC.pdf (accessed on 18 August 2022).
64. Al-Shamma, O.; Ali, R. *A Comparative Study of Cost Estimation*; Technical Report: Coventry University's Repository: Coventry, UK, 2014.
65. ATA. *Standard Method of Estimating Comparative Direct Operating Costs of Turbine Powered Transport Airplanes*; Technical Report; ATA: Albuquerque, NM, USA, 1967.
66. Liebeck, R.H.; Andrastek, D.A.; Chau, J.; Girvin, R.; Lyon, R.R.; Rawdon, B.K.; Scott, P.W.; Wright, R. *Advanced Subsonic Airplane Design and Economic Studies*; Technical Report; NASA CR-195443; NASA: Washington, DC, USA, 1995.
67. AEA. *Short-Medium Range Aircraft—AEA*; Technical Report; AEA: Nashville, TN, USA, 1989.
68. Evans, H. Fuel Cell Price to Drop 70–80% as Production Volume Scales. Available online: https://www.spglobal.com/commodityinsights/en/market-insights/latest-news/electric-power/112020-green-hydrogen-costs-need-to-fall-over-50-to-be-viable-sampp-global-ratings (accessed on 1 December 2022).
69. Mulholland, E.; Rogan, F.; Gallachóir, B.P.Ó. Techno-economic data for a multi-model approach to decarbonisation of the Irish private car sector. *Data Brief* **2017**, *15*, 922–932. [CrossRef] [PubMed]
70. Scott, M.S.; Idriss, H. Market Dynamics to Drag Green Hydrogen to $1.50/kg by 2030. 2022. Available online: https://doi.org/10.1002/9783527628698.hgc022 (accessed on 15 December 2022).
71. Schiavo, M.; Nietvelt, K. How Hydrogen Can Fuel The Energy Transition. Available online: https://www.spglobal.com/ratings/en/research/articles/201119-how-hydrogen-can-fuel-the-energy-transition-11740867 (accessed on 9 December 2022).
72. Johanning, A.; Scholz, D. *Evaluation of Worldwide Noise and Pollutant Emission*; Technical Report; Hamburg University of Applied Sciences: Hamburg, Germany, 2012.
73. European Commission. Flightpath 2050: Europe's Vision for Aviation: Maintaining Global Leadership and Serving Society's Needs. Available online: https://data.europa.eu/doi/10.2777/50266 (accessed on 26 December 2022).
74. HAIQU. HAIQU–Future Regional Aircraft. Available online: https://www.haiqu.de.cool (accessed on 30 October 2022).
75. Niţă, M.F. *Aircraft Design Studies Based on the ATR 72*; Technical Report; Department of Automotive and Aeronautical Engineering—HAW Hamburg: Hamburg, Germany, 2008.
76. ATR-aircraft. ATR 42-600. Available online: https://www.atr-aircraft.com/wp-content/uploads/2020/07/Factsheets_-_ATR_42-600.pdf (accessed on 23 February 2023).
77. Novelli, P. Sustainable Way for Alternative Fuels and Energy in Aviation. Available online: http://large.stanford.edu/courses/2012/ph240/greenbaum1/docs/SW_WP9_D.9.1-July2011.pdf (accessed on 23 February 2023).
78. FlightGlobal. Embraer Shifts 'Energia' Focus to New Hybrid- and Hydrogen-Powered Concepts. Available online: https://www.flightglobal.com/airframers/embraer-shifts-energia-focus-to-new-hybrid-and-hydrogen-powered-concepts/151225.article (accessed on 29 December 2022).
79. Airbus. Airbus Reveals Hydrogen-Powered Zero-Emission Engine. Available online: https://www.airbus.com/en/newsroom/press-releases/2022-11-airbus-reveals-hydrogen-powered-zero-emission-engine (accessed on 29 December 2022).

Article

Enabling Green Approaches by FMS-AMAN Coordination

Nils Ahrenhold *, Izabela Stasicka , Rabeb Abdellaoui , Thorsten Mühlhausen and Marco-Michael Temme

German Aerospace Center (DLR) Braunschweig, Institute of Flight Guidance, Lilienthalplatz 7, 38108 Braunschweig, Germany
* Correspondence: nils.ahrenhold@dlr.de; Tel.: +49-531-295-1184

Abstract: Growing political pressure and widespread social concerns about climate change are triggering a paradigm shift in the aviation sector. Projects with the target of reducing aviation's CO_2 emissions and their impact on climate change are being launched to improve currently used procedures. In this paper, a new coordination process between aircraft flight management systems (FMSs) and an arrival manager (AMAN) was investigated to enable fuel-efficient and more sustainable approaches. This coordination posed two major challenges. Firstly, current capacity-centred AMANs' planning processes are not optimised towards fuel-efficient trajectories. To investigate the benefit of negotiated trajectories with fixed target times for waypoints and thresholds, the terminal manoeuvring area was redesigned for an independent parallel runway system. Secondly, the FMS-AMAN negotiation process plan the trajectories based on time, whereas air traffic controllers guide traffic based on distance. Three tactical assisting tools were implemented in an air traffic controller's working position to enable a smooth transition from distance-based to time-based coordination and guidance. The whole concept was implemented and tested in real-time human-in-the-loop studies at DLR's Air Traffic Validation Center. Results showed that the new airspace design and concept was feasible, and a reduction in flown distance was measured.

Keywords: green approaches; FMS; AMAN; CDA; negotiation process

Citation: Ahrenhold, N.; Stasicka, I.; Abdellaoui, R.; Mühlhausen, T.; Temme, M.-M. Enabling Green Approaches by FMS-AMAN Coordination. *Aerospace* **2023**, *10*, 278. https://doi.org/10.3390/aerospace10030278

Academic Editors: Spiros Pantelakis, Andreas Strohmayer and Jordi Pons-Prats

Received: 2 February 2023
Revised: 8 March 2023
Accepted: 9 March 2023
Published: 11 March 2023

1. Introduction

Addressing environmental challenges, especially global warming, is more than ever a must for the community [1]. This matter is becoming an increasing priority at the regional and global level, which was already investigated by Crompton in 2009 [2]. Europe has made commitments to reduce aviation's environmental footprint [3], not only because of growing political pressure, but also due to the widespread social concern about climate change. This is triggering a paradigm shift in the aviation sector, since the sector is contributing to climate change, increasing noise, affecting local air quality and consequently affecting the health and quality of life of European citizens [4,5]. In 2020, air traffic movements drastically reduced due to the COVID-19 pandemic [6]. Currently, at the end of 2022, the number of movements still being below that of pre-pandemic times [7]. Gudmundsson expects that it will require up to five to ten years to recover to 2019 numbers of air traffic movements [8]. The dramatic reduction in flights is not only considered as negative. On the contrary, it could be seen as an opportunity to rebuild the system and make the air traffic sector greener than before the pandemic. In this context, Brouder and Ateljevic described the extensive economic reset evoked by COVID-19 but also the possibility for a fresh start [9,10]. From a general perspective, the air traffic in Europe was growing until 2019 and is expected to continue increasing significantly in the future again in order to cope with the growing demand for mobility and connectivity [11]. For example, Dube and Gössling assessed the rapid impact of pandemic control measures on the air transport sectors and the prospects for recovery of the global aviation industry [11,12].

The long-term effects on the environment from the aviation sector, mainly caused by aircraft noise and exhaust gases (especially CO_2, nitrogen oxides NO_x and methane), make aviation's environmental footprint a clear target for mitigation efforts. The future growth of air traffic shall go hand in hand with environmentally sustainability policies. Therefore, studies and research are being conducted especially in Europe, exploring possible optimisation of aircraft technologies and air traffic management (ATM) operations. This is investigated by Bolić and Ravenhill within the Single European Sky ATM Research (SESAR) projects [13]. One possible starting point is the optimisation of ground movements at the airport. Within the EPISODE 3 project [14] and the EMMA project [15,16], advanced surface movement guidance systems were developed to design airport movements more efficiently. Furthermore, the use, optimisation and practical implementation of 4D trajectories, including a time parameter, was investigated by [17]. Given the close interdependence between aircraft routing and the resulting impact on the environment, optimisation of flight trajectory design and air traffic control (ATC) operations are appropriate means of reducing emissions in short and medium-term periods. Another target is the analysis of airspace conditions to mitigate delays due to overload. Analysing the air traffic complexity in the approach phase can lead to multi-sector planning operations, which detect overload and reduce delays. The Harmonised ATM Research in Eurocontrol (PHARE) [18] developed a algorithm to calculate the air traffic complexity.

Within the European funded project GreAT (Greener Air Traffic Operations), this paper investigates a new concept for the approach phase with a slightly extended scheduling horizon from 50 to 125 Nautical miles (NM) [19]. The main goal is to enable more sustainable approach procedures in terms of fuel-optimised approaches. This is realised by a new coordination system between aircraft flight management systems (FMSs) and arrival manager (AMAN) for guidance of inbound traffic. For that purpose, a new airspace structure and controller system enhancements for arrival and departure management are investigated within this project. In the following sections, all new concept elements are briefly introduced. In addition, the discrepancy between time-based and currently used distance-based planning is examined. These concept elements and the tactical assisting systems are the basis for the final human-in-the-loop (HITL) validation trials. According to EUROCONTROL's European Operational Concept Validation Methodology (E-OCVM), HITL simulations are appropriate techniques to receive objective and subjective outputs [20]. These outputs are generated by collecting data from simulator logs, observer notes, questionnaires and debriefings. The described study aims to test the system and concept improvements based on the advanced ATM procedures. Furthermore, the aim is to assess the reduction in fuel consumption by operational parameters and therefore, assess the reduction in greenhouse gas emissions too. Aside from the operational parameters directly linked to environmental sustainability, such as flown distance and number of landed aircraft, the air traffic controllers' (ATCOs) mental workload, situation awareness and perceived safety are analysed.

The concept of mental workload is described by Eggemeier and O'Donnell in [21]. An ATCO's mental workload is related to the requirements of the control tasks performed. As a new airspace design was proposed within the GreAT Project, it is important to assess if the ATCOs can handle the traffic within the new airspace structure while maintaining an acceptable level of mental workload. On the one hand, an unknown simulated operational environment with unfamiliar tools, including the radar display and new functionalities in the controller working position (CWP), may induce an additional workload. The extent of the increase will depend on the complexity and measures required to handle the new functions. On the other hand, an automated process of the 4D-FMS aircraft could also cause mental underload. It is furthermore essential that the ATCOs perceive operations as safe and that they maintain an adequate level of situation awareness. According to Endsley [22], situation awareness involves (a) the perception of the elements in the environment, (b) the comprehension of the current situation and (c) the projection of the future status.

For a medium–large-scale airport, such as Munich airport (EDDM), it is assumed that fuel-optimised procedures are not feasible due to the current airspace structure, average traffic flows and applied planning horizon. This hypothesis is underlined by two crucial reasons. Firstly, current AMANs are developed with regard to increased capacity and to support ATCOs at scheduling and sequencing of inbound traffic. Thus, the AMAN's planning process is not optimised towards fuel-efficient trajectories, rather than optimising the capacity. Enabling greener approaches at medium to large sized airports with less CO_2 emissions, such as long-distance independent approaches, depends upon a redesign of the airspace structure and extension of the capacity-centred AMAN calculation. Since these independent long-distance approach procedures start at the top of decent and end on the final decent, aircraft's speed profiles are unknown for ATCOs, which requires more space for coordination with standard approaches [23]. Therefore, a completely new terminal manoeuvring area (TMA), a so-called the extended-TMA (E-TMA), was designed for the independent parallel runway system of EDDM. Additionally, a trajectory negotiation process between the aircraft's FMS and the in-house developed AMAN was established to enable long-distance independent approach procedures. Lastly, tactical supporting tools have been developed and provided for ATCOs to enable a time-based aircraft guidance system instead of the conventional distance-based guidance system, since the guidance of fuel-optimised approach routes follows the time principle in order to meet the negotiated target times, although occasionally, aircraft deviate from their optimum profiles. In the following, each concept will be introduced one after another.

2. Materials and Methods

2.1. Extended Terminal Manoeuvring Area

The redesign of the EDDM TMA consists of four steps, which is simplified in Figure 1. First of all, the EDDM TMA was extended from a range of 50 to 125 NM. Secondly, when crossing the border from an outside sector of the E-TMA, all aircraft were distributed on direct routes leading towards the runway system. Thirdly, aircraft were distinguished by their technical functionality in their onboard FMSs. Thus, approaching traffic was sorted into two categories.

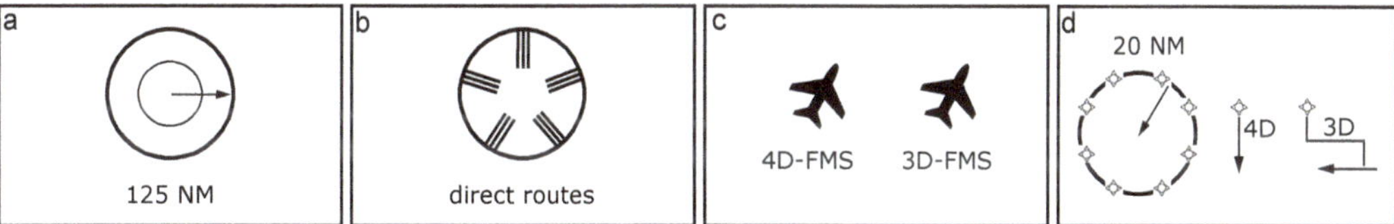

Figure 1. Four fundamental steps to enable green approaches: (**a**) extend the terminal manoeuvring area to 125 NM, (**b**) enable direct routes towards the airport, (**c**) distinguish aircraft by flight management system functionality and (**d**) introduce aircraft separation points and different routes to the end point.

Aircraft equipped with common FMSs, autopilots and no or only simple data links, such as Controller Pilot Data Link Communications (CPDLCs), were categorised. In this concept, these were referred to as 3D-FMS aircraft or non-equipped aircraft. They were able to perform a flight along a calculated trajectory but did not have the ability to meet a target time with less than twenty seconds of reliability, since they cannot sufficiently compensate changing wind conditions with an influence on their own airspeed. Additionally, the limited bandwidth of the data link did not allow a target time negotiation between the FMS and AMAN.

The second category had by aircraft equipped with an advanced FMS or 4D-FMS and a broadband data link. These were referred to as 4D-FMS aircraft and had the ability to perform a long-distance independent approach on a defined route with negotiated target times. With a 4D-FMS, aircraft had the capability to fly along a predefined 4D-trajectory and meet the target times at all points of the way with divergence of less than plus or minus six seconds. Deviations in route, altitude and speed due to changing wind conditions were automatically compensated by the 4D-FMS, even if this may mean a divergence from the optimal approach profile.

In the fourth step to adapt the E-TMA, aircraft separation points (ASP) were introduced. Those ASPs were located around the airport with a distance of around 20 NM to the runways and had nearly the same functionality, such as TMA entry fixes today. The difference from traditional entry fixes is that at this point, the aircraft with differing FMS equipage are split. 4D-FMS aircraft followed a direct route to the direct-only merge points (DOMP), located on the right and left sides of the final approaches. This DOMPs had the task to serve as stream collection points only for the 4D-FMS aircraft from one compass direction. 3D-FMS aircraft were guided conventionally from the ASPs—the downwind transition by the ATCOs.

To separate the inbound streams of 4D-FMS and 3D-FMS aircraft in the area between ASPs and final approaches, the downwind intercept altitude was 8000 ft. In this way, the direct approaches overtook the standard approaches at the possible crossing points. If there was more than one aircraft heading to the same ASP, the wake vortex separation was established with the traffic distribution to nearby ASPs before entering the TMA. If too many aircraft arrived at one ASP at the same time, additional speed and level clearances were advised. In the event of conflicts at the ASPs, the first arrival received its optimal trajectory. If additional aircraft arrived at the ASPs at the same time and could not be guided without conflict due to their optimal target time window, they were automatically treated as conventional aircraft and manually guided over the conventional routes. Figures 2 and 3 illustrate the designed model in three dimensions. The cyan routes correspond to the direct routes towards the airport, purple marks the E-TMA boundaries, direct approaches for 4D-FMS aircraft are displayed in green and the conventional routes for 3D-FMS aircraft are in orange. All remaining blue routes depict designed departure routes.

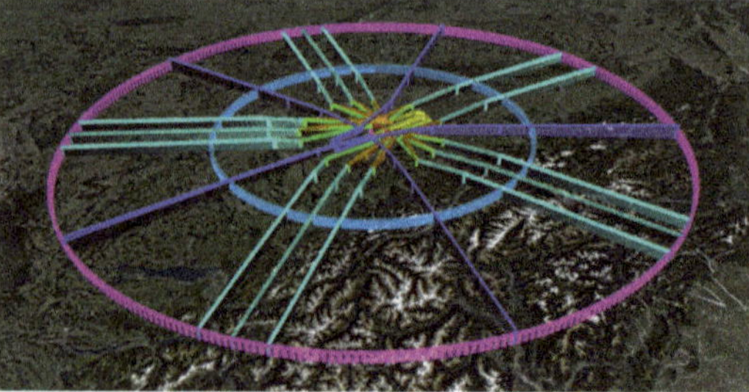

Figure 2. E-TMA: cyan = direct routes, purple = E-TMA boundaries, green = 4D paths, orange = 3D paths, blue = departure, red = direct-only merge points and routes.

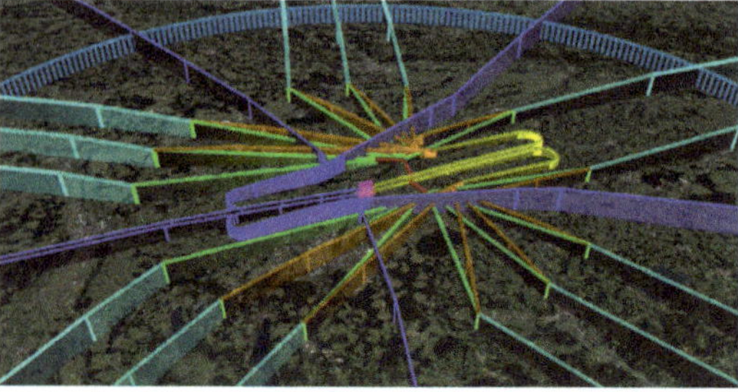

Figure 3. E-TMA with distinction into 3D (orange lines) and 4D (green lines) arrival paths. Red = direct-only merge points and routes, purple = touchdown areas.

2.2. Tactical Assistance Systems

To support approach ATCOs sequencing the inbound traffic in the GreAT airspace structure around an airport, three systems were developed or refined as tactical support systems. In order to obtain an early picture of the target times of 4D-FMS approaches relative to 3D-FMS approaches, the label projection technique "ghosting" was used and extended for continuous decent approaches (CDA). *Ghosting* is the method of projecting an aircraft's label on a radar display on a different route in order to make it easier for the ATCOs to merge two routes at one waypoint [24]. The ghost position is located where, based on current performance, the aircraft would be if flying that route. The visualisation of the arrival slots planned by the AMAN was carried out on the centreline and the final approach with the help of *TargetWindows*. Finally, a precise numerical check of the planned and observed separations was made possible by the *Centerline Separations Visualisation Tool*.

2.2.1. Time-Based Ghosting

Separation between ghost and real aircraft on different routes then showed the actual relative temporal spacing between those objects, as if both aircraft were on the same route [25]. This was originally done for two arrival streams on converging runways simulating a dependent parallel approach [26]. Two different methods can be used to calculate ghost-label positions: Time-based and distance-based ghosting. Distance-based ghosting can be used without problems for regular arrival routes, where two approach streams are merged on which the aircraft move with the same standardised approach procedure and speed [27]. The merging of approach streams with different approach procedures and speeds poses new challenges. These can be partially solved if a time-based "segmented ghosting" with dynamic approach speeds is used for the ghost-label's position calculation [28].

One of the tasks of approach ATCOs in the GreAT study was the merging of manually guided and CDA-performing aircraft with different speed profiles at the late merging point (LMP) onto the jointly used last six nautical miles of the final approach. A particular challenge for ATCOs was that they did not know the speed profiles of the 4D-FMS aircraft. Due to this reason, a time-based form of ghosting was developed, since here, aircraft with significantly different speed profiles had to be merged and therefore projected onto one route [29]. In time-based segmented ghosting, the current ghost-label position is calculated using the negotiated target time of the original aircraft at the late merging point (LMP) and calculating from this point in time back to the actual time to locate the position an aircraft would have if moving with a standard speed profile already on the final approach. The LMP represents the location where the 4D-FMS aircraft and ghost meet on the final approach. For the movement of the ghost on the final approach, a rudimentary flight simulator was implemented in the AMAN, which calculated the aircraft movements along the typical speed profile of a standard approach on the final approach and moved the ghost accordingly (see in Figure 4). Thereby, the phases of speed reduction and constant speeds were tuned for each aircraft so that ghost and real aircraft finally met at the LMP at the negotiated target time. If an aircraft deviates from its negotiated target time, it will also not meet its ghost at the LMP. However, since it will also cause a conflict on the final approach in the event of a deviation, it must then be downgraded to a standard approach and guided conventionally via the downwind and base leg to the final approach. In this case, it loses all the advantages of the direct approach. However, during the validation, we assumed that modern 4D-FMS can accurately maintain a fixed target time for a waypoint with only a few seconds of deviation even under unfavourable wind conditions. During pre-validations, it was shown that it is sufficient for ATCOs, regarding safety aspects, if the ghost label is faded out thirty seconds before reaching the LMP, instead of showing the ghost until aircraft meet at the LMP.

In this way, the approach ATCO was able to implement the distances between manually guided 3D-FMS aircraft and the 4D-FMS aircraft (callsign CDA123, CDA987) on the final approach, while being sure that they can be maintained all the way to the LMP.

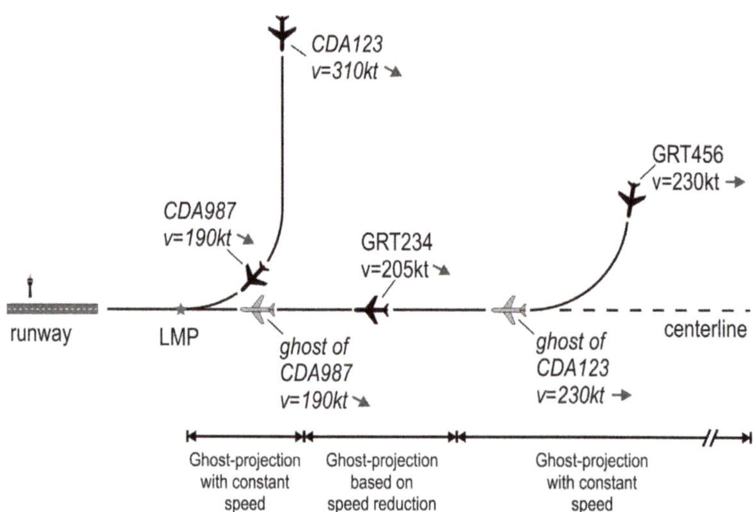

Figure 4. The working principle of 3-segment ghosting. GRT234 and GRT456 are regularly guided aircraft; CDA123 and CDA987 are aircraft (4D-FMS) conducting an long-distance independent approach. The two CDAs are "ghosted" onto the final approach and centreline by adjusting their position calculations in the typical approach procedure of the manually guided aircraft.

2.2.2. TargetWindow

Another optical supporting function used for guidance assistance in operational environments for 3D-FMS aircraft is an indicator in the form of a target circle or arrow on the centreline and final approach. This target circle visualises a position for the merging of two arrival streams. These systems also do consider several turns of an aircraft [30,31]. Another approach is using "slot marker" circles to show the aircraft's expected position along its trajectory if it were conforming to the schedule [32]. Similar target position indicators may also be used for certain waypoints in upper airspace, for wake vortices [33] or in lower airspace for aircraft on several arrival routes, which are mapped onto one centreline [28,34]. For the introduction of the European wake turbulence categories and separation minima on approach and departure (RECAT-EU) I, a new categorisation of the mandatory wake vortex separates them into six. While today most countries use four categories, EUROCONTROL implemented the Leading Optimised Runway Delivery (LORD) display aid for approach ATCOs [35,36]. With two additional triangular symbols for each inbound aircraft moving on the final approach, it follows the principle of DLR's TargetWindow without indicating an additional safe area around the optimal position on the final approach.

A TargetWindow on the ATCOs traffic situation display was a marked interval on the centreline where it was safe for individually guided aircraft to be fed into the planned or established arrival stream by the ATCO [37]. Target positions in this window indicated the optimal positions after a turn-to-base manoeuvre to meet the AMAN's calculated trajectory and touchdown time. When aircraft were flying on downwind, they received a turn-to-base command to perform the base and final leg by the feeder ATCO [38]. It did not matter whether an aircraft was turned in from downwind or guided to the final approach by a direct or a fan approach. The decisive factor was that the aircraft manoeuvres were precisely presented within the TargetWindow when reaching the its last phase of approach.

On the one hand, the task of approach ATCOs was to clear the turn not too early to avoid wake vortex separation violations on final approach. On the other hand, the clearance had to be given early enough, so that the aircraft was not too far behind its predecessor, thereby reducing the capacity and effectiveness of the airport after the turn manoeuvre. A special challenge in this context was the wind, as its influence on the airspeed can change extremely during the 180°-turn from downwind on the final approach.

In the TargetWindows concept, target positions for turning aircraft were indicated by a dotted semicircle on the final approach with the open side facing the for this position by the AMAN scheduled aircraft (Figure 5). The surrounding dotted lined TargetWindow symbolises a safe area around this optimal target position even if the aircraft does not hit its planned position exactly. Furthermore, there is a buffer of half a nautical mile, shown by a tapering of the TargetWindow at both ends. This helps ensuring that ATCOs do not violate separation minima from predecessors and successors.

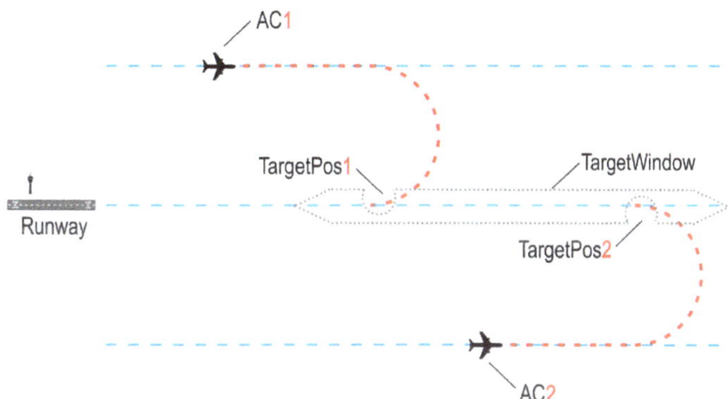

Figure 5. Schematic illustration of the TargetWindow concept displayed on an ATCO's traffic situation display. The dashed pointed area (grey) moves with the time in the direction to the runways. The ATCO's task is to turn the aircraft at the right time, as they fit in the open areas in the TargetWindow to meet their scheduled landing time perfectly. Additionally, ATCOs have the ability to quickly perceive if an aircraft is too fast or too slow. Thus, they can assess if these deviations will have any impact on the wake-vortex safety distances.

With the passage of the time, the TargetWindows moved with the speed of the expected aircraft in the direction of the runway. In this way, the ATCO was shown the current sequence planning and also planned distances of aircraft from each other. From this point of view, the TargetWindow also represented a "ghost", since it projects the position of the corresponding aircraft from another route onto the centreline depending on the distance still to be flown until touchdown, at least as long as the aircraft was moving along its planned trajectory. Unlike a ghost, however, a TargetWindow did not change its movement on the final approach because it represented the ideal position when the corresponding aircraft had to be turned on base and final approach. The ATCOs therefore did use the TargetWindow as an indicator of whether the aircraft was too early or too late at the LMP, and thus at the threshold. Nevertheless, the ATCO retained both the responsibility for the approach guidance and all freedom to follow the AMAN's suggestions or to establish his own sequence. The TargetWindow reacted just as adaptable to traffic changes as the entire AMAN.

2.2.3. Centreline Separation Visualisation Tool

Within the trails, an essential task of an approach ATCO was to set and monitor the separation between aircraft on centreline and final approach. In addition, distance markers (scale) were available on modern primary displays, which allowed a quite fast and reliable estimation of the distances between approaching with subdivision in a nautical mile. For a much more finely graduated distance display, the final approach distance indicator centreline Separation Visualisation Tool (CSVT) was developed (Figure 6). Located in separate windows for each centreline, the aircraft which were currently on final approach approach were represented by defined symbols with callsigns. In addition, the current distances between the aircraft were displayed in NM. In this way, the alphanumeric display enabled the ATCO not only to monitor the current distances, but also to immediately detect any changes in their tendency and to intervene with guidance in the event of imminent separation violations.

Figure 6. The CSVT. The symbols mark the positions of aircraft (triangle), ghosts (square) and TargetWindows (semicircle). The labels' colours represent the aircraft's weight classes (yellow: medium; green: heavy), and the white numbers between the labels indicate the current separation between them.

In addition to actual aircraft, labels for ghosts (squares) and TargetWindows' positions (semicircles) were also displayed, allowing approach ATCOs to estimate how large the separation would be after turning over Base or LMP and before reaching the final approach. Different colours for the weight classes of aircraft allowed more precise differentiation.

2.3. Validation Trials and Setup

The validation was conducted in the form of real-time HITL simulations and focused on the coordination and guidance of arrival streams (especially the 3D-FMS aircraft) within the novel E-TMA concept and supporting tools. The ATCO worked as a director (feeder). All necessary actions normally done by the executive (pick-up) role were fulfilled by the simulation pilots independently. This included, for example, early decent clearances for the 3D-FMS aircraft. Thereby, the 4D-FMS aircraft were untouchable as soon as the negotiation process was completed. This means that the ATCOs were not actively participating in the negotiation process and not directing these aircraft through their assigned flight paths. During the trials, the ATCO was in charge of both independent parallel runways at EDDM. A within-subjects design with the factor "share of 4D-FMS equipped aircraft" was used to examine the dependent variables flown distance, number of landed aircraft, mental workload, perceived safety and situation awareness [39]. Additionally, feedback about the improved assistant tools was received. By evaluating the dependent variables, the targeted environmental impact was recorded.

The trials were structured into two HITL simulations campaigns. This paper focuses on the results of the final second campaign. The first HITL simulation campaign took place in May 2022. The outcome of this campaign provided minor adjustments and improvements for the simulation setup and procedures which were implemented in the second campaign. The second HITL simulation campaign took place in September 2022. Five male ATCOs participated in the second campaign, with an age ranging between 28 and 44 years (M = 36.8, SD = 8.12) and with 2 to 18 years (M = 10, SD = 7.18) of work experience. One of the five ATCOs had used similar supporting tools or concepts before the simulation in the first HITL simulation campaign in May. Nevertheless, the participant's data were included in the final analysis, as no large training effects were expected. Due to the limited sample size, N = 5, the data and results were analysed on a non-parametric level, providing only first indicators.

2.4. Simulation Environment

As simulation environment for the HITL simulation campaigns, the Air Traffic Management and Operations Simulator (ATMOS) [40] of the DLR, Braunschweig Air Traffic Validation Center was selected. The software NARSIM (NLR's Air traffic Management Real-time Simulator) version 8.1 was deployed as generic real-time simulation software [41]. Furthermore, the aircraft's performance was modelled based on the BADA (Base of aircraft data) model version 3.15 by EUROCONTROL [42]. For the HITL simulations, one CWP and three simulation pilot positions were configured. All four working positions were connected via a simulated radio connection (Voice over IP). Simulation pilots were responsible for implementing the aircraft clearances communicated by ATCO via radio connection. Within this study, one simulation pilot controlled up to four aircraft, depending on the traffic situation and flow. A simulation pilot was provided with a slightly different display compared to the ATCO, consisting of the following elements.

(a) Stripview: Listing all flights radioing on simulation pilots' frequency.
(b) Workspace: Displaying flight strips of flights under control of the simulation pilot. Flight strips included aircraft's performance data, such indicated airspeed, heading, flight level or altitude if the aircraft is below the transition level, arrival route and further more.
(c) Radar screen: Providing an overview of the actual traffic picture within the airspace.

2.5. Simulation Setup and Scenarios

The simulation was implemented for EDDM airport with its two parallel runway systems. Both runways have an offset of 1500 m and a length of 4000 m. The distance between the runways is 2300 m, sufficient for independent usage of both runways. During the simulation, runways 26R and 26L were in use. The ATCO was in charge of both independent parallel runways guiding the arrivals streams. Departure was integrated into the simulation but handled by the simulator automatically via a departure manager (DMAN). No limitations regarding the aircraft type or weather restrictions were simulated.

In total, seven different scenarios were developed. Table 1 displays an overview of the seven developed scenarios and their composition. In the following, all scenarios are briefly explained. For the human performance, no baseline was simulated during the simulation campaign. Therefore, the results were compared to the reference scenarios (R1 and R2). The reference scenarios, R1 and R2, serving as the baseline for the objective data, were considered important to ensure comparability for the simulation data. The scenarios were based on real air traffic. Data for the reference scenarios where taken from the OpenSky scientific dataset [43]. These data are not validated and may contain inaccuracies. The data for R1 and R2 were composed by ten operating hours selected from October to December, 2021, consisting of 38–40 (R1) and 18–22 (R2) landings per hour from OpenSky datasets. R2, which provides a smaller number of aircraft, was taken into the analyses too, to see the comparison of simulation results for an even smaller number of arrivals. This was done under the presumption that a greater number of arrivals per hour makes it more difficult to enable direct routes due to safety issues. Hence, more aircraft have to fly conventional routes, which leads to a greater value of distance flown per aircraft.

The simulation scenarios were based on a medium traffic load at EDDM, which equals two thirds of the maximum traffic at EDDM [44]. The number of departures was reduced. The traffic mix by aircraft type was based on typical EDDM traffic conditions in 2022 [45]. Since currently the share of aircraft with advanced FMS varies widely among airlines, the amount of 4D-FMS aircraft included in the scenario was used as decisive parameter to distinguish the scenarios. Thus, the four simulation scenarios were developed with different amounts of 4D-FMS aircraft, starting with 20% for a training scenario and increasing up to 80%. Respectively, in the present paper, the scenarios are referred as R1, R2, T, S30 scenario, S60 scenario and S80 scenario; see Table 1. For the simulation scenarios, only the information from flightradar24 on the real callsigns, aircraft types and departure airports was used [46]. The percentage of heavy aircraft varied between 3 and 19%. Each scenario lasted for 45 min. Aircraft were initialised outside the E-TMA area and flew predefined arrival routes towards the boundaries of the E-TMA.

Table 1. Reference and simulation scenarios: composition and overview (ARR: arrivals).

Scenario ID	ARR per Hour	% of 4D ARR	Traffic Sample	% of ARR Heavy	Time Interval [3]
R1 [1]	38–40	0	2021	0	-
R2 [1]	18–22	0	2021	0	-
T [2]	20	25	2019	13	14:00–14:45
S30 [2]	40	30	02.04.2022	23	07:00–08:00
S60 [2]	40	60	03.03.2022	19	09:00–10:00
S80 [2]	40	80	01.04.2022	3	18:00–19:00

[1] Data taken from OpenSky database. [2] Data taken from Flightradar24 [46]. [3] Time interval in which the data were captured.

Each ATCO participated in a full day of simulations. Each day was scheduled into five sections, starting with a briefing and training session (T) to familiarise the ATCOs with the simulation environment. Thereafter, the simulation scenarios S60, S30 and S80 were run. After each session, the ATCOs were asked to fill in the post-run questionnaire (PRQ). Finally, a second run with the S60 scenario was conducted. This run acted as an explorative simulation run with the ATCO to obtain more in-depth feedback about the system from the ATCOs. During the explorative simulation run, individual components (Ghosts, TargetWindows and CSVT) were deactivated and activated one at a time. Participants received some time to test the system when one of the components was deactivated and were asked how this affected their work as an open question. This was followed by detailed questions about each tool. The debriefing took place after the non-explorative simulation runs and was combined with the explorative simulation run. The debriefing questions were modified from the first to the second HITL campaign, based on ATCOs feedback. Finally after a full day of simulation exercises, ATCOs were prompted to fill in a post exercises questionnaire (PEQ). As the order of simulation runs was kept constant for all participants, training effects or effects of exhaustion cannot be entirely ruled out.

2.6. Validation Methods and Techniques

During the simulations runs, all aircraft data were recorded. This included aircraft performance data, such as velocity, three-dimensional position and actual thrust value. Aircraft performance data were logged. The resulting log files were used for post-analysis to examine the concept's impact on the defined dependent variables flight distance and number of landed aircraft.

The dependent variables mental workload, perceived safety and situation awareness were assessed on the basis of questionnaires and debriefing sessions. Two different sets of questionnaires were administered. The PRQ was used after each simulation run. The PRQ includes the NASA Task Load Index (NASA-TLX) [47,48] and the situation awareness part of the Solutions for Human Automation Partnerships in European ATM (SHAPE) (SASHA) questionnaire, which was developed to assess the effects of system automation and trust on ATCOs' situation awareness [49,50].

NASA-TLX was used to assess the different dimensions of workload [47]. The NASA-TLX includes the subscales mental demand, physical demand, temporal demand, performance, effort and frustration. The subscale physical demand was omitted in the present trials, as no physical demand was expected for the task. Participants were instructed to place a score on slider bars with 21 gradations each, ranging from 0 (low) to 100 (high) (or 0 (good) to 100 (poor) in the case of performance) in steps of 5. Raw TLX ratings were used; i.e., the sub-scales were not weighted. According to Hart [47], this is a common practice and does not reduce sensitivity. A global raw TLX score was computed by calculating the mean of the five subscale ratings.

In order to assess ATCOs' experienced situation awareness, the SASHA questionnaire [50] was administered. SASHA consists of six items on a 7-point Likert-scale from 0 (never) to 6 (always) [51]. By inverting the ratings of items 2, 3, 5 and 6 and then calculating the mean of all item ratings, the overall SASHA score was computed [50]. A higher score represented higher situation awareness and was thus preferable.

The post exercise questionnaire (PEQ) was administered after the ATCOs completed the full simulation day. The PEQ included a bespoke questionnaire. Only selected statements about situation awareness and perceived safety are reported in this paper. Statements were rated on a 5-point Likert-scale from 1 (strongly disagree) to 5 (strongly agree). Means and standard deviations were calculated for the bespoke statements, where mean rating of 3 was used as the success criterion.

In addition to the introduced ATCO radar and supporting tools, the Instantaneous Self Assessment (ISA) measure was integrated into the CWP on a second touchscreen to obtain subjective mental workload ratings [52–54]. The ATCO was prompted to rate their perceived mental workload on a five-point rating scale (1 = under-utilised, 5 = excessively busy) every five minutes [55]. The data were used afterwards to evaluate the ATCOs perceived mental workload in different traffic situations.

3. Results

The aim of the simulation trials was to evaluate whether the GreAT concept, adapted within EDDM terminal airspace, could lead to a reduction in both fuel consumption and greenhouse gas emissions by assessing operational parameters compared to the concept currently applied.

3.1. Traffic and Trajectory Analysis

In the first step of the evaluation, an analysis of the traffic data was carried out in order to determine appropriate research horizon and measure the lengths of the travelled trajectories. Figures 7–9 present in detail the results of the validation trials and effects described above.

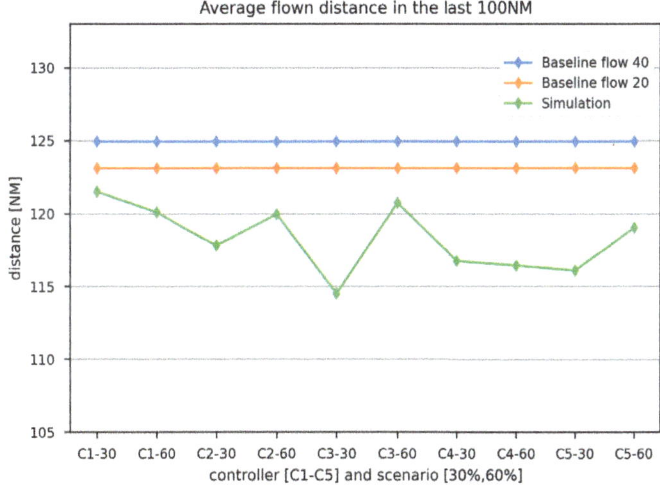

Figure 7. Flight trajectory results obtained under the simulation conditions (green) and compared with real traffic reference values for flows of 40 aircraft per hour (blue) and 20 aircraft per hour (orange).

In Figure 7, the vertical axis presents the average distance flown within the radius of 100 NM from the EDDM reference point for arriving aircraft. The horizontal axis presents the results of validation trials executed by the five ATCOs (C1–C5) testing the two traffic scenarios, differing with distribution of 3D-FMS and 4D-FMS flights, where 30 and 60 correspond, respectively, to the S30 scenario and the S60 scenario. The results obtained during the simulation have been marked as a green line. They can be directly compared with real traffic data for a flow of 40 aircraft per hour (marked in blue—R1) and for a flow of 20 aircraft per hour (marked in orange—R2). These reference flown distances were calculated as averages based on 10 h of arrival traffic at EDDM extracted from the OpenSky database for both.

Treating that as a reference, it can be observed that even with a smaller number of total arrivals in R1 and R2, the introduction of innovative airspace structure, new FMS procedures and ATCOs supporting systems resulted in a reduction in flight distance for all ATCOs and all scenarios. In each case, the simulation results were lower than all reference values.

Figure 8 presents the distances flown by aircraft as cumulative occurrence curve divided into 5 NM lengths, where again the green line refers to the simulation results, and the blue and orange lines present the data of the reference scenarios. Within the simulation scenarios, the numbers of flights covering shorter distances were slightly higher than those in the reference scenarios. This was particularly evidenced by the first two peaks observed in Figure 8, which are substantially higher, representing well over half of scheduled flights (sum of 68.3%) that arrived at the airport in the range of 100–115 NM, in contrast to 36.9% (orange) in one reference. In addition to that, the real traffic data show that a significant amount of flights (corresponding to 25% of occurrences) needed a distance of 125 NM to reach the airport.

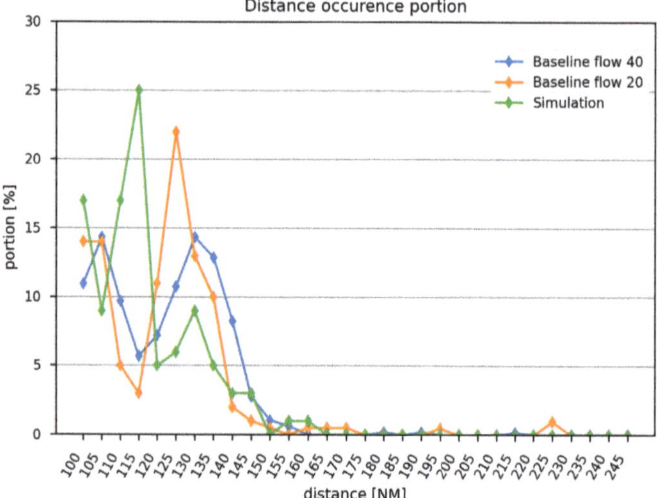

Figure 8. Cumulative occurrence curve for flight-distance results obtained in validation trials (green) and real traffic reference values for flows of 40 aircraft per hour (blue) and 20 aircraft per hour (orange).

The last set of results is related to the number of approach operations performed. This situation is reflected in the results presented in Figure 9. The figure displays a comparison between number of approaches executed in two simulation scenarios, where different distributions of 3D-FMS and 4D-FMS operations are analysed. The blue bars represent the numbers for the S30 scenario, and the orange bars correspond to the S60 scenario. The bars display that for each ATCO, a greater number of landed aircraft was recorded during the S60 scenario in comparison to the S30 scenario.

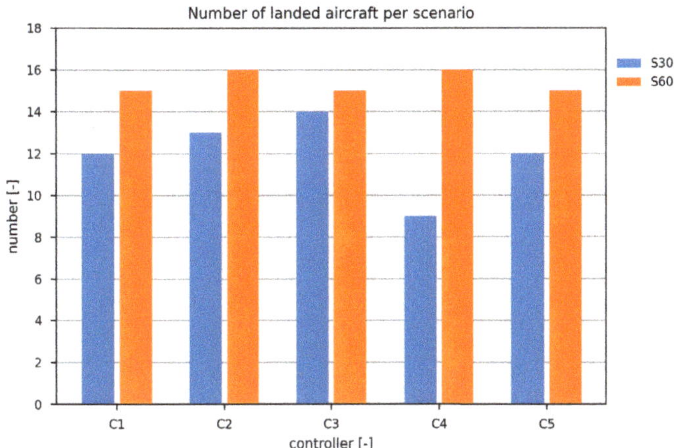

Figure 9. Capacity assessment from validation trials: numbers of landed aircraft listed for five different ATCOs (C1–C5) in each scenario, S30 (blue) and S60 (orange).

3.2. Mental Workload

Alongside the capacity analysis, the mental workload analysis was conducted. Therefore, Figure 10 shows the mean ISA ratings depending on the share of 4D-FMS equipped aircraft (30% vs. 60% vs. 80%). The results were averaged for all participants and assessment times.

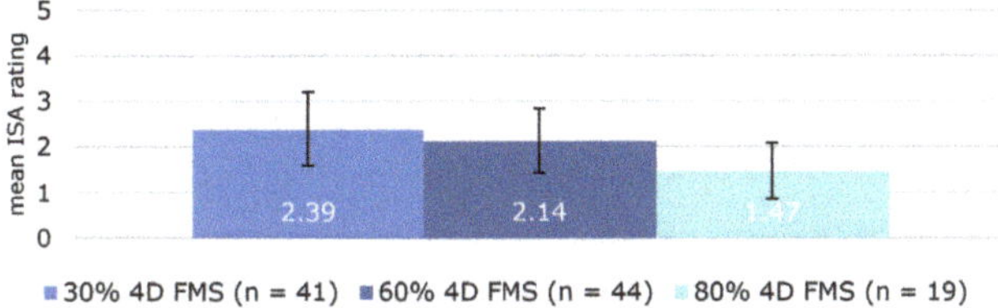

Figure 10. Mean ISA ratings in the different scenarios (S30 vs. S60 vs. S80) summarised over all participants and assessment times. Error bards represent standard deviations.

Mean ISA ratings were the highest in the S30 scenario, followed by the S60 scenario and then the S80 scenario. For the S30 and S60 scenarios, respectively, ISA ratings fell between 2 (relaxed) and 3 (comfortable), indicating a slightly lower than mid-level mental workload. For the S80 scenario, mean ISA ratings were below 2 (relaxed), pointing towards mental underloading. It is worth noting that the S80 scenario was slightly shorter than the others runs, hence the lower sample size.

Moreover, the results from NASA-TLX were assessed. Figure 11 shows the mean raw NASA-TLX scores for scenarios S30 and S60. The mean global score and all mean sub-scores were higher in the S30 scenario than in the S60 scenario, indicating higher overall workload in the S30 scenario than in the S60 scenario on a descriptive level. This is in line with the ISA ratings. Standard deviations were especially high for the sub-scales frustration and performance.

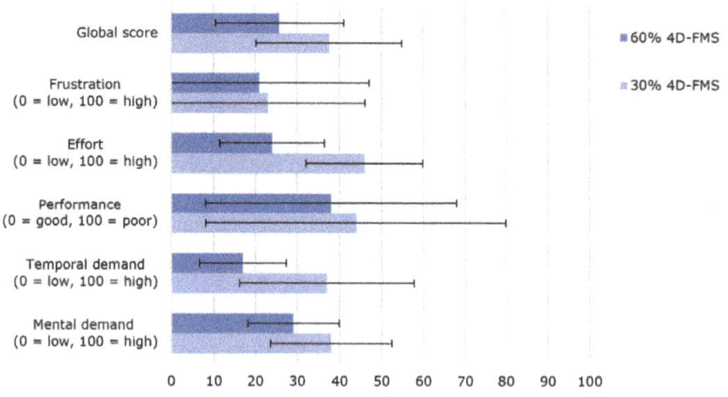

Figure 11. Mean raw TLX scores for the scenarios (S30 vs. S60 vs. S80). Error bars represent standard deviations.

3.3. *Situation Awareness*

Figure 12 depicts the mean SASHA score for the S30 scenario and the S60 scenario. For both conditions, the mean SASHA score was above 4. On a descriptive level, the SASHA scores for the S30 scenario and the S60 scenario differed only slightly. The SASHA score for the S30 scenario was higher than that for the S60 scenario. Participants' mean agreement with the statement "I always had a good mental picture of the situation" was M = 4.00 (SD = 0.71), indicating overall agreement with the statement.

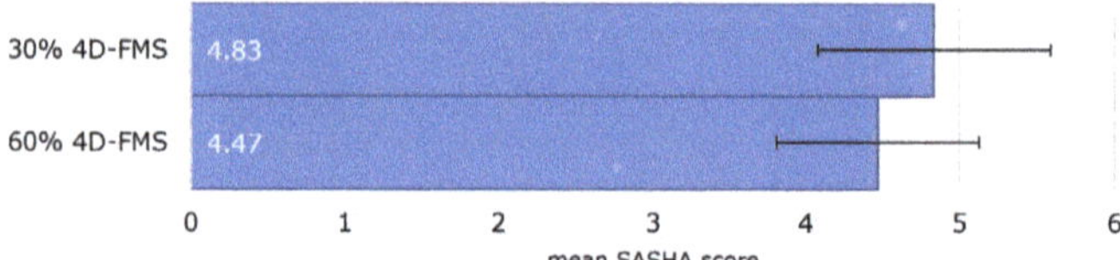

Figure 12. Mean SASHA scores for the scenarios (S30 vs. S60 vs. S80). Error bars represent standard deviations.

3.4. *Safety*

The bespoke statements regarding the perceived safety are shown in Figure 13. All statements regarding the general perceived safety received a mean rating higher than three (neither agree nor disagree) at the minimum. From this rating, it became apparent that the ATCOs seemed to feel in control and safe in controlling the traffic, including the area around the LMP where the traffic was converging as well.

During the debriefing, the ATCOs reported overall safe operation within the new airspace design. Nevertheless, some safety-critical situations were reported, which are also shown in the recorded data. Critical situations were mainly related to the simulation setup, technical issues or lack of experience with the system—for instance, difficulty in judging distances due to lack of measurement tools in unknown airspace. Although aircraft might need to keep adjusting their flying speed constantly to meet the negotiated target times, no further risk was produced. These results reflect the subjective feedback from ATCOs gathered through the questionnaire and debriefing.

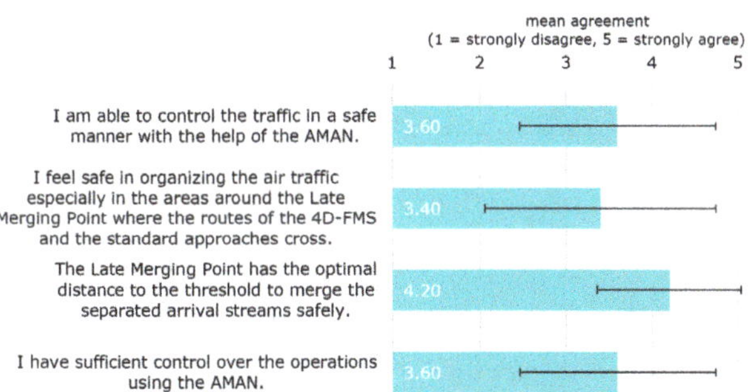

Figure 13. Mean agreement to tailored statements regarding the perceived safety for all scenarios. Error bars represent standard deviations.

3.5. Tactical Assistance Systems

The feedback on the three tactical assistance systems was not free of contradictions. However, the main feedback was positive rather than negative, along with suggestions for possible improvements. All in all, the tactical assistance systems provided the required information about the projected aircraft's positions on the final approach in line with the sequences computed by the AMAN. The provided information was reported by ATCO as helpful to obtaining the whole picture of the traffic situation and to plan ahead. Participants made also several suggestions to further improve the provided data, indicating that there is room for improvements. Some examples of the suggested improvements are listed in the following paragraph.

Based on feedback collected from all ATCOs, the time-based ghosting tool was one of the most used and useful assisting tools made available during the validation trials. It was even considered by most of ATCOs as necessary and crucial for handling the 4D-FMS traffic safety-wise. Participants reported also that they would even like to have it for real operations to be used for other purposes. Nevertheless, some ATCOs felt distracted by the ghost symbol, which could tie up mental capacity. Some would only like to have the sequence number above the ghost symbol or to reduce the label by displaying only the callsign. Others would like to see a special marker when the real aircraft deviates too much from the planned route due to certain environmental conditions. This request could be accommodated by enabling the ATCO to individually customise the ghost symbol using different shapes and colour settings. The ATCOs perceived the TargetWindow as helpful and sufficient for handling the conventional 3D-FMS traffic, making the CSVT superfluous. Consequently, the CSVT was rarely used. Additionally, the ATCOs recorded some minor technical stability issues, which means that in some cases, aircraft did not have a TargetWindow, or the label was shown on the wrong side of the runway system. As an improvement, some features should be made available to ATCOs, enabling them to manually adjust the label, shape and colour settings for the TargetWindow itself. This feedback is in line with feedback on the ghosting.

4. Discussion

The traffic and trajectories analysis results display that during the experimental scenarios, S30 and S60, the trajectories were on average 7.5% shorter than in the baselines R1 and R2. As cumulative events, it was found that the share of 4D-FMS aircraft allowed a greater number of flights to be implemented with shorter approach distances on average. Although the capacity was not planned to be addressed by this solution, it was observed that ATCO assistance tools effectively supported them in guiding the traffic during the validation activities. Taking that into consideration, it can be pointed out that a greater number of FMS equipped aircraft were more efficiently routed for landing by all ATCOs.

Based on ISA measurements, two conclusions for mental workload analysis can be drawn. Firstly, mental workload remained at acceptable levels in the S30 scenario and in the S60 scenario, pointing out that no mental overload arose. This evaluation was also confirmed through ATCOs' feedback during the debriefing sessions. Nevertheless, ISA ratings during the S80 scenario pointed towards mental underload, because increasing automation took away much of the traffic guidance work. Mental workload could be expected to be lower in simulations than in real operations. However, since mental underload is a potential safety risk because crucial events can be missed, this should be tested in further validation campaigns with adjusted preconditions. For example, improved visualisation of the tactical assistance systems and a bigger sample size could be used. Secondly, the experienced ATCO's mental workload seemed to be inversely related to the percentage of 4D-FMS aircraft. Increasing the amount of untouchable 4D-FMS aircraft results in a reduction in the share of 3D-FMS aircraft navigated by the ATCO. Indeed, the number of aircraft a ATCO manages simultaneously at a given time was the most used index to estimate the workload [56]. However, this index is influenced by the way aircraft are spread over space and time [56], and therefore, less aircraft to be managed does not necessarily result in less workload. An alternative explanation could be linked to the main task of the ATCO: Given the route structure (separated by design) and the sequence proposed by the AMAN (considering required separation), the ATCO mainly monitored and guided the 3D-FMS aircraft towards the TargetWindow to meet the optimal position on final approach, unless he decided to choose an alternative path based on direct routing. That being said, the higher the number of 4D-FMS aircraft, the less intervention is required from the ATCO, potentially resulting in lower mental workload.

The results from NASA TLX analysis point out that the mean global score and all mean sub-scores were higher in the S30 scenario than in the S60 scenario, indicating higher overall workload in the S30 scenario than in the S60 scenario on a descriptive level. Those results coincide with the ISA ratings. Additionally, standard deviations were especially high for the subscales frustration and performance. This means that a wide range of answers were attributed to these subscales. This divergence was also indicated during the debriefings and explorative simulation runs. For instance, some ATCOs felt comfortable being in charge of fewer 3D-FMS aircraft, while others pointed out their frustration about the untouchable character of 4D-FMS aircraft. Likewise, some ATCOs tried to further optimise the sequence proposed by the AMAN; others reported strictly following the proposed sequence. The latter might have impacted ATCOs' perceived performance ratings. Nevertheless, the NASA TLX analysis shows that increasing the amount of 4D-FMS aircraft lowers the perceived ATCOs workload. This could result in more spare mental capacity, which can be used for other ATCOs tasks, such as safety monitoring or improving sequence planning.

During the debriefing session, ATCOs named both the ghosts and the TargetWindows as beneficial, if not essential, for increasing and maintaining situation awareness. However, one ATCO raised the concern that situation awareness will be lost if the share of 4D-FMS-equipped aircraft is too high. Communication between ATCOs and 4D-FMS pilots is reduced to a minimum after the initial call. Such a little amount of exchange of information could reduce situational awareness for specific aircraft. In short, it can be concluded that besides the discussed effects, the perceived ATCO situation awareness remained at an acceptable level for a 4D-FMS aircraft percentage of up to 60%. More research will be needed to assess the impact of higher percentages of untouchable 4D-FMS aircraft on situation awareness. Monitoring automated systems and assuming a more passive role instead of actively engaging with a system can impair situation awareness, possibly resulting in an out-of-the-loop performance problem [57]. As a higher share of 4D-FMS aircraft leads to the ATCO passively monitoring more aircraft, an overly large number of 4D-FMS aircraft might result in lowered situation awareness. This possibility should be critically considered in future research.

To sum up, the new airspace design and supporting functions were considered as acceptable from ATCO perspectives in terms of safety. ATCOs felt able to provide the same safety level compared to current operations. Nevertheless, ATCOs addressed potential safety risks, such as the possible loss of situation awareness for overly high shares of 4D-FMS aircraft.

Although the qualitative and quantitative assessments provided promising initial results, it should be added that there exist still some limitations. Those limitations are mainly related to the constraints on the large-scale implementation of such systems. For example, no baseline scenario was used to compare human performance results. Additionally, the human performance data were analysed on a non-parametric, descriptive level; i.e., no statements can be made regarding statistically significant differences due to the sample size of five per iteration. Therefore, a bigger sample size should test the GreAT concept and its impacts on efficiency, mental workload and situation awareness.

5. Conclusions

The development and implementation of the GreAT concept within the DLR real-time simulation environment ATMOS was realised. The HITL validation campaign was conducted with a total of five ATCOs from HungaroControl. The obtained results allow two conclusions to be drawn. The first one indicates that the developed GreAT concept has proved to be supportive for ATCO by sequencing arrival traffic in a safe and efficient manner. This has a proven influence on the mean flight distance per aircraft, and therefore, a positive impact on the reduction of overall fuel consumption. Furthermore, reducing the total fuel consumption mitigates the exhausted gas. These results indicate that through the implementation of the GreAT concept, a important reduction of aviation's environmental impact is feasible. The second conclusion is that the 4D-FMS aircraft guidance may to certain extent increase airport capacity. Increasing the number of aircraft equipped with advanced FMS leads to more landed aircraft with shorter flown distances, while keeping the mental workload, situation awareness and safety to acceptable levels. All in all, the GreAT concept carries the potential to lower aviation's impact on the environment while keeping airport capacity at least on the same level. Further research should focus on the large-scale implementation and gathering of statistically significant results by increasing the sample size. Additionally, a baseline for the human performance results could bring new insights on the impacts of the GreAT concept on mental workload and situation awareness.

Author Contributions: Conceptualisation, N.A., I.S., T.M. and M.-M.T.; methodology, N.A., T.M. and M.-M.T.; simulation and software, N.A.; validation, N.A., I.S., R.A. and M.-M.T.; formal analysis, N.A., I.S., R.A., T.M. and M.-M.T.; investigation, N.A.; resources, T.M.; data curation, N.A.; writing—original draft preparation, N.A., I.S., R.A., T.M. and M.-M.T.; writing—review and editing, T.M. and M.-M.T.; visualization, N.A., I.S. and M.-M.T.; supervision, T.M.; project administration, T.M. and M.-M.T.; funding acquisition, M.-M.T. All authors have read and agreed to the published version of the manuscript.

Funding: This project has received funding from the European Union's Horizon 2020 research and innovation programme under grant agreement n° 875154.

Institutional Review Board Statement: Not applicable.

Informed Consent Statement: Informed consent was obtained from all subjects involved in the study.

Data Availability Statement: Data can be available by contacting the corresponding author, Nils Ahrenhold. Data for reference study were taken from https://opensky-network.org/datasets/states/, accessed on 18 November 2022.

Conflicts of Interest: The authors declare no conflict of interest.

Abbreviations

3D-FMS	Aircraft with conventional FMS equipment
4D-FMS	Aircraft with advanced FMS equipment
AMAN	Arrival Manager
ASP	Aircraft Separation Point
ATC	Air Traffic Control
ATCO	Air Traffic Controller
ATM	Air Traffic Management
ATMOS	Air Traffic Management and Operation Simulator
BADA	Base of Aircraft Data
CSVT	Centreline Separation Visualisation Tool
CDA	Continuous Decent Approach
CPDLC	Controller Pilot Data Link Communications
CWP	Controller Working Position
DLR	Deutsches Zentrum für Luft- und Raumfahrt e.V.
DMAN	Departure Manager
DOMP	Direct-only Merge Point
EDDM	Munich airport
E-OCVM	European Operational Concept Validation Methodology
E-TMA	Extended TMA
FMS	Flight Management System
ft	foot/feet
GreAT	Greener Air Traffic Operations
HITL	Human-in-the-Loop
ISA	Instantaneous Self Assessment
LMP	Late Merging Point
LORD	Leading Optimised Runway Delivery
NARSIM	NLR's Air traffic management Real-time SIMulator
NASA TLX	NASA Task Load Index
NM	Nautical Mile
PEQ	Post Exercise Questionnaire
PHARE	Harmonised ATM Research in Eurocontrol
PRQ	Post Run Questionnaire
RECAT-EU	lEuropean wake turbulence categories and separation minima on approach and departure
SESAR	Single European Sky ATM Research
SASHA	Situation Awareness for SHAPE (SASHA)
SHAPE	Solutions for Human Automation Partnerships in European ATM
TMA	Terminal Manoeuvring Area

References

1. Brazzola, N.; Patt, A.; Wohland, J. Definitions and implications of climate-neutral aviation. *Nat. Clim. Chang.* **2022**, *12*, 761–767. [CrossRef]
2. Crompton, T.; Kasser, T. *Meeting Environmental Challenges: The Role of Human Identity*; WWF-UK: Godalming, UK, 2009; Volume 29.
3. Meltzer, J. Climate Change and Trade—The EU Aviation Directive and the WTO. *J. Int. Econ. Law* **2012**, *15*, 111–156. [CrossRef]
4. Ebi Kristie, L.; Jan, C. Semenza Community-based adaptation to the health impacts of climate change. *Am. J. Prev. Med.* **2008**, *5*, 501–507.
5. Haines, A.; Kovats, R.S.; Campbell-Lendrum, D.; Corvalán, C. Climate change and human health: Impacts, vulnerability and public health. *Public Health* **2006**, *7*, 585–596. [CrossRef] [PubMed]
6. *Effects of Novel Coronavirus (COVID-19) on Civil Aviation: Economic Impact Analysis*; International Civil Aviation Organization (ICAO): Montréal, QC, Canada, 2020. Available online: https://www.aaco.org/Library/Files/Uploaded%20Files/Economics/Corona%20studies/3dec%20ICAO_Coronavirus_Econ_Impact.pdf (accessed on 20 November 2022).
7. EUROCONTROL Data Snapshot: The Path to Recovery for Intra-European and Intercontinental Flights. Available online: https://www.eurocontrol.int/sites/default/files/2023-01/eurocontrol-data-snapshot-38-revised.pdf (accessed on 24 January 2023).
8. Gudmundsson, S.V.; Cattaneo, M.; Redondi, R. Forecasting temporal world recovery in air transport markets in the presence of large economic shocks: The case of COVID-19. *J. Air Transp. Manag.* **2021**, *91*, 102007. [CrossRef] [PubMed]

9. Brouder, P. Reset redux: Possible evolutionary pathways towards the transformation of tourism in a COVID-19 world. *Tour. Geogr.* **2020** 22 484–490. [CrossRef]
10. Ateljevic, I. Transforming the (tourism) world for good and (re) generating the potential 'new normal'. *Tour. Geogr.* **2020**, *22*, 467–475. [CrossRef]
11. Dube, K.; Godwell, N.; Chikodzi, D. COVID-19 pandemic and prospects for recovery of the global aviation industry. *J. Air Transp. Manag.* **2021**, *92*, 102022. [CrossRef] [PubMed]
12. Gössling, S.; Scott, D.; Hall, C.M. Pandemics, tourism and global change: A rapid assessment of COVID-19. *J. Sustain. Tour.* **2020**, *29*, 1–20. [CrossRef]
13. Bolić, T.; Ravenhill, P. SESAR: The past, present, and future of European air traffic management research. *Engineering* **2021**, *7*, 448–451. [CrossRef]
14. *Single European Sky Implementation Support though Validation*; European Comission CORDIS: Brussels, Belgium, 2010. Available online: https://cordis.europa.eu/project/id/37106 (accessed on 18 November 2022).
15. *European Airport Movement Management by A-SMGCS*; European Comission CORDIS: Brussels, Belgium, 2008. Available online: https://cordis.europa.eu/project/id/503192/it (accessed on 18 November 2022).
16. *European Airport Movement Management by A-SMGCS, Part 2*; European Commission CORDIS: Brussels, Belgium, 2008. Available online: https://cordis.europa.eu/project/id/503522 (accessed on 18 November 2022).
17. Kuenz, A.; Mollwitz, V.; Korn, B. Green trajectories in high traffic TMAs. In Proceedings of the IEEE/AIAA 26th Digital Avionics Systems Conference, Columbia, MD, USA, 25–26 September 2007. [CrossRef]
18. Schaefer, D.; Meckiff, C.; Magill, A.; Pirad, B.; Aligne, F. Air traffic complexity as a key concept for multi-sector planning. In Proceedings of the 20th Digital Avionics Systems Conference (Cat. No.01CH37219), Daytona Beach, FL, USA, 14–18 October 2001. [CrossRef]
19. EAD Basic. Available online: https://www.ead.eurocontrol.int/cms-eadbasic/opencms/en/login/ead-basic/ (accessed on 10 January 2023).
20. European Organisation for the Safety of Air Navigation. *European Operational Concept Validation Methodology E-OCVM*; European Organisation for the Safety of Air Navigation: Brussels, Belgium, 2010; Version 3.0 Vol. I.
21. Eggemeier, F.T.; O'Donnell, R.D. *A Conceptual Framework for Development of a Workload Assessment Methodology*; Defense Technical Information Center: Fort Belvoir, VI, USA; Wright State University: Dayton, OH, USA, 1982; Volume ADA289489; pp. 1–12.
22. Endsley, M.R. Situation awareness in aviation systems. In *Handbook of Aviation Human Factors*; CRC Press: Boca Raton, FL, USA, 1999; Volume 257; p. 276.
23. Erkelens, L.J.J. Advanced Noise Abatement Procedures for Approach and Departure. In Proceedings of the AIAA Guidance, Navigation, and Control Conference and Exhibit, Monterey, CA, USA, 5–8 August 2002.
24. Mundra, A.D.; Smith, A.P. Capacity enhancements in IMC for converging configurations with down-link of aircraft expected final approach speeds. In Proceedings of the 20th DASC, 20th Digital Avionics Systems Conference (Cat. No. 01CH37219), Daytona Beach, FL, USA, 14–18 October 2001; Volume 2.
25. Beers, C.S. *Sourdine II: Concept of Operations for Schiphol Airport Simulations*; National Aerospace Laboratory, NLR: Amsterdam, The Netherlands, 2005.
26. Smith, A.P.; Becher, T.A. A study of SPACR ghost dynamics applied to RNAV routes in the terminal area. In Proceeding of the 24th Digital Avionics Systems Conference, Washington, DC, USA, 30 October–3 November 2005; Volume 1.
27. Becher, T.A.; Barker, D.R.; Smith, A.P. Methods for maintaining benefits for merging aircraft on terminal RNAV routes. In Proceedings of the 23rd Digital Avionics Systems Conference (IEEE Cat. No. 04CH37576), Salt Lake City, UT, USA, 28 October 2004; Volume 1.
28. Oberheid, H.; Weber, B.; Temme, M.-M.; Kuenz, A. Visual assistance to support late merging operations in 4D trajectory-based arrival management. In Proceedings of the IEEE/AIAA 28th Digital Avionics Systems Conference, Orlando, FL, USA, 23–29 October 2009; Volume 1.
29. Temme, M.-M.; Kuenz, A.; Oberheid, H. FAGI: Segment Ghosting und Targeting in der Anflugplanung. Internal Report: DLR-IB 112-2009/40, Deutsches Zentrum für Luft- und Raumfahrt e.V. 2010. Available online: https://elib.dlr.de/62345/ (accessed on 18 November 2022).
30. Shepley, J. Near-term terminal area automation for arrival coordination. In Proceedings of the Eighth USA/Europe Air Traffic Management Research & Development Seminar, Napa, CA, USA, 29 June–2 July 2009.
31. Atkins, S.; Capozzi, B. Relative Position Indicator for merging mixed RNAV and vectored arrival traffic. In Proceedings of the IEEE/AIAA 30th Digital Avionics Systems Conference, Seattle, WA, USA, 16–20 October 2011; p. 2A4-1. [CrossRef]
32. Parke, B.; Bienert, N.; Chevalley, E.; Omar, F.; Buckley, N.; Brasil, C.; Yoo, H.-S.; Borade, A.; Gabriel, C.; Lee, P.; et al. Exploring management of arrival spacing using route extensions with terminal spacing tools. In Proceedings of the IEEE/AIAA 34th Digital Avionics Systems Conference (DASC), Prague, Czech Republic, 13–17 September 2015; p. 3E1-1. [CrossRef]
33. Burnett, K.; Scully, G.; Davis, D.; Krause, J.; Cooper, K.; Musclow, R.; Beasley, P. Visual Aircraft Spacing Tool. U.S. Patent App. 12/464,070, 19 November 2009.
34. Uebbing-Rumke, M.; Temme, M.-M. Controller aids for integrating negotiated continuous descent approaches into conventional landing traffic. In Proceedings of the 9th USA/Europe ATM Seminar, Berlin, Germany, 14–17 June 2011.

35. Cappellazzo, V.; Treve, V.; De Visscher, I.; Chalon, C. Design Principles for a Separation Support Tool Allowing Optimized Runway Delivery. In Proceedings of the 2018 Aviation Technology, Integration, and Operations Conference, Hilton Head, SC, USA, 25–29 June 2018; Volume 4237.
36. European Wake Turbulence Categorisation and Separation Minima on Approach and Departure. Available online: https://www.eurocontrol.int/publication/european-wake-turbulence-categorisation-and-separation-minima-approach-and-departure (accessed on 2 January 2023).
37. Ohneiser, O. *Flight Guidance Support to Integrate Conventional Equipped Aircraft into a Time Based Arrival Flow*; Grandt, M., Schmerwitz, S., Eds.; German Society of Aerospace: Bonn, Germany, 2012; pp.175–192.
38. Ohneiser, O.; Temme, M.-M.; Rataj, J. Trawl-Net Technology for Timely Precise Air Traffic Controller Turn-To-Base Commands. In Proceedings of the Eleventh USA/Europe Air Traffic Management Research and Development Seminar, Lisbon, Portugal, 23–26 June 2015.
39. Charness, G.; Gneezy, U.; Kuhn, M.A. Experimental methods: Between-subject and within-subject design. *J. Econ. Behav. Organ.* **2012**, *81*, 1–8. [CrossRef]
40. Air Traffic Management and Operations Simulator (ATMOS). Available online: https://www.dlr.de/content/en/research-facilities/air-traffic-management-and-operations-simulator-atmos.html (accessed on 18 November 2022).
41. Ten Have, J.M. The development of the nlr atc research simulator (Narsim): Design philosophy and potential for ATM research. *Simul. Pract. Theory* **1993**, *1*, 31–39. [CrossRef]
42. European Organisation for the Safety of Air Navigation. *Base of Aircraft Data (BADA) Product Management Document*; EEC Technical/Scientific Report No. 2009-008; European Organisation for the Safety of Air Navigation: Rue de la Fusée, Belgium, 2009. Available online: https://citeseerx.ist.psu.edu/document?repid=rep1&type=pdf&doi=a401e21db54026b17eb4b4bd3ab8121d1bc57649 (accessed on 18 November 2022).
43. The OpenSky Network. Available online: https://opensky-network.org/ (accessed on 18 November 2022).
44. Flughafenkoordination Deutschland, Airport Capacity Parameters. Available online: https://fluko.org/wp-content/uploads/2022/10/Airport-Capacity-Parameters_S23_L3_20221002.pdf (accessed on 21 November 2022).
45. EDDM—Aviation Intelligence Unit. Available online: https://ansperformance.eu/dashboard/stakeholder/airport/db/eddm#traffic (accessed on 22 December 2022).
46. Flightradar24: Live Flight Tracker. Available online: https://www.flightradar24.com (accessed on 19 November 2022).
47. Hart, S.G. NASA-task load index (NASA-TLX); 20 years later. *Proc. Hum. Factors Ergon. Soc. Annu. Meet.* **2006**, *50*, 904–908. [CrossRef]
48. Hart, S.G.; Staveland, L.E. Development of NASA-TLX (Task Load Index): Results of empirical and theoretical research. *Adv. Psychol.* **1988**, *52*, 139–183.
49. Dehn, D.M. Assessing the impact of automation on the air traffic controller: The SHAPE questionnaires. *Air Traffic Control Q.* **2008**, *16*, 127–146. [CrossRef]
50. Situation Awareness for SHAPE (SASHA). Available online: https://ext.eurocontrol.int/ehp/?q=node/1609 (accessed on 28 September 2022).
51. Joshi, A.; Kale, S.; Chandel, S.; Pal, D.K. Likert scale: Explored and explained. *Br. J. Appl. Sci. Technol.* **2015**, *7*, 396. [CrossRef]
52. Kirwan, B.; Evans, A.; Donohoe, L.; Kilner, A.; Lamoureux, T.; Atikinson, T.; MaxKendrick, H. Human factors in the ATM system design life cycle. In Proceedings of the FAA/Eurocontrol ATM R&D Seminar, Saclay, France, 16 June 1997; pp. 16–20.
53. Instantaneous Self Assessment of Workload (ISA). Available online: https://ext.eurocontrol.int/ehp/?q=node/1585 (accessed on 12 December 2022).
54. Tattersall, A.J.; Foord, P.S. An experimental evaluation of instantaneous self-assessment as a measure of workload. *Ergonomics* **1996**, *39*, 740–748. [CrossRef] [PubMed]
55. Brennen, S.D. An experimental report on rating scale descriptor sets for the instantaneous self assessment (ISA) recorder. In *Portsmouth: DRA Maritime Command and Control Division. DRA Technical Memorandum (CAD5)*; 1992; Volume 92017. Available online: https://skybrary.aero/sites/default/files/bookshelf/1963.pdf (accessed on 18 November 2022)
56. Athènes, S.; Averty, P.; Puechmorel, S.; Delahaye, D.; Collet, C. ATC complexity and controller workload: Trying to bridge the gap. In *International Conference on HCI in Aeronautics*; AAAI Press: Cambridge, MA, USA, 2002; pp. 26–60.
57. Endsley, M. R.; Kiris, E. O. The out-of-the-loop performance problem and level of control in automation. *Hum. Factors* **1995**, *37*, 381–394. [CrossRef]

Article

Decarbonised Future Regional Airport Infrastructure

Markus Meindl [1,*], Cor de Ruiter [2], Valerio Marciello [3], Mario Di Stasio [3], Florian Hilpert [4], Manuela Ruocco [5], Fabrizio Nicolosi [3], Nils Thonemann [6], Karen Saavedra-Rubio [6], Louis Locqueville [6], Alexis Laurent [6] and Martin Maerz [1]

1 Institute of Power Electronics, Friedrich-Alexander-Universität Erlangen-Nürnberg, 90429 Nuermberg, Germany
2 Rotterdam the Hague Innovation Airport, 3045 AP Rotterdam, The Netherlands
3 Department of Industrial Engineering, Universitá degli Studi di Napoli Federico II, 80138 Naples, Italy
4 Fraunhofer Institute for Integrated Systems and Device Technology IISB, 91058 Erlangen, Germany
5 SmartUp Engineering srl, 80123 Naples, Italy
6 Section for Quantitative Sustainability Assessment, Department of Environmental and Resource Engineering, Technical University of Denmark, 2800 Kongens Lyngby, Denmark
* Correspondence: markus.meindl@fau.de

Abstract: Sustainability and, especially, emission reductions are significant challenges for airports currently being addressed. The Clean Sky 2 project GENESIS addresses the environmental sustainability of hybrid-electric 50-passenger aircraft systems in a life cycle perspective to support the development of a technology roadmap for the transition to sustainable and competitive electric aircraft systems. This article originates from the GENESIS research and describes various options for ground power supply at a regional airport. Potential solutions for airport infrastructure with a short (2030), medium (2040) and long (2050) time horizon are proposed. This analysis includes estimating the future energy demand per day, month and year. In addition, the current flight plan based on conventional aircraft is adapted to the needs of a 50-PAX regional aircraft. Thus, this article provides an overview of the energy demand of a regional airport, divided into individual time horizons.

Keywords: on-ground energy supply; hybrid-electric aircraft; airport infrastructure; sustainable aviation

Citation: Meindl, M.; de Ruiter, C.; Marciello, V.; Stasio, M.D.; Hilpert, F.; Ruocco, M.; Nicolosi, F.; Thonemann, N.; Saavedra-Rubio, K.; Locqueville, L.; et al. Decarbonised Future Regional Airport Infrastructure. *Aerospace* **2023**, *10*, 283. https://doi.org/10.3390/aerospace10030283

Academic Editor: Jordi Pons-Prats

Received: 23 February 2023
Revised: 9 March 2023
Accepted: 10 March 2023
Published: 13 March 2023

1. Introduction

Sustainability and reducing emissions are significant challenges for airports. Frankfurt airport will reduce CO_2 emissions by around 65% until 2030 and operate in 2045 without any CO_2 emissions. A total of 34% of the vehicles in the airport in Frankfurt are hybrid-electric or hydrogen-based. With this advantage, it was possible to reduce the CO_2 emissions on this German airport by around 35% since 2010 [1]. Aircraft manufacturers such as Airbus plan to introduce hydrogen planes by 2035. With the code ZEROe (short for zero emissions), Airbus plans three types of passenger planes that rely on liquid hydrogen (LH_2) as fuel [2]. Aircraft manufacturers have already initiated a transition to sustainable aviation, which the airports strive to follow. However, there are enormous challenges, such as the generation of hydrogen or electricity from renewable energies, to decarbonising the industry entirely. The electrification of aircraft systems raises the question of whether airports will be among the largest electricity consumers in our infrastructure in the future. At the small Corisco International Airport, Source [3] proposes the renewable energy generation of 307.42 MWh/year with an energy surplus of 41.30% by integrating wind turbines (WTs), photovoltaics and diesel generators. This integration will reduce annual greenhouse gas emissions on the island by 98.50% [4]. The Soekarno-Hatta Airport Railink Project is one of the projects the Indonesian government prioritizes to ensure reliable mass transport to and from Soekarno-Hatta International Airport [5,6]. In this study, the electricity demand for the operation of the Soekarno-Hatta airport railway is discussed and compared with the demand for the existing substation. The results of this study show that the substations will require a small amount of additional capacity. However, overall, the existing substations will remain reliable even though additional capacity will be required in 2030 to maintain the reliability of the electricity supply for two different services. According to the calculations, the cost of the additional capacity is USD 4.3 million. Electricity costs are estimated at USD 85,000 to 100,000/month for the first year, with 89 trips per day [6]. These examples show that both small and very large airports are currently investing heavily in electrification to drive the electrification of the entire aviation industry. The first simulations of energy supply technologies for a regional airport show that the energy demand of a regional airport with 13 gates will increase from 6 GWh to 22.53 GWh by operating 49 hybrid-electric aircraft per day [7]. As part of the Clean Sky 2 ENhanced electrical energy MAnagement (ENIGMA) project, a centralised smart supervisory control (CSS) with enhanced electrical energy management (E2-EM) capability was developed for an Iron Bird electrical power generation and distribution system (EPGDS) [8]. These projects show that this is a very current and important research topic.

The Rotterdam The Haque Airport (RTHA) is a subcontracting partner of the Clean Sky 2 GENESIS project (Gauging the ENvironmEntal Sustainability of electrIc and hybrid aircraft Systems) and seeks to change its infrastructure to adopt hybrid-electric aircraft (HEA). To accomplish this, RTHA plans to electrify around 70% of its aircraft traction fleet from 2030 onwards and aims to replace the remaining 30% with hydrogen-powered aircraft by 2050 [9]. Therefore, they must develop and adapt their ground power supply strategies to meet the demand for HEA (Hybrid-Electric-Aircraft) traffic. The study aims to determine the energy requirements for a regional airport's operation and the expected emissions. This paper is framed in the context of GENESIS, which corresponds to the EU theme JTI-CS2-2020-CFP11-THT-13 under the Clean Sky 2 programme for Horizon 2020 and presents a forward-looking view focusing on the assessment of appropriate energy supply technologies for ground energy storage, grid connection and power transmission to aircraft. Based on these technologies, a flight plan and the design of a 50 PAX HEA developed in the project, the energy requirements for operations at a regional airport can be estimated [10]. The fuel types for HEA change depending on the time horizon. The energy demand of the developed HEA was used to classify the energy demand of a conventional aircraft (ATR 42 with a Pratt and Whitney PW127 engine). In the short-term (2025–2035) and medium-term (2035–2045), a direct comparison between kerosene and a mixture of kerosene and sustainable aviation fuels (SAF) can be made. This study also assumes that LH_2 and a battery in the medium-term can power HEA. In the long-term (2045–2055+) horizon, hybrid-liquid–hydrogen aircraft are assumed exclusively. Based on the energy requirements of the aircraft, which were provided by two partners of the consortium UniNa (Università degli Studi di Napoli Federico II) and SmartUp Engineering, the flight plan and the number of take-offs and landings, the emissions can be estimated. In addition, a flight plan is being developed to replace conventional aircraft with HEA and thus enable more environmentally friendly air traffic. Table 1 provides the key figures of the conventional aircraft and HEA designed with GENESIS, with information on the amounts of fuel (kerosene, SAF and LH_2) and battery energy consumed per kilometre. Results are presented in this table for two separate missions: a 600 nmi mission, which was used to size both conventional aircraft and HEA concepts, and a shorter 200 nmi mission, more representative of the typical mission for a regional turboprop aircraft.

Table 1. Overview of fuel and battery energy consumptions per km based on calculations performed by UNINA and SmartUp.

2025–2035	kg kerosene/km	kg SAF$_{(4)*}$/km	kWh/km	kg LH2/km
Short-term 200 nmi$_{(1)*}$				
Short-term ICE$_{(ref,2)*}$	1.25	1.23		
Short-term ICE$_{(2)*}$ + Battery	0.87	0.96	2.23	
Short-term 600 nmi $_{(1)*}$				
Short-term ICE$_{(ref,2)*}$	0.98	0.96		
Short-term ICE$_{(2)*}$ + Battery	0.86	0.94	0.76	
2035–2045	**kg kerosene/km**	**kg SAF$_{(4)*}$/km**	**kWh/km**	**kg LH2/km**
Medium-term 200 nmi$_{(1)*}$				
Medium-term ICE$_{(ref,2)*}$	1.14	1.12	0.00	
Medium-term ICE$_{(2)*}$ + Battery	0.64	0.63	2.39	
Medium-term PEMFC$_{(3)*}$ + Battery			2.58	0.16
Medium-term 600 nmi$_{(1)*}$				
Medium-term ICE$_{(ref,2)*}$	0.90	0.88		
Medium-term ICE$_{(2)*}$ + Battery	0.46	0.20	1.61	
Medium-term PEMFC$_{(3)*}$ + Battery			1.75	0.21
2045–2055	**kg kerosene/km**	**kg SAF$_{(4)*}$/km**	**kWh/km**	**kg LH2/km**
Long-term 200 nmi$_{(1)*}$				
Long-term ICE$_{(ref,2)*}$	1.11	1.09		
Long-term PEMFC$_{(3)*}$ + Battery			2.37	0.16
Long-term 600 nmi$_{(1)*}$				
Long-term ICE$_{(ref,2)*}$	0.88	0.86		
Long-term PEMFC$_{(3)*}$ + Battery			1.60	0.19

(ref)*: Reference aircraft with conventional gas turbine engines as power plant technology. (1)*: Nautical mile. (2)*: Internal Combustion Engine. (3)*: Polymer Electrolyte Membrane Fuel Cell. (4)*: Sustainable Aviation Fuel.

2. Methodology

2.1. Supply Technologies under Consideration

The electrification of aircraft comes with immense stress to the airport's electrical system due to the vastly increased power demand for charging airplanes. In addition, a mid-to-long-term infrastructure should aim to improve the airports' entire energy system, including information technology (IT) and control systems, lighting, and general-use low voltage supply. Furthermore, building a hydrogen infrastructure with a new generation of tanks, pipelines, and supply possibilities to refuel the aircraft is also necessary. Instead of overhauling the existing system, the short-term (2025–2035) analysis will focus on an electrical grid whose sole purpose is to charge aircraft on the airfield. The medium-term (2035–2045) time perspective includes more fast-charging stations and the possibility of including a hydrogen infrastructure. As assessed for the regional airport, the needed hydrogen infrastructure onsite will provide only storage on wheels and tanks. The hydrogen production will be off-site in a nearby harbour. The components currently limiting the airport infrastructure and required to be installed and/or revised are (i) connection to the main grid, (ii) local photovoltaic supply, (iii) charger for aircraft, (iv) charger for airfield support vehicles, (v) local battery storage and (vi) local hydrogen storage.

Even with a local supply of electricity from the airside solar park and storage, the grid feed-in for an infrastructure capable of charging multiple aircraft and support vehicles will need to be connected to the medium-voltage grid (6 kV to 60 kV) due to the high-power demand. The distribution network on the airport premises should then be realised on a low-voltage level (or according to the low voltage directive 2014/35/EU) due to safety reasons. The AC (analog current)-concept utilises a common AC bus for the distribution system. It is a standard electrical installation, as it is common in most of the world [11].

Photovoltaic power generation, fuel cell technology, stationary battery storage, and mobile batteries in aircraft and escort vehicles are DC (direct current)-based by nature. A second concept links all components to a low-voltage DC grid (<1500 V). This DC grid must be rebuilt entirely at an airport [11]. This possibility is mentioned in this article for completeness but is not considered in more detail in the scenarios, as the RTHA airport operates with an AC-based grid. Sections 2.1.1 and 2.1.2 briefly describe the type of supply voltage and the basic connection of a Fast-Charging-Station to the airport grid.

2.1.1. Voltage on the Grid

Compared with the DC-based method, the AC based approach benefits from being the currently established technology at the airport side. Therefore, a large selection of installation components is available, as are trained technical staff to work with standard grids. However, from a technical perspective, the DC approach has multiple benefits. For example, integrating additional battery storage units and photovoltaic systems is simplified by eliminating the AC/DC or DC/AC conversation stages for all DC-based devices. Therefore, fewer conversation stages are needed between local battery storage, photovoltaic generators, and consumers (e.g., DC chargers). In addition, a DC-based grid increases the system's efficiency and decreases complexity. Furthermore, DC-connected systems are easier to control than AC-connected systems [11].

There is also an advantage when it comes to power distribution cables. An AC system uses a four-wire setup for the power distribution network. As such, the transported power results in Equation (1).

$$P_{AC} = \sqrt{3} \cdot U_{peak} \cdot I_{eff} \cdot \cos(\varphi) \tag{1}$$

with the effective current (I_{eff}), the peak voltage (U_{peak}) and the power factor ($\cos(\varphi)$). Using the same four-wire setup for a DC system with two wires used for positive and negative, the maximum current per wire matches the effective AC to not overload the conductor. For AC systems, the insulation rates for the peak voltage. As such, the nominal voltage in a DC system can be U_{peak}, and the power in a four-wire DC system results in Equation (2).

$$P_{DC} = 2 \cdot U_{peak} \cdot I_{eff} \tag{2}$$

Comparing the DC and AC power results in Equation (3):

$$\frac{P_{DC}}{P_{AC}} = \frac{2 \cdot U_{peak} \cdot I_{eff}}{\sqrt{3} \cdot U_{peak} \cdot I_{eff} \cdot \cos(\varphi)} = \frac{2}{\sqrt{3} \cdot \cos(\varphi)} \tag{3}$$

Even with a power factor $\cos(\varphi)$ of 1, the same wire system can transport about 15% higher power using a DC system. Using the reciprocal reduces the required copper cross-section of the wiring system to approximately 85% for the same power. Despite the many advantages of a DC network, however, the AC network is considered in this study because, as already mentioned, the AC network is an established technology, trained specialists are available, and the airport is equipped with an AC network. However, developing a DC network could also become a key element in the future.

2.1.2. Fast-Charging-Stations

According to safety standards [12], galvanic isolation must be guaranteed between the main distribution 3-phase AC-grid and the charging station. This can either be realised by utilizing a low frequency (LF-) transformer or an isolating DC-DC converter (MF-transformer), which operates in the medium frequency (MF) range. These two possibilities are visualised in Figure 1 as "Topology 1" and "Topology 2". Higher frequencies allow for lower material effort, and hence the size of the passive devices. Therefore, Topology 2 provides benefits over Topology 1 in the form of a less "bulky" transformer. However, a transformer connects the regional airport to the medium-voltage grid. Accordingly, "Topology 1" is considered in this study.

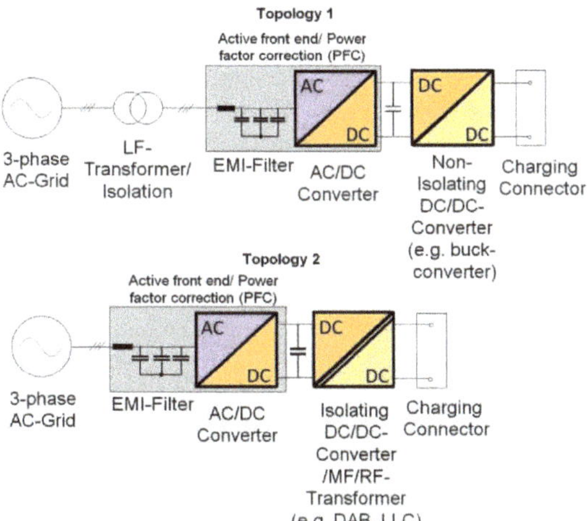

Figure 1. Basic topologies for DC-Fast-Charging of electric vehicles.

2.1.3. Hybrid-Electric-Aircraft Configurations

First, Figure 2 visualises an overview of a 50 PAX hybrid-electric aircraft with a gas turbine and battery as an energy source and the drivetrain's arrangement, valid for two different reference entry-into-service (EIS) years 2030 and 2040 [10].

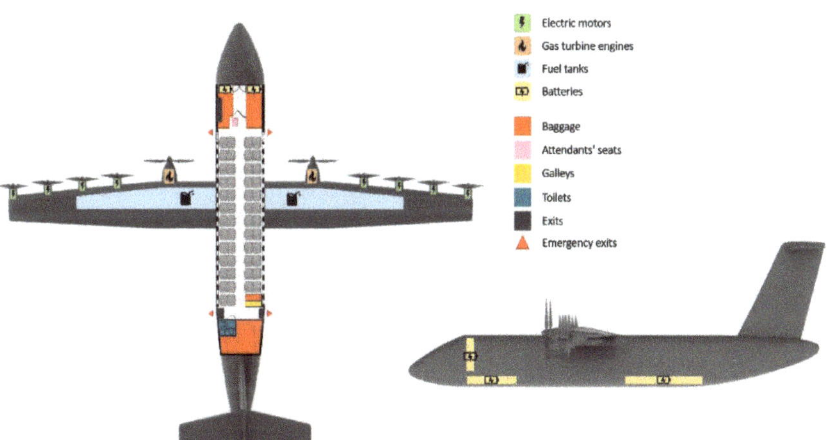

Figure 2. Schematic view of an HEA (EIS 2030/2040, ICE + Battery) - Reprinted/adapted with permission from Ref. [10]. 2023, Marciello, V.; Di Stasio, M.; Ruocco, M.; Trifari, V.; Nicolosi.

The propulsive architecture adopted for the short- and medium-term scenarios was based on a serial/parallel partial hybrid configuration with two distinct propulsive lines. This choice made it possible to use the distributed electric propulsion during the ground phases to increase the lifting capabilities of the aircraft, compensating for the increased mass due to the advanced powerplant. At cruise, in light of the lower efficiency of distributed propellers, delivering all the shaft power through the primary line is preferable, redirecting the energy from the electric storage. New secondary electric machines were designed based on nominal RPMs equal to 8000 since it was found that it is optimal to delegate to gearboxes the task of adapting the number of revolutions to that of the propellers. The specific fuel consumption reflects the usage of pure hydro-processed esters and fatty acids synthetic paraffinic kerosene (HEFA-SPK) as fuel [10].

The second HEA configuration deals with a Proton Exchange Membrane Fuel Cell and a battery (PEMFC + Battery). Figure 3 gives a short introduction and overview of the aircraft design and the arrangement of the components. The main difference concerning the Internal combustion engine (ICE) + Battery scenario is that there is only a single-drivetrain, referred to as primary in the present context, with five engines of equal power attached to each semi-wing. For this reason, the electric machines will generate thrust through the propeller for the aircraft in all phases of flight. Since there is no distinction between primary and secondary propulsion lines, the production costs, as well as the maintenance costs of the aircraft, would benefit from having installed electric machines all rated at the same power. The propulsive architecture with Proton Exchange Membrane Fuel Cell (PEMFC) is based on a full-electric configuration, where part of the electricity is produced directly by the fuel cells through the reaction of hydrogen with air. The atmospheric air is supposed to be supplied through suitable air intakes and compressed up to the operating pressure of the fuel cells using a centrifugal compressor. Based on the power and energy requirements, Li-S batteries were identified to be the best choice for this application.

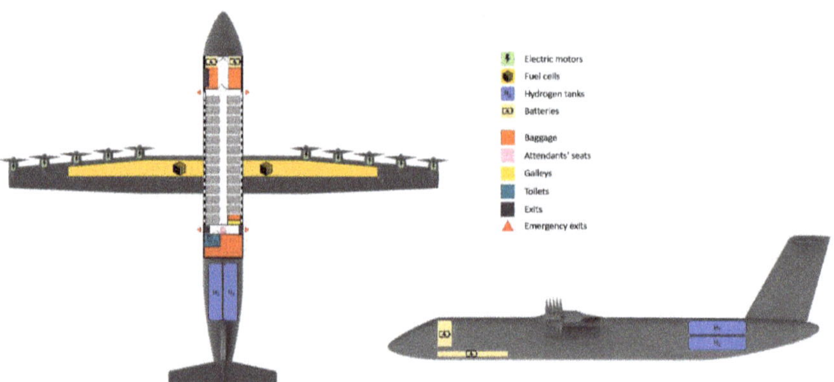

Figure 3. Schematic view of the aircraft (EIS 2040/2050; PEMFC + Battery) - Reprinted/adapted with permission from Ref. [10]. 2023, Marciello, V.; Di Stasio, M.; Ruocco, M.; Trifari, V.; Nicolosi.

2.1.4. Airport Infrastructure Scenarios

This paper is about the energy requirements of an airport for the operation of hybrid-electric aircraft, so only the most necessary supply technologies are briefly presented here. In principle, a Photovoltaic (PV) system, battery storage, wind power and an on-ground electrolyser/fuel cell also make sense to include in an airport network. However, this would go beyond the scope of this paper, which is therefore limited to the necessary components to describe the methodology, which applies to both presented scenarios in the following. Figure 4 shows possible airport infrastructure for a conventional fuel supply and fast-charging stations for the HEAs. This configuration is used with ICE + Battery HEA for the short-term and medium-term horizon. A 10 kV/400 V transformer connects the medium-voltage and 400 V AC airport grid. The left side of Figure 4 shows this connection to the medium-voltage grid. The AC/DC converters and DC/DC boost converters are connected to the airport AC grid, as described in simplified form in Section 2.1.2, to provide the necessary high charging currents. For the sake of simplicity, the regular 3-phase AC grid is shown here with a line for a better overview. In addition, the airport is connected to the AC grid as an electric load to supply the terminals, lightning, etc., with electric energy. Furthermore, charging possibilities for baggage cars, busses and other airport vehicles are connected to the airport. Finally, the lower part of Figure 4 shows the kerosene and SAF mixture ratio supply as fuel.

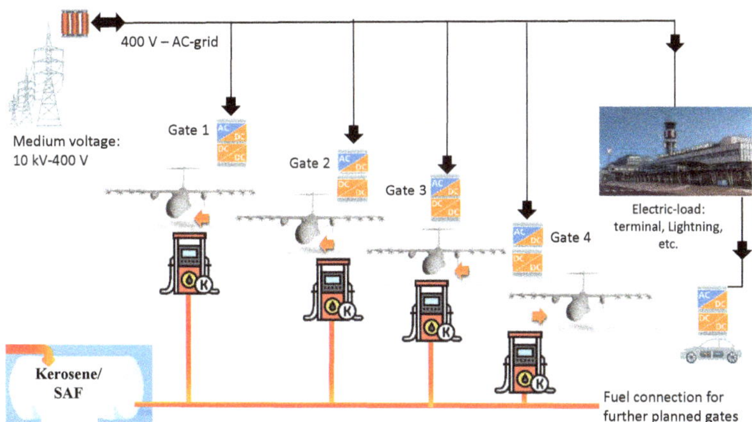

Figure 4. Airport infrastructure for the ICE + Battery HEA for medium– and short–term horizons.

Figure 5 shows possible airport infrastructure for a LH_2 and battery hybrid electric aircraft. This configuration is used for the medium-term and long-term horizon with PEMFC + Battery HEA. The connection to the medium voltage grid on the left side above is shown in Figure 5. The AC lines are shown with one single line similarly to Figure 4. Furthermore, the AC/DC and DC/DC boost converters for the fast charging stations are shown in the upper part of Figure 5. The hydrogen production for the airport hydrogen supply is done off-site at a port near the airport. Wind energy is converted into electrical energy. This electrical energy is converted into hydrogen via electrolysers and liquefied (i, ii). Trailer trucks transport liquid hydrogen to the airport, and the aircraft can be refuelled directly (iii). This process can be seen on the bottom right-hand side of Figure 5.

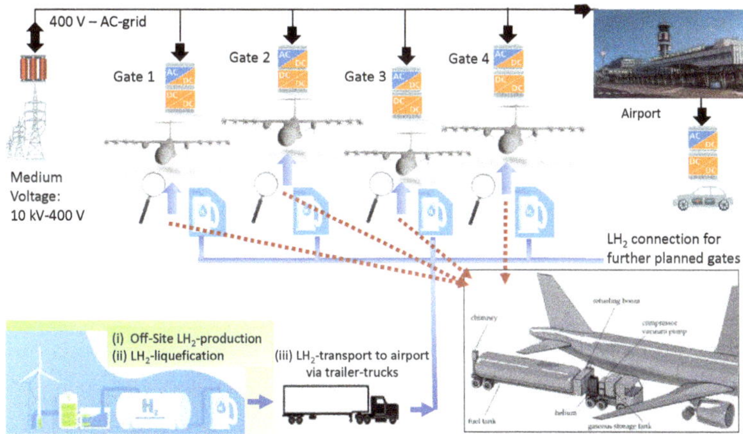

Figure 5. Airport infrastructure for the LH_2 powered PEMFC + Battery HEA for medium– and long–term horizons.

2.2. Scenario Definitions

This section introduces the individual scenarios for consideration at a regional airport of the future. These scenarios give an insight into which assumptions were made.

Based on the configurations and assumptions prepared by UNINA for the aircraft, these were further developed for the benefit of the regional airport. The flights of the HEA have been fitted into RTHA's flight schedule based on the following assumptions formulated by UNINA. Due to the COVID-19 pandemic, a flight plan from 2019 was used because air traffic was not restricted in 2019. These data form the basis for the operation of the hybrid-electric aircraft in the scenarios of a regional airport of the future.

- Analysis of the Air Traffic Data for 2019, collected from RTHA and elaborated to collect the necessary key figures;
- The flight data from RTHA 2019 were used to select the flights relevant to the HEA: commercial flights with a destination with of maximum distance of 1111.2 km;
- These data were used to replace the fossil flights with HEA with max 50 PAX and a maximum distance of 1111.2 km;
- HEA flights replace fossil flights with 50 passengers or fewer, with a maximum of 50 passengers;
- Two HEA flights replace fossil flights with more than 50 passengers but fewer than 100, with a maximum of 50 passengers;
- Three HEA flights would replace a flight with more than 100 passengers but fewer than 150, with a maximum of 50 passengers each;
- Four HEA flights would replace all other flights with more than 150 passengers.

To calculate the energy required, an extra 330 km is added to the distance of 1111.2 km for any possible calamities (holding/diversion, etc.). Partly in connection with this choice, the gradual transition of aviation will accelerate the introduction of smaller but environmentally friendly aircraft with smaller passenger capacity. This may mean that aircraft such as a Boeing 737 will eventually be replaced by 2–4 environmentally friendly aircraft with a capacity of 50 passengers, followed by hybrid/electric aircraft with 100–150 passengers. For the first period, this means a severe increase in aircraft and flight movements at the airport. This will also have a massive impact on the infrastructure and organisation at the airport, with the comment that this transition will be gradual to allow the airport to prepare for it. Airport infrastructure is geared to existing aircraft for take-off, landing, taxiing, refuelling, loading/unloading, etc. Thus, it will change when adopting other disruptive aircraft propulsive technologies. The future regional airport infrastructure design focuses on the HEA, suitable for 50 PAX with a flight range of 1111.2 km.

- For 2025–2035, aircraft with combustion engines fuelled by 90% kerosene and 10% SAF and electric engines powered by batteries are assumed (ICE + Battery-2030). The SAF fuel will be HEFA (Hydroprocessed Esters and Fatty Acids), briefly explained in [13];
- In 2035–2045, aircraft with combustion engines fuelled by 75% kerosene and 25% SAF and electric engines powered by batteries are assumed (ICE + Battery-2040). In addition, some flights will be replaced by HEA with electric engines and PEMFC powered by liquid hydrogen (PEMFC + Battery-2040);
- In 2045–2055+, all HEA flights will use fuel cells and electric motors. There will be a further developed PEMFC (PEMFC + Battery-2050) installed.

In calculating the energy requirements in the different scenarios, a distinction has been made between typical mission (200 nmi) and design mission (600 nmi) flights. For all flights, the possibility of diversions, etc., corresponding to 330 additional km per flight has been included.

2.3. *Emmisions*

UNINA and SmartUp presented in [14] a general approach for emissions estimation, based on the results, produced with a gas turbine engine performance calculation tool, the semi-empirical approach illustrated in [15] and average data included in [16]. This emission model requires a manageable amount of input data: the engine's overall pressure ratio, fuel flow rate, operating conditions and ambient conditions are the only information required. The output of this model consists of the emission indices (EIs) for the following species: nitrogen oxides (NO_x), carbon monoxide (CO), hydrocarbons (HC), carbon dioxide (CO_2), water vapor (H_2O) and sulphur dioxide (SO_2). For the EIs of CO_2, H_2O and SO_2, constant average values listed in Table 2 were assumed.

Table 2. Average EIs for CO_2, H_2O and SO_2 for conventional (Jet A-1) aviation fuel, assumed according to [16].

Pollutant	EI (g/kg)
CO_2	3149.0
H_2O	1230.0
SO_2	0.84

It can be assumed that the approach adopted by UNINA and SmartUp can be considered reliable as long as the fuel is conventional jet fuel (i.e., Jet A-1). For this reason, in the case of HEFA-SPK blends, calibration factors to correct the EI of the above species were considered in [14]. Specifically:

- A correction for CO_2 EI as a function of HEFA-SPK mixtures has been established. According to [14], a reduction in CO_2 EI of 0.78% can be obtained for 50% blends. For pure HEFA-SPK, the reduction doubles (−1.56%);
- For the water vapor EI, a linear regression law was obtained for the percentage changes as a function of the HEFA-SPK mixing ratio. According to this law, an increase of 10% can be expected for pure HEFA-SPK [17];
- For CO and SO_2 EIs, linear regression laws were determined instead using the data collected by [18]. According to the equations obtained from these data, the use of pure HEFA-SPK could lead to a reduction of −22% in CO-EI and −74% in SO_2-EI;
- Finally, for NO_x and HC-EIs, [18] suggests a negligible impact of biofuel blends. Consequently, no percentage changes were considered.

These deviations were all applied to the reference EI values calculated with the original approach established for conventional aviation fuel. With the help of these methods and estimates, the emission impact in the different scenarios can be estimated [14].

2.4. Economic Aspect

In addition to the environmental and conservation benefits, switching from fossil fuels to sustainably produced fuels is also financially attractive for airports, airlines and travellers. The results of several studies indicate a reduction in fuel costs ranging from 15% to 40%, depending on the study assumptions. The results of a recent Swedish study [19] show the total costs for routes and aircraft in Table 3 below in euros (EUR). Table 3 shows that costs increase with distance. Previous studies concluded that the operating costs of electric aircraft are between 30 and 40% lower. In contrast to the previous study, the table shows the difference between the conventional Jetstream JS31 (19 PAX) aircraft and the Swedish ES-19 (19PAX) aircraft. The difference is between 15 and 22% per route.

Table 3. Total emission cost comparison of two aircrafts according to [19].

Route	Distance [km]	Jetstream JS31 [EUR]	ES-19 [EUR]
Linköping-Visby	178	EUR 2306	EUR 1805
Pajala- Luleå	194	EUR 2380	EUR 1886
Umeå-Östersund	296	EUR 2976	EUR 2378
Sälen-Arlanda	326	EUR 3253	EUR 2636

Combining the cost difference from Table 3 with other data from the literature in Table 4 and internal information from RTHA, costs can be estimated and determined for the HEA and are listed in Table 4.

Table 4. Cost estimation for HEA emissions.

Energy	Unit	Cost in EUR
Carbon tax ETS [20]	kg CO_2	0.25
Kerosene [21]	kg	1.20
SAF [21]	kg	2.00
Electric [22]	kWh	0.60
LH_2 [21]	kg	6.00

Based on the fundamental literature research on the costs of the individual energy parameters and carbon tax for the HEA aircraft, an estimate can be made for future typical mission (200 nmi) and design mission (600 nmi) flights. For now, only the lower energy requirements for the different scenarios were used in these cost estimates. Fuel costs and landing fees account for 30% of the ticket price. For this reason, the fuel cost is calculated on the price per kilometre. Then, the fuel saving per kilometre can be calculated on an aircraft basis. Finally, ticket price savings can be calculated and stated on an aircraft basis, with 30% of the total ticket price.

3. Results

Sections 3.1 and 3.2 will describe the procedure in the medium-term scenario in more detail. The approach to determining energy demand, emissions and ticket prices is similar for all time horizons. Therefore, the methodology described in Section 2 is carried out once here. However, the different aircraft configuration already indicated in Table 1 was used. The fuel mix ratio in the medium-term (ICE + Battery) is 75% kerosene and 25% SAF, and the infrastructure is shown in Figure 4.

For the aircraft configuration PEMFC + Battery, the infrastructure was considered as in Figure 5. The results for an HEA with FC and battery are also described in Sections 3.1 and 3.2. A PEMFC is used for the fuel cell technology. This allows a direct comparison between the operation of an ICE + Battery and a PEMFC + Battery HEA.

3.1. Determining Energy Demand ICE + Battery and PEMFC + Battery HEA

Based on information from the relevant commercial flight and the configuration of the HEA, an overview of the fuel requirements for the eligible flights was made. To determine the maximum fuel and electricity supplies, one of the busiest days was selected for flights up to 1111 km to the destination.

For the medium term, based on these assumptions, the amount of electricity which the HEA flights would potentially require on a busy day at a regional airport is reported in Table 5. The departure airport in this study is Rotterdam. Table 5 lists the destinations and the amount of kerosene, SAF and electrical energy or LH_2 and electrical energy required for the HEA in parentheses. The electrical energy demand for ICE + Battery HEA is listed as "Electric ICE [kWh]". The electrical energy demand for PEMFC + Battery HEA is listed as "Electric ICE [kWh]" in Table 5. These destinations are determined from the number of PAX and the distance, as already described in Section 2.3. It can be seen that several flights would have to take off at the same time. This is, of course, not possible, but it should serve here as an introduction to show the potential energy demand of the HEA. Energy consumption per flight is high in the morning and evening for the London destination and average for European flights. The total capacity required is highest in the late afternoon, shown in Figure 6. This high capacity is because three flights with many passengers depart in the afternoon. Figure 6 illustrates the high capacity of the period from 16:30 to 17:04. The initial calculations and simulations show that the short-term scenario requires a daily kerosene demand of 13.37 tonnes, an SAF demand of 4.39 tonnes and an electrical energy demand of 46.68 MWh (yellow line).

Table 5. Daily fuel and electricity amount per hybrid-electric flight, 2040.

Day 2030	PAX	Distance [km]	GMT [Time]	Kerosene [kg]	SAF [kg]	Electric ICE [kWh]	LH_2 [kg]	Electric PEMFC [kWh]
London (2—HEA)	76	308	05:19	612	200	3049	209	3288
London (2—HEA)	78	308	08:38	612	200	3049	209	3288
Bergerac (3—HEA)	147	844	12:04	1743	572	5670	740	6128
Pula (4—HEA)	164	1049	16:30	2730	896	8881	1158	9598
Wien (3—HEA)	135	953	16:49	1905	625	6197	808	6697
Montpellier (4—HEA)	179	925	17:04	2485	816	8082	1054	8735
Pisa (4—HEA)	186	1021	19:56	2675	878	8700	1135	9403
London (2—HEA)	97	308	20:08	612	200	3049	209	3288
TOTAL				**13,374**	**4389**	**46,678**	**5523**	**50,425**

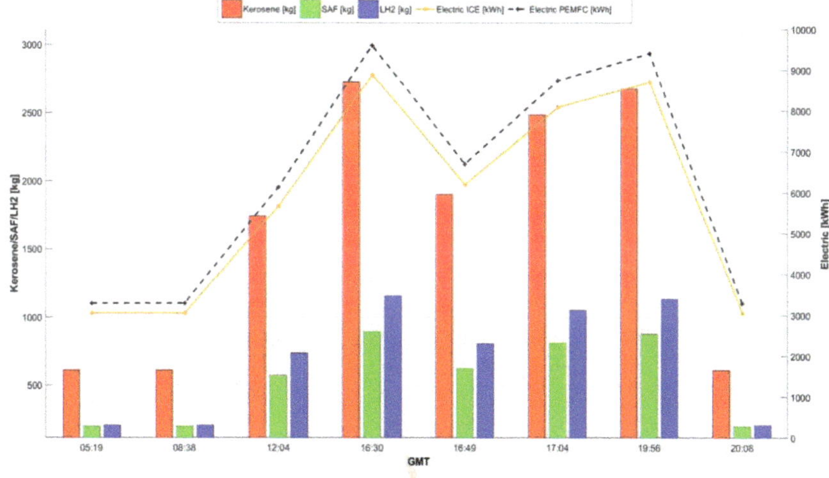

Figure 6. Daily fuel and electricity amount per hybrid-electric flight, 2040.

To compare the impact of LH_2, based on the combination of information from the respective RTHA traffic flight and the configuration of the newly developed medium-term HEA with PMFC + Battery, an overview of the fuel requirements for the considered flights is shown in Table 5 as "LH_2" and "Electric PEMFC". Initial calculations and simulations show that the HEA in the medium-term scenario with PMFC + Battery no longer requires the daily kerosene demand of 23.5 tonnes (short-term) and 13.37 tonnes (medium-term-ICE + Battery). Similarly, the SAF demand of 2.55 tonnes (short-term) and 4.39 tonnes (medium-term-ICE + Battery) is no longer needed. Instead, a liquid hydrogen requirement of 5.523 tonnes is now determined to fuel the aircraft. In addition, the PEMFC + Battery medium-term HEA will be fitted with a battery of higher capacity and power, increasing the demand for electrical energy from 26.05 MWh (short-term) and 46.68 MWh (medium-term-ICE + Battery) to 50.425 MWh (black line).

In order to replace these flights with hybrid-electric flights, a new flight schedule with new departure times must be created. This new flight schedule is presented in Table 6. In this table, the old departure times of the original flight plan are listed again. New departure times are introduced in the column to the right with the destination abbreviation. These new departure times are based on the original time, and an average time of 10 min assumed between the departure times. These 10 min are for taxiing from the gate to the runway and subsequent take-off. The kerosene, SAF and electrical energy consumption of each HEA is given and composed of typical mission (200 nmi) and design mission (600 nmi) flights. The maximum number of passengers per flight is 50. The maximum range of the potential flights was kept below 1111.2 km to represent a realistic scenario.

Table 6. Possible replacements through HEAs, 2040.

Destination	Time—Old	Time—New	PAX	km	Kerosene [kg]	SAF [kg]	Electric ICE [kWh]	LH_2 [kg]	Electric PEMFC [kWh]
London	05:19	LO 05:19	50	308	306	100	1.525	105	1643.94
London	05:19	LO 05:29	26	308	306	100	1.525	105	1643.94
London	08:38	LO 08:38	50	308	306	100	1.525	105	1643.94
London	08:38	LO 08:48	28	308	306	100	1.525	105	1643.94
Bergerac	12:04	BE 12:04	50	844	581	191	1.890	247	2042.76
Bergerac	12:04	BE 12:24	50	844	581	191	1.890	247	2042.76
Bergerac	12:04	BE 12:44	47	844	581	191	1.890	247	2042.76
Pula	16:30	PU 16:00	50	1049	683	224	2.220	290	2399.46
Pula	16:30	PU 16:10	50	1049	683	224	2.220	290	2399.46
Pula	16:30	PU 16:20	50	1049	683	224	2.220	290	2399.46
Pula	16:30	PU 16:30	14	1049	683	224	2.220	290	2399.46
Vienna	16:49	VIE 16:39	50	953	635	208	2.066	269	2232.42
Vienna	16:49	VIE 16:49	50	953	635	208	2.066	269	2232.42
Vienna	16:49	VIE 16:59	35	953	635	208	2.066	269	2232.42
Montpellier	17:04	MO 17:09	50	925	621	204	2.021	264	2183.7
Montpellier	17:04	MO 17:19	50	925	621	204	2.021	264	2183.7
Montpellier	17:04	MO 17:29	50	925	621	204	2.021	264	2183.7
Montpellier	17:04	MO 17:39	29	925	621	204	2.021	264	2183.7
Pisa	19:56	PI 19:36	50	1021	669	220	2.175	284	2350.74
Pisa	19:56	PI 19:46	50	1021	669	220	2.175	284	2350.74
Pisa	19:56	PI 19:56	50	1021	669	220	2.175	284	2350.74
Pisa	19:56	PI 20:06	36	1021	669	220	2.175	284	2350.74
London	20:08	LO 20:16	50	308	306	100	1.525	105	1643.94
London	20:08	LO 20:26	47	308	306	100	1.525	105	1643.94
TOTAL					**13,374**	**4389**	**46,678**	**5523**	**50,425**

As in the scenarios before, PEMFC + Battery HEA should replace conventional aircrafts in this medium-term scenario. These energy requirements are also listed in Table 6 on the right side as "LH_2" and "Electric PEMFC".

Figure 7 shows the new flight plan's results and the energy required. It is apparent that in the early morning, for the flights to London (LO), 1525 kWh of electrical energy is required to charge the aircraft and refuel them for the flight. The kerosene quantity is 306 kg, and the SAF quantity is 100 kg, with the previously defined specifications of 75% kerosene and 25% SAF. The equalisation of the flights to Pula (PU), Vienna (VIE) and Montpellier (MO) show an electrical energy demand of 2220 kWh to 2021 kWh. The flight schedule was equalised, and the electrical energy required from 16:00 to 17:29. The kerosene/SAF requirement of a maximum of 683 kg/flight can also be easily provided. Four take-offs to Pisa (PI) are required in the evening, with an electrical energy quantity of 2175 kWh and a kerosene quantity of 669 kg/flight. As soon as the last flight at 20:26 to London has taken off with an electrical energy quantity of 1525 kWh and 306 kg of kerosene, the electrical energy consumption of the airport can be reduced again.

As in the scenarios before, the HEA's kerosene, SAF and electrical energy consumption are now eliminated. Figure 7 shows the results of the new flight plan for required LH_2 (blue) and electrical energy (black). Early morning flights to London (LO) require 1644 kWh of electrical energy in the medium term to recharge the aircraft and refuel for the flight. This is because a more powerful battery is installed in the PEMFC aircraft than in the previous time horizon. The liquid hydrogen quantity is 105 kg instead of the paraffin quantities of 306 kg (medium-term) and 557 kg (short-term). The reconciliation of the flights to Pula (PU), Vienna (VIE) and Montpellier (MO) resulted in an electrical energy demand of 2399 kWh to 2184 kWh, which is significantly higher than in the previous scenarios, as expected. The flight schedule was adjusted, and electrical energy is required from 16:00 to 17:29. The kerosene/SAF requirement of a maximum of 1191 kg/flight in the short term and a maximum of 683 kg/flight (medium-term) is now also omitted here. A maximum of 290 kg of liquid hydrogen is required for the flight to Pula. Four take-offs to Pisa (PI) are required in the evening, with an electrical energy quantity of 2351 kWh. The paraffin amounts of 1167 kg/flight in the short-term horizon and 669 kg/flight in the medium horizon with ICE + Battery are omitted, and 284 kg liquid hydrogen per flight is required. Once the last flight has taken off at 20:26 to London with an electrical energy quantity of 1644 kWh and 105 kg of liquid hydrogen, the electrical energy consumption of the airport can be reduced again. In the long term, storing electrical energy not needed in large batteries or converting it into liquid hydrogen can be considered.

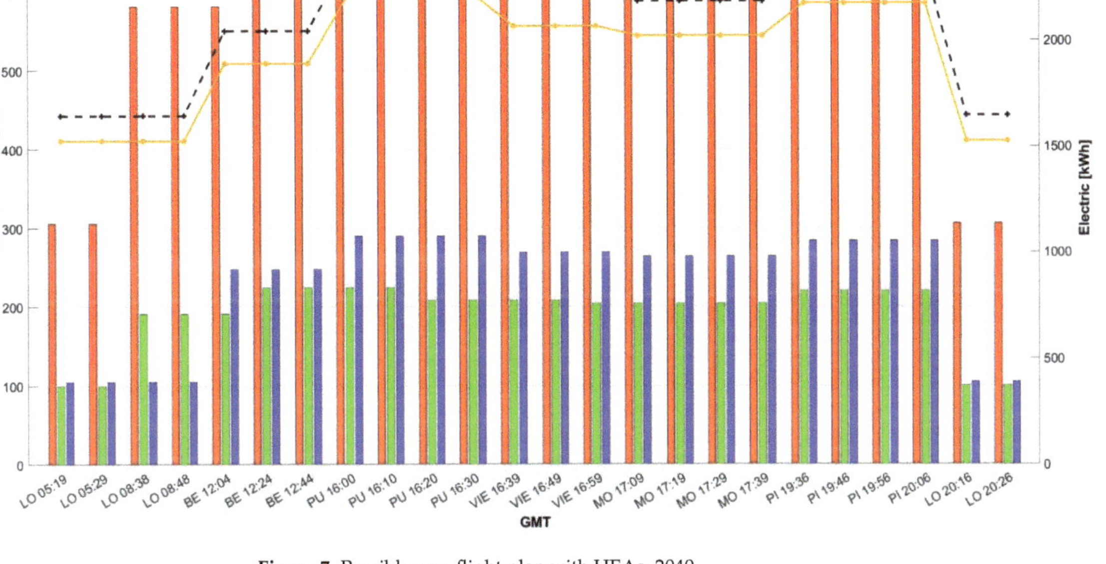

Figure 7. Possible new flight plan with HEAs, 2040.

Finally, the annual energy demand for the short-term scenario is given in Table 7. Table 7 shows the energy demand for hybrid-electric flights in 2030 per month to determine the loading and refuelling energy for one year in 2030. It was concluded that 3215 tonnes of kerosene, 1056 tonnes of SAF and 11.704 GWh of electrical energy would be required in the short-term horizon to operate the HEA. These calculations were made with a fuel mix ratio of 75% kerosene, 25% SAF and an HEA configuration.

Table 7. Charging and refuelling HEA energy requirements per month, 2040.

Month	Kerosene [Tons]	SAF [Tons]	Electric ICE [MWh]	LH2 [Tons]	Electric PEMFC [MWh]
January	209	69	762	84	823
February	268	88	966	108	1043
March	355	117	1269	144	1371
April	261	86	975	103	1053
May	305	100	1111	123	1200
June	324	106	1171	131	1265
July	345	113	1221	141	1319
August	332	109	1165	136	1258
September	314	103	1128	127	1218
October	208	68	798	81	862
November	130	43	529	49	571
December	162	53	610	64	658
TOTAL	**3215**	**1056**	**11,704**	**1291**	**12,640**

For the PEMFC + Battery aircraft, the energy demand is listed on the right side of Table 7 as "LH2 [tons]" and "Electric PEMFC" [MWh]. This table shows the energy demand for HEA flights in 2040 per month to determine the loading and refuelling energy for one year in 2040. It was found that, instead of 3215 tonnes (medium-term-ICE + Battery), 1291 tonnes of liquid hydrogen are now required to operate the HEA in the medium-term scenario. The requirement of 1056 tonnes of SAF in the medium-term horizon are eliminated accordingly. The demand for electrical energy of 11.704 GWh (medium-term-ICE + Battery) increases to 12.640 GWh (black line). The demand for electrical energy is 74% higher than for the short-term horizon (Table 13). The demand for electrical energy in the medium-term with PEMFC is almost 8% higher than in the medium-term scenario with ICE + Battery. This is due, on the one hand, to the increased battery capacity in the PEMFC aircraft, and on the other hand to the use of liquid hydrogen. The charging and refuelling energy for the HEA is shown in Figure 8.

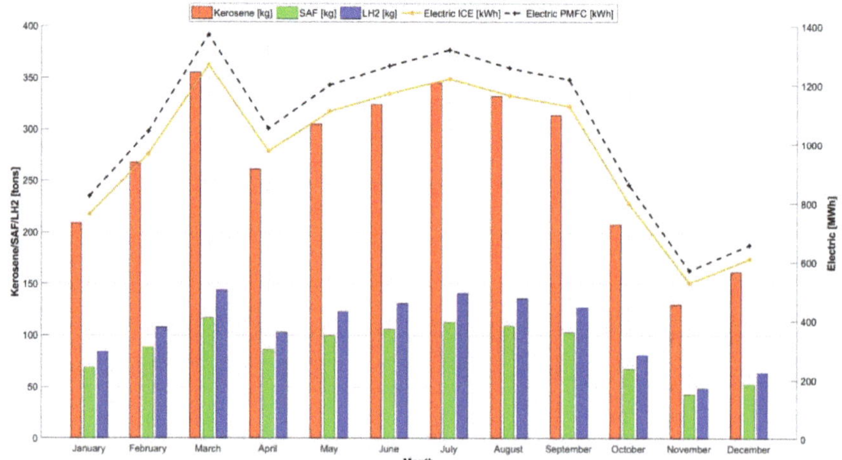

Figure 8. Charging and refuelling HEA energy requirements, 2040.

3.2. Determining Cost Estimations and Emissions for the Short-Term

In this section, a cost and emission forecast for the period 2025–2035 will be given. First, it should be explained how the data were obtained. It is important to read Sections 2.2 and 2.3 first. Gilbarco Tritium RT175-S DCFC Fast Charge Single Electric Vehicle 175 kW Charging Stations have a list price of USD 105,000 each. For a charging station with double capacity, an investment of USD 175,000 is considered [23]. Costs for maintenance have not yet been released.

Capex and Opex of the Maeve Recharge 30-ft container with 8 MW battery capacity and control module have also not yet been released. The battery pack cost will be lower than the market price for new batteries because it is reused from electric aircraft. The final megawatt charging system (MCS) standard is expected to be published in 2024 [24,25].

For the ticket price calculation in the short-term scenario, the data and calculations in Section 2.3 were used as a base Then, using the flight distance, the information from RTHA and the composition of the current ticket price, the price of a passenger per km can be given. It is further assumed that fuel costs and landing fees account for 30% of the ticket price. Furthermore, three possible environmental price increases offered by Lufthansa [26,27] were included and applied to the ERJ 190 and B737. A number is given in the brackets after the respective conditions, indicating which scenarios were considered in the following tables. These three environmental price increases amount to:

- A 100% climate project subsidy (100% describes that, with this selection, the full 2.6%, which is additionally paid by the client, goes into climate projects)—2.6% → (1);
- An 80% climate project subsidy and 20% SAF fuel—21% → (2);
- A 100% SAF fuel and CO_2 emissions reduced—by 96% → (3).

In addition, an average inflation rate of 2.44% was assumed, which resulted over the last 50 years in Germany [28]. This inflation rate is also included in the ticket prices, to give a realistic estimate of the prices for different time horizons.

- 2.44% inflation rate in terms of 2040 → (4).

For the price comparison per ticket with the GENESIS flight, Scenario 2 was assumed in the short term. Therefore, this scenario is considered with 20% SAF fuel and is comparable with the HEA case study. The calculated costs for the short-distance flight are shown in Table 8, and the costs for the medium-distance flight are in Table 9. These calculations and data show that HEA ticket prices are somewhat higher than conventional ticket prices for typical mission flights such as to London. However, in the medium-term scenario, the ticket price for a flight with ICE + Battery HEA is 1.8 below the comparable ticket price with 20% SAF. The expected ticket price for design mission flights is 11.8% below the comparable price when using an HEA. As soon as HEA flights with PEMFC + Battery can be offered, the ticket price difference is considered very attractive purely on the basis considered: a price saving of 46.7% is expected for typical mission flights and 40.4% for design mission flights.

Table 8. Costs for typical mission flights, medium-term (2040) forecast.

London 308 km	Scenario	Delta/km [EUR]	Cost/km and PAX [EUR]	Delta/km and PAX [%]	Cost/km (PAX; Env.; inflation) [EUR]	Delta/Ticket HEA [%]
ERJ190	(4)	23.33	0.048		0.24	
ERJ190	(1.4)	23.94	0.049		0.24	
ERJ190	(2.4)	28.31	0.058		0.29	
ERJ190	(3.4)	45.77	0.094		0.47	
HEA-ICE	(4)	14.18	0.057	−1.8%	0.28	−1.8%
HEA-PEMFC	(4)	12.44	0.050	−46.7%	0.25	−46.7%

Table 9. Costs for design mission flights, short-term (2040) forecast.

Pula 1049 km	Scenario	Delta/km [EUR]	Cost/km and PAX [EUR]	Delta/km and PAX [%]	Cost/km (PAX; Env.; inflation) [EUR]	Delta/Ticket HEA [%]
B737	(4)	24.40	0.04		0.20	
B737	(1.4)	25.05	0.04		0.20	
B737	(2.4)	29.64	0.05		0.24	
B737	(3.4)	40.11	0.077		0.38	
HEA-ICE	(4)	10.46	0.042	−11.8%	0.21	−11.8%
HEA-PEMFC		11.43	0.046	−40.4%	0.23	−40.4%

Table 10 shows an estimate of the ticket development for 2040, which can be derived using the presented method. This table illustrates very well the impact of inflation and the environmental bonus in the categories on different routes. According to this, the EIS of PEMFC + Battery HEA results in competitive ticket prices for HEA PEMFC tickets. The tickets for the flight to Pisa are 17% more expensive than the expected ticket prices without subsidy (4). As soon as customers want to fly with "80% climate project subsidy and 20% SAF fuel (2,4)", the ticket PEMFC HEA is already 4% cheaper. Nonetheless, it should always be mentioned that the calculation was made without the high investment research and operating costs.

Table 10. Ticket price forecast, 2040.

Destination	Distance [km]	GMT [Time]	Ticket (4) [EUR]	Ticket (1.4) [EUR]	Ticket (2.4) [EUR]	Ticket (3.4) [EUR]	Ticket HEA ICE [EUR]	Ticket HEA PEMFC [EUR]
London	308	05:19	73.31	75.24	88.97	143.83	87.33	76.63
Bergerac	844	12:04	164.78	169.15	200.11	323.92	176.49	192.90
Pula	1049	16:30	204.81	210.23	248.71	402.59	219.35	239.76
Wien	953	16:49	186.06	191.00	225.95	365.75	199.28	217.81
Montpellier	925	17:04	180.60	185.38	219.31	355.00	193.42	211.42
Pisa	1021	19:56	199.34	204.62	242.07	391.85	213.50	233.36

Nevertheless, the savings on the expected ticket price per passenger offer a first estimate to make these investments lucrative for airlines and to justify the initial investments with a long view into the future. This fact confirms the previously established thesis that HEA flights have the potential to be financially attractive and environmentally friendly.

Finally, the HEA flights' estimated emissions for the short-term scenario are given for an average day, month and year. The calculation basis was the methods described in Section 2.3. The results are presented in Table 11. The HEA produce daily emissions of almost 58 tonnes of CO_2. Annual emissions of nearly 13 863.65 tonnes of CO_2 are expected. The NO_x values are 49.619 tonnes per year, whereas 20.04 tonnes of CO are expected to be emitted annually. The values were estimated according to the procedure presented in Section 2.3. These high emissions indicate the urgency of transitioning towards sustainable hybrid-electric aviation.

Table 11. Emissions forecast of HEA flights in the medium-term (2040) scenario.

	Fuel [ton]	CO_2 [ton]	NO_x [ton]	HC [ton])	CO [ton]	H_2O [ton]	SO_2 [ton]
			Day—ICE + Battery				
Kerosene	13.374	43.769	0.156	0.006	0.067	17.096	0.012
HEFA-SPK	4.389	13.890	0.050	0.002	0.017	6.062	0.001
TOTAL		**57.659**	**0.206**	**0.008**	**0.083**	**23.158**	**0.013**
			Month—ICE + Battery				
Kerosene	267.917	876.807	3.126	0.123	1.334	342.481	0.234
HEFA-SPK	88.000	278.497	1.009	0.040	0.336	121.551	0.020
TOTAL		**1155.304**	**4.135**	**0.163**	**1.670**	**464.032**	**0.253**
			Year—ICE + Battery				
Kerosene	3215.000	10,521.679	37.515	1.481	16.009	4109.770	2.807
HEFA-SPK	1056.000	3341.968	12.104	0.475	4.034	1458.609	0.235
TOTAL		**13,863.647**	**49.619**	**1.956**	**20.044**	**5568.379**	**3.041**
			Day—PEMFC + Battery				
LH_2	5.523	0.000	0.000	0.000	0.000	9.849	0.000
			Month—PEMFC + Battery				
LH_2	107.583	0.000	0.000	0.000	0.000	191.849	0.000
			Year—PEMFC + Battery				
LH_2	1291.000	0.000	0.000	0.000	0.000	2302.192	0.000

For further classification and comparison purposes, a conventional aircraft from D1.2 [13] was used in Table 12. These flights were considered with kerosene only. By comparing the emissions of Tables 11 and 12, it can be deduced that, by flying with PEMFC + Battery HEA, 49.5% CO_2, 51.1% NO_x and 48% H_2O saving can be achieved. Flying with a PEMFC + Battery HEA, 100% CO_2, 100% NO_x and 77.9% H_2O savings can be achieved.

Table 12. Emissions of a comparable short-term conventional aircraft (with new gas turbine engines installed) in combination with flight schedule.

	Fuel [ton]	CO_2 [ton]	NO_x [ton]	H_2O [ton]
		Day		
Kerosene	26.091	114.168	0.423	44.594
		Month		
Kerosene	519.224	2272.003	8.411	887.449
		Year		
Kerosene	6230.696	27,264.041	100.935	10,649.384

3.3. Results over All Time Horizons

This section summarises all data for the operation of a regional airport for the different time horizons and aircraft configurations. The results for the short-term scenario (ICE + Battery—2030) follow the procedure described in Sections 3.1 and 3.2, but here the fuel composition is, as already mentioned, 90% kerosene and 10% SAF. In addition, a lower powerful battery is installed. The results for the long-term scenario (PEMFC + Battery—2050) are obtained according to the procedure also described in Sections 3.1 and 3.2. Here, a further developed PEMFC and further developed battery are included in the aircraft configuration. For more detailed information on the aircraft configuration, please refer back to [10] or [13].

The already-presented results of the medium-term scenario (ICE + Battery) and medium-term scenario (PEMFC + Battery) are taken up in the following tables. They can be classified as short-term (ICE + Battery—2030) and long-term (PEMFC + Battery—2050). Table 13 shows the annual energy demand for the process of the HEA in different time horizons. It was found that, instead of 5608 tonnes of paraffin (short term), 3215 tonnes (medium term ICE + Battery) and 1291 tonnes of liquid hydrogen (medium term PEMFC + Battery), 1234 tonnes of liquid hydrogen would now be required to operate the HEA in the long-term scenario. The electrical energy demands of 7233 GWh (short term), 11,704 GWh (medium term -ICE + Battery) and 12,640 GWh (medium term -PEMFC + Battery) are now 11,622 GWh. The demand for electrical energy is 60% higher than in the short term. The demand for electrical energy in the medium-term scenario with PEMFC is almost 0.7% lower, and thus almost identical to the medium-term scenario ICE + Battery. Overall, the demand for electrical energy in the medium-term scenario with PEMFC + Battery is 8.1% lower than in the short-term scenario.

Table 13. The yearly amount of energy for HEA 2025–2055.

	Kerosene [Tons]	SAF [Tons]	Electric [MWh]	LH2 [Tons]
2030—ICE + Battery	5608	610	7233-	
2040—ICE + Battery	3215	1056	11704	
2040—PEMFC + Battery			12,640	1291
2050—PEMFC + Battery			11,622	1234

Table 14 shows the expected and extrapolated ticket prices for the different time horizons. The approach was the same as in Sections 2.2, 3.1 and 3.2. The HEA ticket price is expected to be 49.4% cheaper for typical mission flights and 45.7% for design mission flights in the long-term PEMFC + Battery scenario. The list was compiled without the high investment, research and operating costs. As described in the respective sections, the price calculations considered environmental aspects and expected inflation rates.

Table 14. Costs for typical mission flights—forecast summary.

London 308 km	Scenario	Delta/km [EUR]	Cost/km and PAX [EUR]	Delta/km and PAX [%]	Cost/km (PAX; Env.; Inflation) [EUR]	Delta/Ticket HEA [%]	Delta/km [EUR]
				2030			
ERJ190	(4)	19.50	0.05	-	0.20	-	61.29
ERJ190	(1.4)	20.02	0.05	-	0.20	-	62.91
ERJ190	(2.4)	23.67	0.06	-	0.24	-	74.38
ERJ190	(3.4)	38.26	0.09	-	0.39	-	120.25
ICE + Battery	(4)	11.72	0.06	−3.00%	0.23	−3.00%	72.18
				2040			
ERJ190	(4)	23.33	0.05	-	0.24	-	73.31
ERJ190	(1.4)	23.94	0.05	-	0.24	-	75.24
ERJ190	(2.4)	28.31	0.06	-	0.29	-	88.97
ERJ190	(3.4)	45.77	0.09	-	0.47	-	143.83
ICE + Battery	(4)	14.18	0.06	−1.80%	0.28	−1.80%	87.33
PEMFC + Battery	(4)	12.44	0.05	−46.70%	0.25	−46.70%	76.63
				2050			
ERJ190	(4)	27.15	0.05	-	0.28	-	85.33
ERJ190	(1.4)	27.87	0.05	-	0.28	-	87.58
ERJ190	(2.4)	32.95	0.06	-	0.34	-	103.56
ERJ190	(3.4)	53.27	0.09	-	0.54	-	167.42
PEMFC + Battery	(4)	13.75	0.05	−49.40%	0.28	−49.40%	84.71
				Pula1049 km			
				2030			
B737	(4)	20.40	0.04	-	0.16	-	171.22
B737	(1.4)	20.94	0.04	-	0.17	-	175.76
B737	(2.4)	24.78	0.05	-	0.20	-	253.33
B737	(3.4)	40.11	0.08	-	0.32	-	336.58
HEA	(4)	7.88	0.04	−20.50%	0.16	−20.50%	165.37
				2040			
B737	(4)	24.40	0.04	-	0.20	-	204.81
B737	(1.4)	25.05	0.04	-	0.20	-	210.23
B737	(2.4)	29.64	0.05	-	0.24	-	248.71
B737	(3.4)	47.97	0.08	-	0.38	-	402.59
ICE + Battery	(4)	10.46	0.04	−11.80%	0.21	−11.80%	219.35
PEMFC + Battery	(4)	11.43	0.05	−40.40%	0.23	−40.40%	239.76
				2050			
B737	(4)	28.41	0.04	-	0.23	-	238.39
B737	(1.4)	28.41	0.04	-	0.23	-	244.71
B737	(2.4)	28.41	0.05	-	0.28	-	289.49
B737	(3.4)	28.41	0.08	-	0.45	-	468.61
PEMFC + Battery	(4)	12.12	0.04	−45.70%	0.24	−45.70%	254.36

However, the high savings in the expected ticket price per passenger offer an excellent field to make these investments lucrative for airlines and passengers through hybrid-electric typical and design mission flights. This fact confirms the previously established thesis that hybrid-electric flights have the potential to be financially attractive and environmentally friendly. The assumed costs for CO_2 compensation are justified here, as more and more institutions, such as FAU, are obliged to pay CO_2 compensation on ticket prices for business trips.

Table 15 summarises the extrapolated and expected emissions of the different time horizons and aircraft types. The mentioned reference aircraft (ATR 42 with a Pratt and Whitney PW127 engine) is listed first under the 2012 category for comparison purposes.

Table 15. The yearly missions HEA flights forecast estimation compared to the reference aircraft—Summary 2025–2050.

	Fuel [Ton]	CO_2 [Ton]	NO_x [Ton]	HC [Ton])	CO [Ton]	H_2O [Ton]	SO_2 [Ton]
			2012—Reference aircraft				
Kerosene	6230.696	27,264.041	100.935	-	-	10,649.384	-
			2030—HEA				
Kerosene	5608	18,353.211	65.438	2.583	27.926	7168.768	4.896
SAF (HEFA-SPK)	610	1930.493	6.992	0.275	2.33	842.568	0.136
TOTAL	6218	20,283.704	72.43	2.858	30.256	8011.336	5.031
			2040—HEA—ICE + Battery				
Kerosene	3215	10,521.679	37.515	1.481	16.009	4109.77	2.807
SAF (HEFA-SPK)	1056	3341.968	12.104	0.475	4.034	1458.609	0.235
TOTAL	4271	13,863.647	49.619	1.956	20.044	5568.379	3.041
			2040—HEA—PEMFC + Battery				
LH_2	1291	-	-	-	-	2302.192	-
			2050—HEA—PEMFC + Battery				
LH_2	1234	-	-	-	-	2025.9	-

4. Conclusions

This paper presents the results of an energy demand analysis for a future regional airport over three different time horizons. This study presents different options for the ground power supply of a regional airport and possible solutions for the airport infrastructure with a short (2030), medium (2040) and long (2050) time horizon. The results include estimating the future energy demand per day, month and year and the energy demand. To accommodate the increasing number of flights, the flight plan was adapted to the needs of a 50-PAX regional aircraft. This new flight plan provides the opportunity to present an overview of the results for the energy demand of a regional airport, broken down by individual time horizons. The result of this work describes the energy demand for the airport's operation, the expected emissions and an estimate of ticket prices. The findings confirm that airports will require an enormous amount of electrical energy due to the electrification of air traffic. Accordingly, the infrastructure of airports will also have to change. Furthermore, the study shows that the transition to sustainable hybrid-electric aviation is attractive due to lower emissions and adjusted ticket prices.

In future work, a full-fledged prospective Life Cycle Assessment (LCA) in accordance with the methodology proposed by [29] needs to be performed to consider all relevant life cycle stages and additional environmental impacts besides climate change. The inclusion of additional emerging propulsion systems (e.g., direct H_2 use in the gas turbine), aircraft types (besides the regional HEA), and other means of reducing airport/aircraft emission (e.g., air traffic management) would broaden the scope and enrich the discussion of the transition of airports.

Author Contributions: This section reports the main contributions of each author according to the Contributor Roles Taxonomy (CRediT) (http://img.mdpi.org/data/contributor-role-instruction.pdf). Conceptualisation: M.M. (Markus Meindl), C.d.R. and F.H. Methodology: M.M. (Markus Meindl), C.d.R., V.M., M.D.S., M.R., F.H. and L.L. Data collection: M.M. (Markus Meindl), C.d.R., V.M., M.D.S. and M.R. Validation: M.M. (Markus Meindl), C.d.R., V.M. and M.D.S. Data maintenance: M.M. (Markus Meindl), C.d.R., V.M. and M.D.S. Writing—original draft preparation: M.M. (Markus Meindl). Writing—review and editing: M.M. (Markus Meindl), C.d.R., V.M., M.D.S., M.R., F.N., F.H., N.T., K.S.-R., A.L. and M.M. (Martin Maerz). Supervision: M.M. (Martin Maerz). Project administration: A.L. and M.M. (Martin Maerz). Funding acquisition: M.M. (Martin Maerz). All authors have read and agreed to the published version of the manuscript.

Funding: This study is part of the GENESIS project (https://www.genesis-cleansky.eu/ (9 March 2023)). The GENESIS project has received funding from the Clean Sky 2 Joint Undertaking (JU) under Grant Agreement n° 101007968. The JU receives support from the European Union´s Horizon 2020 research and innovation programme and the Clean Sky 2 JU members other than the Union. This study only reflects the authors' views; the JU is not responsible for any use that may be made of the information it contains.

Data Availability Statement: Not applicable.

Acknowledgments: This study was conducted as part of the GENESIS project (https://www.genesis-cleansky.eu/ (9 March 2023)). The authors are grateful to all project partners for the fruitful discussions and expert advice about the main elements of hybrid-electric architectures, which formed a solid theoretical foundation for the studies presented in this work.

Conflicts of Interest: The authors declare no conflict of interest. The funders had no role in the design of the study; in the collection, analyses, or interpretation of data; in the writing of the manuscript; or in the decision to publish the results.

Abbreviations

The following abbreviations are used in this manuscript:

Abbreviation	**Meaning**
AC	Alternating Current
CO_2	Carbon Dioxide
DC	Direct Current
EI	Emission Indices
FAU-LEE	Friedrich Alexander University—Lehrstuhl für Leistungslektronik
FC	Fuel Cell
GENESIS	Gauging the ENvironmental sustainability of electrIc aircraft Systems
HEA	Hybrid-Electric-Aircraft
ICE	Internal Combustion Engine
LCA	Life Cycle Assessment
LF	Low Frequency
LH2	Liquid hydrogen
IT	Information Technology
MCS	Megawatt Charging System
MW	Megawatt
NOx	Nitrogen Oxide
PAX	Persons approximately
PEMFC	Proton Exchange Membrane Fuel Cell
PV	Photovoltaic
RTHA	Rotterdam The Hague Airport
SAF	Sustainable Aviation Fuel
TLAR	Top-level Aircraft Requirements
UNINA	Universita degli Studi di Napoli Federico II

References

1. Steinike, S. Die grüne Basis: Flughäfen sind bei den Themen Lärmschutz und Emissionssenkung schon lange aktiv. Für die Umstellung auf neue Antriebsenergien, wie batterieelektrisches FLiegen und Brennstoffzellen, stehen ihnen aber ein aufwendiger Umbau der Infrastruktur vor. *Flug Revue* **2022**, 32–37.
2. Airbus. ZEROe. Available online: https://www.airbus.com/en/innovation/zero-emission/hydrogen/zeroe (accessed on 15 February 2023).
3. Metar-Taf. Corisco International Airport | OCS | Piloten-Infos. Available online: https://metar-taf.com/de/airport/OCS (accessed on 22 February 2023).
4. Kilimi, M.G.R.; Motjoadi, V.; Bokoro, P.N. Improvement of an Off-grid Electricity Supply System: A Case Study in Corisco International Airport. In Proceedings of the 2021 International Conference on Electrical, Computer, Communications and Mechatronics Engineering (ICECCME), Flic en Flac, Mauritius, 7–8 October 2021; pp. 1–8.

5. Jakarta Airport Guide—Jakarta Soekarno-Hatta International Airport. Available online: https://www.jakartaairportonline.com/ (accessed on 22 February 2023).
6. Untari, I.; Raharjo, H.S.; Nayusrizal, N.; Hudaya, C. Analysis of Electricity Power Consumption and Cost Impact of Soekarno-Hatta Airport Railink Project. In Proceedings of the 2019 IEEE 2nd International Conference on Power and Energy Applications (ICPEA), Singapore, 27–30 April 2019; pp. 28–32.
7. Meindl, M.; Weber, K.J.; Maerz, M. Ground-Based Power Supply System to Operate Hybrid-Electric Aircraft for Future Regional Airports. In Proceedings of the ESARS-ITEC Europe-Venice (Italy), Vinece, Italy, 29 March 2023; 29 March 2023.
8. Sumsurooah, S.; He, Y.; Torchio, M.; Kouramas, K.; Guida, B.; Cuomo, F.; Atkin, J.; Bozhko, S.; Renzetti, A.; Russo, A.; et al. ENIGMA—A Centralised Supervisory Controller for Enhanced Onboard Electrical Energy Management with Model in the Loop Demonstration. *Energies* **2021**, *14*, 5518. [CrossRef]
9. Rotterdam the Hague Airport. *A Vision for Hydrogen Flying at Rotterdam The Hague Airport*; Rotterdam the Hague Airport: Rotterdam, The Netherlands, 2022.
10. Marciello, V.; Di Stasio, M.; Ruocco, M.; Trifari, V.; Nicolosi, F.; Meindl, M.; Lemoine, B.; Caliandro, P. Design Exploration for Sustainable Regional Hybrid-Electric Aircraft: A Study Based on Technology Forecasts. *Aerospace* **2023**, *10*, 165. [CrossRef]
11. Srdic, S.; Lukic, S. Toward Extreme Fast Charging: Challenges and Opportunities in Directly Connecting to Medium-Voltage Line. *IEEE Electrific. Mag.* **2019**, *7*, 22–31. [CrossRef]
12. Deutsche Kommission Elektrotechnik Elektronik Informationstechnik. Konduktive Ladesysteme für Elektrofahrzeuge. DIN EN IEC 61851-1 (VDE 0122-1):2019-12. Available online: https://www.vde-verlag.de/normen/0100544/din-en-iec-61851-1-vde-0122-1-2019-12.html (accessed on 18 January 2023).
13. Marciello, V.; Nicolosi, F.; Di Stasio, M.; Ruocco, M. GENESIS D1.2-Scenarios & Requirements for Future Electric/Hybrid Propulsion and Conventional A/C: Università degli Studi di Naples "Federico II" (UNINA), Neapel. 2022. Available online: https://www.genesis-cleansky.eu/deliverables/ (accessed on 9 March 2023).
14. Mario Di Stasio (UNINA), Manuela Ruocco (SMARTUP), Valerio Marciello (UNINA), Fabrizio Nicolosi. Turbine/ICE Generator Set Technology Analysis: GENESIS D2.5. 2023. Available online: https://www.genesis-cleansky.eu/deliverables/ (accessed on 9 March 2023).
15. Filippone, A.; Bojdo, N. Statistical model for gas turbine engines exhaust emissions. *Transp. Res. Part D Transp. Environ.* **2018**, *59*, 451–463. [CrossRef]
16. Eurocontrol. Forecasting Civil Aviation Fuel Burn and Emissions in Europe—Interim Report. Available online: https://www.eurocontrol.int/publication/forecasting-civil-aviation-fuel-burn-and-emissions-europe-interim-report (accessed on 22 February 2023).
17. Narciso, M.; de Sousa, J.M.M. Influence of Sustainable Aviation Fuels on the Formation of Contrails and Their Properties. *Energies* **2021**, *14*, 5557. [CrossRef]
18. National Academies of Sciences, Engineering, and Medicine. *ACRP Web-Only Document 41: Alternative Jet Fuels Emissions: Quantification Methods Creation and Validation Report*; Transportation Research Board: Washington, DC, USA, 2019.
19. Shahwan, K. Operating Cost Analysis of Electric Aircraft on Regional Routes. Bachelor's Thesis, Lingköping University, Norrköping, Sweden, 2021.
20. Blair Comley, Paul Boulus, Pip Best, Fiona Hancock, Steve Hatfield-Dodds, Emma Herd, Mathew Nelson, Matthew Cowie, Andrew Lee and Jarrod Allen. Essential, Expensive and Evolving: The Outlook for Carbon Credits and Offsets: An EY Net Zero Centre Report, EY, Sydney. 2022. Available online: https://assets.ey.com/content/dam/ey-sites/ey-com/en_au/topics/sustainability/ey-net-zero-centre-carbon-offset-publication-20220530.pdf (accessed on 9 March 2023).
21. Kos, J.; Peerlings, B.; Lim, M.N.A.; Lammen, W.F.; Posada Duque, A.; van der Sman, E.S. Novel Propulsion and Alternative Fuels for Aviation towards 2050: Promising Options and Steps to Take (TRANSCEND Deliverable D3.2). 2022. Available online: https://reports.nlr.nl/server/api/core/bitstreams/af600792-5757-48d8-b153-45513acf97d9/content (accessed on 6 January 2023).
22. Petrichenko, L.; Sauhats, A.; Diahovchenko, I.; Segeda, I. Economic Viability of Energy Communities versus Distributed Prosumers. *Sustainability* **2022**, *14*, 4634. [CrossRef]
23. JME Ellsworth. Electric Vehicle Charging Stations. Available online: https://www.jmesales.com/electric-vehicle-charging-stations/?sort=pricedesc (accessed on 3 January 2023).
24. MCS Task Force; the Charging Connections Focus Group. CharIN Whitepaper Megawatt Charging System (MCS): Recommendations and Requirements for MCS Related Standards Bodies and Solution Suppliers. 2022. Available online: https://www.charin.global/media/pages/technology/knowledge-base/c708ba3361-1670238823/whitepaper_megawatt_charging_system_1.0.pdf (accessed on 7 December 2022).
25. Maeve. Flight Time: 1 Hour 30 Minutes. Charging Time: 35 Minutes. Available online: https://maeve.aero/recharge (accessed on 3 January 2023).
26. Lufthansa. Lufthansa Booking. Available online: https://www.lufthansa.com/de/de/homepage (accessed on 6 January 2023).
27. Lufthansa. Nachhaltiger Fliegen. Available online: https://www.lufthansa.com/de/de/flug-kompensieren (accessed on 20 February 2023).

28. Statistisches Bundesamt. Inflationsrate in Deutschland von 1950 bis 2022. 2022. Available online: https://de.statista.com/statistik/daten/studie/4917/umfrage/inflationsrate-in-deutschland-seit-1948/ (accessed on 9 March 2023).
29. Thonemann, N.; Schulte, A.; Maga, D. How to Conduct Prospective Life Cycle Assessment for Emerging Technologies? A Systematic Review and Methodological Guidance. *Sustainability* **2020**, *12*, 1192. [CrossRef]

Article

Determination of the Parachute Harness Critical Load Based on Load Distribution into Individual Straps with Respect of the Skydiver's Body Position

Robert Grim *, Robert Popela, Ivo Jebáček *, Marek Horák and Jan Šplíchal

Institute of Aerospace Engineering, Brno University of Technology, 60190 Brno, Czech Republic
* Correspondence: robert.grim@vutbr.cz (R.G.); jebacek@fme.vutbr.cz (I.J.)

Abstract: This article evaluates the redistribution of forces to the parachute harness during an opening shock load and also defines the ultimate limit load of the personal parachute harness by specifying the weakest construction element and its load capacity. The primary goal of this research was not only to detect the critical elements but also to gain an understanding of the force redistribution at various load levels, which could represent changes in body mass or aerodynamic properties of the parachute during the opening phase. To capture all the phenomena of the parachutist's body deceleration, this study also includes loading the body out of the steady descending position and asymmetrical cases. Thus, the result represents not only idealized loading but also realistic limit cases, such as asymmetric canopy inflation or system activation when the skydiver is in a non-standard position. The results revealed a significant difference in the strength utilization of the individual components. Specifically, the back webbing was found to carry a fractional load compared to the other webbing used in the design in most of the scenarios tested. Reaching the maximum allowable strength was first achieved in the asymmetric load test case, where the total force would be equal to the value of 7.963 kN, which corresponds to the maximum permissible strength of the carabiner on the measuring element three. In the same test case, the second weakest point would reach the limiting load force when the entire harness is loaded with 67.89 kN. This information and the subsequent analysis of the individual nodes provide a great opportunity for further strength and weight optimization of the design, without reducing the load capacity of the harness as a system. The findings of this study will be used for further testing and possible harness robustness optimization for both military and sport parachuting.

Keywords: parachute harness; opening load; limit load

Citation: Grim, R.; Popela, R.; Jebáček, I.; Horák, M.; Šplíchal, J. Determination of the Parachute Harness Critical Load Based on Load Distribution into Individual Straps with Respect of the Skydiver's Body Position. *Aerospace* **2023**, *10*, 83. https://doi.org/10.3390/aerospace10010083

Academic Editors: Spiros Pantelakis, Andreas Strohmayer and Jordi Pons-Prats

Received: 22 November 2022
Revised: 5 January 2023
Accepted: 11 January 2023
Published: 14 January 2023

1. Introduction

In order to follow the modern trend of using lightweight materials and optimized product design, it is necessary to focus on the individual construction elements and dimension them exactly according to the expected requirements. Nowadays, proven designs of parachute harnesses are commonly used, which do not differ greatly in the materials and construction elements used. [1]. The effective sizing of individual elements is only possible when detailed information about their loading is available. This is the aim of this research, which in principle, can be divided into three main stages. The first stage is to obtain the opening shock force, which characterizes the aerodynamic parameters of the canopy. This force is then applied to the harness worn by a dummy fixed in different positions. The final stage is to identify the decomposition of this total force into the individual structural elements. This information is then used for evaluating the load capacity of the elements and to gain insight into the possibility of subsequent optimization of the structure. In the results section of this research, it is shown that the safety margins of the separate construction elements differ significantly. This suggests a unification of the safety coefficients of the individual components when the maximum required load is reached. The outcome would be a lighter overall structure. However, the intention is not just to analyze the case for one specific loading force representing a particular canopy during its opening shock load test. The objective is to obtain a percentage value of the total force that will be transmitted to each node and to prove or disprove that this redistribution is constant over a certain range of loading. Expressing this dependence would imply the possibility of applying the presented results without restriction on the magnitude of the total force, in other words, for arbitrarily chosen canopies and activation parameters.

Nowadays, structural overload tests, which include the entire sequence of canopy activation, are considered standard [2,3]. The procedure is performed based on the Technical Standard 135 published by the Parachute Industry Association (PIA) [4]. These types of tests evaluate whether a harness shows signs of mechanical damage after activation under specific conditions. Capture 4.3.6, named "Structural Overload Tests", defines the general conditions for a drop test that the complete parachute assembly or separate components must withstand. This approach has a major disadvantage for the proposed force redistribution analysis, since dropping a dummy from an airplane or helicopter does not guarantee an exact position when the rescue system is activated. As will be shown in the results section of this document, not only does the opening shock load play a major role but also the direction of it does. In fact, the proposed test methodology is a combination of dynamic and static tests that are primarily used for the certification of paragliding harnesses and work at heights for safety harnesses [5–7]. They have a common main feature, namely, the fact that both force and position conditions are precisely defined.

The possibility to divide the parachute assembly test into separate tests of individual components allows one to obtain the dynamic opening force of the canopy first, and then apply this force to the harness as a static load. This makes it possible to position the dummy in the different setups and evaluate the required influence of the skydiver's body position, asymmetry during parachute activation, and the tightening or loosening of each strap.

To achieve the above goals, the following tasks were completed. The first step was the development of the drop test laboratory and methodology. The most important factor was reaching the activation speed exactly according to the specification. This was achieved using a real-time measurement of the speed together with a backup timer. Data logging of the forces in the connection between the parachute and ballast is performed at a frequency of 200 Hz. The opening force is recorded by measuring the carabiners, so it is possible to reach force on both attachment points. A similar logging of the force has been presented in this publication [8].

The next challenging step was to measure the force in flexible straps. Few studies have focused on the measuring tension in flexible structures, such as parachute fabric, but for the purpose of the proposed aim, this could not be used [9–11]. The requirement for strain gauges developed, especially for the purpose of this research, was that they be versatile enough to be used regardless of the exact type of strap on with which they were installed. The second important goal was to minimize the influence of structural rigidity. Preferably, the measuring components were made of parts from which the harness itself is assembled. The main advantage of this research is that it provides a very detailed analysis of the distribution of forces in the individual parts of the harness. By achieving dimensionally small load cells, it was possible to install the strain gauges in all the necessary places so that a load from the individual nodes was captured during one harness load cycle. In order to obtain comprehensive data, the different dummy positions that may occur when the rescue system is activated were also investigated.

Information about the forces in the individual parts of the harness gives the possibility to analyze the dimensioning of each part according to the required load. Thus, by evaluating the data, it was possible to identify the critical elements for different dummy configurations and harness settings. Subsequently, identifying non-uniformity in the sizing of the individual elements based on their actual loading is also very important information. The outcome is, therefore, a vision of significant weight savings without affecting the load capacity of the whole system, once the above results are incorporated into the design of the new version.

2. Materials and Methods

2.1. Design of the Tested Harness

For the purpose of the test, a fully articulated harness from serial production has been chosen. However, all non-load-bearing elements that increase pilot comfort were intentionally removed. The aim was to extract only the structural frame to obtain more variability in the positioning of the measuring elements. Because the removed elements do not affect the strength of the harness, similarity to a fully equipped harness intended for real use is guaranteed. The complete scheme of used materials is highlighted in Figure 1.

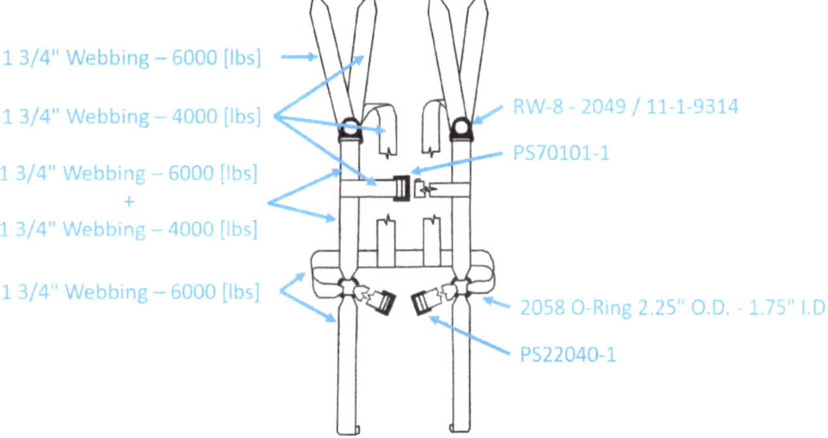

Figure 1. Scheme of used materials.

Straps of the same width were used for the entire harness. The only differences are in their declared strengths. Regarding the buckles, the exact types are also shown. The names stated in Figure 1 are the trade names of the buckles. It is therefore possible to trace their exact parameters, which will be discussed further in the following section.

2.2. Equipment for Measuring Forces in Webbing

Measuring the forces in a flexible structure, such as parachute webbing, is a very specific issue. The main reason is that straps in the final configuration are not uniformly loaded in most cases. It is necessary to consider, for example, the partial loading of the sides of the strap due to the seating around the dummy's body. Hence, there was a need to develop a custom measuring element. The major request was also to use elements with minimum dimensions to have the possibility of implementing them into each webbing. The minimum dimensions of the feature also ensured the smallest possible influence on the harness' structural characteristics. In addition, the aim was to eliminate any undesirable loads, for example, from bending. When selecting the positions for the load cells, consideration was given to placing them in positions where only tension could be expected, and the bending component was eliminated as much as possible.

Regarding the design of the load cells, modification of the buckle frames was determined to be the most appropriate and least costly option. The aim was to ensure that the stiffness of the structural node was not adversely affected by the incorporation of the individual measuring components. Therefore, the same carabiners were always selected for the design of the strain gauges as those already used in the design chain. For the purpose of this study, only two types of buckle bases were required. The stronger type is a carabiner with the trademark PS 22040-1 and a declared strength of 2500 lbs, while the weaker type is for use in chest webbing and has a declared strength of 500 lbs. Figure 2 shows the initial geometry of the buckles used before any modifications. Figure 2 highlights the initial geometry of the used buckles before any modification. To obtain the basic frame for strain gauge installation, it was necessary to remove the moving part used to fix the strap in the tightened position.

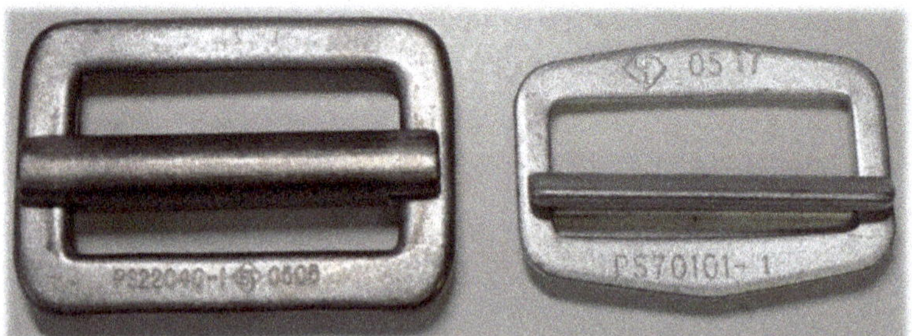

Figure 2. Selected buckles for the frame preparation to create the measuring load cell.

Once a clear rectangular base plate was prepared, four strain gauges were installed on the degreased surface, one on each side of the buckle. The intention was to reach a full bridge connection that will ensure accurate measurement of the force, regardless of whether the buckle is loaded symmetrically or not. This assumption has been confirmed during the calibration, which has been performed for all six load cells. In the procedure, the carabiners were loaded evenly and unevenly, in the sense that one side of the carabiner was loaded more than the other. This process verified that the total measured force did not vary from case to case. Repeating the above approach for each load cell separately guarantees accurate measurement of all elements, regardless of any manufacturing tolerances of the buckle or inaccurate placement of the strain gauge on its surface. The specific regression curves of these load cells are shown in Figure 3. The regression dependence is linear over the entire applied load range, and therefore, a very high accuracy of force measurement can be assumed. The dependence of the material tension on the total loading force generated by the tensile test machine is shown.

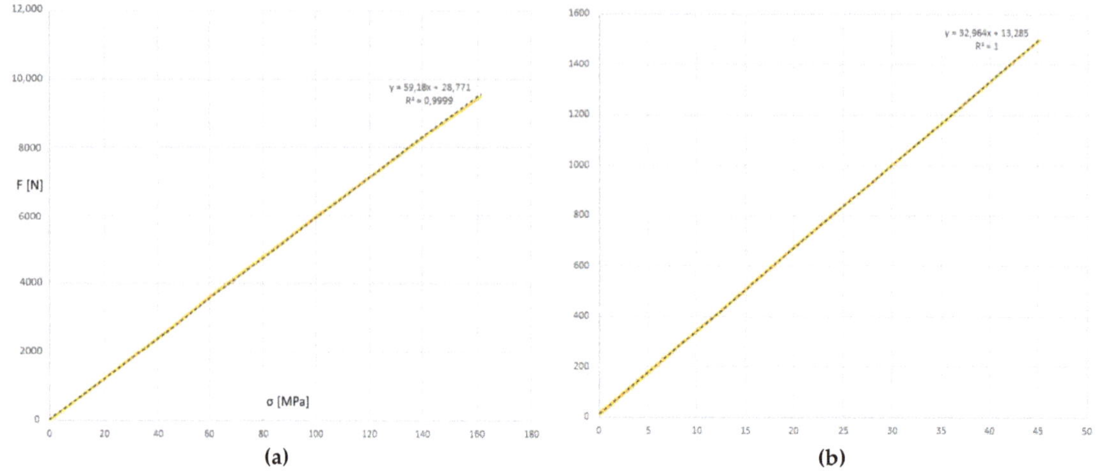

Figure 3. Exact buckle plate calibration curve: (**a**) curve related to baseplate PS22040-1 and (**b**) curves related to baseplate PS 70101-1.

The final load cell implemented into the harness structure is highlighted in Figure 4. It was necessary to provide mechanical protection around the strain gauge against damage when the harness settles on the metal dummy. During the first moments of loading, significant movements occur. The area around the strain gauges was sealed with hot melted glue. This method of protection proved to be sufficient as no damage occurred during the test.

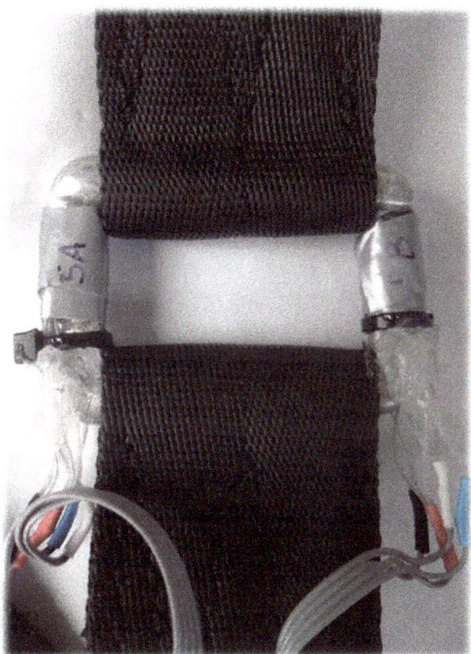

Figure 4. Final load cell manufactured on PS22040-1 baseplate sewn into structure.

2.3. Design of the Harness Fitted with the Load Cells

When fitting the load cells to the harness, symmetry was utilized. Measurement elements were sewn only on the right side of the harness, so only six of them were needed. To obtain complete load decomposition into individual elements during the asymmetric load cases specified in the next section of the publication, it was only necessary to perform one additional test.

Despite the fact that it was proven during calibration that the laterally uneven loading of the measuring element does not affect the accuracy of the total read-out forces, it was the intention to place the elements only in locations where there would be only tensile loading without influence from bending. It can be expected that when placing the measuring element in a position where bending occurs, measurement inaccuracies will begin to appear. The main reason for this would be the contribution of friction in the contact with "dummy–webbing". With this knowledge, the optimum position for each element has been found. The location of the load cells sewn into the harness structure is highlighted in Figure 5. The numbering of the elements given in the diagram matches the following naming of the forces in the evaluation of the results.

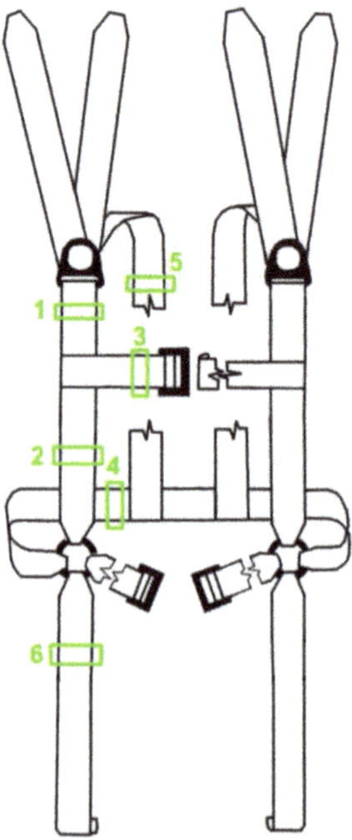

Figure 5. Load cell layout diagram, including markings.

2.4. Fitting the Harness to the Test Dummy

Great care was taken in fitting the harness to the test dummy. The position of the individual structural points was set to fully correspond with the real position of the human body. To achieve this, the harness was specially adjusted in length for the real dimensions of the dummy. In the area of the legs and buttocks, the measuring devices were lined with felt sheets. This method eliminated bending of the strain gauges while also reducing friction. Detailed placement of the loading cells once the harness is fitted to the dummy is displayed in Figure 6. The numbering of individual load cells is also highlighted. Based on this marking, the redistribution of forces was subsequently evaluated, which is presented in the following sections.

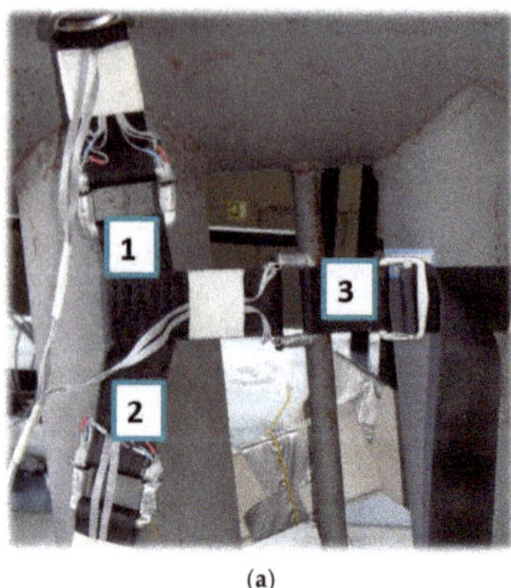

(a)

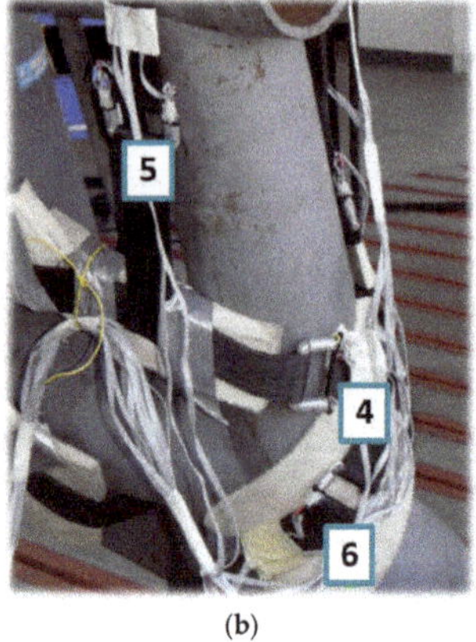

(b)

Figure 6. Alignment of measuring elements: (**a**) Front side view of the dummy. (**b**) Back side view of the dummy.

2.5. Tested Configurations

The number of test configurations and their setup were designed considering the parameters identified as important for the survey. The aim was to perform a series of tests in different spatial positions of the dummy so that it would then be possible to evaluate the effect of force redistribution on the individual elements, based not only on the geometry of the structure itself but also on the different spatial body positions in which the rescue system would be activated. This will replicate the use of emergency parachutes, which are assumed to be activated in positions other than those for ideal and stable skydivers. Another important objective was to determine the effect of chest strap tightening. Nevertheless, the main feature of all tests will be the evaluation of the critical element in terms of its structural capacity. In other words, the critical element may not be the same in all cases. Based on the mentioned requirements, six different test cases have been established:

1. Symmetrical load;
2. Unsymmetrical load–$F_{RIGHT} = 2 \cdot F_{LEFT}$;
3. Unsymmetrical load–$F_{LEFT} = 2 \cdot F_{RIGHT}$;
4. Symmetrical load–dummy, fixed at about 15° face down;
5. Symmetrical load–dummy, fixed at about 15° back down;
6. Symmetrical load–loose chest strap.

The meaning of "symmetrical loading" is the fact that a force of the same magnitude is applied to the left and right sides of the harness. The opposite is the case for the second and third tests, as their aim is to obtain information on the redistribution of the forces in the case of opening the canopy asymmetrically. Regarding test cases 1, 2, 3, and 6, the steel dummy has only one fixation point in the bottom between its legs. The position of equilibrium is determined just by the center of gravity and by adjusting the individual straps. Tilted cases 4 and 5 have generally fixed positions in the space defined. The setup of the configurations in the unloaded state is shown in Figure 7.

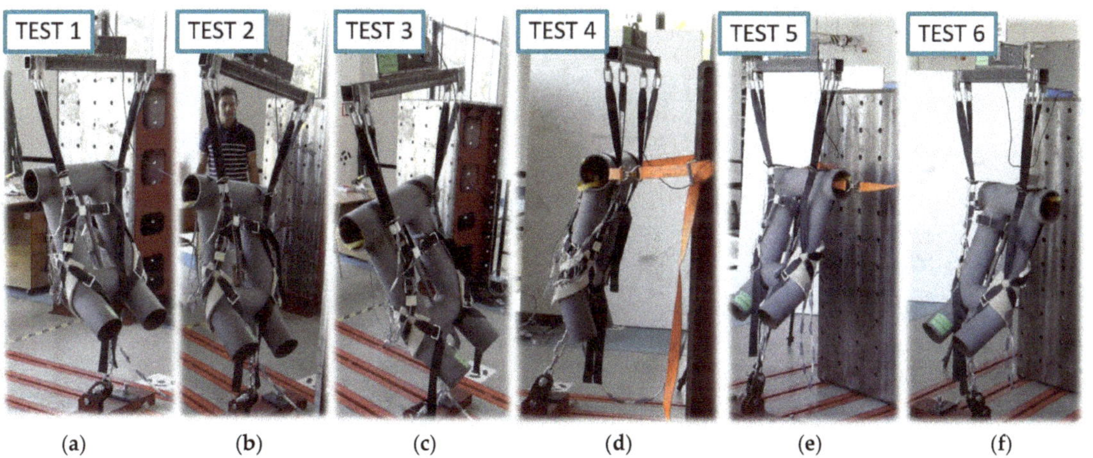

Figure 7. Equilibrium position in the unloaded state for each test case: (**a**) test case 1, (**b**) test case 2, (**c**) test case 3, (**d**) test case 4, (**e**) test case 5, and (**f**) test case 6.

2.6. Drop Test Laboratory Setup for Reaching the Opening Shock Load

In order to reach the inputs in the harness load test, it is necessary to obtain data from the actual drop, which reflects parachute aerodynamic characteristics during the opening phase. For this purpose, a special drop test laboratory has been developed. The scheme, with a description of the basic components, is highlighted in Figure 8. The test laboratory consists of electronics that record the most important information regarding the entire drop in real time. The high recording frequency allows the system to be activated with high accuracy and is also very important for the next postprocessing of the data. Not only speed-related information is recorded, but also the duration of the freefall, forces in separate connections between parachute and laboratory, position in the coordinate system, and G-force. Data from the indicated speed and free fall time are used to activate the system at the required speed. Exceeding the design speed should mean the destruction of the parachute, and in most cases, damage to the laboratory caused by high-speed impact with the ground.

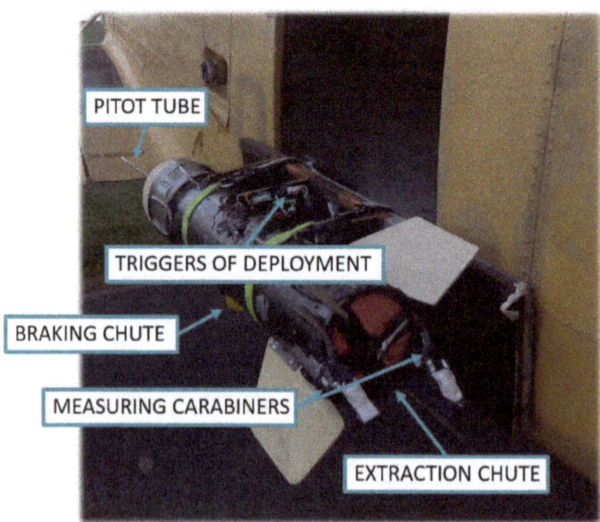

Figure 8. A scheme of a fully equipped laboratory mounted on the AN-2 before takeoff.

Information related to G-Force and load in separate connections is the value that needed to be identified for further analysis. Even though the laboratory records the complete progress of the quantities over the whole drop test sequence, for the present research, only one point is important, which is the maximum peak related to the opening shock load. The amount of asymmetry read from the measuring carabiners on the attachment points can be considered as additional information and input to the proposed tests 2 and 3.

It should be highlighted that the activation of the parachute must be executed in the same manner as if it were activated by a skydiver while wearing the serial harness. That means that the folding of the parachute into the storage area needs to be executed in the same way and deploying the drogue chute equipped with the serial extracting spring is also required. There should be free space in the area of deployment that would cause the elements to be trapped during the pulling sequence. This ensures that there will be no unwanted delay in the deceleration of the ballast. It is an important parameter for drop tests that use gravitation to accelerate the laboratory to the final speed. Any delay in activation thereafter means exceeding the design speed, which is unacceptable.

3. Results

3.1. Drop Test Evaluation

To obtain information about the forces, the test conditions were designed so that they corresponded to the conventional values at which the skydiving or rescue systems are activated. It was intended to determine the G factor generated by the canopy at the typical terminal speeds in a stable belly-to-earth position. The conditions of the drop test were specified as:

- activation speed $v_{activation}$ = 200 [km/h].
- weight of the ballast $m_{laboratory}$ = 130 [kg].

Information related to parachute geometry will not be presented, as it is considered the property of the company that provided the canopy for the test. Nevertheless, for the purpose of this research, that information is not important. Generally, a gliding parachute equipped with a slider to reduce the opening shock load was used. The results presented in this study define the maximum aerodynamic force generated by the canopy for the subsequent analysis. The point corresponding with the maximum opening shock load is highlighted in Figure 9 by point three.

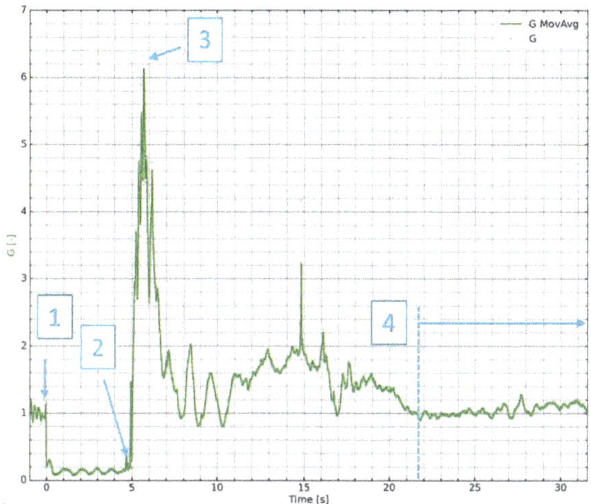

Figure 9. Overload record during reference drop test. 1. release of the system from the airplane; 2. reaching the activation speed–releasing the drogue chute; 3. point of the maximum system overload; and 4. initial point of the steady descent.

The drop test identified the maximum overload as G = 6.3 [–]. By converting the overload to a force based on the input mass using Equation (1), the opening shock load is obtained:

$$F_{shock\_load} = G \cdot g \cdot m_{laboratory} = 6.3 \cdot 9.81 \cdot 130 = 8034.4\ [N] \quad (1)$$

3.2. Results of Harness Loading

The aim of the test was to progressively load the harness up to the force corresponding with the maximum opening shock load generated by the canopy during the activation phase. The gradual loading process will help to obtain information on whether the individual straps are taking the same percentage of the total applied force during all processes. If the test results show that there is a uniform percentage redistribution above a certain load value that does not change further, the results of this study can be used without regard to the opening shock load magnitude. In other words, the loading of the individual elements would then only be determined by the total applied force and the redistribution factor based on this research. As the result shows, this will play a major role in the investigation of the element's safety margin.

Based on the load output from the drop test, the maximum required force has been established as 8000 N. In order to not distort the data due to the initial settling of the harness on the dummy, a preload of 1500 N was applied before each test. This level of preload was set by an estimation. Above this value, there was no further movement of the harness on the dummy's body.

It must be noted that the values of the force recorded by the station also include the weight of the dummy and hanging devices. Hence, some postprocessing was necessary to have comparable results. The presented values of the forces are zeroed at the beginning of the test. This ensures that the weight of the equipment and the initial tension of the webbing are no longer present. As a result, the data only represent the value increment gained from main loading force redistribution into individual segments.

The maximum applied forces are not totally identical between the separate cases. The reason for this is that the readout of the forces was performed manually. Once the operator saw the desired load on the display, further loading was stopped. These small differences do not affect the evaluation of the individual tests. Only the parameters from the separate test entered the analysis of the load capacity, followed by a recalculation of the critical element's theoretical load capacity at a given configuration. The subsequent comparison in Table 1 includes a conversion to the theoretical critical force, which can already be used to compare the harness load capacity for different configurations.

Table 1. Complete results of the maximum forces during loading.

	Test One	Test Two	Test Three	Test Four	Test Five	Test Six	
Applied Force	7665	7699	7599	7735	7748	7787	[N]
Force 1_{MAX}	3920	5045	2719	3544	3788	3869	[N]
Force 2_{MAX}	3806	4560	3350	3432	3816	3856	[N]
Force 3_{MAX}	1910	2151	1884	1892	1886	936	[N]
Force 4_{MAX}	2	1	22	1	345	0	[N]
Force 5_{MAX}	3	4	9	2	241	2	[N]
Force 6_{MAX}	2001	2238	1697	1715	1923	2105	[N]

In general, all test cases have the same evolution, where four main stages can be identified:

1. Dummy rotation into the steady position;
2. Settling the harness–force redistribution;
3. Gradual loading with straight slope;
4. Limit force for gradual loosening of the buckle.

The first stage is, in general, caused by the difference in position of the center of gravity of the steel dummy and the point where the dummy is fixed to the ground of the test room. During the first seconds of loading, the dummy is rotating to the new equilibrium state, which is not changing significantly during further loading. The position change is highlighted in Figure 10.

During the second phase, the system stops rotating and only the harness itself begins to show signs of slight movement on the dummy's body. This is clearly visible in the chart showing the load profiles of the individual elements. Once the load exceeds a certain value, the harness is already static, and there is a steady increase in force. This phenomenon allows the referenced generalization of the published results to use this procedure for a different opening shock load. The assumption is that if a higher load was applied, it could be expected to increase in separate positions with respect to the obtained redistribution ratio. This is one of the most important findings. This idea is possible to apply to all six tests, which enables the calculation of the theoretical limit load of the harness for all the configurations.

The fourth point of the observation showed undesirable conditions. The buckle located on the right leg webbing started to gradually and irregularly loosen once the overall load exceeded the value of around 3000 N. This could be caused by the wear of the buckle, the hard base under the buckle, or a combination of both phenomena. The relaxation was gentle and did not affect the final results presented.

The load redistribution of the total applied load into individual load cells is shown for all six tests in Figures 11–16.

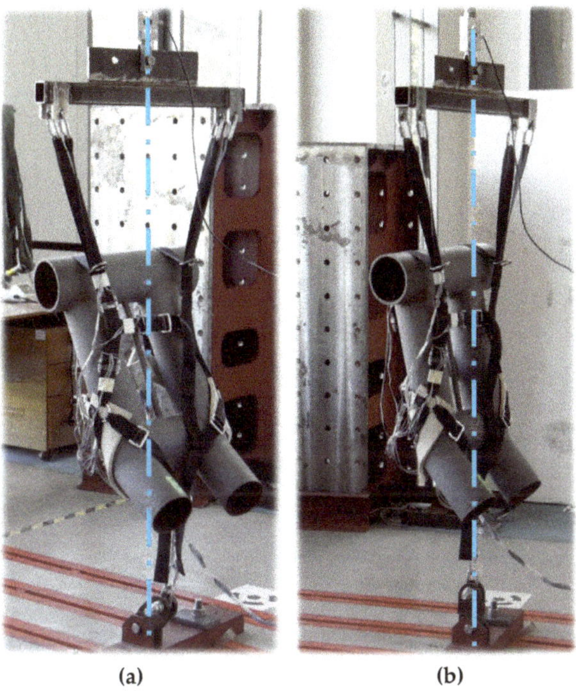

Figure 10. Rotation of the dummy during the first stage of test case one: (**a**) Unloaded, and (**b**) loaded.

Figure 11 displays the load profile of the test configuration one. Progressive loading began at force 0 N, which represents a steady state in which the harness is only loaded by gravity and tightening the webbings to fit the dummy. Loading was stopped at the maximum overload value of 7665 N. Above the value of 4000 N, forces are already steadily distributed, and the same slope of the curves is observable. The load in elements one and two differs by 120.9 N. It can be assumed that this difference in load is transferred by the chest strap, which was tightened.

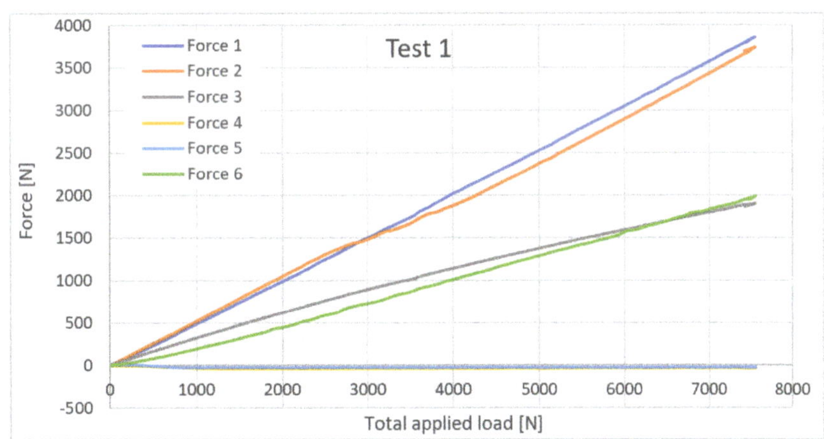

Figure 11. Total force redistribution into individual load cells during test one.

Figure 12 displays the load profile of the test configuration two. During this test, the force applied to the right connection points of the harness was two times higher than to the left side. This resulted in higher loading of the chest webbing and bigger differences between the loads in positions one and two Loading begins at 0 N and increases to a maximum of 7699 N, as in the previous test. Unlike the first test, the forces were already evenly distributed once the total applied load reached a value of 1000 N.

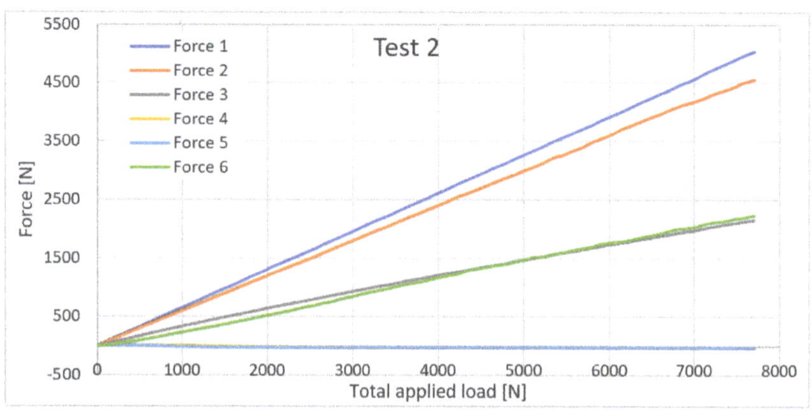

Figure 12. Total force redistribution into individual load cells during test two.

Figure 13 displays the load profile of the test configuration three. During this test, the force applied to the left connection points of the harness was two times higher than to the right side. This resulted in a different redistribution pattern. The load cell one measured a lower load than the load cell two. Loading started at 0 N and reached a maximum of 7599 N. The forces were evenly distributed once the total applied load reached 3000 N.

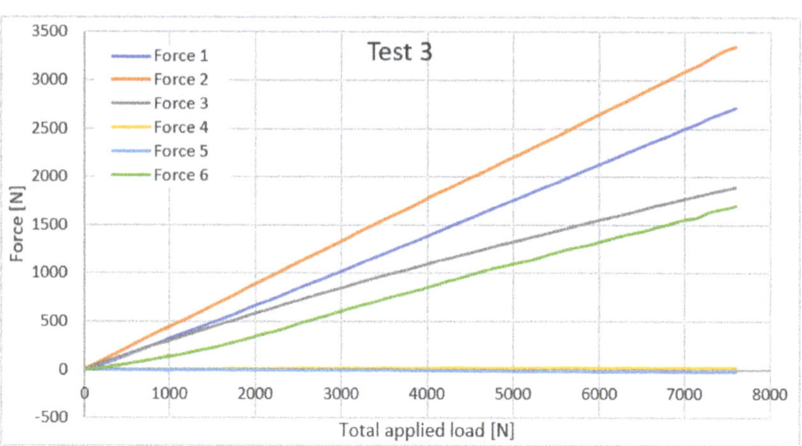

Figure 13. Total force redistribution into individual load cells during test three.

Figure 14 displays the load profile of the test configuration four. The dummy was fixed in place and rotated face down during this test. This resulted in a loosening of the back straps, which were not carrying any load. Loading started at 0 N and reached a maximum of 7735 N. During this test, load cells in positions one and two measured similar forces throughout the whole process. As in the second test, once the total applied load reached 1000 N, the forces were already evenly distributed.

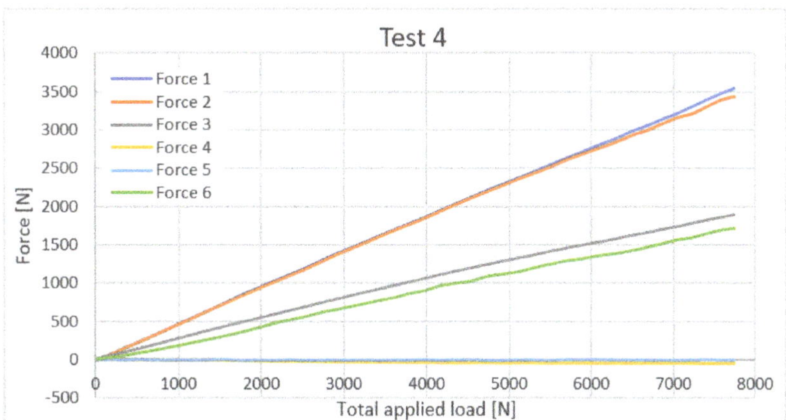

Figure 14. Total force redistribution into individual load cells during test four.

Figure 15 displays the load profile of the test configuration five. The dummy was fixed in place and rotated backwards during this test. This increased the loading on the back straps. Loading started at 0 N and reached a maximum of 7748 N. During this test, load cells in positions one and two measured similar forces throughout the whole process. The forces were evenly distributed once the total applied load reached 2000 N.

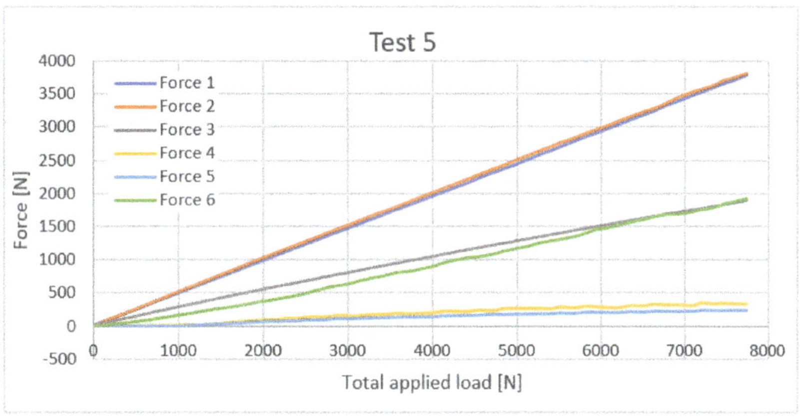

Figure 15. Total force redistribution into individual load cells during test five.

Figure 16 displays the load profile of the test configuration six. During this test, the chest webbing has been loosened. This resulted in lower loading of the chest webbing compared to test one, which has the same setup but with tightened chest webbing. Loading started at 0 N and reached a maximum of 7787 N. The forces were evenly distributed once the total applied load reached 1000 N, the same as in the second test.

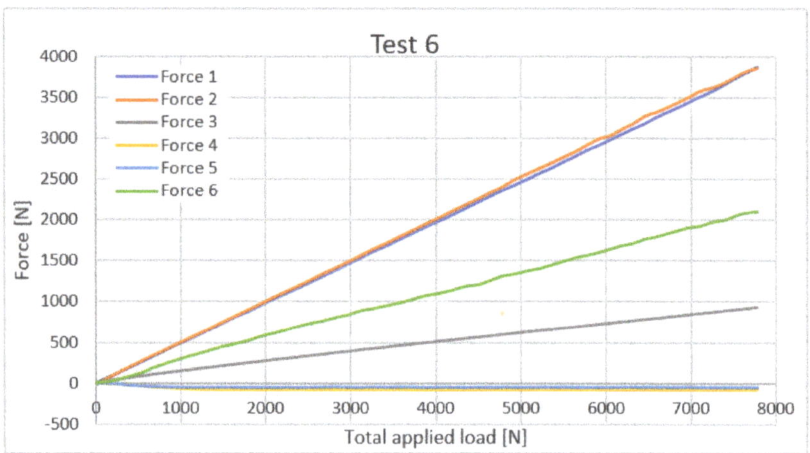

Figure 16. Total force redistribution into individual load cells during test six.

Table 1 summarizes the measured forces in separate load cells with respect to the maximum applied force, which is also highlighted. The marking of the separate elements is performed according to the established convention. As already discussed, the peak of the applied force is not identical for all test cases because of the delay between switching off the load hydraulic cylinder, which was mechanically operated. Nonetheless, the evaluation of the results within the single tests has no effect on the results of the subsequent load capacity analysis.

3.3. *Determination of the Theoretical Load Capacity of Separate Configurations*

Once the ratio of the redistribution of the total applied force to the individual structural elements is known, it is possible to calculate the load capacity of the individual components. For this operation, it is important to know the strength of each component declared by the manufacturer [12–14]. In Table 2, these values for single parts are listed. For clarity, Figure 17 shows the position of the items marked in the table. The mark C indicates the carabiner (buckle) and mark S indicates the strap (webbing). Its number corresponds to the specific designation of the measuring element.

Table 2. Declared strength of the elements.

Element	F_{limit}	
C1	2225	[N]
C2	11,121	[N]
C3	11,121	[N]
S1 and S2	44,482	[N]
S3 and S5	17,793	[N]
S4 and S6	26,689	[N]

With this knowledge, it is possible to use Equations (2)–(3) to determine the limiting force during the canopy activation stage, at which the maximum allowable component load is reached. The given force calculation will thus show not only the critical element but also the strength margin for the other components. This makes the uniformity or non-uniformity of the strength margin of each element visible at first sight. As an example, test case one is analyzed according to the mentioned procedure. With the use of Equations (2)–(3), the results defining the critical applied force to reach the limit force in separate construction elements for test one are as stated in Table 3. The calculations shown in Equations (2)–(3) are demonstrated on the critical element for test case one.

b_1 = percentage value of the force carried by the element, relative to the loading force.
F_{c1} = force at a particular position C1.
$F_{resultant}$ = total loading force that represents the opening load of the parachute.
F_{limit} = the manufacturer's declared element limit force.
F_{crit} = loading force at which the maximum permitted force value is reached.

$$b_{1(C1)} = \frac{F_{c1}}{F_{resultant1}} \cdot 100 = \frac{1910}{7665} \cdot 100 = 24.9\ [\%] \tag{2}$$

$$F_{crit\_1(C1)} = \frac{F_{limit(C1)}}{\left(\frac{b_{1(C1)} \cdot F_{resultant1}}{100}\right)} \cdot F_{resultant1} = \frac{F_{limit(C1)}}{b_{1(C1)}} \cdot 100 = \frac{2224.91}{24.92} \cdot 100 = 8928.3\ [N] \tag{3}$$

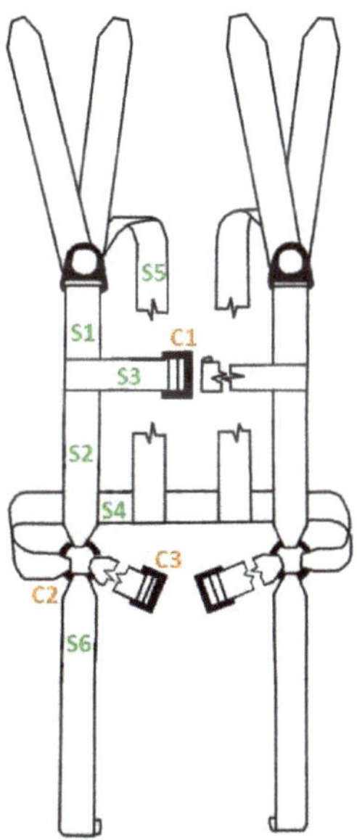

Figure 17. The highlighted position of the evaluated elements.

Table 3. Recalculation of critical force for test case one.

Element	b_1 [%]	F_{limit} [N]	F_{crit1} [kN]
S1	51.1	44,482	86.971
S2	49.7	44,482	89.590
S3	24.9	17,793	71.401
S4	0.03	26,689	89,134
S5	0.04	17,793	46,058.5
S6	26.1	26,689	102.231
C1	24.9	2225	8.9283
C2	49.7	16,014	32.252
C3	26.1	11,121	42.596

The results show that while the theoretical strength of element C1 is achieved at 8.928 kN, element S2 can withstand a value more than ten times higher. By evaluating all the remaining tests, it was found that C1 is a critical element for all configurations. The other elements indicate a significant difference in the safety margin compared to element C1.

Table 4 summarized the critical force calculated according to Equation 1 for all the test setups. This procedure also highlights the most vulnerable position of the dummy, where the limit load of element C1 first appeared.

Table 4. Extracting the limit force at the weakest point of the harness for all test configurations.

	F_{crit_C1} [kN]
Test one	8.9283
Test two	7.963
Test three	8.972
Test four	9.098
Test five	9.143
Test six	18.51

It is evident from the results that the asymmetric load is the first case in which the limit load is exceeded. In other words, test configuration two. However, the differences in the strengths of individual cases are not so significant. It varies up to 14.8% for tests one-five.

The evaluation of element C1 brings another important piece of information related to the influence of chest strap tightening. This can be achieved based on the comparison of the results from test one, where the chest strap was tightened, and test six, where it was loose. The difference in maximum value to reach the limit load is about 107% higher in the case of a loosening chest strap.

4. Discussion

The present study focuses on two major chapters. First, is a test of the aerodynamic characteristics of the canopy in its activation phase to obtain the opening shock load during its activation. The second part is the application of this specific load to the harness and finding the redistribution of the total applied load to the different structural nodes of the harness. The force application is extended to loads in different configurations.

In the canopy testing phase, it was important to maintain the design parameters, which were determined based on the terminal speeds at which a parachutist falls in free fall. For the purpose of the test, the speed was set at 200 km/h. The progress of the test confirmed the normal canopy function, and the results from the first test were used for the postprocessing. The logging of the individual parameters proved to be sufficient to provide the data for the following harness analysis, which is the main goal of the research. The G-force data were used to calculate the opening shock load. As a supplement, measuring carabiners on each side of the canopy hinge were also used. The existing measuring carabiners showed non-standard behavior during the test, and therefore, the results were not included in the evaluation. However, they are the subject of additional internal development, and further practical use is planned for future tests. The maximum value of the measured opening shock overload was $G = 6.3$. With a weight mass of $m = 130$ kg, this overload was converted to an opening shock load of $F_{shock\_load} = 8035$ N. For subsequent analysis, it was stated that the testing force could be rounded up to the value of $F_{shock\_load} = 8000$ N. The provided methodology proved to be adaptable and can be used for a variety of activation speeds and ballast masses or types of parachutes. This brings the possibility of filling the certification requirements defined by the Technical Standard 135 in the section "Structural Overload Tests" [4].

The second section of this paper focuses on a comprehensive loading study of the fully articulated parachute harness structure. The decomposition and incorporation of each element's strength limits, as well as the determination of the applied force's redistribution into separate parts, shed light on the sizing of individual elements. Load cells based on the minimum dimension parameters were developed to measure forces in separate webbings. With this benefit, serial production parts were used for the strain gauge installation. This procedure ensured minimum costs and, due to the load cell's small dimensions, minimal structural influence once it was sewn into the final harness. During calibration, it was verified that laterally uneven loading of the measuring feature has no effect on the measurement precision. It was one of the biggest concerns when considering the location of the separate features. It was expected that, because of the required number of load cells, there would not be enough space to locate them all in ideal positions. The element on the webbing heading to the back of the dummy was, despite all efforts, loaded partly by bending. The aim of future work will be to develop an element that will not be sensitive to the bending component. A very important observation was identified. Once the harness settles on the dummy body, the redistribution of the forces related to the applied force does not change significantly. This opened the possibility of generalizing the research. It can be expected that by increasing the applied force above the presented value of 8000 N, the same trend of redistribution will follow. This means that the use of a different canopy or changes in activation speed resulting in a different opening shock load can be applied for recalculation.

In all cases, the buckle located on the chest webbing was identified as the weakest element. Based on the results, the fixed spatial position that loads this element the most is an asymmetrically loaded case. This is because the chest strap took some of the force from the opposite side through the cross-connection. However, the amount of load transferred by this carabiner is strongly dependent on how tight the strap is. It was identified that the difference in critical force can vary by up to 107%. This result is also expected because the loosening of the chest webbing causes the main load to pass from the pilot's buttocks directly to the upper hinge points. When tightened, the webbing tends to create two triangles that tend to expand under the load. When using the harness, it is therefore a good idea to tighten the harness, but in the case of the chest strap, tighten it only to the point that the harness cannot come loose on the pilot. Any overtightening will not bring any benefit. It will only cause overloading of the structural node.

It can be expected that the maximum allowed load declared by the manufacturer uses some safety factors. This means that even though elements would reach their maximum limit, no visible damage would be observable. Compared to the test of a complete assembly according to TS-135 [4], exceeding the recommended limit of single structural nodes is not controlled. After the test, the harness structure must only be inspected for visible damage. Hence, the testing procedure proposed in this study is considered safer as it allows for the direct monitoring of each component.

For further extension of this study, the strength test of individual components up to visible damage is suggested. Incorporating the maximum strength limits of the parts assumes a significant increase in the maximum load limit of the entire harness.

5. Conclusions

During the research, a methodology was developed to optimize the structure of the parachute harness with the purpose of lightening it. In order to efficiently design the harness for the intended load, an extensive study was conducted to consider the loading of the individual structural elements. A test of the parachute has been performed to reach the required aerodynamic characteristics, which were interpreted as opening shock load. The maximum force obtained from the drop test was directly used to test the harness itself. The static loading test of the harness was designed to gradually increase the load from zero to the maximum load. Different load cases were also incorporated to capture the effect of non-ideal pilot positions during system activation. As a result, the study presents a complete analysis of the separate cases together with a percentage redistribution of the forces during gradual loading. The results evaluation also provided an overview of the sizes of the individual elements. The chest strap is significantly the weakest point, and most of the elements show several times higher strength. Therefore, the intent is to focus on these elements and create a harness with an adapted design. The vision is to create a lightweight harness without affecting the overall load capacity.

Author Contributions: Conceptualization, R.P. and R.G.; methodology, R.P. and R.G.; software, I.J.; validation, M.H., I.J. and R.G.; formal analysis, R.G.; investigation, R.G., J.Š. and M.H.; resources, I.J.; data curation, R.G.; writing—original draft preparation, R.G.; writing—review and editing, R.G.; visualization, R.G. and M.H.; supervision, R.P. and I.J.; project administration, R.P.; funding acquisition, R.P.; All authors have read and agreed to the published version of the manuscript.

Funding: The work of Robert Grim, Robert Popela and Marek Horák on this paper has been supported by the Technology Agency of Czech Republic grant project no. TN01000029 National Competence Centre for Aeronautics and Space under the National Centres of Competence 1 programme—support programme for applied research, experimental development and innovation.

Data Availability Statement: The data presented in this study are available on request from the corresponding authors.

Acknowledgments: Robert Grim wishes to thank to the BUT Institute of Aerospace Engineering for providing the laboratory and the equipment that was necessary to perform the tests in such range.

Conflicts of Interest: The authors declare no conflict of interest.

References

1. *Parachute Rigger Handbook (Change 1)*; LULU PR: Oklahoma City, OK, USA, 2015; ISBN 0359096352.
2. Zhang, S.Y.; Li, Y. Pierangelo MASARATI a Bo Wen QIU. New general correlations for opening shock factor of ram-air parachute airdrop system. In *Aerospace Science and Technology*; Elsevier: Amsterdam, The Netherlands, 2022; p. 129. ISSN 1270-9638.
3. Maydew, R.C.; Peterson, C.W. *Design and Testing of High-Performance Parachutes La Conception et Les Essais Des Parachutes à Hautes Performances*; Advisory Group for Aerospace Research: Neuilly sur Seine, France, 1991; ISBN 92-835-0649-9.
4. Performance Standards for Personnel Parachute Assemblies and Components: Technical Standard 135. Available online: https://www.pia.com/wp-content/uploads/TS-135v1.4.pdf/ (accessed on 2 May 2022).
5. Para-Test. Available online: https://para-test.com/ (accessed on 21 October 2022).
6. Prostakishin, D.; Nam, T.P. Dynamic test method for full body harnesses exploited in cold climate. *IOP Conf. Ser. Mater. Sci. Eng.* **2020**, *945*, 012027.
7. *EN 361:2002 Full Body Harness*; ATRA Technology Centre, Wyndham Way, Telford Way: Kettering, UK, 2022; Available online: https://www.satra.com/ppe/EN361.php (accessed on 19 December 2022).
8. Kalavsky, P.; Rozenberg, R.; Tobisova, A.; Antosko, M. Fall testing of the personal airborne equipment backpack: Ground and flight testing. *Appl. Sci.* **2022**, *12*, 3671. [CrossRef]
9. WAGNER; Peggy, M. *Experimental Measurement of Parachute Canopy Stress During Inflation*; National Technical Information Service (NTIS): Springfield, VA, USA, 1978; Available online: https://apps.dtic.mil/sti/pdfs/ADA058474.pdf/ (accessed on 13 January 2023).
10. Cochrane, C.; Lewandowski, M.; Koncar, V. A flexible strain sensor based on a conductive polymer composite for in situ measurement of parachute canopy deformation. *Sensors* **2010**, *10*, 8291–8303. [CrossRef] [PubMed]
11. Min, A.O.; Jin, Y.; Zhao, M.; Xu, S. Research and Application of Heavy-Equipment Parachute Rope Tension Sensor. *J. Sens.* **2018**, *2018*, 1–10.
12. PS22040-1: QUICK-FIT Adapter. Available online: https://catalog.dj-associates.com/item/military-hardware/adapters/ms22040-1 (accessed on 21 November 2022).
13. PS 70101: Adapter, Reversible Friction. Available online: https://catalog.dj-associates.com/item/special-hardware/adapter-reversible-friction/ms-ps-70101 (accessed on 2 May 2022).
14. 2058-2.25: O-Ring. Available online: https://catalog.bourdonforge.com/item/forged-metal-safety-o-rings/2058-o-ring-2-25-o-d-1-75-i-d-/2058 (accessed on 2 May 2022).

Article

Methodology for Determining the Event-Based Taskload of an Air Traffic Controller Using Real-Time Simulations

María Zamarreño Suárez [1,*], Rosa María Arnaldo Valdés [1], Francisco Pérez Moreno [1], Raquel Delgado-Aguilera Jurado [1], Patricia María López de Frutos [2] and Víctor Fernando Gómez Comendador [1]

1 Department of Aerospace Systems, Air Transport and Airports, School of Aerospace Engineering, Universidad Politécnica de Madrid (UPM), 28040 Madrid, Spain
2 ATM Research and Development Reference Centre (CRIDA), 28022 Madrid, Spain
* Correspondence: maria.zamsuarez@upm.es

Abstract: The study of human factors in aviation makes an important contribution to safety. Within this discipline, real-time simulations (RTS) are a very powerful tool. The use of simulators allows for exercises with controlled air traffic control (ATC) events to be designed so that their influence on the performance of air traffic controllers (ATCOs) can be studied. The CRITERIA (atC event-dRiven capacITy modEls foR aIr nAvigation) project aims to establish capacity models and determine the influence of a series of ATC events on the workload of ATCOs. To establish a correlation between these ATC events and neurophysiological variables, a previous step is needed: a methodology for defining the taskload faced by the ATCO during the development of each simulation. This paper presents the development of this methodology and a series of recommendations for extrapolating the lessons learnt from this line of research to similar experiments. This methodology starts from a taskload design, and after RTS and through the use of data related to the subjective evaluation of workload as an intermediate tool it allows the taskload profile experienced by the ATCO in each simulation to be defined. Six ATCO students participated in this experiment. They performed four exercises using the SkySim simulator. As an example, a case study of the analysis of one of the participants is presented.

Keywords: human factors; air traffic management; air traffic controllers; subjective workload assessment; air traffic control events; simulation platform; taskload; real-time simulations

Citation: Zamarreño Suárez, M.; Arnaldo Valdés, R.M.; Pérez Moreno, F.; Delgado-Aguilera Jurado, R.; López de Frutos, P.M.; Gómez Comendador, V.F. Methodology for Determining the Event-Based Taskload of an Air Traffic Controller Using Real-Time Simulations. *Aerospace* **2023**, *10*, 97. https://doi.org/10.3390/aerospace10020097

Academic Editors: Jordi Pons-Prats and Michael Schultz

Received: 9 November 2022
Revised: 16 January 2023
Accepted: 16 January 2023
Published: 18 January 2023

1. Introduction

Aviation combines great technological development with key activities performed by humans. Human factors are dedicated to better understanding how humans can safely and efficiently integrate with technology in aviation [1]. At the tactical level, air traffic controllers (ATCOs) are at the core of today's air traffic management (ATM) [2].

As their work has a direct influence on air traffic safety, ATCOs must be highly trained and skilled to provide air traffic control (ATC) services [3]. Due to the great responsibility associated with the activities performed by these professionals, identifying which situations cause greater difficulty in their decision-making process so as to be able to model these situations to limit their workload is of great interest. In particular, the study of air traffic controllers' workload is a topic of significant relevance within the air traffic industry.

The environment in which ATCOs work is inherently dynamic and requires them not only to perform their tasks safely and efficiently, but also to interact with a multitude of systems and coordinate with other people. Exploration of digitalisation and the possibility of automating some of these tasks is increasingly prevalent in current research. As an example, ref. [4] presents an experiment conducted with six ATCOs to analyse the possibilities and challenges of automation in terms of teamwork in a realistic ATC en route phase scenario. The authors in [5] presented a methodology for predicting if and when

ATCOs would react to the presence of conflicts, which was developed through the use of deep learning techniques.

The use of simulations in the research and development stages has many advantages. One of them is the ability to create highly realistic exercises with controlled ATC events in order to understand their influence on the decision-making processes of air traffic controllers when resolving such events. In addition, it also allows for the testing and validation of new functionalities prior to large-scale implementation. These simulations also allow for the recreation of unusual or emergency situations in a controlled manner. In summary, current needs and future trends make simulators invaluable tools for both ATCO training and ATM system development [6].

1.1. CRITERIA Project

The CRITERIA project (atC event-dRiven capacITy modEls foR aIr nAvigation) is a collaborative project between Universidad Politécnica de Madrid (UPM) and CRIDA, the Spanish ATM Research and Development Reference Centre. The objective of this project is to establish capacity models based on ATC events. In the same way, another fundamental objective is to integrate the study of the human factors associated with air traffic controllers into these models.

ATC events are a model of the different actions carried out by air traffic controllers. From the beginning of the project, it was defined as one of the requirements that the data used in the analysis should be obtained during the development of the project itself so as to improve the traceability of the data and the explainability of the models.

For this purpose, an en route flight simulator has been used for the development of simulations. The simulator chosen was SkySim, which was developed by SkySoft-ATM. Within this project, SkySim has been used for the development of real-time simulations. However, the simulator can also be used for procedure and airspace design, ATC training, or the testing of new systems and functionalities. SkySim consists of simulation, user interface, management, recording, and replay modules [7].

Two ATC positions are available within the simulation platform. The user interface module consists of a radar display with a full graphical presentation of all ATC information and data [8].

Before this project started (though after the setup of the platform and a set of preliminary exercises had been designed), test simulations were carried out to study the feasibility of starting a project focused on the study of the neurophysiological variables of ATCOs and their relationship with ATC events. The methodology and preliminary results validating interest in the launch of a more detailed and robust project can be found in [9].

The ATC events mentioned in the above reference are conceptually the same as those that will be defined in later sections. The naming of the events has changed as a consequence of the development of the test simulations. Similarly, the exercises presented in the present study are practically the same as the first four exercises described in the previous reference. Following the development of the test simulations, improvements were made and minor errors were identified and corrected.

During the execution of the simulation and data acquisition campaign, the following data were recorded:

- Subjective records of the workload perceived by participants during the course of the simulations.
- Data on neurophysiological variables, in particular electroencephalography (EEG) and eye-tracking data.
- Information on the actions carried out by the participants during the exercises.

Once the first simulations had been developed, it was necessary to define a methodology for assessing the impact of the ATC events that occurred during the development of the simulations on the ATCOs. This methodology is one of the main contributions of this paper.

Studying ATC events separately is not sufficient in itself. A variable is required that relates these events to the difficulty perceived by the ATCOs.

1.2. Taskload and Workload

Taskload is a measure associated with the challenge and difficulty faced by a person in completing a task [10]. ATCOs are subject to multiple task demand loads, or taskloads, over time [11]. Another key concept is mental workload. Mental workload reflects the subjective experience of individuals when performing specific tasks in specific environments and under specific time constraints [12].

Taskload and workload are not synonyms. While the concept of taskload refers to external duties, the amount of work, or the number of tasks to be performed by the ATCO, the concept of workload refers to the individual effort made by a person and his or her subjective experience under given conditions [13].

In [14], the authors identify some of the factors that influence the variables of taskload and workload. Factors such as airspace demand, interface demand, and procedure demand influence taskload. On the contrary, some factors that influence workload are skills, strategy, and expertise.

There are multiple approaches to defining the taskload faced by ATCOs. As an example, in [15], a study was conducted to determine whether air traffic control communication events would predict subjective estimates of controller workload and controller taskload measures. There were four taskload components considered in this study: two principal components related to the number and duration of communication-related events and two principal components related to the content of voice communications. This study is an example of the fact that, despite the generic definition of taskload, the way it is quantified and its components vary according to the specific objective of the study.

The authors in [16] discuss a comparison of different complexity metrics related to the ability to match the subjective workload results obtained in a simulation. The definitions of taskload and workload in this reference coincide with the definitions of these terms in this line of research. In the above reference, it is also explained that one of the simplest ways to quantify taskload is to count the number of aircraft present in a sector. Similarly, it is noted that this approach presents limitations, as it does not consider the evolution of aircraft in the sector. The study presented in this paper aims to use a more comprehensive measure of taskload than aircraft counts. To do so, the experiment described in this paper uses event-based taskload as a metric, which results from considering the taskload contribution of the aforementioned set of ATC events.

Based on the definitions of taskload and workload, it can be concluded that taskload is the part of the work demand imposed on the controller purely due to the tasks he/she has to perform [17]. The taskload value of the ATC events included in the research line of this paper is specific to each event. However, the workload perceived by the participants when facing such events will be specific to each person.

1.3. Hypothesis and Obejctives of the Study

To meet the objectives of the research line, tasks include a massive analysis of the data on neurophysiological variables so that correlation between their evolution over time and the ATC events that have taken place during the simulation can be defined. The aim is to establish general patterns and set limits on certain combinations of events so that the workload of ATCOs is within acceptable levels.

The starting hypotheses of the study are as follows:

Hypothesis 1: *It is possible to determine a taskload distribution profile based on data recorded in a control position that can serve as a reference for the subsequent analysis of the evolution of the neurophysiological variables of ATCOs.*

Hypothesis 2: *Based on a taskload distribution profile, variables related to subjective workload assessment can be used to establish whether this baseline profile is the best reference for the subsequent study of the evolution of neurophysiological variables.*

To test the hypotheses of this study, the development of a laboratory experiment with controlled ATC events, the use of an ATC simulator, and the development of a campaign of real-time simulations were chosen. Initially, when the exercises were designed, the starting point was a designed taskload profile. The research question to be answered now is whether this designed taskload is a good reference or whether it is necessary to define a more adequate taskload profile. Only by answering this question will it be possible to continue the study of events and the analysis of the combinations of events that induce a higher workload on the participants.

Additional information about the participants will be presented in later sections. The sample was selected in such a way that participants were as homogeneous as possible in relation to their age, training, and skills. In this first part of the research line, it was important to invest as many hours as possible in the simulator to obtain a robust methodology. For this reason, the chosen participants are ATCO students.

The study presented in this paper has two main objectives: (i) to establish a suitable taskload profile, which will serve as a reference for further studies in the project and determine whether the designed taskload profile can fulfil this function or whether it is necessary to define a new one; and (ii) once this reference is established, to study which events or combinations of events cause the most complex situations within the sector.

To the authors' knowledge, this study is original as it includes as taskload references a series of specific events to this line of research. Similarly, the methodology used and the process of analysing the data collected during the experiment are also novel.

Studies focused on the taskload and workload concepts sometimes present the important limitation of using the two terms interchangeably. In this paper, right from the introduction, the aim is to eliminate this limitation by clarifying the differences between the two concepts. In the same way, another risk that can arise from these studies focused on such specific topics is that they have a limited application and cannot be extrapolated to other research. To overcome this limitation, this study, in addition to generalising the methodology to other similar experiments, includes a series of recommendations when discussing the results obtained. These recommendations are intended to serve as a guide for applying the lessons learnt during this research to similar subsequent studies.

The remainder of this paper aims to present the steps that have been taken to achieve these objectives. The structure is as follows. In Section 2, Materials and Methods, the steps followed in the study are presented, as well as some details on the development of the simulations and the data recorded. Section 3 presents the results obtained after the simulations were carried out by the six participants. Section 4 presents two subsections. In the first, the results are explained using a case study of one of the participants and the key findings are identified. The second subsection presents some recommendations for future research based on the lessons learnt in this study. Finally, Section 5 summarises the results obtained and the next steps to be taken within the line of research.

2. Materials and Methods

Before discussing the results obtained, it is necessary to present the methodology followed to achieve them. This section includes general information on the methodology followed in the first subsection, as well as some details on the events considered in the design of the simulations, useful information on the simulations carried out, and an explanation of the data recorded, respectively, in the following subsections.

2.1. Methodology

This subsection summarises the methodology followed throughout the study, from the first steps taken to setup the simulation platform to detailed analysis of the event combinations and their influence on the taskload and subjective workload of ATCOs. The methodology followed in this study can be seen in the form of a flow chart in Figure 1. Four stages have been defined:

- Previous steps: Includes all the activities that needed to be carried out in order to get the simulation platform up and running and to start working with the participants.
- Simulation campaign and data recording: Includes the process of running the experiment, as well as the recording of all the data associated with the simulations.
- Definition of the best taskload profile: To be able to study the relationship between the evolution of the neurophysiological variables and the ATC events in the simulation, it is necessary to determine the best reference taskload profile. If this profile proves to be different from the designed profile, it will be necessary to define a new methodology to obtain a baseline that considers the actual events that occurred during the simulations. This stage is a decision point represented by a diamond with two possible outputs in Figure 1. Once this baseline is established, the first objective of the paper mentioned in Section 1 will be achieved.
- Event analysis: Once the baseline has been established, the next step is to study which events or combinations of events induce the greatest difficulty in the exercise and the highest workload on the controller. This step addresses the second objective of the paper.

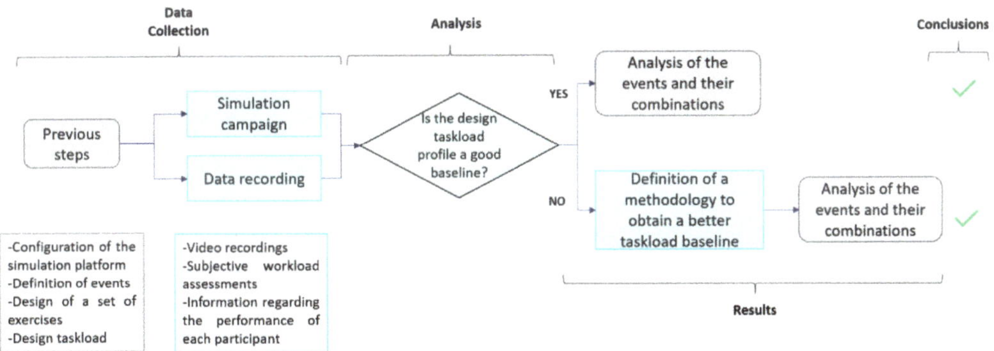

Figure 1. Methodology to follow to obtain an event-based taskload profile of reference and to study ATC events and their combinations.

Before simulations could be performed, several preliminary steps had to be taken. First, the simulation platform was configured. Subsequently, the ATC events that would serve as the basis for exercise design were defined. For this first stage of the project, a total of four exercises of increasing difficulty were designed. When events were introduced at specific moments of the exercise, a unique designed taskload profile was obtained for each of the exercises.

Once the exercises had been designed, the next step was to develop the simulation campaign. In total, six participants participated in the simulations. Each of them simulated each of the four exercises. Several data were recorded during the development of the exercises.

For this line of research, the data of interest were the video recordings of the simulations, which provided information on the actual events and actions carried out by the participants, and subjective workload assessment data. For this purpose, the Instantaneous Self-Assessment (ISA) method was implemented in the simulator through the use of a window that appeared on the radar screen every two and a half minutes to ask participants to evaluate their perceived workload.

The ISA method is based on the idea of asking the operator to assess their workload at regular intervals. At each assessment, the operator is required to select a value on a scale of 1–5. On this scale, 1 means under-utilised and 5 means excessively busy [18]. The ISA method was chosen because it is considered less intrusive than other subjective workload assessment methods [19]. Furthermore, it can be run during the progression of exercises.

The first objective was to find an event-based taskload baseline. In fact, the ultimate goal was to create a profile with the taskload values for each minute of the simulation.

When the exercises were created, a designed taskload profile was defined. Therefore, the first question to be solved is whether this designed taskload can be used as a reference. This question is represented by the diamond in Figure 1. There are two options: 'yes' and 'no'. The first option is to demonstrate that the answer is 'yes'. This would be the simplest possible situation. In that case, since the designed events of the simulations are known in advance, the designed taskload profile could be used directly as a reference when studying the evolution of neurophysiological variables.

The other alternative is that the answer to the question is negative. If it is not possible to use the designed taskload profile as a reference, it will be necessary to establish a methodology for obtaining a better reference based on the events that actually took place during the development of the simulations before addressing the second objective of the study and moving towards analysis of the events and their relationship with the recorded subjective workload values.

2.2. Events Considered in the Design of Exercises

To design the exercises, the first step was to agree on the ATC events to be considered when creating a designed taskload profile that would characterise each exercise. For this purpose, a series of workshops were organised that brought together ATM experts as well as people with previous experience using SkySim.

Within the group of experts, the vision of experienced working controllers was highly valued. On the other hand, researchers with previous experience in the design of simulation exercises and in the development of validations in research projects at the national and European level were included in the working group.

In particular, there were two previous studies that were very useful for defining the final ATC events. On the one hand, at the national level, UPM had previous experience working with the Spanish Aviation Safety Agency.

On the other hand, the events considered in the SESAR AUTOPACE project were also of great interest [20].

The results of the AUTOPACE project provide a better understanding of how cognition and automation coexist, thus supporting new strategies for training and interface design [21]. Although the objectives are not aligned with those of the present line of research, the events defined in this project have been considered as a reference, as has the definition of difficulty in the design of the exercises.

Figure 2 shows, in schematic format, the activities that were identified as key ATC tasks (shaded in blue), the designed ATC events associated with each of them (shaded in green), and the base score for each of the events (rectangles with a white background).

The first aspect that was defined was the key activities carried out by ATCOs in nominal traffic situations. The result can be seen in the six blue rectangles shown in Figure 2, including the identification of an aircraft, the takeover/handover process, the identification of a conflict and its resolution, and, finally, the monitoring of the traffic present in the sector. On the basis of these activities, at least one event associated with each of them was defined. The twelve events considered can be seen in the green rectangles in the figure above. For each event, an average duration and a base score were defined.

Eleven of the events have an absolute value associated with the event. The monitoring event is the only one that is relative. It is associated per minute with each of the aircraft within the sector at a given time.

In defining the twelve events, complexity factors aligned with those identified by other authors have been considered. Specifically, the authors of [22] conducted a Principal Component Analysis (PCA) on 24 complexity factors defined in the literature to reduce them. The final result was a set of eight complexity factors. Several of these factors have been considered when defining the events listed in Figure 2, in particular, aircraft count, aircraft vertical transitioning, and conflict sensitivity. In this line of research, the complexity factor of aircraft count has been considered by defining three scenarios of traffic density that condition the base values of taskload. Similarly, in the case of conflicts, the contribution of aircraft climbing or descending has been taken into account by giving conflicts where one or both aircraft are changing their flight level a higher taskload value than conflicts where aircraft are at cruise level.

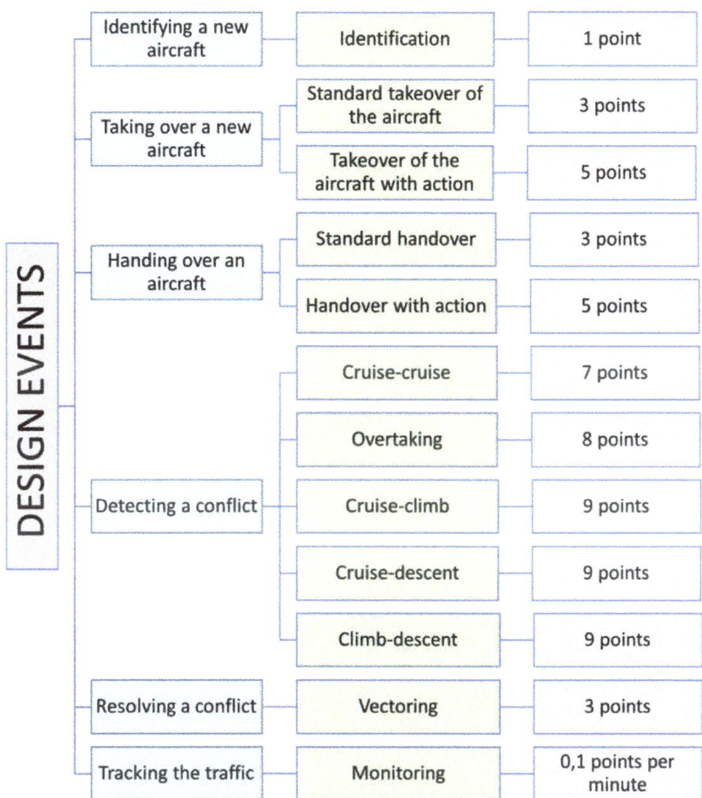

Figure 2. Identification of the key activities carried out by ATCOs (blue rectangles) and the ATC events associated with them (green rectangles). In total, twelve ATC events form the basis of the simulations in the experiment conducted.

Air traffic control is a service task whose duty is to prevent conflicts between aircraft [23]. The events associated with conflicts were discussed in detail. It was decided that conflicts should be categorised according to the state in which the aircraft were: in cruise flight or climbing or descending. When two aircraft are on the same trajectory and one of them starts to approach the other, until the separation minima are infringed, an overtaking event occurs.

ATCOs are required to ensure that minimum separation standards are complied with at all times in terms of horizontal and vertical separation between aircraft [24]. Regarding the level of automation of the platform, the configuration used in this experiment is very similar to the so-called attention-guided mode in [25]. In this configuration, conflict detection is automatically performed by the simulator. However, the ATCO retains the role of controlling the aircraft and is not assisted in resolving the conflict. In the context of this experiment, a conflict is defined as a situation where the minima of 5.0 NM in the horizontal plane and 1000 ft in the vertical plane are infringed.

In the simulator's conflict detection tool, the detection threshold in the horizontal plane is set to 7.9 NM. This allows ATCOs to receive information in advance of a conflict situation occurring. The conflict detection tool helps them identify which aircraft are involved in the conflict situation, how close they will be at the closest point of approach

(CPA), and the time until this point is reached. However, the decision-making process to resolve the conflict remains in the hands of the ATCO, without guidance. In short, conflict detection is automated on the simulation platform, but resolution is not.

Although there are many factors that can increase the complexity of an event for a controller, traffic density is one of the key factors that increases perceived workload [26].

To account for this contribution, the base score of each event increases as the number of simultaneous aircraft in the sector increases. For this purpose, three traffic density scenarios were defined: low (less than five aircraft), medium (between five and nine aircraft), and high (more than ten aircraft).

Additional information on the definition of events and the assignment of taskload values can be found in [9]. This reference also includes a table that compares the basic characteristics of the four exercises used in this study, including the number of aircraft in each exercise, the number of events, or the sector in which the simulations were conducted.

2.3. Details of the Simulation Campaign

To ensure safe and efficient traffic flow, ATCOs must predict future flight paths based on their perception and interpretation of multiple data on the radar display [27]. In this study, it was considered from the outset that one of the cornerstones of the experiment should be data collected in a especially designed simulation campaign. In this way, in addition to the numerical data, it is also possible to access all the information concerning the radar display, as well as the actions taken by ATCOs.

The exercises simulated in this research reproduce realistic en route scenarios where aircraft are established at certain flight levels. En route ATCOs are responsible for monitoring, controlling, and managing aircraft and traffic flows in the ATC sectors they have been assigned [28]. In the exercises simulated in the experiment described in this paper, the sectors of responsibility varied throughout the exercises, though sectors within Madrid Area Control Centre (ACC) airspace were always used.

In the simulations, each participant simulated four exercises. Each exercise lasted 45 min and was designed to be simulated in parallel in two sectors. The taskload value for the first exercise was 141.40, and this value continued increasing until Exercise 4, which had a value of 228.55. Table 1 presents a comparison of the designed taskload values for each of the exercises.

Table 1. Comparison of the designed values of the four exercises in the simulation program.

Exercise	Total Designed Taskload	Maximum Taskload	Minute Maximum Taskload
1	141.40	7.425	0:11:00
2	165.20	10.400	0:29:00
3	193.25	10.875	0:30:00
4	228.55	11.400	0:31:00

As mentioned above, to introduce subjective assessment of the workload by participants, the ISA method was used. This method was implemented on the platform through a Python program that was run in parallel to the simulations.

A window asking the participants to evaluate their perceived level of workload appeared every two and a half minutes in a fixed place on the radar display. To do so, they had to press one of the five buttons available under the question, with '1' being the lowest value and '5' the highest value. They had 20 s to select one of the options before the window automatically closed.

The data presented in this paper relate to six participants. All participants were ATCO students with an average age of 21 years and previous knowledge in the field of air traffic management. The ATCO students who participated in the study were selected on the basis of their performance in other practical tests developed during their training.

Throughout their training, participants were trained in concepts related to airspace management, conflict resolution strategies, and the operation of a control position. Although they had previously performed different exercises and simulations with other simulation platforms, this experiment was their first contact with the SkySim platform. For this reason, prior to the test exercises, they underwent specific training on the platform.

The test simulations for each participant took place on two days, with an interval of one week between them. Each day the participant performed simulations, they would simulate two exercises with a one-hour break between. Before starting the simulations, the participants were informed of the aim of the project, and all agreed to their data being analysed as part of the research.

2.4. Data Registered after the Simulations

During the course of the simulations, a multitude of data were recorded. Specifically, for the purposes of this work, two categories of data were of interest:

- Firstly, the video recordings of the radar screen during the simulations. From these recordings, it is possible to obtain information about the events that actually took place, the actions taken by the participants, and the conflict resolution strategies followed.
- Secondly, information that was obtained about the subjective workload values evaluated and the minute of simulation in which each of the values was selected.

In the study of human factors, the exclusive use of subjective measures of workload assessment has certain limitations. On the contrary, some of its main advantages are the relatively low effort required to acquire data and high user acceptance [29]. These advantages were considered decisive for the implementation of subjective measures in this study.

As mentioned above, neurophysiological data were also recorded during the simulations and will be studied in later stages of this line of research. Physiological measures have been shown to be sensitive to differences in taskload and task demand in a variety of domains [30]. For this reason, they are of interest in the study. However, to be able to compare the variation in these variables against the taskload, the first step is to have a good baseline for that taskload.

The use of subjective workload values is a preliminary step. It is assumed that the most complex traffic situations and the most difficult combinations of existing events will lead participants to evaluate these traffic situations with the highest workload values. The idea is to use these values to define the best baseline taskload profile for future use in determining other workload indicators.

From the data related to the subjective assessment of workload, two variables are of interest:

- The first variable is the subjective workload value selected in each query by the participant. The possibilities are that a numerical value (1–5) is recorded or, in case the participant did not respond, a "not assessed" is recorded.
- The second variable is reaction time. This variable can take values in the range of 0–20 s, as this was the time that the participant had to select one of the values of the ISA scale before the window closed. Reaction time is calculated as the difference between the time in the simulation when the participant selects one of the values and the time in the simulation when the ISA method window appears.

These two variables will be used as an intermediary step in the establishment of a methodology that can obtain the best reference taskload profile. The results of this analysis and its implications are presented in the following section.

3. Results

In the safety-critical area of ATC, workload remains a dominant consideration when seeking to improve the performance of ATC systems [31]. As mentioned above, subjective workload data will be used as an indicator to establish the best event-based taskload baseline.

The starting point is to try to assess the suitability of the design profile as a baseline. This would be the simplest situation, since this profile is available from the beginning of the creation of the exercises. In the following two subsections, the results obtained from the combined representation of this design profile and reaction time data and workload values will be presented.

3.1. Analysis of the Reaction Time Variable

To establish a relationship between the reaction time variable and the designed taskload of each exercise, it was decided that a combined graph should be created. The same graph shows the designed taskload profile, which is different for each of the exercises, and superimposed on it are the reaction time values for each of the six participants.

A representation of the four exercises can be seen in Figure 3. Each participant is represented by a different geometric shape and colour, as can be seen in the legend that appears under the four plots.

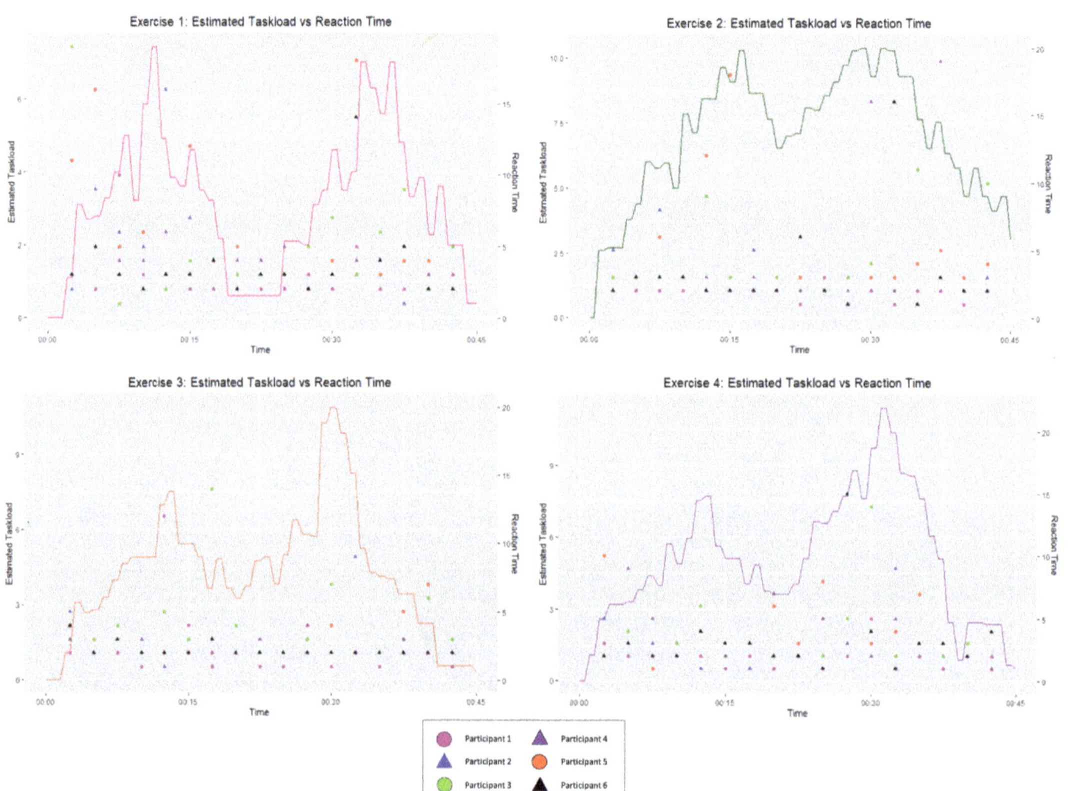

Figure 3. Combined representation of the designed taskload profile of the different exercises together with the reaction time values for the six participants. The first row of graphs presents the two simplest exercises, with Exercise 1 on the left and Exercise 2 on the right. The second row presents the data from Exercise 3 on the left and Exercise 4 on the right.

Each of the exercises has been associated with a different colour to facilitate the interpretation of the plots. The taskload profiles of Exercise 1 appear in magenta, those of Exercise 2 in green, those of Exercise 3 in orange, and finally those of Exercise 4 in purple.

To correctly interpret Figure 3, the following aspects should be considered.

- The x-axis of the four graphs represents the time elapsed since the start of the simulation. All exercises lasted 45 min.

- Each graph has two vertical axes: the left vertical axis is associated with the taskload profile and the secondary axis on the right is associated with the reaction time variable.
- The left vertical axis indicates the value of the designed taskload per minute of simulation. In each of the graphs, the upper limit of this axis is different considering that the difficulty increases progressively from the first exercise to the last.
- The vertical axis on the right, i.e., the secondary axis, indicates the reaction time value for each of the participants. In all graphs, the values on this axis range from 0 to 20 s. There are reaction time values every two and a half minutes, as they are recorded at the moments when the participants assessed their workload.

As can be seen in the figure above, the taskload distribution profile for each exercise is different. In the exercise design process, the starting point was a specific shape of designed taskload and a total reference taskload score. Specifically:

- The taskload profile of Exercise 1 was designed to be symmetric with two cycles of taskload and a low event valley in the central part. In each of the cycles, the taskload was intended to progressively increase to a maximum and then decrease again.
- The taskload profile for Exercise 2 was designed to have two taskload cycles separated, again, by a valley. In this case, the area with fewer events did not have taskload values as low as in the case of the previous exercise.
- The taskload profile of Exercise 3 was designed to be non-symmetric. In this case, the first cycle would reach a taskload maximum lower than the second cycle.
- The profile of Exercise 4 has characteristics similar to those of the previous exercise. The difference is that in this exercise the maximum taskload values are higher.

Table 1 shows a summary of the design characteristics of each of the exercises to compare their design values. The second column presents the total taskload value for each of the exercises. The next two columns present the maximum value of the designed taskload and the minute of simulation at which this value was expected to be reached.

Taking all the above into account, the obtained values for reaction time and the designed taskload for each of the exercises were represented in a combined graph.

The initial hypothesis was that reaction time would increase as the taskload faced by the ATCO increased. Given that the situation in the sector is more complex, it would be expected that the controller would take longer to assess the workload.

However, such a correlation was not observed in any of the four exercises. Contrary to what might be expected, the highest reaction time values appear at the beginning of Exercise 1. The explanation for these values is not that the situation in the sector was more complex, but that the participants were not yet familiar with the platform, the radar screen, or the additional windows.

Some isolated cases that confirm the initial hypothesis are the values of Participants 3, 5, and 6 in Exercises 1, 2, and 4, where the reaction time values increase at times with high taskload values. However, in general, the expected generalised relationship is not observed. Based on the analysis conducted, it was considered that, in the case of the registered data from participants, reaction time was not a significant variable in the study, and it was decided to discard it.

3.2. Analysis of Subjective Workload Scores

Once the variable associated with reaction time had been discarded, the approach was repeated while considering the workload values recorded by the participants. A combined representation of the exercise design profile and superimposed subjective workload values was created.

Figure 4 presents a combined representation of the designed taskload profile and superimposes the workload values assessed by each participant.

The rationale behind the plots is the same as in the case of reaction time. The only difference is that, in this case, the secondary axis presents the workload values on a scale of 1 to 5.

In addition to the series identifying each of the participants, an additional series of data points has been included in the graphs. The values indicated with an orange star correspond to the mean values of the six participants in each of the moments where the workload is evaluated. As can be seen, in general, the values evaluated by each of the participants are closer to the mean in the first evaluations of the exercises and in the final minutes. In the middle minutes of the exercise and in the intermediate minutes of the taskload cycles, the values assessed by the participants are more dispersed due to the different actions implemented by the participants, especially in the conflict resolution processes.

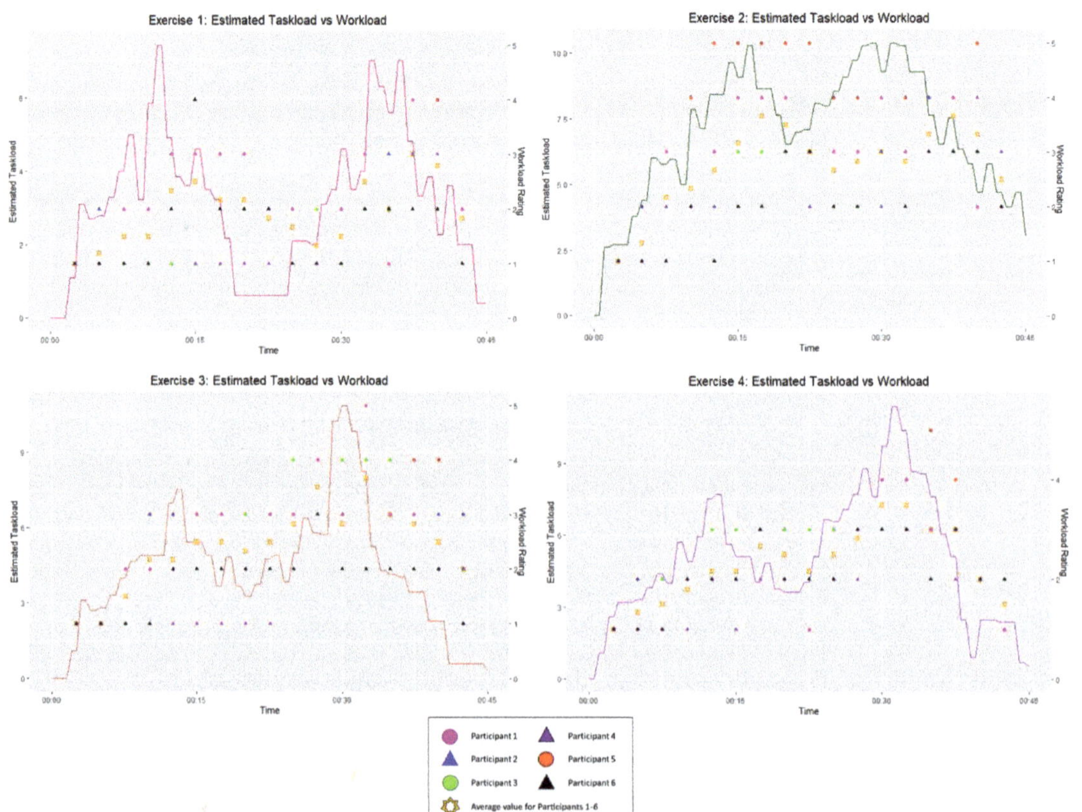

Figure 4. Combined representation of the designed taskload profile of the exercises together with subjective workload assessments for the six participants. The first row of graphs presents Exercise 1 on the left and Exercise 2 on the right. The second row presents the data from Exercise 3 on the left and Exercise 4 on the right.

Unlike what happened in the case of the reaction time variable, the workload values evolve throughout the exercise. The initial hypothesis in this case is that the highest subjective workload values are reached in minutes when the designed taskload is highest. However, in general, this relationship is not observed.

The fact that workload values evolve over the course of the exercise makes them a variable of interest in the study. After the combined graph analysis, the results obtained are as follows:

- The general tendency of the participants is to assess the highest workload values out of phase with the designed taskload. This is particularly clear in the graphs of Exercise 1 and Exercise 3. The reason for this is that events that have a higher value of taskload associated with them appear to have a longer duration than in the designed taskload profile. Therefore, participants must implement actions to deal with these events for longer periods of time. Depending on the actions selected by each participant, especially in conflict resolution processes, the taskloads of more complex events can influence the workload of ATCOs for longer periods of time.

- Considering that the purpose of the study is to identify the situations associated with the highest workload values assessed by ATCOs, it is necessary to define a reference profile capable of explaining the events that take place at the moments when the workload values are at their maximum.
- In general, the trend in workload assessments does not follow the designed taskload profile. To continue with the study, this designed taskload is not a good baseline.
- It is necessary to establish a taskload profile based on the actual situation experienced by each participant during the simulations.

All of the above leads to the conclusion that the designed taskload is not a good reference. This profile was unique for each of the exercises. However, the decision-making process of each ATCO and the conflict resolution strategies used are specific to each controller.

Therefore, in order to explain the workload values evaluated by each of them, it is necessary to compare these values with a specific taskload profile for each participant and each exercise.

Analysis of the graphs in Figure 4 leads to the conclusion that the actual taskload experienced by each controller was different from the designed taskload. Therefore, it is necessary to determine the actual events that took place in the simulation for each participant so that a taskload reference representative of what actually happened may be obtained.

To find out which events took place during the simulations and at what time, the radar screen recordings of each controller were examined. The steps in the methodology to obtain the actual profile were the following:

1. For each exercise, the minutes of the simulation in which the absolute events occurred were recorded. The taskload values associated with each event were the same as those shown in Figure 2. The aim is to enable a comparison between the design and actual taskload profiles.
2. In addition to the taskload of absolute events, there is the taskload associated with aircraft monitoring. The time interval in which an aircraft is monitored is calculated as the difference between the time at which the identification event starts and the time at which the event associated with the handover ends.
3. During analysis of the recordings, two new events were identified that had not been considered during the design of the exercises.
   a. The first event is the change of flight level. Some participants, upon identifying that two aircraft were about to encounter a conflict, would anticipate the situation and change the flight level of one of the aircraft before being alerted by the conflict detection tool. This event was assigned a base score of 2 points.
   b. The second event is the change of speed. As in the previous case, some participants detected in advance that an overtaking conflict was going to occur. In this case, some participants considered that the easiest way to resolve it was to change the speed of one of the aircraft involved. Since the taskload induced is similar to that of flight level changes, this event was also scored with a base value of 2 points.
4. The events with the highest associated taskload are conflicts. These situations were analysed in great detail. The greatest differences with respect to the designed profile were found to occur as a consequence of conflict resolution. In the design of the exercises, for each of the designed conflicts, a generic vectoring event was assigned for conflict resolution. In actual simulated exercises, several participants needed to try different conflict resolution strategies before resolving a conflict. This was especially acute in the case of the second cycle of Exercises 3 and 4, where two conflicts were designed to take place with a short time interval between.

5. Taking all of the above into account, a bar chart was designed that represents the taskload associated with each minute of the simulation. This actual taskload can differ from the designed taskload in terms of the number of events, as well as in the start and end times of some events. To understand whether the participant had perceived the exercise as easier or more difficult than initially designed, a combined representation of the actual and designed taskload profiles was made, and tables were created comparing the score values of the two profiles to understand how the actual profile differed from the designed one.
6. The actual taskload profile obtained for each exercise and for each participant was compared with the subjective workload values assessed by each participant. In this way, it was finally possible to identify which event or set of events led to the highest workload values.

4. Discussion

This section is structured in two parts. The first presents a case study to reflect the implementation of the steps of the methodology listed in the previous section. It then presents a series of general results obtained by repeating this analysis for all exercises and all participants.

The second part includes a series of recommendations for future research in the field based on the results of this study. The results presented in this study are specific to this. The idea of including a set of recommendations here is to justify the interest in the methodology and to highlight the lessons learnt for the benefit of other researchers in the field who might consider developing a similar experiment.

4.1. Case Study and Generalised Results

4.1.1. Case Study

Following all the steps in the methodology, the actual taskload profiles of each of the participants were constructed and analysed one by one. As an example, this subsection explains the detailed analysis of one of these actual taskload profiles. Exercise 1 of Participant 5 (ID5) was selected to be the case study.

After reviewing the radar screen recording of Exercise 1 and studying the decisions made by Participant 5, a total of 47 absolute events were identified, as well as the start and end times of each event.

For these absolute events, the taskload derived from aircraft monitoring was added. From these data, a bar chart representing the actual taskload profile was constructed and compared to the designed taskload profile.

The combined plot of both diagrams can be seen in Figure 5. For each minute of simulation, the green bars represent the designed taskload and the blue bars represent the actual taskload.

When comparing the two bar charts in the previous figure, it can be seen that the taskload distribution is different in the simulated exercise. In addition to the fact that the taskload values per minute of simulation are different, it can also be seen graphically that the taskload distribution in each of the cycles is not maintained.

The designed taskload profile was symmetric. However, symmetry has been lost in the actual taskload, with the highest taskload values being reached in the second part of the exercise.

Table 2 shows a comparison of the most relevant data for each of the designed and actual profiles.

The first row of the table above presents the total taskload values for each of the cases. The taskload profile of the actual exercise is higher than the designed profile. The first conclusion is that Participant 5's simulation was a more difficult Exercise 1 than the one that had been initially designed.

The next two rows compare the scores associated with the absolute events and the monitoring event (defined per minute for each aircraft).

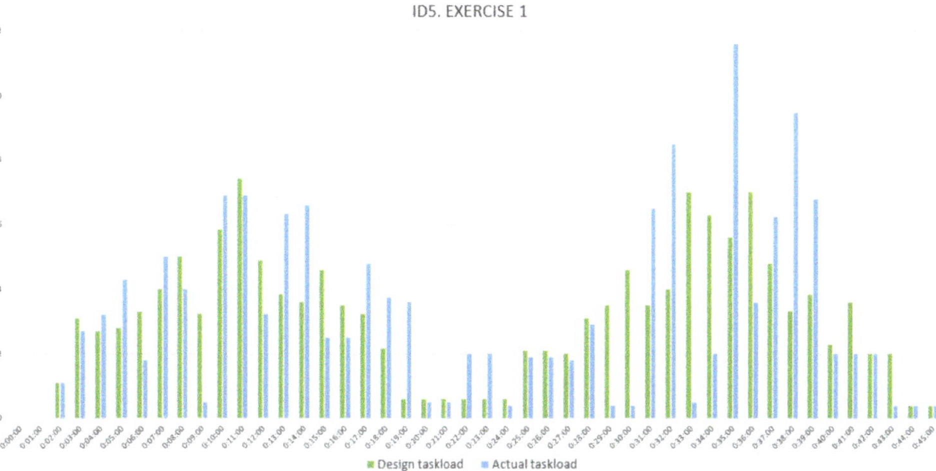

Figure 5. Combined representation of the designed taskload profile (green bars) and the actual taskload profile (blue bars) for Exercise 1 of Participant 5.

Table 2. Comparison of the most relevant data that characterise the designed and actual taskload profiles.

	Designed	**Actual**
Total taskload	141.400	146.800
Event taskload without monitoring	118.000	124.600
Monitoring taskload	23.400	22.200
Maximum taskload	7.425	11.600
Minute of maximum taskload	00:11:00	00:35:00
Number of absolute events	44	47

As can be seen, in the case of this participant, the taskload associated with the absolute events is higher than that of the design. This is due to the fact that a greater number of events appear in the simulation than those initially designed, fundamentally due to the resolution of conflicts. Specifically, the conflict that occurs at 00:10:41 is resolved by changing the flight level of one of the aircraft. As the exercise progresses, this aircraft must be returned to its original flight level in order to comply with the flight plan of its flight progress strip. In the same way, in order to resolve the conflict that takes place at 00:35:00, several vectoring events are required, as the first one is insufficient in terms of respecting the separation minima between aircraft.

In contrast, the monitoring score is slightly lower. This is explained by the fact that the participant handed over some aircraft earlier than planned in the design. Therefore, they spent less time in the sector.

The maximum designed taskload was 7.425 points. The most significant difference in the table is that the maximum taskload that occurred in the exercise was 11.600 points.

In the design, the minute with the maximum taskload was foreseen to be minute 11. Minutes 33 and 36 had a similar taskload associated with them, although slightly lower than the maximum. These high taskload values are associated with the occurrence of design conflicts in the exercise.

In the case of this specific participant's taskload, the highest value was reached at minute 35. This high value is explained by an accumulation of events. Some of them have a taskload value that is not too high. This is the case with respect to the identification and takeover of an aircraft. The problem is that in this exercise, they gather in the same minute in which the participant identifies and starts to resolve a conflict.

Finally, the last row compares the number of absolute events. In the case of the simulated exercise, three additional events occurred compared to the design. Specifically, they were events associated with conflict resolution, as the participant had to try different strategies due to the first not being effective.

Once the actual profile of the participant has been obtained and analysed, it needs to be compared with the subjective workload values assessed by the participant. A combined graph of the participant's actual profile and the subjective workload values assessed can be seen in Figure 6.

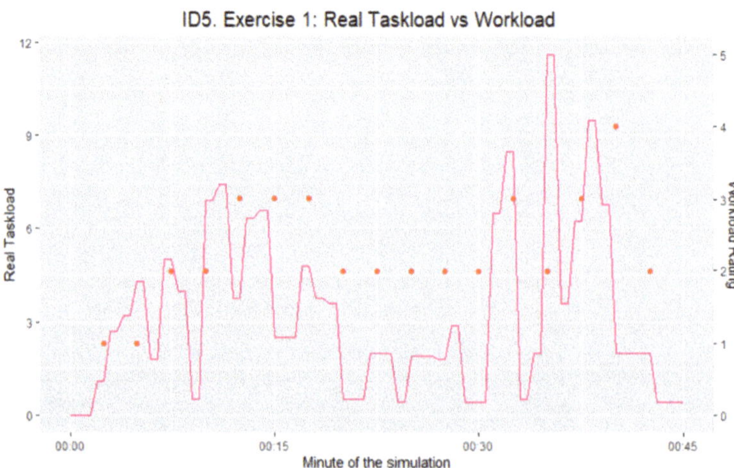

Figure 6. Comparison between the actual taskload profile obtained for Exercise 1 of Participant 5 and the workload values assessed by the participant during the simulation.

4.1.2. Generalised Results

From analysis of the above graph for the different participants, some general conclusions can be drawn about the relationship between the assessed workload values and the identified ATC events.

- When comparing the workload values with the actual profile, it is possible to explain some values that seemed unusual compared to the design profile.
- In general, there is a correlation between the moments in which the participant evaluates the workload to be at the highest level and the highest taskload values of the exercise.
- The effect of events that produce an increased workload is spread over time. Even if the subsequent taskload is lower, participants evaluate the workload in a sustained way over time. An example of this can be clearly seen in Figure 6 in the workload assessments that take place in the central part of the exercise.
- A common phenomenon is that, once participants have assessed the workload with a value higher than '1', it is very rare that they assess it with the minimum value, except in the valley moments of the exercises or in the last workload assessment at 00:42:30.
- The most complex events are those associated with conflicts and their resolution.
- Events evaluated with a higher workload value include conflicts that are not resolved during the first attempt.
- Simpler events, such as aircraft handovers, are perceived to be more difficult after conflict resolution and with a larger number of aircraft in the sector.
- When comparing the workload assessment of the participants for a given time, differences are especially found in the middle part of the exercise and in the intermediate minutes of the two taskload cycles. These differences are due to the different actions performed by each participant. This fact, once again, justifies the definition of a taskload profile for each participant and each exercise. These differences in actions are

particularly remarkable in conflict resolution processes. Some participants implement actions to resolve the conflict that are successful on the first attempt. On the contrary, other participants must implement several conflict resolution strategies if the first attempt is not satisfactory. For this reason, depending on where the participant is in the decision-making process when the workload assessment question is asked, the recorded values may vary.

The results presented in this study were obtained with a small sample of participants. Therefore, caution should be exercised when extrapolating these conclusions to all air traffic controllers or potential participants in the simulations. In the case under study, this sample of participants has been used to define the methodology and these steps are applicable to other subjects, although the results obtained by introducing more participants may vary. Therefore, these first participants are considered a validation group for the methodology to be followed. In future stages of research, a larger number of participants will be included in the study.

Another limitation is the use of ATCO students instead of air traffic controllers with operational experience. Future work in this research line aims to overcome this limitation. These ATCO students have been involved so as to obtain a methodology that is as robust as possible. In later stages, in-service ATCOs will be included as participants.

These limitations are common to other studies conducted in the study of human factors associated with air traffic controllers. The authors of [32] focused on assessing the effects of the number of crossings, traffic flows, and aircraft separation on the mental workload and perceived emotion of ATCOs. One of the limitations identified in the study was the caution necessary in interpreting the results, as only two ATCOs were included as participants.

Another study in which the number of participants is considered a limitation is [33]. In this case, the aim of the study was to monitor the heart rate variability of the controllers and to determine the suitability of the methods used. To overcome this limitation, including a larger number of participants in the study is suggested. However, the results obtained with the study sample are considered beneficial, as they provide an indication of the suitability of the analysis methods used in this type of research.

In the study documented in [34], the future work reported is very much in line with that proposed in this line of research. In their study, evaluating the behavioural response in ATCOs in terms of the use of the procedural control bay and the electric flight strip bay using human-in-the-loop simulations was proposed. Two experts and two trained subjects were used as participants. Future work includes the development of a study with a larger number of participants, as well as on-site replication in a real operational scenario.

Another possible solution to help overcome the limitation of all participants being ATCO students is to consider a mixed group where some participants have previous experience as ATCOs. As an example, this logic is applied in [17]. In this study, the objective was to present the development and evaluation of a 3D space-based metric solution for air traffic control workload. Participants were part of two different expertise groups: four were retired ATCOs and the other six were researchers in the ATM domain or participants who had completed an ATC course. The study presents comparative results between the two groups of participants.

4.2. Recommendations

The results presented in this study, as well as the baseline ATC events, are specific to this research. The input parameters to the methodology will vary in other studies depending on many variables, such as the number of aircraft introduced in the simulation, the characteristics of the participants in the study, the simulator used, the ATC events defined, etc.

However, the methodological process for obtaining the results can be extrapolated to other studies. Once the methodology has been validated within the CRITERIA project, this methodology will be used by CRIDA in other ongoing research projects.

Based on the experience gained in extrapolating the methodology to other research projects, a series of recommendations of interest have been identified based on the results of this study. These five recommendations could be considered by researchers who wish to implement a similar experiment using an ATC simulation platform.

- From the beginning of the experiment's design, it is very important to clearly document the ATC events to be considered. Based on previous work, it is advisable to at least define the mean duration parameter and a designed taskload value. Researchers are advised to keep these values realistic and to consider the opinion of experts with experience in the simulator being used for their definition.
- Based on the values defined in the previous step, in addition to a list of the events to be included in the study, it is recommended to represent the designed taskload profile by, for example, using a bar chart and defining one bar for each minute of simulation. This will allow researchers to have a reference of what is expected to happen during the simulation before they start developing them. If an observer is watching the simulations with this reference profile in front of them, they can already draw some initial conclusions about the performance of the participants.
- Whenever possible, researchers are recommended to register radar screen recordings during the exercises. As demonstrated in this study, they have been vitally important in understanding what actually happened during the exercise and in comparing the airspace conditions during the simulation to the design parameters.
- Before going directly toward a study of the temporal evolution of neurophysiological parameters, it is necessary to invest some time in determining a good profile that considers the actual situation of what happens in simulation exercises. This study has proposed a methodology to determine the best reference profile.
- In order to obtain the best reference profile for the study of neurophysiological variables, it is recommended to use an intermediate tool that allows researchers to conclude graphically whether the profile considered is consistent with the perception of the participants during the simulation. In this study, values related to the subjective assessment of workload that were obtained via the ISA method were used.

5. Conclusions and Future Work

The study of human factors in aviation integrates the human component with the advanced technology used in air traffic management.

Within this discipline, the use of real-time simulations presents many advantages, including the ability to design exercises with known ATC events and the ability to study their influence on the response of air traffic controllers.

This line of research aims to establish capacity models that consider the neurophysiological variables recorded during the development of ad hoc real-time simulation exercises designed for the project.

The problem is that, in order to develop a detailed and valid study of these variables, it is necessary to compare their evolution with a valid taskload profile, which represents the actual difficulty faced by the participant during performance of the exercises.

The results of this study show that the two initial hypotheses were correct. In this experiment, a methodology has been established to define an event-based taskload profile that is suitable for studying the evolution of neurophysiological variables. The future objective is to extrapolate the study to the real operation of a control unit.

It has been demonstrated that, in the case of the data analysed in this study, the subjective workload values obtained after implementing the ISA method on the platform have been a good intermediate tool for assessing the suitability of considering the different taskload profiles as a reference. In this paper, the methodology that was followed to obtain this reference taskload has been presented, as well as the main conclusions obtained after comparing the actual taskload profile obtained to the subjective workload values assessed by the participants.

From analysis of the designed taskload profile and the subjective workload values, it has been shown that the designed taskload profile of the exercises is not the best baseline for studying the combinations of events that generate the greatest difficulty. To solve this problem, a methodology has been defined to obtain the actual event-based taskload profile.

This reference will be used in later research investigating the behaviour of neurophysiological variables related to brain activity and eye-tracking.

When comparing the actual profile to subjective workload assessments, it is possible to explain values that previously seemed unusual. By being aware of the events occurring in each minute of the simulation, it is possible to draw a number of general conclusions about the difficulty of the events perceived by the participants.

Some of these conclusions are in line with what might be expected, based on previous work:

- The events with the greatest associated difficulty are conflicts and their resolutions.
- Difficulty increases as the number of simultaneous conflicts in the sector increases.
- As the number of aircraft in the sector increases, certain events initially considered simple are perceived as more difficult due to the different aircraft that need to be monitored.

However, the results of the study have also revealed some interesting trends:

- At first, it was suspected that the factor of greatest difficulty was the number of simultaneous conflicts. However, it has been shown that a factor that causes the difficulty to increase greatly is the conflicts that are not resolved at the first attempt and which require a new resolution strategy, i.e., those situations in which the participant tries a resolution strategy, it does not work, and it is thus necessary to change the strategy.
- Those situations perceived as more difficult are not necessarily those where two conflicts occur in parallel, but those where the resolution of a conflict is prolonged over time, especially when the participant has tried different forms of resolution that are not effective and that worsen the situation of the first conflict detected.
- In relation to the above, it was found that events that were initially assigned a low taskload value were perceived as more difficult if they occurred while one or more conflicts were present in the sector and when the participant had to deal with them immediately after the resolution of a conflict.
- The general tendency in the assessment of workload in Exercises 2, 3, and 4 is to assess only the lowest value before the occurrence of the first conflict. Once the participants have to resolve the first conflict, even if the situation in the sector is under control and no additional events occur, it is very rare that the value chosen in the ISA scale is '1'.

The results obtained meet the objectives defined in the study and have allowed the establishment of a methodology that can be used to obtain the actual taskload profile of each participant, which can subsequently be used to compare the evolution of neurophysiological variables. On the basis of these findings, the following future work is defined:

- Once the methodology has been shown to work and is of interest, it will be necessary to extend the process to a larger number of participants.
- Taking the event-based taskload as a reference, the evolution of neurophysiological variables will be related to the ATC events recorded and the relationship between these variables and the traffic conditions in the sector that are established.

Two main limitations of the work presented in this paper have been identified:

- Firstly, the small number of participants included in the study.
- Secondly, the participants were ATCO students and therefore did not have the experience of real controllers. The results obtained could vary when repeating the study with ATCOs in service.

Future work will be organised in order to address these two limitations. The first set of participants was reduced so that the methodology could be validated before extending the study to a larger group. As demonstrated in this work, the methodology has been validated. Therefore, a simulation campaign has already been carried out that involves a larger number of participants. Work is currently in progress to review videos of the simulations and to determine the taskload profiles of what happened in the simulations according to the methodology described in this paper.

In this first stage of the process, ATCO students have been selected as participants with the aim of investing as much time as possible in the simulator. Through these tests with students, the idea was to obtain a methodology and relationships between ATC events and neurophysiological variables, as well as to be able to carry out a multitude of tests with the data obtained. Based on all the experience accumulated with the ATCO students' simulations, the data collection and analysis process will already be optimised. Once the results obtained are sufficiently robust, the next step is to replicate the experiment with ATCOs in service or participants with operational experience. This phase of the study will be developed in close collaboration with CRIDA.

Author Contributions: Conceptualization, M.Z.S. and R.M.A.V.; methodology, M.Z.S. and V.F.G.C.; software, M.Z.S., F.P.M. and R.D.-A.J.; validation, R.M.A.V., F.P.M., P.M.L.d.F. and V.F.G.C.; data curation, M.Z.S. and P.M.L.d.F.; writing—original draft preparation, M.Z.S.; writing—review and editing, R.M.A.V., P.M.L.d.F. and V.F.G.C.; project administration, R.M.A.V. and P.M.L.d.F.; funding acquisition, P.M.L.d.F. All authors have read and agreed to the published version of the manuscript.

Funding: This research received no external funding.

Institutional Review Board Statement: Not applicable.

Informed Consent Statement: Informed consent was obtained from all subjects involved in the study.

Data Availability Statement: Data available on request.

Acknowledgments: The development of this paper is framed within the CRITERIA project (atC event-dRiven capacITy modEls foR aIr nAvigation). This project is framed within a research, development (R&D), and innovation partnership between CRIDA and Universidad Politécnica de Madrid. The authors would like to thank CRIDA for their support in the development of this research line.

Conflicts of Interest: The authors declare no conflict of interest.

References

1. Yazgan, E.; Sert, E.; Şimşek, D. Overview of Studies on the Cognitive Workload of the Air Traffic Controller. *Int. J. Aviat. Sci. Technol.* **2021**, *2*, 28–36. [CrossRef]
2. Isufaj, R.; Koca, T.; Piera, M. Spatiotemporal Graph Indicators for Air Traffic Complexity Analysis. *Aerospace* **2021**, *8*, 364. [CrossRef]
3. Juričić, B.; Antulov-Fantulin, B.; Rogošić, T. Project ATCOSIMA—Air Traffic Control Simulations at the Faculty of Transport and Traffic Sciences. *Eng. Power Bull. Croat. Acad. Eng.* **2020**, *15*, 2–9.
4. De Rooji, G.; Borst, C.; van Paassen, M.M.; Mulder, M. Flight Allocation in Shared Human Automation En-Route Air Traffic Control. In Proceedings of the 21st International Symposium on Aviation Psychology, Corvallis, OR, USA, 1 May 2021; pp. 172–177.
5. Bastas, A.; Vouros, G. Data-Driven Prediction of Air Traffic Controllers Reactions to Resolving Conflicts. *Inf. Sci.* **2022**, *613*, 763–785. [CrossRef]
6. Çetek, C.; Aybek, F.; Çinar, E.; Cavcar, A. New Directions for Air Traffic Control Simulators: A Discussion to Guide the Selection and Renovation of Simulators. *Aeronaut. J.* **2013**, *117*, 415–426. [CrossRef]
7. Simulation. SkySim | Skysoft ATM Solutions. Available online: https://www.skysoft-atm.com/air-traffic-management/simulation/ (accessed on 22 December 2022).
8. SkySim Datasheet. Available online: http://783910.web05.swisscenter.com/wp-content/uploads/datasheet_skysim.pdf\T1\textbar{} (accessed on 22 December 2022).
9. Zamarreño Suárez, M.; Arnaldo Valdés, R.M.; Pérez Moreno, F.; Delgado-Aguilera Jurado, R.; López de Frutos, P.M.; Gómez Comendador, V.F. How Much Workload Is Workload? A Human Neurophysiological and Affective Cognitive Performance Measurement Methodology for ATCOs. *Aircr. Eng. Aerosp. Technol.* **2022**, *94*, 1525–1536. [CrossRef]
10. Jazzar, A.; Alharasees, O.; Kale, U. Assessment of Aviation Operators' Efficacy in Highly Automated Systems. *Aircr. Eng. Aerosp. Technol* ahead-of-print. **2022**, *95*, 302–311. [CrossRef]
11. Abdul, S.M.B.; Borst, C.; Mulder, M.; van Paassen, M.M. Measuring Sector Complexity: Solution Space-Based Method. In *Advances in Air Navigation Services*; Magister, T., Ed.; InTech: London, UK, 2012; ISBN 978-953-51-0686-9.
12. Tao, D.; Tan, H.; Wang, H.; Zhang, X.; Qu, X.; Zhang, T. A Systematic Review of Physiological Measures of Mental Workload. *Int. J. Environ. Res. Public. Health* **2019**, *16*, 2716. [CrossRef]
13. Lean, Y.; Shan, F. Brief Review on Physiological and Biochemical Evaluations of Human Mental Workload. *Hum. Factors Ergon. Manuf. Serv. Ind.* **2012**, *22*, 177–187. [CrossRef]

14. Hancock, P.A.; Desmond, P.A. (Eds.) *Stress, Workload, and Fatigue*, 1st ed.; CRC Press: Boca Raton, FL, USA, 2000; ISBN 978-1-4106-0044-8.
15. Manning, C.A.; Mills, S.H.; Fox, C.M.; Pfleiderer, E.M.; Mogilka, H.J. *Using Air Traffic Control Taskload Measures and Communication Events to Predict Subjective Workload*; Office of Aerospace Medicine: Washington, DC, USA, 2002.
16. Rahman, S.M.B.A.; Borst, C.; van Paassen, M.M.; Mulder, M. Cross-Sector Transferability of Metrics for Air Traffic Controller Workload. *IFAC-Pap.* **2016**, *49*, 313–318. [CrossRef]
17. Somers, V.L.J.; Borst, C.; Mulder, M.; van Paassen, M.M. Evaluation of a 3D Solution Space-Based ATC Workload Metric. *IFAC-Pap.* **2019**, *52*, 151–156. [CrossRef]
18. Instantaneous Self Assessment of Workload (ISA) | HP Repository. Available online: https://ext.eurocontrol.int/ehp/?q=node/1585 (accessed on 22 December 2022).
19. Marinescu, A.; Sharples, S.; Ritchie, A.; Lopez, T.; McDowell, M.; Morvan, H. Physiological Parameter Response to Variation of Mental Workload. *Hum. Factors* **2018**, *60*, 31–56. [CrossRef]
20. SESAR JU 2017. "D3.2 competence and training requirements". In *AUTOPACE Project. H2020-SESAR2015-1.*
21. De Frutos, P.L.; Parla, E.P.; Ballestín, L.; Cañas, J.J.; Ferreira, P.; Comendador, F.G.; Lucchi, F. Quantitative Prediction of Automation Effects on ATCo Human Performance. In Proceedings of the 8th International Conference on Research in Air Transportation, Castelldefels, Spain, 25–29 June 2018.
22. Djokic, J.; Lorenz, B.; Fricke, H. Air Traffic Control Complexity as Workload Driver. *Transp. Res. Part C Emerg. Technol.* **2010**, *18*, 930–936. [CrossRef]
23. Triyanti, V.; Azis, H.A.; Iridiastadi, H. Yassierli Workload and Fatigue Assessment on Air Traffic Controller. *IOP Conf. Ser. Mater. Sci. Eng.* **2020**, *847*, 012087. [CrossRef]
24. De Reuck, S.; Donald, F.; Siemers, I. Factors Associated with Safety Events in Air Traffic Control. *Ergon. SA* **2014**, *26*, 18.
25. Wang, Y.; Hu, R.; Lin, S.; Schultz, M.; Delahaye, D. The Impact of Automation on Air Traffic Controller's Behaviors. *Aerospace* **2021**, *8*, 260. [CrossRef]
26. Edwards, T.; Martin, L.; Bienert, N.; Mercer, J. The Relationship Between Workload and Performance in Air Traffic Control: Exploring the Influence of Levels of Automation and Variation in Task Demand. In Proceedings of the Human Mental Workload: Models and Applications; Longo, L., Leva, M.C., Eds.; Springer International Publishing: Cham, Switzerland, 2017; pp. 120–139. [CrossRef]
27. Vallesi, A. The Case of Air Traffic Control. In *Theory-Driven Approaches to Cognitive Enhancement*; Colzato, L.S., Ed.; Springer International Publishing: Cham, Switzerland, 2017; pp. 293–303. ISBN 978-3-319-57505-6.
28. Jakšić, Z.; Janić, M. Modeling Resilience of the ATC (Air Traffic Control) Sectors. *J. Air Transp. Manag.* **2020**, *89*, 101891. [CrossRef]
29. Fürstenau, N.; Radüntz, T. Power Law Model for Subjective Mental Workload and Validation through Air Traffic Control Human-in-the-Loop Simulation. *Cogn. Technol. Work.* **2022**, *24*, 291–315. [CrossRef]
30. Nixon, J.; Charles, R. Understanding the Human Performance Envelope Using Electrophysiological Measures from Wearable Technology. *Cogn. Technol. Work* **2017**, *19*, 655–666. [CrossRef]
31. Pham, D.-T.; Alam, S.; Duong, V. An Air Traffic Controller Action Extraction-Prediction Model Using Machine Learning Approach. *Complexity* **2020**, *2020*, 1659103. [CrossRef]
32. Trapsilawati, F.; Liu, Y.; Wee, H.J.; Subramaniam, H.; Sourina, O.; Pushparaj, K.; Sembian, S.; Chun, P.; Lu, Q.; Chen, C.-H.; et al. Perceived and Physiological Mental Workload and Emotion Assessments in En-Route ATC Environment: A Case Study. *Transdiscipl. Eng. A Paradig. Shift* **2017**, *5*, 420–427. [CrossRef]
33. Socha, V.; Hanáková, L.; Valenta, V.; Socha, L.; Ábela, R.; Kušmírek, S.; Pilmannová, T.; Tecl, J. Workload Assessment of Air Traffic Controllers. *Transp. Res. Procedia* **2020**, *51*, 243–251. [CrossRef]
34. Rahman, S.; Sidik, M.; Shukri, M.; Nazarudin, M. IOP Controller Response Behaviour during Procedural Control with Surveillance Information. *IOP Conf. Ser. Mater. Sci. Eng.* **2018**, *405*, 012004. [CrossRef]

 *aerospace* 

Article

Comparative Analysis of Flame Propagation and Flammability Limits of CH_4/H_2/Air Mixture with or without Nanosecond Plasma Discharges

Ghazanfar Mehdi , Maria Grazia De Giorgi *, Sara Bonuso , Zubair Ali Shah, Giacomo Cinieri and Antonio Ficarella

Department of Engineering for Innovation, University of Salento, Via per Monteroni, 73100 Lecce, Italy
* Correspondence: mariagrazia.degiorgi@unisalento.it

Abstract: This study investigates the kinetic modeling of CH_4/H_2/Air mixture with nanosecond pulse discharge (NSPD) by varying H_2/CH_4 ratios from 0 to 20% at ambient pressure and temperature. A validated version of the plasma and chemical kinetic mechanisms was used. Two numerical tools, ZDPlasKin and CHEMKIN, were combined to analyze the thermal and kinetic effects of NSPD on flame speed enhancement. The addition of H_2 and plasma excitation increased flame speed. The highest improvement (35%) was seen with 20% H_2 and 1.2 mJ plasma energy input at $\phi = 1$. Without plasma discharge, a 20% H_2 blend only improved flame speed by 14% compared to 100% CH4. The study found that lean conditions at low flame temperature resulted in significant improvement in flame speed. With 20% H_2 and NSPD, flame speed reached 37 cm/s at flame temperature of 2040 K at $\phi = 0.8$. Similar results were observed with 0% and 5% H_2 and a flame temperature of 2200 K at $\phi = 1$. Lowering the flame temperature reduced NO_x emissions. Combining 20% H_2 and NSPD also increased the flammability limit to $\phi = 0.35$ at a flame temperature of 1350 K, allowing for self-sustained combustion even at low temperatures.

Keywords: flame propagation; nanosecond plasma discharges; lean burning; ZDPlaskin; CHEMKIN

Citation: Mehdi, G.; De Giorgi, M.G.; Bonuso, S.; Shah, Z.A.; Cinieri, G.; Ficarella, A. Comparative Analysis of Flame Propagation and Flammability Limits of CH_4/H_2/Air Mixture with or without Nanosecond Plasma Discharges. *Aerospace* **2023**, *10*, 224. https://doi.org/10.3390/aerospace10030224

Academic Editors: Spiros Pantelakis, Andreas Strohmayer and Jordi Pons-Prats

Received: 31 January 2023
Revised: 19 February 2023
Accepted: 23 February 2023
Published: 25 February 2023

1. Introduction

Combustion is a crucial factor in air transportation due to the high energy density of liquid fuels. However, the low efficiency of current aeroengines and the production of harmful emissions contributing to climate change are pressing issues. To comply with strict emission regulations set by CAEP (Committee on Aviation Environmental Protection) and improve fuel efficiency, various international organizations are exploring the concept of lean combustors.

Lean fuel burning is an effective solution for reducing NO_x emissions by lowering flame temperature. However, these low temperature flames are prone to critical instabilities that can lead to re-ignition and flame blowout issues [1,2]. To address issues with methane combustion, the addition of a more reactive and cleaner fuel such as hydrogen could be a practical solution [3]. Blending methane with hydrogen has been shown to enhance performance and reduce emissions without modifying existing combustors [4]. Hydrogen is a carbon-free fuel with low ignition energy, a wide flammability range, fast flame propagation, and high reactivity [3]. Several studies in the past [3–6] have focused on the impact of hydrogen on the flame speed of CH_4/H_2 mixtures. Halter et al. [5] studied the effect of hydrogen content and inlet pressure on the laminar flame speed of CH_4/H_2 flames, with results indicating that the laminar flame speed improved with increasing hydrogen content and decreased with increasing inlet pressure.

Mandilas et al. [6] studied the impact of hydrogen on iso-octane-air and methane mixtures in both laminar and turbulent conditions. They found that using hydrogen led to earlier flame instabilities but improved laminar flame speed at lean limits in turbulent combustion. Adding hydrogen to methane slightly improved reactivity at lean conditions, but also increased complexities, safety issues, and thermoacoustic instabilities [3–6]. Flame speed was slightly better at lean compared to rich conditions [4]. Non-thermal plasma combustion can improve flame stability, flame speed, and lean blowout limits. NTP enhances combustion through kinetic, thermal and momentum effects [7]. NTP improves combustion through three mechanisms: kinetic (creation of active particles from fuel decomposition), thermal (increased fuel/air mixture temperature), and momentum (ionic wind and flow motion from electro-hydrodynamic forces) [8].

Among NTP technologies, nanosecond plasma discharge (NSPD) has gained attention due to its ability to effectively produce excited states and active particles [9,10]. NSPD also rapidly heats the gas, which accelerates combustion [11,12]. Despite numerous research studies on NTP combustion [1], commercialization is only possible with the development of accurate numerical models for plasma chemistry in combustion. Our group [12–15] has studied CH4/air mixtures with NSPD for flame propagation and ignition enhancement. We compared ignition delay, flame speed, and flammability limits under different conditions and found that NSPD improved ignition, flame propagation, and flammability limits due to the production of neutral radicals and increased mixture temperature. The improvements were primarily due to NSPD's kinetic effects. Prior studies on NSPD have individually considered H2/air and CH4/air mixtures.

While initial studies have explored the kinetics of NSPD, a comprehensive understanding of plasma mechanisms for CH_4/H_2/air mixtures is still lacking. It has been shown that the evolution of active particles over time provides the most accurate analysis of plasma kinetics [16,17].

This paper presents a study of CH_4/H_2/air with nanosecond plasma discharge. There is currently no numerical study available on methane blended hydrogen plasma-assisted combustion. Both plasma and combustion kinetics were analyzed using validated mechanisms and compared to previously published experimental data. The impact of NSPD and hydrogen content on flame propagation and flammability limits in methane/air mixtures was studied. A comparative analysis of flame speed enhancement with and without plasma actuation was performed using different methane blended hydrogen ratios.

2. Numerical Procedure and Kinetic Modelling

2.1. Numerical Procedure

Numerical analyses were conducted using two solvers: ZDPlasKin (0D Plasma kinetic solver) [18] and CHEMKIN (Chemical kinetic solver) [19]. The methodology is shown in Figure 1 and explained in [13]. ZDPlasKin was used to analyze the kinetic and thermal effects of NSPD in CH_4/H_2/Air mixture. BOLSIG+ was linked to ZDPlasKin to predict the temporal evolution of excitation states and the reactions producing free radicals/active particles. It has been assumed that the non-equilibrium plasma created from a CH_4/H_2/air mixture at atmospheric pressure is uniformly distributed, which is a similar assumption to what was previously executed in [13]. Although the nanosecond pulsed plasma combustion process is three-dimensional and not homogeneous, we used a simplified homogeneous model. To investigate the effects of plasma CH_4/H_2/air products on flame speed and flammability limits, we used the plasma products of CH_4/H_2/air as the inlet domain of the reactor.

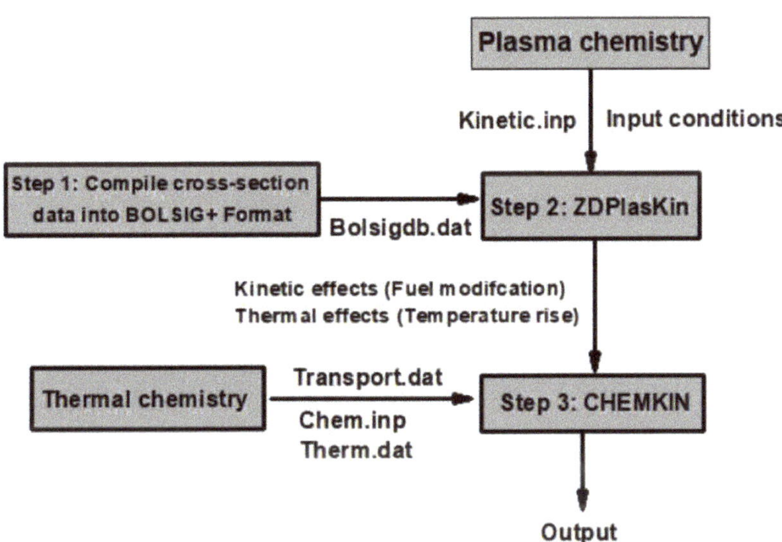

Figure 1. Flowchart for numerical analysis.

ZDPlasKin boundary conditions were set as ambient temperature and pressure, fixed EN and electron number density, and initial CH_4/H_2/Air composition. The simplified homogeneous model was used as in [20]. ZDPlasKin simulation was performed using the integral mean value of E_N obtained from experiments, about 200 Td over 10^{-6} s, as shown in Figure 2. Experimental setup and E_N estimation are described in [13]. The gas temperature was predicted using equations from [18]. The adiabatic gas temperature was calculated from the energy conservation equation and reallocation of electrical power P_{ext} to electron translational degree P_{elec}, gas internal degree P_{chem}, and gas P_{gas}:

$$Pext = Pgas + Pelec + Pchem \tag{1}$$

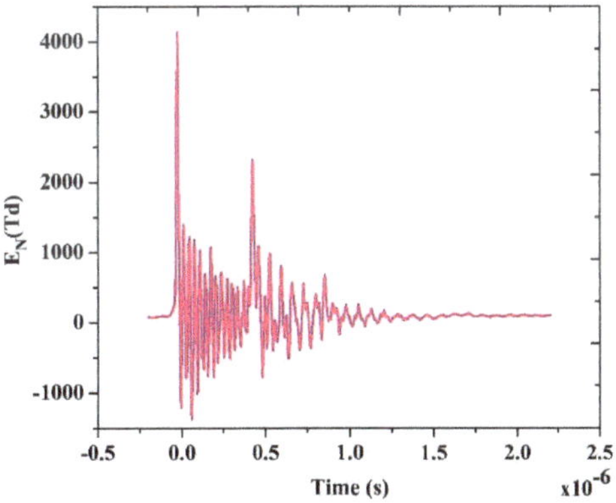

Figure 2. Experimental E_N value used for numerical analysis [13].

The above equation can be described below.

$$Pext = e[Ne]veE \tag{2}$$

$$Pgas = \frac{1}{\gamma - 1} + \frac{d(NTgas)}{dt} \tag{3}$$

$$Pelec = \frac{3}{2} + \frac{d([Ne]Te)}{dt} \tag{4}$$

$$Pchem = \sum_{i}^{n} Qi + \frac{d[Ni]}{dt} \tag{5}$$

where E is the reduced field, e is the elementary charge, T_e is the electron temperature, ve is the drift velocity of electrons $[N_e]$ is the electron density, N is the total gas density, $\gamma = 1.2$ is the specific gas heat ratio and Q_i is the potential energy of species i.

The results gained from the ZDPlaskin solver in terms of neutral and excited species (at a time of 0.5 ms because the residence time is too short that autoignition chemistry does not significantly influence the reactants compositions), and the gas temperature of the activated region were introduced into the CHEMKIN solver to investigate the combustion process.

A 1-D premixed laminar flame speed reactor was employed to analyze combustion characteristics, considering thermal diffusion and multicomponent diffusion options. The adaptive mesh parameters were set as CURV = 0.5 and GRAD = 0.05, with absolute and relative error criteria of $A_{TOL} = 1 \times 10^{-9}$ and $R_{TOL} = 1 \times 10^{-5}$, respectively. The total number of grid points used was typically 350–400. In this study, we have established that the calculation domain of the CHEMKIN reactor ranges from −2.0 cm upstream to 4.0 cm downstream with respect to the reactor and is sufficient to attain adiabatic equilibrium. Numerical analyses were performed at various fueling conditions based on H_2/CH_4 ratio (x_{H2}) with or without plasma actuation. Table 1 shows the mole fraction of CH_4, and H_2 reactants at equivalence ratio of 1.

Table 1. Reactants mole fraction of CH_4/H_2/Air flames at plasma on and off conditions.

Case No.	H_2 (%)	CH_4	H_2	O_2	N_2	Plasma (ON/OFF)
1	0	0.0950	/	0.1900	0.7149	OFF
2	5	0.0935	0.0049	0.1894	0.7122	OFF
3	10	0.0917	0.0101	0.1885	0.7094	OFF
4	20	0.0879	0.0219	0.1869	0.7031	OFF
5	0	0.0950	/	0.1900	0.7149	ON
6	5	0.0935	0.0049	0.1894	0.7122	ON
7	10	0.0917	0.0101	0.1885	0.7094	ON
8	20	0.0879	0.0219	0.1869	0.7031	ON

2.2. Plasma Kinetic Model

A comprehensive literature review was conducted to develop an extended plasma kinetic mechanism for CH_4/H_2/Air mixture. It consists of 161 species and 1382 plasma and gas-phase reactions, and includes ionization reactions, charged transfer reactions, dissociation reactions, excited species reactions, recombination reactions, relaxation reactions, and three-body recombination reactions. The mechanism also included 38 exciting species and 35 charged species. The relevant reactions were taken from [21–24]. The collision cross-sectional data were taken from the LXCat data source [13]. Further information can be found in the previously published study [13].

2.3. Combustion Kinetic Model

The NSPD generated kinetic effects (neutral radicals, active particles, excited species) and thermal effects were used to study the effect on flame speed in a CHEMKIN combustion model. The CH_4/H_2/Air combustion kinetic model was created with an expanded version of the combustion mechanism, incorporating thermodynamics and transport data. The mechanism was updated from GRI-Mech v3.08 with ozone reactions [25] and updated hydrogen combustion mechanism including the excited species O(1D), OH(2+), $O_2(a^1g)$ [26]. A sub-model of the excited species OH* and CH* has also been added [27]. Furthermore, the reaction mechanism of ions and excited species of CH_4/Air mixture was also considered [28].

3. Validation of Kinetic Models

The plasma kinetic model was validated using an experimental study in [29] by comparing the mole fraction of the decay process of O atoms. The plasma kinetic model was validated using an experimental study in [13]. The model accurately predicts the O atom mole fraction, in good agreement with experimental data.

The combustion kinetic model was validated using experiments by Coppens [30], Hermanns [31], the Konnov mechanism [32], and the San Diego mechanism [33]. The combustion kinetic mechanism was tested with H_2 and CH_4 fractions equal $x_{H2} = 0.05$ and $x_{CH4} = 0.95CH_4$ at different equivalence ratios. Figure 3 shows the validation results for burning velocities of CH_4/H_2/Air mixture with 5%, 30%, and 40%. H_2 content. The model shows good agreement with experimental data compared to other mechanisms, with slightly lower values at rich burning conditions as seen in [30,31].

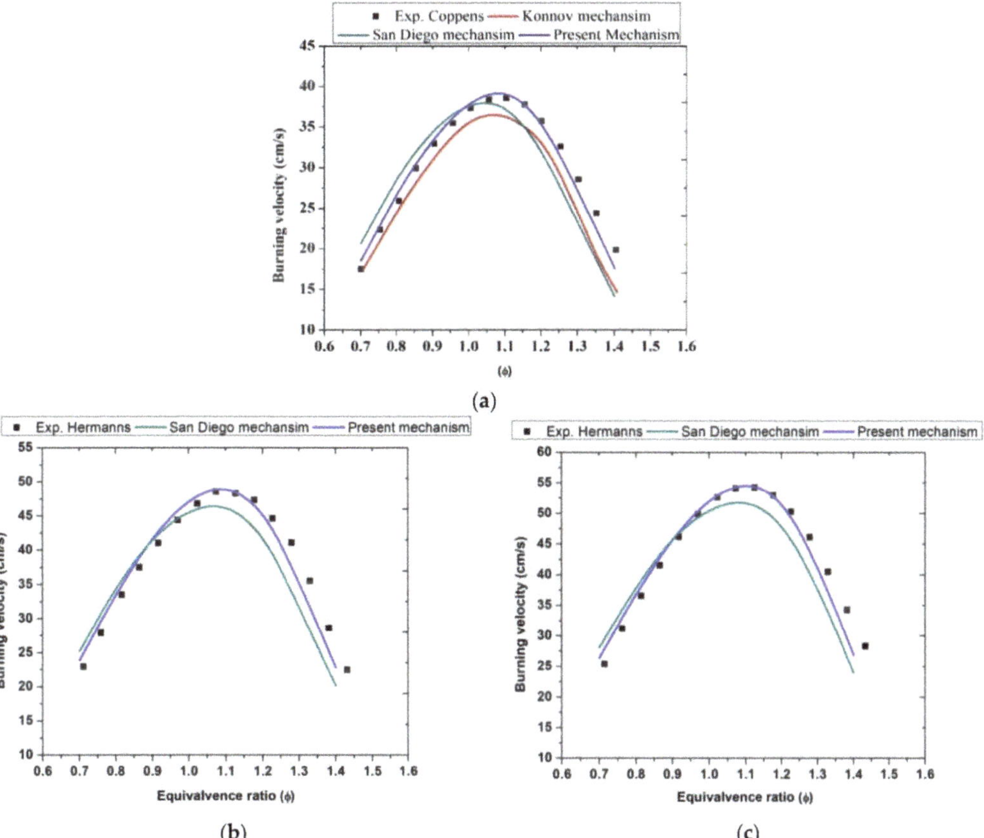

Figure 3. Validation: predicted values of burning velocities of CH_4/H_2/Air mixture with a H_2 content of (**a**) 5%, (**b**) 30% and (**c**) 40%.

4. Results and Discussions

The present analysis was conducted under fixed plasma conditions: $E_N = 200$ Td, repetition frequency equal to 1000 Hz, and electron number density equal to 10^7 cm^{-3}. The effect of varying H_2 contents (x_{H2} from 0 to 0.2) on active particle production was studied in methane/air. Figure 4 shows the temporal evolution of active particles (H, OH, CH, and CH_3) as predicted by ZDPlaskin simulations under fixed plasma actuation conditions, only changing the H_2 content in the methane/air mixture. An increase in H_2 concentration led to a significant improvement in the mole fraction of active species. The improvement in active particles production was linearly proportional to the rise in H_2 content, with the maximum concentration observed at 20% H_2. The rapid decomposition of H_2^+ into H, (E + H_2^+ => H + H) due to its simple molecular structure and high reactivity and the subsequent reactions with other intermediate species, led to increased concentration of active particles. The maximum mole fraction of H was 0.00704 (Figure 4a), which was almost twice the OH species equal to 0.0038 (Figure 4b). The maximum mole fraction of CH_3 was 0.006 (Figure 4c), slightly less than H but two orders higher than the molar fraction of CH (0.000041, Figure 4d).

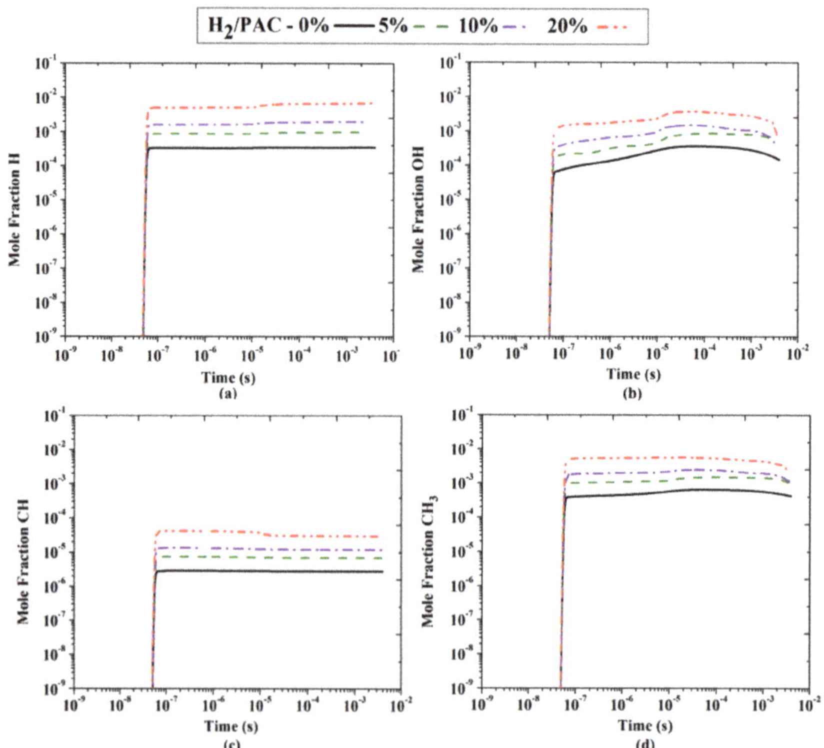

Figure 4. Temporal evolution of (**a**) H, (**b**) OH, (**c**) CH, (**d**) CH_3 concentrations at different contents of H_2 in methane/air mixture using fixed plasma actuation conditions.

The H atoms were produced during the decomposition process when electrons reacted with the ions of CH^+, CH_2^+, CH_3^+, CH_4^+, and H_3O^+. The primary reactions contributing to the production of H, CH, and CH_3 were E + CH_4^+ = CH_3 + H and E + CH_3^+ = CH + H + H (reaction rates: 10^{20} and 10^{21} cm^3 s^{-1}, respectively). The OH radicals were produced through the reaction O(1D) + CH_4 = CH_3 + OH (reaction rate: 10^{24} cm^3 s^{-1}).

As shown in Figure 5, the mole fraction of active species was significantly improved with an increase in H_2 content in the methane/air mixture. The highest mole fraction of O atoms was observed at a 20% H_2 content (x_{H2} = 0.2) with a value of 0.0158 (Figure 5b). This was due to the decomposition of excited O_2 species when reacting with H atoms and H_2O molecules, which increased in concentration due to the presence of H2 molecules in the methane/air mixture. The dominant reaction path was H + $O_2(V_4)$ => O + OH (reaction rate: 1023 cm^3/s). The O atoms produced began to reduce after 10^{-4} s, likely due to the short reactive time of O atoms leading to their consumption during recombination and intermediate reactions. Similarly, ozone concentration improved as shown in Figure 5c, with a mole fraction of 0.00729, close to that of H atoms (0.00704) at a 20% H_2 content.

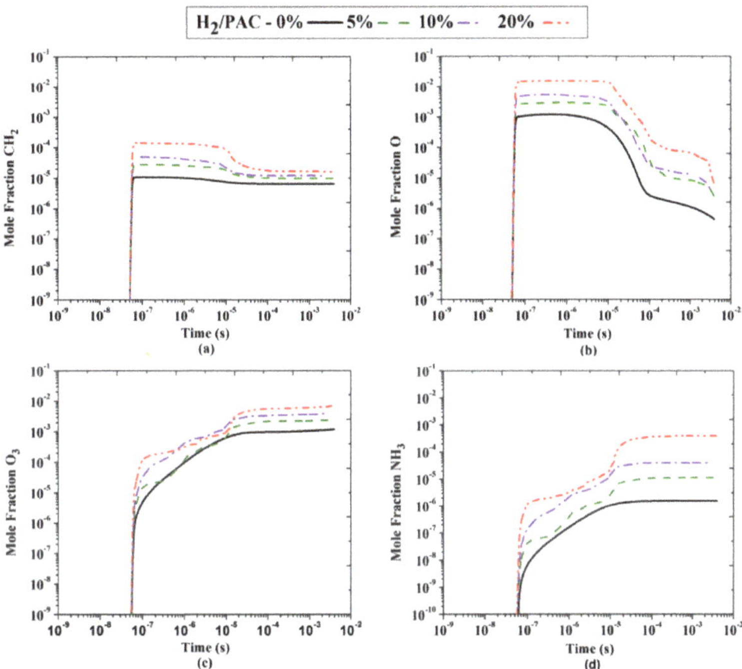

Figure 5. Temporal evolution of (**a**) CH_2, (**b**) O, (**c**) O_3, (**d**) NH_3 concentrations at different contents of H_2 in methane/air mixture using fixed plasma actuation conditions.

Ozone is primarily generated through the reaction of O atoms with molecular oxygen or with excited species of oxygen. The most significant reaction is $O + O_2 + N_2 = O_3 + N_2$, as it has a reaction rate of about 1023 cm^3/s. Nitrogen acts as a third body, removing excess energy. Ozone can also improve combustion analysis as it increases flame speed [25]. The temporal evolution of ammonia was also found to improve when nitrogen seed particle was added to methane-blended hydrogen.

The kinetic and thermal effects predicted by ZDPlasKin were introduced into Chemkin to investigate the flame speed and maximum flame temperature. The reactant mole fraction of active particles and excited species predicted by ZDPlasKin were added to Chemkin to account for kinetic effects. This was executed with a 0.5 ms residence time, which is too short to affect the autoignition chemistry and the reactant composition. The flame speed and peak flame temperature were investigated using a pre-mixed laminar flame speed reactor at different methane-blended hydrogen mixture compositions with or without NSPD.

Figure 6a showed that flame speed improved with increasing hydrogen content and plasma excitation. At stoichiometric mixture, adding hydrogen ($x_{H2} = 0.2$) to the methane/air mixture resulted in a 14% increase in flame speed (Δs_L). A further improvement of 35% was achieved with plasma discharge. At leaner condition ($\phi = 0.6$) and same H_2 fraction, Δs_L was 16.7% without and 52% with plasma actuation. However, the same flame speed was observed for both cases of $x_{H2} = 0.2$ and $x_{H2} = 0.05$ with PAC at lean and stoichiometric conditions, similarly in case of $x_{H2} = 0.1$ and $x_{H2} = 0$ with PAC. It means the same range of flame speed could be reached by varying both hydrogen fraction and plasma discharge.

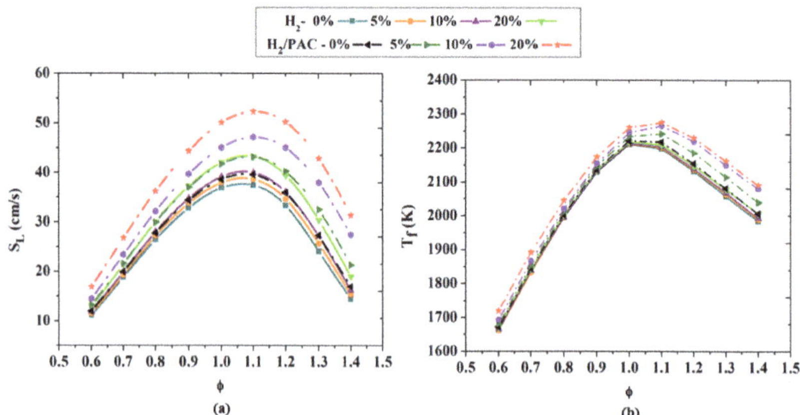

Figure 6. Comparison of (**a**) flame speed and (**b**) flame temperature at various equivalence ratios using different H_2 contents with or without NSPD.

Figure 6b shows the predicted peak flame temperature T_f with or without plasma for various H_2 fractions. The results showed that increasing H_2 did not affect T_f without plasma discharges. However, with plasma, T_f was affected by H_2 at fixed operating conditions, especially for the rich mixture ($\Phi > 1$). A slightly increase in T_f was found at lean and stoichiometric conditions.

The study found that lean conditions at low flame temperature resulted in significant improvement in flame speed. With 20% H_2 and NSPD, flame speed reached 37 cm/s at flame temperature of 2040 K at $\phi = 0.8$. Similar results were observed with 0% and 5% H_2 and a flame temperature of 2200 K at $\phi = 1$. It was observed that the same flame speed can be achieved at lean conditions by reducing T_f, leading to reduced NO_x emissions.

Figure 7 compares the improvement in flame speed (%) at lean, stochiometric, and rich conditions with $x_{H2} = 0.2$ with or without NSPD. It was observed that the improvement trend was $\phi = 1.4 > \phi = 0.6 > \phi = 1$. At lean conditions ($\phi = 0.6$), adding $x_{H2} = 0.2$ improved flame speed by 15%, however, using both $x_{H2} = 0.2$ and plasma resulted in a more than 50% improvement. At rich conditions, the largest improvement was seen with plasma due to the increased fuel causing more active particles to be produced during NSPD".

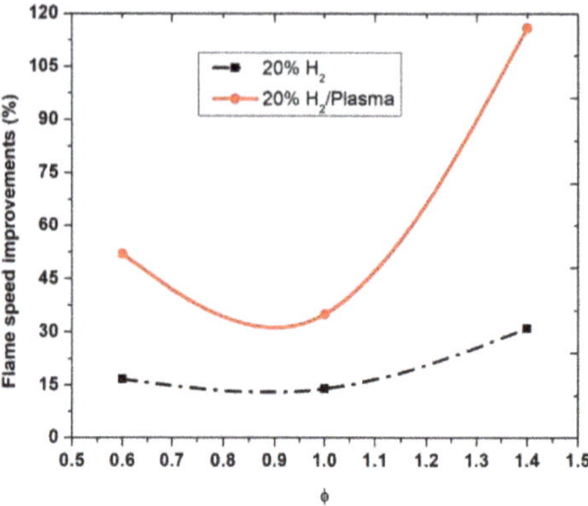

Figure 7. Comparative behavior of flame speed improvements (%) at lean, stoichiometric, and rich conditions.

Literature [26] showed that the molecular excited species oxygen $O_2(1\Delta)$ increased burning velocity by 1% without plasma. The reaction path $H_2 + O_2(1\Delta) = H + HO_2$ was found to play a significant role. More than 5% of O_2 was converted to $O_2(1\Delta)$ in the presence of electric discharge at ambient pressure [34]. Thus, plasma discharge could produce significant amounts of $O_2(1\Delta)$. Figure 8 analyzed the role of $O_2(1\Delta)$ using hydrogen blends with or without plasma discharge. The results showed no change in $O_2(1\Delta)$ production with hydrogen blends alone at various equivalence ratios. However, with the use of plasma discharge, there was a significant rise in excited species production, especially at higher hydrogen content.

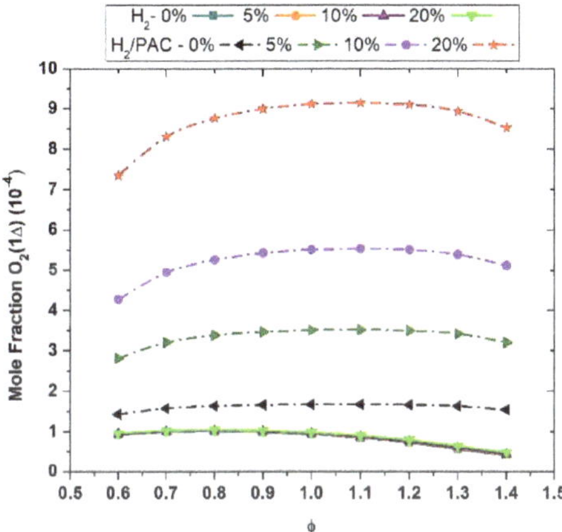

Figure 8. Comparison of molecular excited species $O_2(1\Delta)$ at various equivalence ratios using different H_2 contents with or without NSPD.

Impact of atomic excited species O(1D) on flame speed was studied using plasma and hydrogen blends (Figure 9). Results showed low O(1D) concentration increased with hydrogen and plasma, but still had minimal effect on combustion. However, it could be increased with the increase in plasma amplitude.

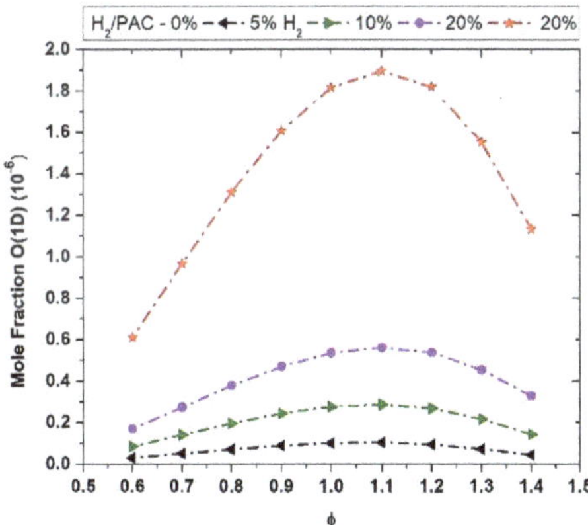

Figure 9. Comparison of atomic excited species O(1D) at various equivalence ratios using different H_2 contents with NSPD.

Free radicals such as O, H, and OH are active due to unpaired electrons and short lived in combustion [35]. They initiate chain reactions and branching. Figure 10a–d show mole fraction profiles of O, H, OH, and CH_3 using hydrogen blends with/without plasma discharge. Adding hydrogen increased O, H, and OH mole fractions, but decreased CH_3

slightly (Figure 10d). Using NSPD in hydrogen blends raised O, H, and OH concentrations and moved the reaction region upstream. CH_3 mole fraction was also slightly increased with plasma discharge. OH particles had the highest concentration at 0.009 mole fraction. Main reactions producing O, H, and OH particles are described as follows in Equations (6) and (7).

$$OH + H_2 = H + H_2O \tag{6}$$

$$H + O_2 = O + OH \tag{7}$$

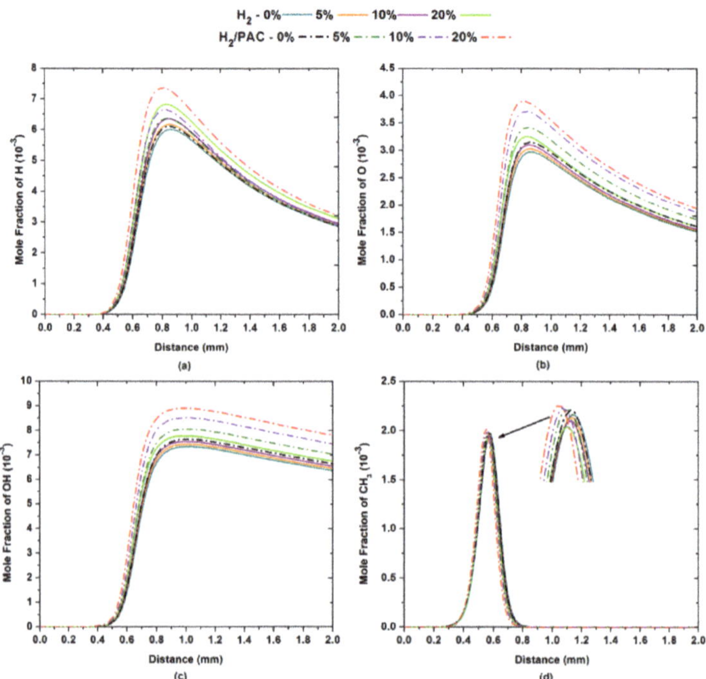

Figure 10. Mole fraction profiles of (**a**) H, (**b**) O, (**c**) OH, and (**d**) CH_3 with different blends of hydrogen without or with NSPD.

Figure 11 shows the production rate of Equations (6) and (7) with $x_{H2} = 0$ and $x_{H2} = 0.2$ with/without NSPD. The rate increased and the peak shifted upstream with hydrogen addition, but with $x_{H2} = 0.2$ and plasma, a significant impact was seen.

H_2 and O_2 mole fractions change with hydrogen blends and NSPD, shown in Figure 12. H_2 transforms from intermediate species to initial reactant in methane flames with $x_{H2} \geq 0.2$ and NSPD. H_2 starts reacting upstream in $x_{H2} = 0.2$, confirmed by [36]. H_2 promotes combustion and moves the reaction region towards upstream due to its higher reactivity than CH_4.

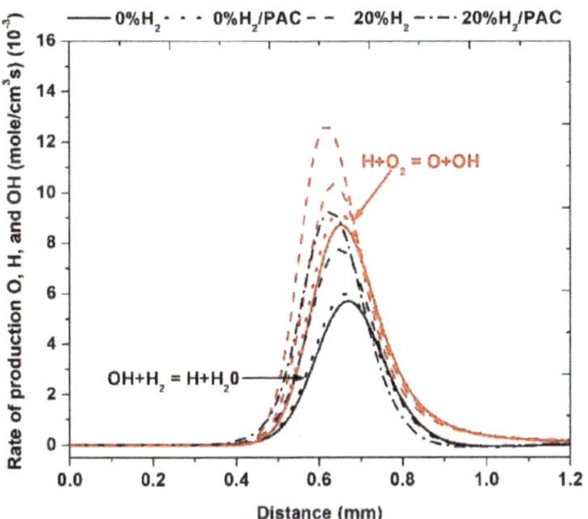

Figure 11. Rate of production of O, H, and OH with different blends of hydrogen without or with NSPD.

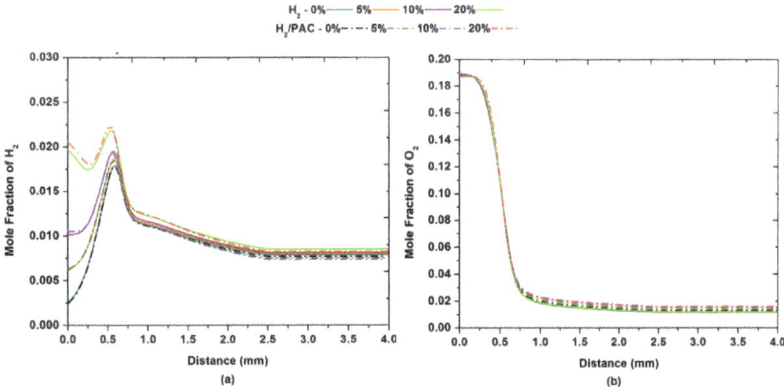

Figure 12. Mole fraction profiles of H_2 and O_2 with different blends of hydrogen without or with NSPD.

Figure 13a illustrates the mole fractions of CH_4 in different H_2 blends with or without the NSPD. The addition of H_2 and NSPD leads to a decrease in CH_4 mole fraction, possibly due to the high reactivity of H_2 and lower CH_4 concentration. The oxidation of CH_4 greatly increased and its profiles were shifted towards the upstream sides. CH_4 was mainly consumed by reactions with active particles O, H, and OH. The dominant CH_4 consumption reactions are listed below.

$$OH + CH_4 = CH_3 + H_2O \quad (8)$$

$$H + CH_4 = CH_3 + H_2 \quad (9)$$

$$O + CH_4 = OH + CH_3 \quad (10)$$

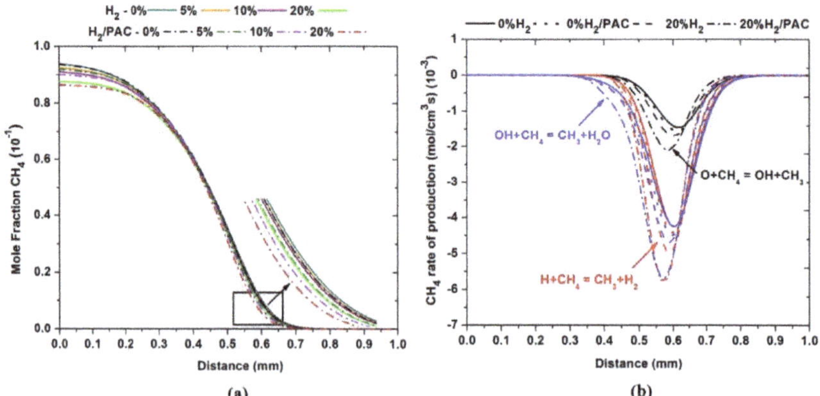

Figure 13. (**a**) Mole fraction profile of CH_4 (**b**) Rate of production of CH_4 with different blends of hydrogen without or with NSPD.

The rate of production of Equations (8)–(10) using hydrogen contents $x_{H2} = 0$ and $x_{H2} = 0.2$ with or without NSPD is shown in Figure 13b. CH_4 consumption was increased for reaction Equations (8)–(10) and the peak of the reaction region was shifted towards the upstream with the addition of hydrogen contents with or without plasma. However, when combining the H_2 blends of $x_{H2} = 0.2$ with NSPD, a noticeable impact was observed. It was because hydrogen is more reactive, which promoted methane combustion. The concentration of active particles O, H, and OH were increased when methane was blended with hydrogen, mainly due to the chemical effects. Moreover, the NSPD further improved the combustion process due to the thermal (moderate gas heating) and kinetic effects (excitation, ionization and decomposition of fuel and air molecules occurred, which resulted in the production of intermediate fuel fragments and active particles).

Finally, the lean flammability limit is discussed as the minimum equivalence ratio for flame propagation. Figure 14 shows the lean flammability limit using hydrogen contents $X_{H2} = 0$ and $X_{H2} = 0.2$ with or without NSPD. The flammability limit remained at $\phi = 0.6$ without H_2 and plasma but improved to $\phi = 0.5$ with the addition of $X_{H2} = 0.2$. Plasma discharge had a significant impact on the flammability limits, with $\phi = 0.45$ at flame temperature about 1500 K.

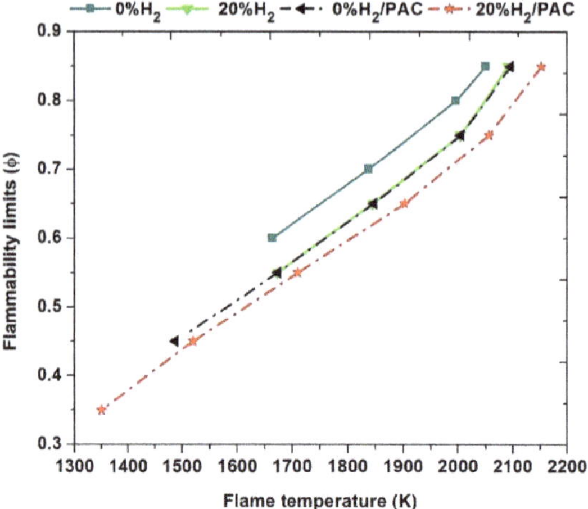

Figure 14. Flammability limits with different blends of hydrogen without or with NSPD.

Combining X_{H2} = 0.2 and NSPD increased the flammability limit to ϕ = 0.35 at 1350 K, allowing self-sustained combustion at lower flame temperatures and reduced NO_x emissions. The improved flammability limits reduce fuel consumption due to the enhanced reactivity and chemical effects of H_2 and thermal and kinetic effects of NSPD.

5. Conclusions

This paper investigated the impact of NSPD on enhancing the flame propagation of CH_4/H_2/air mixture under ambient temperature and pressure. A reduced electric field experimentally estimated was used for numerical investigation. An extended version of the plasma and combustion kinetic mechanism was applied and validated using available experimental and numerical data. ZDPlasKin was used to predict the temporal evolution of active particles and the results were integrated into CHEMKIN to enhance the flame speed. The numerical study was carried out with varying H_2 contents from 0 to 20% in methane/air with or without plasma actuation. It was noticed that with the enrichment of H_2 concentration in the methane/air mixture at fixed plasma, the mole fraction of active species was significantly improved. However, the improvements in the production of active particles were linearly increased with the increase in H_2 contents. The highest improvement in flame propagation was observed at 20% H_2/Plasma reaching 35%.

Flame speed improvement was significantly higher at lean conditions and low flame temperatures. For instance, at an equivalence ratio of 0.8, 20% H_2/Plasma resulted in a flame speed of 37 cm/s at a flame temperature of around 2040 K. This same flame speed was also observed in the case of 0% and 5% H_2 with a flame temperature close to 2200 K, meaning that high flame speed can be achieved at lean conditions and low flame temperatures, reducing NO_x emissions. Figure 7 shows the comparison of flame speed improvement at lean, stoichiometric, and rich conditions with x_{H2} = 0.2 with or without NSPD. The improvement in flame speed was higher at lean conditions (equivalence ratio of 0.6) with the addition of x_{H2} = 0.2, reaching 15%. Combining x_{H2} = 0.2 with plasma discharge significantly increased flame speed by more than 50%. Furthermore, the combination of H_2 blend (x_{H2} = 0.2) and NSPD improved the flammability limit to equivalence ratio 0.35 at a flame temperature of 1350 K, allowing for reduced fuel consumption.

Author Contributions: Conceptualization, M.G.D.G.; Methodology, M.G.D.G. and G.M.; Software, G.M.; Validation, G.M.; Formal analysis, G.M.; Investigation, G.M. and S.B.; Data curation, G.M., G.C., Z.A.S. and S.B.; Writing—original draft preparation, G.M.; Writing—review and editing, M.G.D.G.; Supervision, M.G.D.G. and A.F.; Project administration, M.G.D.G. and A.F.; Funding acquisition, G.M. All authors have read and agreed to the published version of the manuscript.

Funding: The work was supported and funded by the PON R&I 2014–2020 Asse I "Investimenti in Capitale Umano" Azione I.1 "Dottorati Innovativi con caratterizzazione industriale"—Corso di Dottorato in "Ingegneria dei Sistemi Complessi" XXXV ciclo—Università degli Studi del Salento"—Borsa Codice: DOT1312193 no. 3. This project is also received funding from the Clean Sky 2 Joint Undertaking (JU) under the grant agreement no. 831881 (CHAiRLIFT). The JU received support from the European Union's Horizon 2020 research and innovation program and the Clean Sky 2 JU members other than the Union.

Conflicts of Interest: The authors declare no conflict of interest.

References

1. Ju, Y.; Sun, W. Plasma Assisted Combustion: Dynamics and Chemistry. *Prog. Energy Combust. Sci.* **2015**, *48*, 21–83. [CrossRef]
2. Mehdi, G.; Bonuso, S.; De Giorgi, M.G. Plasma Assisted Re-Ignition of Aeroengines under High Altitude Conditions. *Aerospace* **2022**, *9*, 66. [CrossRef]

3. Jackson, G.S.; Sai, R.; Plaia, J.M.; Boggs, C.M.; Kiger, K.T. Influence of H_2 on the response of lean premixed CH_4 flames to high strained flows. *Combust. Flame* **2003**, *132*, 503–511. [CrossRef]
4. Hawkes, E.R.; Chen, J.H. Direct Numerical Simulation of Hydrogen-Enriched Lean Premixed Methane— Air Flames. *Combust. Flame* **2004**, *138*, 242–258. [CrossRef]
5. Halter, F.; Chauveau, C.; Djebaili-Chaumeix, N.; Gokalp, I. Characterization of the effects of pressure and hydrogen concentration on laminar burning velocities of methane–hydrogen–air mixtures. *Proc. Combust. Inst.* **2005**, *30*, 201–208. [CrossRef]
6. Mandilas, C.; Ormsby, M.P.; Sheppard, C.G.W.; Woolley, R. Effects of hydrogen addition on laminar and turbulent premixed methane and iso-octane–air flames. *Proc. Combust. Inst.* **2007**, *31*, 1443–1450. [CrossRef]
7. De Giorgi, M.G.; Bonuso, S.; Mehdi, G.; Shamma, M.; Harth, S.R.; Zarzalis, N.; Trimis, D. Enhancement of Blowout Limits in Lifted Swirled Flames in Methane-Air Combustor by the Use of Sinusoidally Driven Plasma Discharges. In Proceedings of the Active Flow and Combustion Control 2021: Papers Contributed to the Conference "Active Flow and Combustion Control 2021", Berlin, Germany, 28–29 September 2021; King, R., Peitsch, D., Eds.; Springer: Cham, Switzerland, 2022; pp. 66–82. [CrossRef]
8. Meng, Y.; Gu, H.; Chen, F. Influence of Plasma on the Combustion Mode in a Scramjet. *Aerospace* **2022**, *9*, 73. [CrossRef]
9. De Giorgi, M.G.; Mehdi, G.; Bonuso, S.; Shamma, M.; Harth, S.; Trimis, D.; Zarzalis, N. Characterization of Flame Behavior and Blowout Limits at Different Air Preheating Temperatures in Plasma Assisted Stabilized Combustor. In Proceedings of the Turbo Expo: Power for Land, Sea, and Air, Rotterdam, The Netherlands, 13–17 June 2022; American Society of Mechanical Engineers: New York, NY, USA, 2022; 86007, p. V03BT04A048. [CrossRef]
10. Mehdi, G.; Bonuso, S.; De Giorgi, M.G. Development of plasma actuators for re-ignition of aeroengine under high altitude conditions. In *IOP Conference Series: Materials Science and Engineering*; IOP Publishing: Bristol, UK, 2022; Volume 1226, p. 012034.
11. Ju, Y.; Lefkowitz, J.K.; Reuter, C.B.; Won, S.H.; Yang, X.; Yang, S.; Sun, W.; Jiang, Z.; Chen, Q. Plasma Assisted Low Temperature Combustion. *Plasma Chem. Plasma Process.* **2016**, *36*, 85–105. [CrossRef]
12. Mehdi, G.; Bonuso, S.; De Giorgi, M.G. Effects of Nanosecond Repetitively Pulsed Discharges Timing for Aeroengines Ignition at Low Temperature Conditions by Needle-Ring Plasma Actuator. *Energies* **2021**, *14*, 5814. [CrossRef]
13. Mehdi, G.; Fontanarosa, D.; Bonuso, S.; De Giorgi, M.G. Ignition thresholds and flame propagation of methane-air mixture: Detailed kinetic study coupled with electrical measurements of the nanosecond repetitively pulsed plasma discharges. *J. Phys. D Appl. Phys.* **2022**, *55*, 315202. [CrossRef]
14. Fontanarosa, D.; Mehdi, G.; De Giorgi, M.G.; Ficarella, A. Assessment of the impact of nanosecond plasma discharge on the combustion of methane air flames. *E3S Web Conf.* **2020**, *197*, 10001. [CrossRef]
15. Mehdi, G.; De Giorgi, M.G.; Fontanarosa, D.; Bonuso, S.; Ficarella, A. Ozone Production With Plasma Discharge: Comparisons Between Activated Air and Activated Fuel/Air Mixture. In Proceedings of the ASME Turbo Expo 2021: Turbomachinery Technical Conference and Exposition, Virtual, 7–11 June 2021; American Society of Mechanical Engineers: New York, NY, USA, 2021; Volume 3B, p. V03BT04A036. [CrossRef]
16. Starikovskiy, S.M. Topical review: Plasma-assisted ignition and combustion. *J. Phys. D Appl. Phys.* **2006**, *39*, 265–299. [CrossRef]
17. Ruma, M.; Ahasan, H.; Ranipet, H.B. A Survey of Non-thermal plasma and their generation methods. *Int. J. Renew. Energy Environ. Eng.* **2016**, *4*, 6–12.
18. Pancheshnyi, S.; Eismann, B.; Hagelaar, G.J.M.; Pitchford, L.C. ZDPlaskin Zero-Dimensional Plasma Kinetic Solver. 2008. Available online: http://www.zdplaskin.laplace.univ-tlse.fr/ (accessed on 8 September 2021).
19. Lutz, A.E.; Kee, R.J.; Miller, J.A. SENKIN: A FOR- TRAN. In *Program for Predicting Homogeneous Gas Phase Chemical Kinetics with Sensitivity Analysis*; Report No. SAND87-8248; Sandia National Laboratories: Livermore, CA, USA, 1988.
20. Aleksandrov, N.L.; Kindysheva, S.V.; Kukaev, E.N.; Starikovsjaya, S.M.; Starikovskii, A.Y. Simulation of the Ignition of a Methane-Air Mixture by a High-Voltage Nanosecond Discharge. *Plasma Phys. Rep.* **2009**, *35*, 867–882. [CrossRef]
21. Capitelli, M.; Ferreira, C.M.; Gordiets, B.F.; Osipov, A.I. *Plasma Kinetics in Atmospheric Gases*; Springer: Berlin/Heidelberg, Germany, 2000.
22. Flitti, A.; Pancheshnyi, S. Gas heating in fast pulsed discharges in N2–O2 mixtures. Eur. *Phys. J. Appl. Phys.* **2009**, *45*, 21001. [CrossRef]
23. Mao, X.; Rousso, A.; Chen, Q.; Ju, Y. Numerical modeling of ignition enhancement of CH4 /O2 /He mixtures using a hybrid repetitive nanosecond and DC discharge. *Proc. Combust. Inst.* **2019**, *37*, 5545–5552. [CrossRef]
24. Mao, X.; Chen, Q.; Guo, C. Methane pyrolysis with N_2/Ar/He diluents in a repetitively pulsed nanosecond discharge: Kinetics development for plasma assisted combustion and fuel reforming. *Energy Convers. Manag.* **2019**, *200*, 112018. [CrossRef]
25. Halter, F.; Higelin, P.; Dagaut, P. Experimental and detailed kinetic modelling study of the effect of ozone on the combustion of methane, Energy Fuels, 2011, 25 2909–16. *Energy Fuels* **2011**, *25*, 2909–2916. [CrossRef]
26. Konnov, A.A. On the role of excited species in hydrogen combustion. *Combust. Flame* **2015**, *162*, 3755–3772. [CrossRef]
27. Walsh, K.T. Quantitative Characterizations of Coflow Laminar Diffusion Flames in a Normal Gravity and Microgravity Environment. Ph.D. Thesis, Yale University, New Haven, CT, USA, 2000.
28. Cámara, C.F.L.; Éplénier, G.; Tinajero, J.; Dunn-Rankin, D. Numerical Simulation of Methane/Air Flames Including Ions and Excited Species. *Combust. Inst. Provo UT USA* **2015**.
29. Uddi, M.; Jiang, N.; Mintusov, E.; Adamovich, I.V.; Lempert, W.R. Atomic oxy- gen measurements in air and air / fuel nano-second pulse discharges by two photon laser induced fluorescence. *Proc. Combust. Inst.* **2009**, *32*, 929–936. [CrossRef]

30. Coppens, F.H.V.; De Ruyck, J.; Konnov, A.A. The effects of composition on burning velocity and nitric oxide formation in laminar premixed flames of $CH_4 + H_2 + O_2 + N_2$. *Combust. Flame* **2007**, *149*, 409–417. [CrossRef]
31. Hermanns, R.T.E.; Kortendijk, J.A.; Bastiaans, R.J.M.; De Goey, L.P.H. Laminar burning velocities of methane-hydrogen-air mixtures. *Submitt. Combust. Flame* **2007**. [CrossRef]
32. Konnov, A.A. Detailed Reaction Mechanism for Small Hydrocarbons Combustion. 2022. Available online: http://homepages.vub.ac.be/~akonnov/ (accessed on 8 September 2021).
33. Williams, F. San Diego Mechanism. 2010. Available online: http://maeweb.ucsd.edu/combustion/cermech/index.html (accessed on 8 September 2021).
34. Kozlov, V.E.; Starik, A.M.; Titova, N.S. Enhancement of combustion of a hydrogen-air mixture by excitation of O_2 molecules to the $a\ ^1\Delta_g$state. *Combust. Explos. Shock Waves* **2008**, *44*, 371–379. [CrossRef]
35. Law, C.K. *Combustion Physics*; Cambridge University Press: New York, NY, USA, 2006.
36. Ying, Y.; Liu, D. Detailed influences of chemical effects of hydrogen as fuel additive on methane flame. *Int. J. Hydrogen Energy* **2015**, *40*, 3777–3788. [CrossRef]

Article

Innovative Integration of Severe Weather Forecasts into an Extended Arrival Manager

Marco-Michael Temme [1,*], Olga Gluchshenko [1], Lennard Nöhren [1], Matthias Kleinert [1], Oliver Ohneiser [1], Kathleen Muth [1], Heiko Ehr [1], Niklas Groß [1], Annette Temme [1], Martina Lagasio [2], Massimo Milelli [2], Vincenzo Mazzarella [2], Antonio Parodi [2], Eugenio Realini [3], Stefano Federico [4], Rosa Claudia Torcasio [4], Markus Kerschbaum [5], Laura Esbrí [6], Maria Carmen Llasat [6], Tomeu Rigo [7] and Riccardo Biondi [8]

1 German Aerospace Center (DLR), 38108 Braunschweig, Germany
2 CIMA Research Foundation, 17100 Savona, Italy
3 Geomatics Research & Development srl (GReD), 22074 Lomazzo, Italy
4 National Research Council of Italy, Institute of Atmospheric Sciences and Climate (CNR-ISAC), 00133 Rome, Italy
5 Austro Control, 1030 Vienna, Austria
6 Department of Applied Physics, Universitat de Barcelona, 08007 Barcelona, Spain
7 Meteorological Service of Catalonia, 08029 Barcelona, Spain
8 Dipartimento di Geoscienze, Università degli Studi di Padova, 35131 Padova, Italy
* Correspondence: marco.temme@dlr.de

Abstract: In the H2020 project "Satellite-borne and INsitu Observations to Predict The Initiation of Convection for ATM" (SINOPTICA), an air traffic controller support system was extended to organize approaching traffic even under severe weather conditions. During project runtime, traffic days with extreme weather events in the Po Valley were analyzed, an arrival manager was extended with a module for 4D diversion trajectory calculation, two display variants for severe weather conditions in an air traffic controller primary display were developed, and the airport Milano Malpensa was modelled for an air traffic simulation. On the meteorological side, three new forecasting techniques were developed to better nowcast weather events affecting tactical air traffic operations and used to automatically organize arrival traffic. Additionally, short-range weather forecasts with high spatial resolution were elaborated using radar-based nowcasting and a numerical weather prediction model with data assimilation. This nowcast information was integrated into the extended arrival manager for the sequencing and guiding of approaching aircraft even in adverse weather situations. The combination of fast and reliable weather nowcasts with a guidance support system enables severe weather diversion coordination in combination with a visualization of its dynamics on traffic situation displays.

Keywords: air traffic management and airports; air traffic control; arrival manager; controller support system; severe weather visualization; nowcasting; weather research and forecasting

Citation: Temme, M.-M.; Gluchshenko, O.; Nöhren, L.; Kleinert, M.; Ohneiser, O.; Muth, K.; Ehr, H.; Groß, N.; Temme, A.; Lagasio, M.; et al. Innovative Integration of Severe Weather Forecasts into an Extended Arrival Manager. *Aerospace* **2023**, *10*, 210. https://doi.org/10.3390/aerospace10030210

Academic Editor: Spiros Pantelakis

Received: 24 January 2023
Revised: 21 February 2023
Accepted: 21 February 2023
Published: 24 February 2023

1. Introduction

In some regions of the world, adverse weather conditions such as thunderstorms and hailstorms (convective cells) are one of the biggest challenges in commercial aviation as they can have major impacts on air traffic control (ATC) and airlines in terms of safety and capacity [1]. Climate change is intensifying the water cycle [2], thus bringing more intense rainfall and associated flooding, as well as more intense drought in many regions. Current studies suggest that aviation contributes to between 2% and 3% of global warming in the long term through CO_2 and in the short term through methane and contrails [3]. It is expected that climate change, through its impact on atmospheric processes, especially on short-lived and highly localized phenomena (thunderstorms, hailstorms, etc.), will also affect air traffic management activities. This phenomenon, also commonly known as thundercloud, is a main safety risk and can require circumnavigation, thus disrupting air traffic flow, increasing delays, and lowering cost-efficiency. At many major airports, capacity potential is already largely exhausted as global air traffic has continued to grow in recent years [4]. Despite economic fluctuations and the influence of the COVID-19 pandemic, a recovery and further steady increase in flight movements is also predicted for the future [5]. With narrow airspaces and the high density of movements, the probability of flight delays increases in adverse weather conditions.

For more than 9% of the first half of 2018, Munich airport was affected by thunderstorms around or close to the Alps [6]. In 2019, EUROCONTROL reported that 21% of the delayed flights in Europe were caused by adverse meteorological conditions [4]. Around 10% of all European departure flight delays result from adverse weather, varying between 2% at Charles de Gaulle in Paris and 20% in Istanbul [7]. In Austria, 95% of regulated airport traffic delays were caused by weather in 2018 [8]. In addition to reduced visibility [9], snow, short-term de-icing operations, and other strong weather events sometimes make areas of airspace impassable. Unfortunate weather situations are also responsible for the complete closure of airports [10]. The cost of a hub airport closure, for example, can exceed three million EUR per hour for all stakeholders combined [11]. The enormous economic significance that adverse and extreme weather events can have on aviation is also becoming apparent.

An early and coordinated avoidance of adverse weather conditions can make en-route flights and approaches more efficient; however, there are currently no appropriate solutions that are operationally available for a guidance system that takes dynamic and convective weather into account. Consequently, ATC requires innovative solutions for avoiding emerging severe weather during flight. Adverse weather is often limited to regional areas; however, when blocking main flight routes, this can have a significant impact on overall air traffic. Thus, it is of utmost importance to support air traffic stakeholders both on board and on the ground. If appropriate nowcast and forecast weather data are available, the process of circumnavigation can be improved by implementing computer-based air traffic controller and pilot support systems. These systems support ATCOs with route information, altitude and speed advisories, and enhanced traffic displays with weather cells and affected aircraft visualizations. To avoid areas with flashes and hail, diversions should be designed in a way that pilots and ATCOs are able to choose their routes according to their training and experience. This paper describes the technical implementation of an air traffic controller support system for adverse weather situations in the vicinity of airports. In addition, an initial validation of the prototypical system and its results is described, which will be incorporated into the further development of the used arrival manager.

1.1. Current Situation

ATCOs are responsible for maintaining separations between aircraft. Today, spacing to severe weather is the responsibility of the pilot in the event of occurring convective cells. When adverse weather occurs on the aircraft's planned or assigned route, the pilot decides whether to fly through it, fly over it, or avoid it to the left or right. In addition to the onboard weather radar, pilots can also use external live weather applications such as eWAS, which can transmit current weather data to the cockpit via a fast SatCom data link [12].

The ATCO clears this avoidance maneuver and ensures that it does not result in critical separation infringements with other traffic. Center controllers, responsible for approach and lower and upper airspace, have no weather radar available for their sectors. The systems, visualizations, and procedures developed by the German Aerospace Center's (DLR's) Institute of Flight Guidance presented in this paper use available weather information to initiate four-dimensional route rescheduling and arrival sequencing. These tools provide appropriate rerouting advisories for ATCOs at an early stage using forecasts from 30 to 60 min in advance so that pilots do not have to avoid adverse weather situations at short notice. This relieves both the pilot and the ATCO and allows them to react in sufficient time in advance and at a reasonable distance so that rerouting trajectories stay safe and efficient [13].

1.2. Related Work

Dealing with weather in ATC has multiple dimensions, such as accurate weather data, on-ground and on-board systems, visualizations of weather for pilots, ATCOs and supervisors, the determination of weather effects, the rerouting of aircraft, and the monitoring of weather-affected air traffic using deviations from standard routes.

To develop controller support systems for adverse weather conditions, in addition to meeting ATCOs' requirements from these systems, it is also necessary to understand the factors influencing pilots' decisions and their situational awareness when facing predicted or suddenly arising severe weather on their planned flight route [14]. Air traffic management decision making in thunderstorm situations is affected by the uncertainty of such situations [15,16]. Studies have shown that pilots pursue different strategies when avoiding severe weather conditions in general [17,18], even with onboard support functionalities that highlight unobtrusive and important weather characteristics [19,20]. One of the outcomes of Ahlstrom's validations was the discovery that there is still a lack of actual research on the needs of Terminal Maneuvering Area (TMA) ATCOs concerning weather information and support functions [21]. Recent studies also show that air traffic controller support systems may reduce their workload in adverse weather conditions in the TMA [22,23]. The use of decision support tools along with input data considering current and forecasted weather for a user display was also proposed by Evans for the decision loop including impact analysis and mitigation plans [24].

Although weather is one of the most important factors for aviation, ATCOs with no outside view have only selective information about the meteorological conditions in their sector, which they receive from the pilots. When available, they also use general weather information from the Internet, but this is not specifically adapted for aviation. So far, there is no integrated situation-sensitive weather view in their traffic situation display. In 2000, a detailed study on weather impact awareness with respect to the next five minutes that included 20 ATCOs showed that, in almost 40% of the used scenarios, the ATCOs were not fully able to identify the effects of weather on air traffic [25]. Since that time, comprehensive systems such as weather radar have been further developed and are now available at many ATC centers; thus, the meteorological situational awareness of ATCOs should be much greater today. However, detailed studies considering new support systems are not available.

2. User Requirements

Assimilated nowcasting data of convective areas should be visualized and used for arrival scheduling to support ATCOs and reduce their workload in these kinds of non-nominal conditions. For the definition of user requirements, a survey with nine ATCOs was conducted within the project to obtain an overview about their demands and preferences in order to enhance the acceptance and usability of adverse weather visualizations [26]. The results of the survey facilitate suggestions to improve the presentation of adverse weather areas (such as convective cells) on a traffic situation display for aircraft guiding and 4D flight trajectory calculation with target times for significant waypoints [27]. These insights were used for the development of an individually configurable primary display with adverse weather presentation possibilities in order to increase acceptance of the support system by controllers.

2.1. The Requirements Inquiry

A twelve-page questionnaire divided into ten different structured sections was provided to the participants. There were questions about display variants, in which the participants could answer using a seven-point Likert scale, and questions that could be answered with either "yes" or "no". The latter ones had the additional option of specifying one's answer or limiting its validity and scope via an associated comment field. After general questions about the person, five presentation variants of convective cells were introduced, for which the participants were asked to give an estimate of the support quality of the display. The participants were also asked to provide a possible order regarding their personal acceptance of the display modes. The third section concerned to what extent the activation of the display should be automated or manual and whether they wanted a weather display at all. After questions about dynamic and static weather visualizations, as well as questions about a possible forecast period when integrating nowcasts, the final section addressed ideas about the use of additional symbols on the aircraft labels on the radar display to mark aircraft affected by adverse weather and also focused on the integration of additional safety zones around measured and predicted convective cells in the airspace.

There were, in total, nine survey responses that were returned, though not all controllers answered the complete set of questions. The number of participants was not extensive. However, all of them were professional and licensed air traffic controllers. Hence, their answers provide meaningful and valuable insights into the requirements for adverse weather visualizations on traffic situation displays. The respondents were male and their ages were uniformly distributed between the considered age groups (Figure 1).

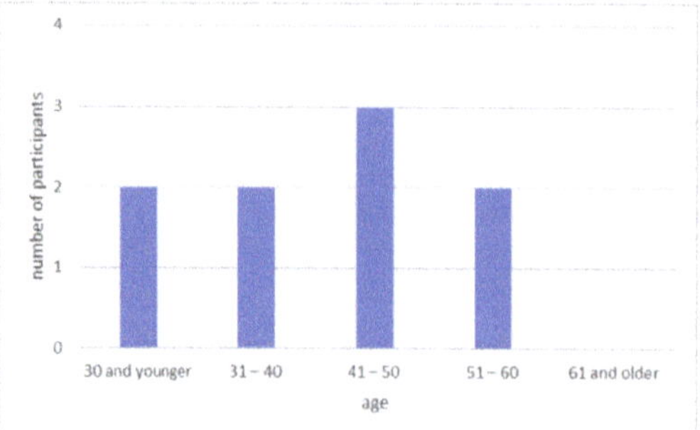

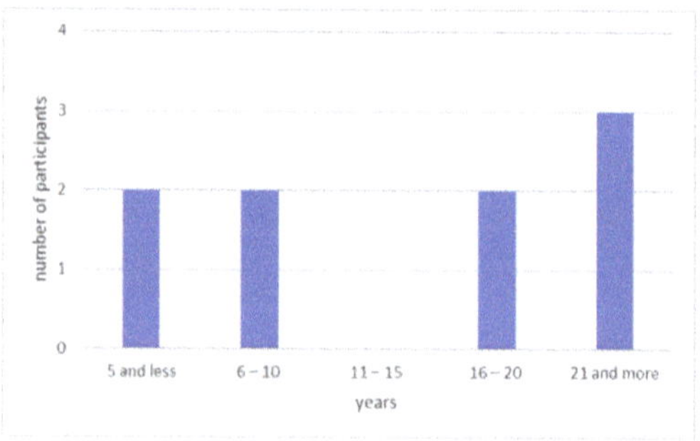

Figure 1. Age distribution and professional experience of the participating controllers.

2.2. *The Requirement Wishes of the Air Traffic Controllers*

Although the integration of adverse weather is intended as a support for advanced scheduling and sequencing, it was usually perceived by controllers as a taking over of additional responsibility. Some of the respondents emphasized that it is the responsibility of the pilot to ensure a safe flight, including evasive maneuvers due to adverse weather [26]. One of the most common comments regarded a possible overload of the controller's display, either with additional information or with its colored presentations. The next point was that the controller's display may only represent the actual state at any time and not a forecasted one. On the other hand, the respondents stressed that displaying severe weather would be beneficial for better planning and less interference with traffic flows. Due to the large differences in the reactions of the pilots, respondents assumed that meaningful and realistic categorization with regard to the dangerousness of a weather situation was impossible. Experience shows that one aircraft can fly on the left side of a convective cell, the next one on the right side, and the last one through some severe weather area depending on the experience of the pilot and the interpretation of available onboard weather radar. Generally, this supporting tool is conceivable for a planning controller. For executive controllers, there might be a risk of visual overload on the traffic display. In principle, it was noted that it is a very good approach to show current weather data in the radar traffic image. However, there should be the possibility of manually switching this information on or off with a button. This should be used by ATCOs if necessary for better planning of traffic in relation to sequence creation. The represented information should always match the actions of the controller. For instance, airspace being "usable" or "not usable" means that a display with two possible states should be used without detailed information about meteorological airspace conditions.

The essence of the results from the requirements analysis for the development of the controller support system in the project can be summarized very well with the phrase *"less is more"*. Overall, eight out of the nine controllers that participated in the survey welcomed a way of quickly and directly accessing weather information that is relevant to them. One of their basic requirements is that any type of information, whether it is provided graphically or numerically, must be able to be activated and deactivated quickly and easily on the display. These requirements build the basis for further developments in the SINOPTICA project that can visualize adverse weather on a controller's traffic situation display.

3. AMAN Air Traffic Controller Support

Arrival Manager (AMAN) systems have been developed and deployed in Europe over the course of the last 25 years. They are primarily designed to provide automated sequencing support for ATCOs handling traffic arriving at an airport by continuously calculating arrival sequences and times for flights while considering the locally defined landing rate, the required spacing for flights arriving to the runway, and other criteria. The AMAN has to generate flyable 4D trajectories with target times for all significant waypoints, including runway thresholds.

During the last twenty-five years, the DLR's Institute of Flight Guidance in Braunschweig has developed arrival management systems for different kinds of scientific applications at various international airports. The latest version of DLR's previously developed arrival manager tools "COMPAS" [28] and "4D-Planner" [29] is called the 4-dimensional Cooperative Arrival Manager (4D-CARMA). Both previous versions were the result of research projects in close cooperation with the Deutsche Flugsicherung GmbH (DFS). Considering different constraints, such as weight classes, runway separation criteria, or runway allocation, 4D-CARMA uses radar data and additional information, such as the flight plans of all arriving aircraft, for sequencing and trajectory calculation. AMANs are pure suggestion systems and have a planning horizon of around one hour. They ease ATCOs' tasks by taking over the particularly difficult planning and optimization of approach sequences while considering all given constraints. This technical support in approach planning can have a clearly positive influence on the effectiveness of ATCOs' work since approaching aircraft are integrated at an early stage, the required distances are precisely considered, throughput is slightly increased, and approach trajectories are more direct and thus shorter [30]. This is especially true when the decision support systems are combined with speech recognition [31]. In recent years, developments in Arrival Managers have gone in two main directions: On the one hand, the planning horizon has been systematically extended to several hundred miles, giving aircraft a precise target time for the threshold and, thus, a position in the landing sequence already in the en-route phase [32]. On the other hand, attempts are being made to use machine learning (ML) methods to build pilot support systems that are based on trajectories that have actually been flown and can thus better represent the typical actions of controllers and pilots in certain situations than classic deterministic algorithms [33,34]. With the help of ML, attempts are also being made to support more environmentally friendly approach planning and thus further reduce the environmental impact of aviation [35].

The first arrival managers already developed the systematic base for ATCO support, and this has not changed in principle. Accordingly, the tasks of an AMAN can be divided into different levels: "Sequence Planning" calculates optimal landing sequences based on airspace structure, current air traffic situations, and performance criteria for all aircraft in the airspace. "Trajectory Calculation" creates optimal 4D routes for every individual aircraft to fulfil the planned sequence. "Advisory Generation" deduces the required instructions from air traffic controller to pilot to follow the calculated trajectory, and "Conformance Monitoring" tracks if aircraft follow the planned trajectories. In order to support adverse weather avoidance, the assistant functionalities of an AMAN have to be adapted and new assistant functionalities need to be provided. The most important elements of an AMAN capable of supporting severe weather area diversions are conflict detection between routes and weather areas, alternate route finding, sequence calculation, and trajectory generation.

3.1. Diversion Route Finding

The AMAN 4D-CARMA used in SINOPTICA first creates an individual waypoint list for the 4D trajectory calculation for each aircraft, which consists of both physical navigation points, such as Very High Frequency Omnidirectional Radios (VORs) or Distance Measurement Equipment (DME) and virtual Aeronautical Information Publications (AIPs), or Flight Management System (FMS) waypoints with all their altitudes and speed constraints. Usually, this waypoint list is based on the defined standard approach routes and transitions of the airport and the approach direction. If a conflict between a trajectory and a polygon of severe weather has been detected, the waypoints located inside the polygon are first determined. Afterwards, the points directly in front of and behind the polygon are searched for.

The points inside the polygon are deleted from the waypoint list, and the points before and after it are set as the start and end points for the diversion. For the route, points to the right and left of the weather are now calculated in the direction of flight, which would allow for a triangular diversion (for example, waypoints IP $[P_1 P_2]$ left or IP $[P_1 P_2]$ right in Figure 2). If it turns out that the waypoints in front of or behind the polygon are too close to it that an aircraft would have to fly a steep turn, these points are accordingly moved further away from the polygon along the route (from waypoint P_1 to P_0 in Figure 2).

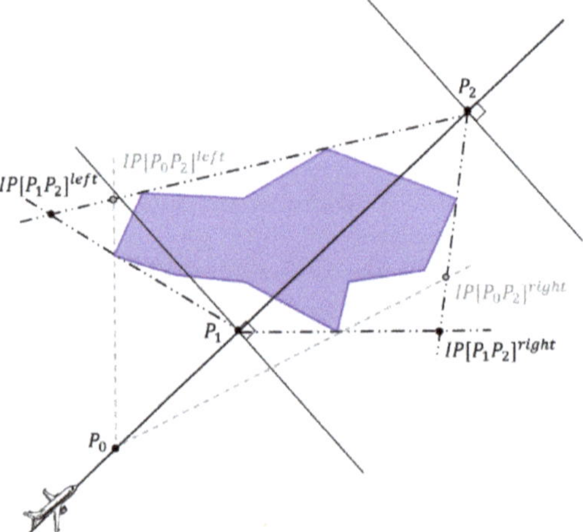

Figure 2. The principle of diversion route calculation. P_0, P_1, and P_2 are points outside a weather polygon and represent the possible start and end points of the diversion. All waypoints on the waypoint list between them will be dismissed. Additionally, selected diversion waypoints such as $IP\ [P_1 P_2]^{right}$ will be integrated and used for 4D trajectory generation.

The next step is to determine the route lengths that involve a right- or left-side diversion. In addition, the moving direction and speed of the severe weather is considered when choosing the diversion's direction. Subsequently, the newly determined waypoints are integrated into the current waypoint list and used for arrival sequencing and trajectory calculation.

3.2. Arrival Sequencing

For sequence planning, the shortest possible approach routes with the highest possible approach speeds are determined, as well as the longest routes with the lowest speeds. These result in an earliest and a latest landing time for each aircraft, and a target time window can thus be planned. The sequence is now determined by using the shortest possible approach times for each aircraft. If two aircraft fall below their weight class-dependent minimum separation, the succeeding aircraft is moved back in time until the minimum separation is reached on final and for touchdown. If an aircraft is shifted behind its latest possible landing time, it must be guided into a holding. This finally results in the landing times for the precise trajectory calculation.

3.3. Trajectory Generation

The basis of the trajectory calculation is flight performance data and a waypoint list with local constraints regarding speed and altitude limits. This trajectory must then be subjected to two screenings. Firstly, whether the trajectory for the aircraft is feasible is tested. This includes, for example, checking the radii of curves with respect to the approach speed planned there. Another test criterion is conflicts with other aircraft. A conflict test with polygons representing severe weather areas is a new addition. Conflict detection is implemented by comparing the planned trajectories and the current or predicted weather areas. If both conditions are met, the estimated landing time can be calculated from the new trajectory.

Several functions are available in 4D-CARMA for calculating descent and reduction rates, which are calculated on the basis of EUROCONTROL's Base of Aircraft Data (BADA) [36]. Therefore, the rates depend also on the current altitudes and speeds of aircraft. Thus, if the calibrated air speed (CAS) is reduced at the same time, the descent rate is reduced by almost 50% compared to a descent rate with a constant CAS to indicated air speed (IAS) ratio. However, there is also the possibility of using correction parameters for the BADA data resulting from simulations and flight tests.

Initiated by the planned target time, the trajectory is calculated backwards starting from the threshold into the air. The forward calculation always begins at the current position of the aircraft. It takes place for at least 25 s from the actual time because it is assumed that no AMAN-advised flight state changes are possible during this short period of time due to the operations of a pilot or controller. However, if advisories that are already displayed to the controller are known, it is assumed that these clearances are also given and executed at the scheduled time.

The reverse calculation is normally carried out from the Final Approach Fix (FAF) in the direction of the current aircraft position to the end point of the forward calculation. It receives the waypoint list until where the forward calculation has been performed as input, with all points between the Final Approach Fix (FAF) and the threshold being removed.

When calculating the trajectory coordinates, it is assumed that the current position and track (flight direction) are known. Furthermore, the 2D route of the aircraft is described with constraints through a list of predetermined waypoints $P_1, \ldots, P_N$. These constraints are the maximum and minimum values for the flight levels and CAS speeds to be maintained. The waypoint P_0 corresponds here to the current aircraft position, and the waypoint P_N corresponds to the runway threshold. If the position of the aircraft now approaches the next waypoint on the route (P_i) by less than the predetermined distance L of 2 NM, a track change to the next point (P_{i+1}) with a constant radius is started. Of course, this applies only if the flight to the next waypoint is connected with a significant change in direction. After validation trials with air traffic controllers in different projects, directional changes of more than 0.5 degrees are considered as a significant route change in DLR's AMAN.

In both cases—the forward and the reverse calculation—an attempt is made to place an arc from the current position onto the following segment so that the end of the arc will track in the direction of the following waypoint or the segment that can be flown. If this is not possible, a direction change of a maximum of three degrees per second takes place per integration step until the new track runs directly to the next waypoint. At the end, the track is always flown to the next waypoint, but it can happen that the track then deviates somewhat from the segment between the last waypoint and the next waypoint.

This deviation is not critical in most phases of the approach. In this way, however, it can always be ensured that deviations around severe weather areas always take place in a soft arc and thus enable a realistic flight path.

3.4. Dynamic Severe Weather Visualization on a Traffic Display

Since convective cells should be visualized on a 2D display, the current and forecasted cells can be represented by discrete sets of simple polygons. Each polygon of a set that is valid for some time period has an identification number that provides unique correlation between sets of cells. The polygons have to be displayed without covering essential traffic information, such as aircraft labels or additional flight plan information. For animation of actual weather development, two methods for comparison were implemented with different demands in terms of the level of detail for weather information. In addition to a linear extrapolation of the last weather development, this also includes a procedure based on a morphing algorithm which, in addition to the last meteorological measurement, also considers nowcasts in the animated presentation.

3.4.1. Algorithm for Advanced Weather Visualization

The mentioned polygons are simple, closed, and defined by the ordered set of its vertices consecutively linked by line segments without intersections that bound a connected interior area. Polygons with self-intersections, as well as polygons with holes, were not considered since they are not suitable for the representation of areas that aircraft should not fly through.

The main goal of the intended visualization is to perform a realistic interpolation of the transformation of a polygons' set at the current time t into a set of forecasted polygons at the time moment $t + \Delta t$ over the time period Δt. The interpolation should have small time steps of less than five seconds in order to avoid gaps between consecutive visualization pictures and to follow the update rhythm of aircraft radar data.

Since convective cells move and can appear, disappear, merge together, or split into small parts over time, it is necessary and sufficient to develop approaches for performing the following main interpolations:

1. A transformation of one polygon into another polygon;
2. A transformation of one polygon into several polygons and vice versa;
3. A transformation of one polygon in simultaneous splitting and merging cases of this polygon.

The formulated task belongs to the class of 2D polygon morphing problems. This class is well studied, especially for complex shapes, because of their wide applicability in computer graphics and animation. Therefore, these solutions imply significant computational effort or are instead developed for some initial objects with properties irrelevant for the visualization of weather. A detailed overview of the related literature is out of the scope of this paper. An overview on the existing methods is given in [37]. Here, we refer to works related to polygon morphing.

Approaches for polygon morphing consist of two main steps: the mapping of polygons by some characteristic or feature points and the specification of methods of interpolation or curves. The last step is more complicated because retaining some of the characteristic features of the considered objects during interpolation is often desirable. One of the main goals in defining interpolation methods is avoiding the local self-intersection of polygon boundaries.

There are many heuristics presented in the literature. Guaranteed intersection-free polygon morphing, described in [38], relies on an analytical basis. However, the approach uses a significant number of interior points and exterior Steiner vertices that increase its complexity. The morphing of simple polygons with the same number of edges that are correspondingly parallel is explored in [39]. Usually, morphing algorithms require user assistance to relate the morphing objects. Malkova introduced intuitive polygon morphing; however, the source and destination polygons must spatially overlap [40]. Moreira dealt with the application of 2D polygonal morphing techniques to create spatiotemporal data representations of moving objects continuously over time [41]. The movement of icebergs in the Antarctic seas was used as a case study and the data sources were sequences of satellite images capturing the position and shape of the icebergs on different dates. The authors applied a perception-based approach in [42], in which the so-called feature points of the morphing objects were determined. The main challenges were the determination of feature points and correspondences between feature points. Since this work investigated the visualization of moving gaseous objects, rotations and similarities had an insignificant role compared to the solid objects in [41].

Considering relatively small time-steps between consequent sets of convective cells compared to their movement speed and transformation, linear interpolation appears to be the best choice for the defined approximation requirements.

3.4.2. Moving and Morphing Algorithm Description

The first mentioned transformation of one polygon into another provides the basis for the two remaining types of transformations, as well as the subsequent morphing of the set of polygons at time t into the set of polygons at time $t + \Delta t$ over the time period Δt. Therefore, this should be primarily considered.

Decomposition of Polygon Boundaries

To morph one polygon into another, the boundaries of considered polygons are decomposed into four corresponding parts to obtain a more natural interpolation of the transformation that avoids significant turns and rotations of the boundary during interpolation. For this purpose, four sets of vertices with minimal and maximal abscissas and with minimal and maximal ordinates have to be found for the current and for the forecasted polygon. These sets can contain more than one point, and some sets can coincide. If the considered set consists of more than one point, the simplest solution is to take the first point from each set of vertices in the vertex index increase direction. Another option is to choose a point from this set, which provides balanced length for the boundary parts. Boundary parts obtained in this manner are illustrated in Figure 3. Here, Pol^t and $Pol^{t+\Delta t}$ are polygons at time moments t and $t + \Delta t$, respectively. The points E_i^t, $i = 1, \ldots, 8$ and $E_j^{t+\Delta t}$, $j = 1, \ldots, 9$ are the vertices and E^{*Top}, E^{*Right}, $E^{*Bottom}$, E^{*Left} are subsets of vertices with minimal or maximal first or second coordinates of the polygons Pol^t and $Pol^{t+\Delta t}$.

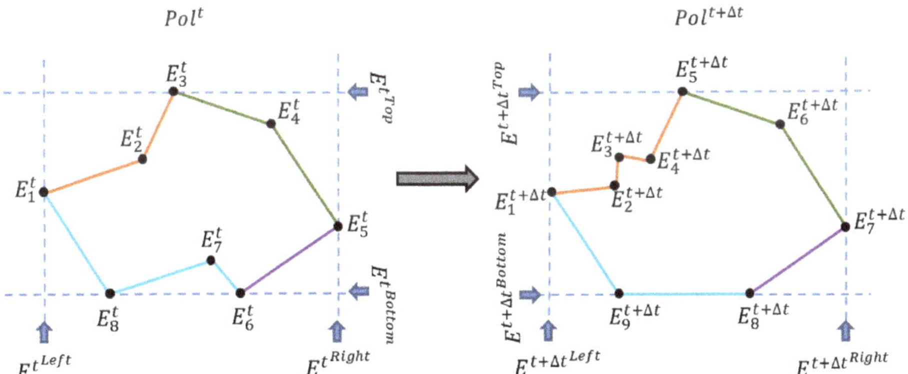

Figure 3. Decomposition of boundaries into four parts (marked by different colors) for the actual polygon Pol^t and the next (forecasted) one $Pol^{t\,+\,\Delta t}$.

After boundary decomposition, morphing of the boundary parts is performed through the linear interpolation of the vertices in s discrete time steps of the duration $\frac{\Delta t}{s}$. To perform the interpolation, the boundary parts are balanced in order to establish one-to-one correspondence between them. This means that the boundary part with the smallest number of vertices receives additional points uniformly distributed along the edges so that, on each edge, the quotient number of additional points plus one additional point on the remaining number of edges in the vertex index increase direction is selected. The described approach is illustrated in Figure 4. Here, the piecewise linear curve A_1A_3 is transformed into the piecewise linear curve B_1B_{10}. The quotient of the division of ten by three is equal to three and the remainder is equal to one. Therefore, the edge A_1A_2 gets three plus one additional points, and three additional equidistant points are selected on the A_2A_3 edge. They are marked red in Figure 4.

Moving and Morphing of a Single Polygon

The described approach provides an identical number of "vertices" on both linear curves to perform one-to-one linear interpolation in s discrete steps. The middle polyline in Figure 4 illustrates the interpolated curve at the time moment $t + \frac{\Delta t}{2}$. Morphing of the polygon Pol^t into $Pol^{t+\Delta t}$ is realized through interpolation of the transformation of four corresponding boundary parts in accordance with the approach visualized in Figure 4. The interpolated polygon at time $t + \frac{\Delta t}{2}$ is illustrated in Figure 5.

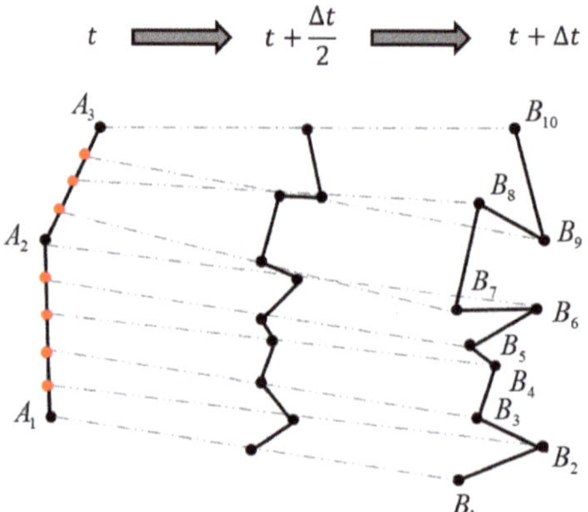

Figure 4. Additional points facilitating one-to-one transformation (marked by red color).

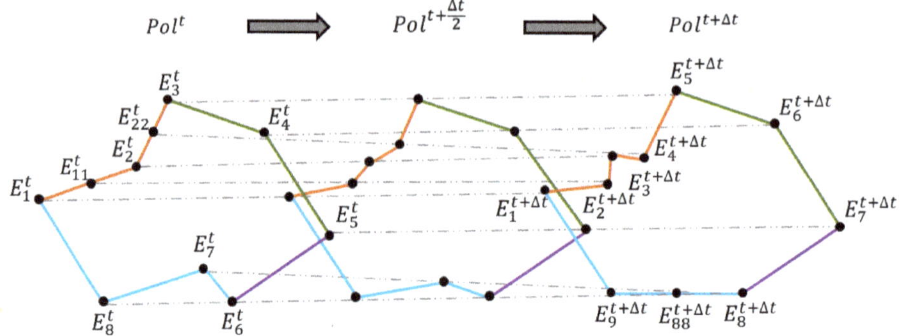

Figure 5. Visualization of one morphing intermediate step at the time moment $t + \frac{t_0}{2}$ accordingly to decomposition of boundaries into four parts (marked by different colors).

The input data to perform interpolation represent the current and forecasted locations of cells, i.e., only the current situation and a forecasted result of transformation after time period Δt are available. Although there is a correlation of identification numbers between sets of polygons reproducing current and forecasted situations, there is no information in the input data in a merging or splitting case, with part of a pre-image corresponding to a transformed polygon and vice versa. There is no way to exactly retrace the decomposition process for available input data.

In the case of an interpolation approach where the forecasted decomposed polygons are mapped to some parts of the current polygon, the free corridors appearing in the visualization may not reflect the real-life situation and can deviate from it significantly. Therefore, from a safety point of view, aircraft should not be directed through these corridors. Additionally, the time periods between the forecasts are relative short (up to ten minutes). Based on these facts, the mapping of polygons without decomposition of the current polygon in the splitting case and without decomposition of the forecasted polygon in the merging case can be taken as a reasonable approach for transformation. This approach is explained and illustrated in the next subsection.

Moving and Morphing Polygons with Splitting and Merging

During the development of occluded fronts, the splitting and merging of individual convective areas always occurs, as well as a shift in the main precipitation areas within the front. For this reason, it is important that an animated presentation format on a traffic situation display can also reproduce these developments as realistically as possible. The mapping of polygons in the splitting case to perform the interpolation of their transformation is shown in Figure 6. The polygon $Pol^{t|1}$ at the current time t decomposes over the time period Δt in four parts, namely $Pol^{t+\Delta t|1^{(t|1)}}, \ldots, Pol^{t+\Delta t|4^{(t|1)}}$, as illustrated in Figure 6. However, the splitting contours are unknown and, as a consequence, the space between appearing parts is not safe. Therefore, as was mentioned in the previous subsection, the current polygon $Pol^{t|1}$ is mapped to all four forecasted polygons $Pol^{t+\Delta t|1^{(t|1)}}, \ldots,$ $Pol^{t+\Delta t|4^{(t|1)}}$. The morphing of the polygon $Pol^{t|1}$ into each forecasted polygon $Pol^{t+\Delta t|1^{(t|1)}},$ $\ldots, Pol^{t+\Delta t|4^{(t|1)}}$ is then performed simultaneously in the way described above. Figure 6 shows interpolation lines, additional points, and the constellation of the polygons at the time period $t + \frac{\Delta t}{2}$. Colors correspond to the mapping of the polygons. An approach to transform one polygon into several polygons, and vice versa, is hereby developed.

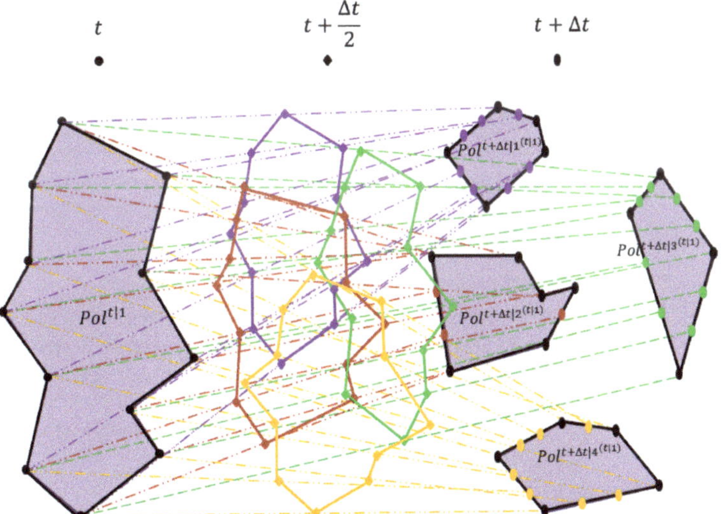

Figure 6. Illustration of interpolation without decomposition of the current polygon. On a radar screen, only the envelope curve is presented. Corresponding mapping is visualized in different colors.

The mapping of whole polygons can be also used to perform a transformation in the case of the simultaneous splitting and merging of the current and forecasted polygons, i.e., when a part of one polygon splits from the polygon and another part merges with some other polygon at the same time. Figure 7 illustrates this mapping example without decomposition.

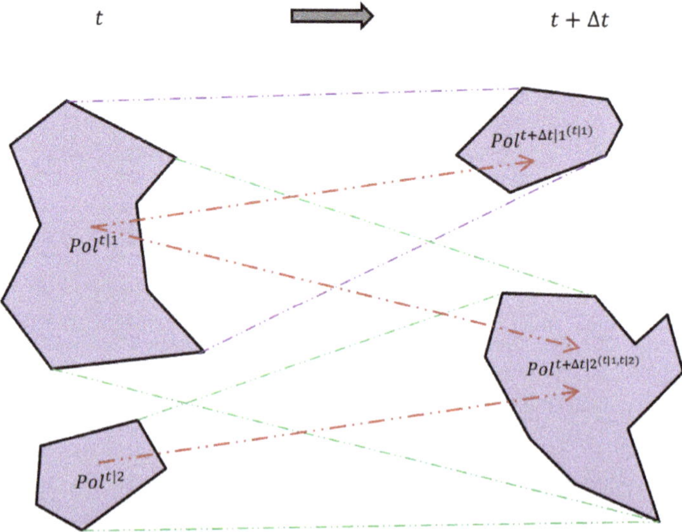

Figure 7. Assignment of polygons for interpolation in the case of simultaneous splitting and merging.

Consequently, the presented morphing approach is applicable for general interpolation of the transformation between two sets of closed polygons representing current and forecasted adverse weather areas over a short period of time.

4. Meteorological Modeling

The three new meteorological forecasting techniques, i.e., Weather Research and Forecasting (WRF), the PHAse-diffusion model for STochastic nowcasting (PhaSt), and the Radar Nowcasting Density of the Vertical Integrated Liquid (RaNDeVIL) model, were developed and extended to better nowcast severe weather events affecting tactical approach operations. For this purpose, short-range severe weather forecasts with very high spatial resolution were elaborated, starting from radar images, through an application of nowcasting techniques combined with the Numerical Weather Prediction (NWP) model and data assimilation.

4.1. Weather Models

WRF model v3.8.1 [43] was used for the numerical simulations. It is a next-generation mesoscale numerical weather prediction system designed to serve both operational forecasting and atmospheric research needs that was developed at the National Center for Atmospheric Research (NCAR) in collaboration with several institutes and universities for operational weather forecasting and atmospheric science research. The model represents the atmosphere with a fully compressible non-hydrostatic set of equations, which is discretized over a staggered Arakawa C-grid. The WRF simulations are produced in two phases: the first configures the model domains, ingests the input data, and prepares the initial conditions, and the second runs the forecast model, which is performed by the forecast component that contains the dynamics solver and physics packages for atmospheric processes (e.g., microphysics, radiation, and the planetary boundary layer). The forecast model components operate within the WRF's software framework V3.8.1 from National

Center for Atmospheric Research (Boulder, Colorado), which handles input/output (I/O) and parallel-computing communications. WRF is written primarily in Fortran, can be built with several compilers, and runs predominately on platforms with UNIX-like operating systems, from laptops to supercomputers. The WRF model is applied extensively under both real-data and idealized configurations for research activity, but it is also used operationally at governmental centers around the world, as well as by private companies. Three domains with a grid spacing of 22.5 km, 7.5 km, and 2.5 km and parametrization schemes used for the WRF operational chain at the CIMA Foundation [44–46] were adopted for the numerical experiments of this work.

PhaSt is a spectral-based nowcasting procedure based on the precipitation fields provided by radar measurements and the stochastic evolution of the transformed fields in spectral space [47]. In the framework of the SINOPTICA project, the PhaSt algorithm has been applied to Vertically Integrated Liquid (VIL) fields, a precursor of convective activity [48]. The algorithm takes an empirical nonlinear transformation of the two precipitation fields used as initial conditions. The method uses the Fourier transform of the two Gaussianized initial fields and their Fourier spectra to obtain the Fourier phase for each wavenumber. The latter is then evolved in time by a stochastic process, while Fourier amplitudes are kept fixed. A Langevin-type model is used to evolve the Fourier phases and to generate a nowcasted Gaussian field. For the SINOPTICA project, the PhaSt algorithm is applied in a sort of Rapid Update Cycle (RUC). For this purpose, three different approaches are evaluated:

- No restart, namely a forecast every 5 min up to one hour;
- A restart every 10 min;
- A restart every 20 min.

The statistical analysis performed in terms of correlation coefficients and Continuous Ranked Probability Score (CRPS), as well as through an object comparison between observed and forecasted cell clusters, proves that the restart every 10 min approach provides the best results; thus, it was decided to apply this approach to all case studies.

RaNDeVIL is a new method for identifying 2D convective cells with the potential to produce nowcasts for severe weather. It is based on previous techniques but considers the Density of the Vertically Integrated Liquid (DVIL) instead of reflectivity fields. By considering only one level instead of several in volumetric data, the new method has the advantage of greater speed, an important capability when air traffic controllers have to make decisions in a very short period of less than a few minutes. Considering that thunderstorms can develop close to airports and the TMA, the algorithm is a valuable tool that can support ATCs in organizing approaching traffic under severe weather conditions [26].

The algorithm can effectively approximate future thunderstorm dynamics [49], with particularly good performance for predictions up to 30 min. Distance differences of less than 4 km between the centres of the observed and predicted areas were observed for more than 85% of cells in the first 5 min nowcasts and between 2 km and 16 km for most 30 min nowcasts. Beyond this time, the results are more unreliable but can still provide valuable information to end users regarding storm propagation direction and possible impacts in future scenarios.

4.2. Data Assimilation and Numerical Simulations

Data assimilation was performed using a 3-Dimensional Variational (3D-Var) technique with the WRF Data Assimilation (WRFDA) system [50] in version 3.9.1. The 3D-Var scheme aims to improve initial conditions by minimizing a penalty or cost function that reduces the misfit between the background forecast and observations [51]. Several experiments that assimilated different datasets were carried out to assess the benefit of data assimilation for the nowcasting of severe convective events that can impact aviation operations. Radar reflectivity, temperature observations from in situ weather stations, and Global Navigation Satellite System (GNSS)-derived data were assimilated every 3 h within a 6 h assimilation window, whereas lightning data were assimilated every 15 min through a nudging technique [52] that increased model instability by adding water vapor when flashes were observed. The results, widely discussed in [46], show that the assimilation of lightning data plays a key role in improving forecast skill. Indeed, the numerical experiment with radar and lightning data was able to produce reliable forecasts at high spatial and temporal resolution that are suitable for ATM purposes.

5. Case Studies

In the framework of the SINOPTICA project, three different severe events were selected to be analyzed. These events were identified during the summer season, when temperatures are higher and a larger amount of energy is available for the triggering of thunderstorms and the development of deep convective systems. The selected events affected the major airports of the Po Valley in northern Italy, with Milano Malpensa (MXP) having 28.8 million passengers affected in 2019 and Bergamo Orio al Serio (BGY) having 13.9 million passengers affected:

1. 11 May 2019 in MXP: A squall line hitting the Malpensa airport producing a large amount of hail. There was no landing for over 1 h at the airport and eight flights were diverted;
2. 7 July 2019 in BGY: General instability affected the Po Valley with intense thunderstorms. Strong downbursts caused some diversions and huge flight delays;
3. 6 August 2019 in MXP and BGY: High atmospheric instability with thunderstorms and hail hitting the Bergamo and Malpensa airports in the late afternoon. Several cancellations and diversions occurred during the event.

5.1. Synoptic Analysis

On 11 May 2019, the Po Valley was affected by strong convective activity that caused economic damage and seriously injured a person in the Lombardy region. The synoptic scenario at 500 hPa was characterized by a stretched trough extending from northern Europe to the Mediterranean basin. In the afternoon hours, the cold air mass at 500 hPa reached the Po Valley with values around −26 °C and the winds consequently shifted to the northwest [46]. On the other hand, the strong south-westerly flow at low levels moved a large amount of water vapor from the Ligurian Sea to the inland, thus increasing convective instability. These meteorological factors produced favorable conditions for the triggering of convective cells over northern Italy. In this context, a squall line hit MXP, producing intense precipitation and a large amount of hail on aprons and runways. The intense thunderstorm also caused several floods in the city of Milan, where some underpasses and metro stations were closed. Instead, the strong downburst winds caused the felling of trees and billboards that required the intervention of firefighters.

On 7 July 2019, the Po Valley was affected by precipitation with a prevalent character of thunderstorms, which was locally very intense and associated in various cases with hail and strong gusts of wind. From a synoptic point of view, in the north of Italy, an anticyclonic field of subtropical nature was retreating towards central-southern Italy under the pressure of more unstable North Atlantic currents at high altitude (Figure 8, left panel). The huge energy reservoir in the lower layers of the atmosphere (humid heat conditions) that had accumulated in the previous days favored the triggering of the first thunderstorms over the Alps as early as midday of 7 July 2019. The intense westerlies moved the first storms towards the plain around 13:00 UTC, but a second group of organized cells hit the central and eastern Po Valley between the late afternoon and evening, producing various forms of damage: numerous garages, basements, roads, and underpasses were flooded, and some buildings were evacuated.

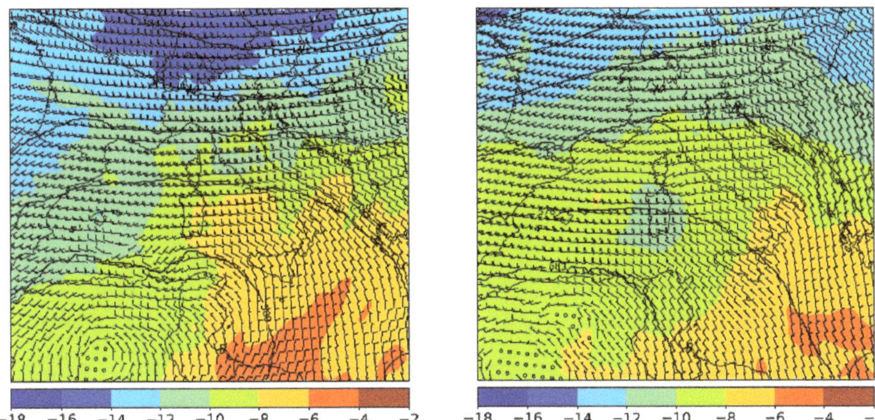

Figure 8. European Centre for Medium-Range Weather Forecasts—High Resolution Forecast (ECMWF-HRES) analysis: 500 hPa temperature (°C), wind (barbs), and geopotential height (dm, contours) on 7 July 2019 (**left panel**) and 6 August 2019 (**right panel**) at 12:00 UTC.

On 6 August 2019, an anticyclone circulation characterized the synoptic scenario over the Mediterranean basin and ensured atmospherically stable conditions and sunny weather over the southern regions of the Italian peninsula (Figure 8, right panel). On the other hand, a deep trough located over the British Isles affected central Europe by touching the Alpine chain during the evening and causing a worsening of the weather conditions over the pre-Alps. In this area, the presence of cold air at 500 hPa, with values around −13 °C, increased the instability of the air column, thus promoting the development of convective cells. In addition, the high values of specific humidity, namely the content of water vapor, at 850 and 950 hPa, in combination with wind shear conditions, easterly wind at 950 hPa, and southerly wind at 850 hPa, produced a very unstable environment. From the early evening, some convective cells affected the Bergamo and Milan provinces with moderate rainfall. Later, the passage of moisture advection over the Alps between 18:00 UTC and 00:00 UTC provided further energy for the convection. Consequently, a more intense convective storm hit the central sector of the Po Valley and produced heavy precipitation, strong wind gusts, and small hail.

5.2. Traffic Analysis: Historical Traffic under Adverse Weather Conditions at Milano Malpensa and Bergamo Airports

In order to evaluate the influence of adverse weather conditions on air traffic in the extended TMA around MXP and BGY, located at ca. 77 km distance from each other, FlightRadar24 datasets from three different selected days with severe events—11 May 2019, 7 July 2019, and 6 August 2019—were analyzed.

In the case of MXP, there were two available days—7 July 2019 and 6 August 2019—without airport closure under disturbed weather conditions and one day—11 May 2019—when adverse weather conditions caused temporal closure of the airport. For BGY, there were two days—11 May 2019 and 7 July 2019—in the dataset where adverse weather did not lead to an airport closure and one day—6 August 2019—when the airport was temporarily closed. As the airports are located adjacent to each other, it is of interest to investigate the datasets in pairs in order to analyze the influence of the changing capacity of one airport on the second one. Therefore, the following pairs of data were considered:

- On 7 July 2019, both airports operated under adverse weather conditions without closures;
- On 6 August 2019, BGY was temporally closed; however, MXP was open all the time;
- On 11 May 2019, MXP was closed for some period of time due to bad weather conditions, whereas BGY continued operations.

The first two datasets recorded on 7 July 2019 contain 322 arrivals at MXP and 144 arrivals at BGY. The travel time of arrivals in the extended TMA and touchdown time are illustrated in Figure 9 in orange for MXP and in blue for BGY. In order to detect peak times while smoothing the influence of outliers, a moving mean value over five neighbor travel times was calculated. These mean values are shown as dark blue and dark orange lines in Figure 9. In the case of MXP, the time until landing in the extended TMA varied between 693 and 2913 s with an average time of 1246 s and five peak periods, the most noticeable of which was between 19:30 UTC and 21:30 UTC (Figure 9). The maximal and the average flown distances in the extended TMA were approximately 302 and 168 km, respectively, whereas the average flown distance at the mentioned peak period was equal to 225 km (i.e., 57 km longer than the average flown distance over the whole day).

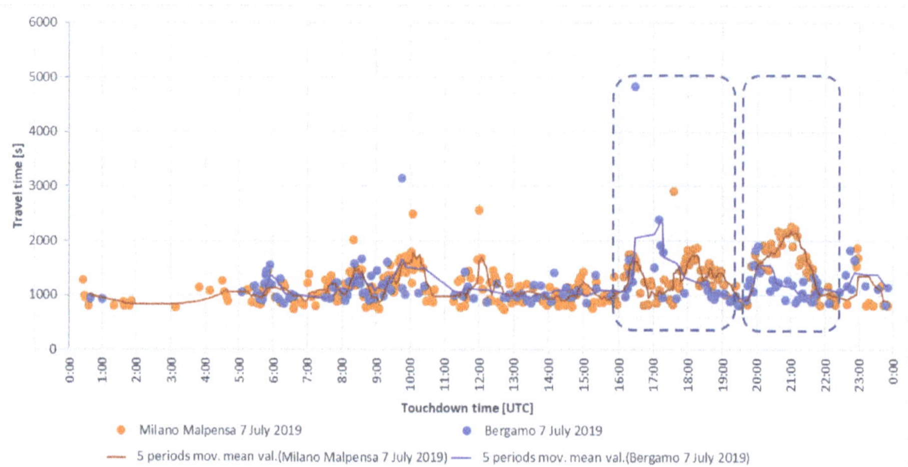

Figure 9. Visualization of travel time in the extended TMA and touchdown time for historical traffic at MXP (orange color) and at BGY (blue color) on 7 July 2019.

In the same day, there were two outlier arrivals at BGY in terms of travel time in the extended TMA. Taking them into account, the time until landing varied between 817 and 4836 s, with an average travel time of 1186 s. The detectable peak period is between 16:15 UTC and 17:30 UTC (Figure 9). The maximal and the average flown distances in the extended TMA are approximately 718 and 160 km, respectively. The average flown distance at the peak time amounts to 247 km with the outlier arrival and 194 km without it. Even without the outlier, the average flown distance during the peak period is 34 km longer than the corresponding value over the whole day.

Summing up, there are two possible coherent peaks at the considered airports that may have been caused by weather conditions. These peaks are pairwise-marked by the areas bounded by the dotted lines in Figure 9.

The second pair of datasets collected on 6 August 2019 contains 421 arrivals. It includes 99 more arrivals than the traffic scenario on 7 July 2019 for MXP and 144 arrivals for BGY. The travel time in the extended TMA and the touchdown time of arrivals are shown in Figure 10. The time until landing in MXP's extended TMA was between 715 and 2436 s, with an average travel time of 1161 s. The maximal and the average flown distances in the extended TMA were approximately 352 and 161 km, respectively. In spite of the fact that this scenario is larger than the previous one for MXP, it has lower values for maximal and average travel time and, as a consequence, average travel distance.

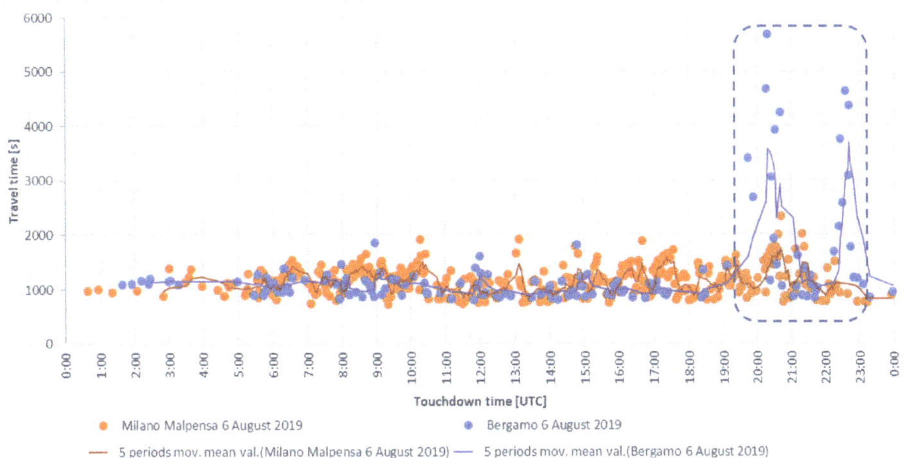

Figure 10. Travel time in the extended TMA vs. touchdown time for historical traffic at MXP (orange color) and at BGY (blue color) on 6 August 2019.

In the same day, the time until landing in the extended TMA around BGY was between 703 and 5679 s, with an average of 1332 s. There are two peaks with respect to travel time in the extended TMA between 19:30 UTC and 23:15 UTC, which are marked by the areas bounded by the dotted lines in Figure 10. The maximal and the average flown distances in the extended TMA were approximately 739 and 185 km, respectively.

The average flown distance during peak time amounts to 310 km and is, through temporal closures of runways, at least 63 km greater than the corresponding average flown distance during the peak time on 7 July 2019, when BGY was not closed. Figure 11 illustrates the accumulated rainfall per ten minutes on 6 August 2019 at BGY, which correlates with the peaks for travel time shown in Figure 10. However, there are no significant peaks between 19:30 UTC and 23:15 UTC at the nearby MXP airport, as illustrated in orange in Figure 10.

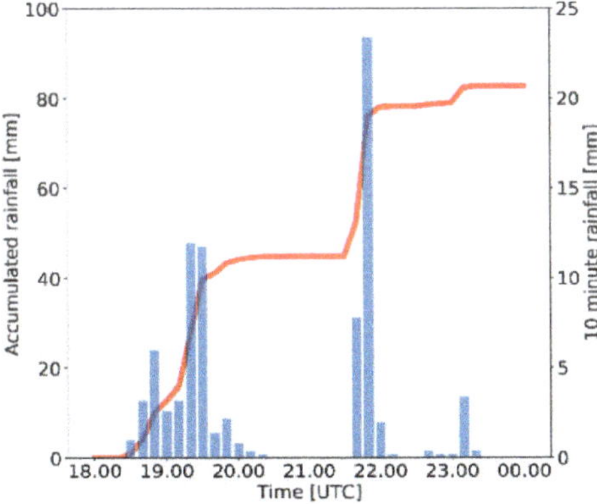

Figure 11. Histogram of the rainfall (blue bars) and accumulated rain (red line) at the Bergamo airport on 6 August 2019 [46].

The last pair of datasets was recorded on 11 May 2019 and contains 233 arrivals to MXP and 122 arrivals to BGY. On this day, a squall line hit MXP between 14:00 UTC and 16:00 UTC causing intense precipitation and heavy hail formation [46]. These adverse weather conditions (Figure 12) led to the airport's closure for about 40 min and some flight delays.

In addition, nine aircraft were diverted to other airports. Figure 13 illustrates travel time at the extended MXP TMA, which was between 719 and 4693 s with an average of 1196 s. This distribution indicates that controllers had to guide many aircraft into holdings. According to Figure 13, the airport was closed at around 16:00 UTC. The maximal and average flown distances in the extended TMA were approximately 676 and 165 km, respectively. The average flown distance during the peak time was equal to 349 km.

Figure 12. Apron of a runway at MXP on 11 May 2019 [53].

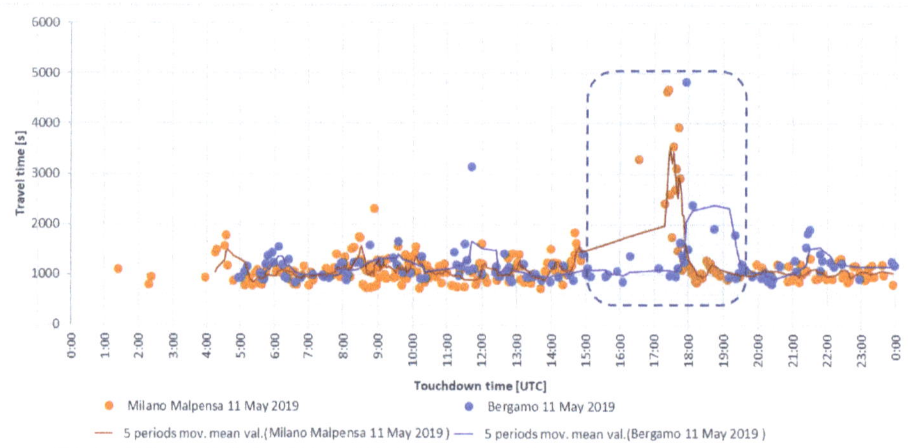

Figure 13. Visualization of travel time in the extended TMA and touchdown time for historical traffic at MXP (orange color) and at BGY (blue color) on 11 May 2019.

Historical traffic at BGY on 11 May 2019 indicated a time until landing in the extended TMA of between 782 and 2985 s, with an average travel time of 1197 s (Figure 13). There was a peak period with respect to travel time between 17:30 UTC and 19:30 UTC, probably as a consequence of the weather conditions at MXP airport described above. The maximal and average flown distances at the BGY extended TMA were approximately 415 and 109 km, respectively, and the average flown distance during the peak period is equal to 215 km. Although the weather conditions at BGY did not cause a runway closure, one can notice the direct influence of the weather conditions at MXP on operations at BGY. Therefore, the historical dataset for MXP on 11 May 2019 was chosen to construct simulation traffic scenarios for investigation of the influence of adverse weather on arrival traffic. Table 1 summarizes the mentioned parameters for the three considered case studies.

Table 1. Data summary of available historical traffic.

Airport	MXP			BGY		
Date	**11 May 2019**	**7 July 2019**	**6 August 2019**	**11 May 2019**	**7 July 2019**	**6 August 2019**
number of arrivals	233	322	421	122	147	144
time until landing in extended TMA (sec)	719–4693	693–2913	715–2436	782–2985	817–4836	703–5679
average time in extended TMA (sec)	1196	1246	1161	1197	1186	1332
maximal flown distance in extended TMA (km)	676	302	352	415	718	739
average flown distance in extended TMA (km)	165	168	161	109	160	185
average flown distance in extended TMA at peak period (km)	349	225	n/a	215	247 (194 without outliers)	310

5.3. *Traffic Scenario Definition*

The simulations involved three traffic scenarios for the TMA at MXP, which were based on FlightRadar24 datasets of the three days (11 May 2019, 7 July 2019, and 6 August 2019). There were two days when weather conditions did not cause airport closure (7 July 2019 and 6 August 2019) at MXP, which were used as baseline scenarios, and one day with an airport closure for some period of time (11 May 2019), which was used as the disturbed scenario.

5.3.1. Scenarios

The real datasets contain 322, 421, and 233 Malpensa flights. They were manually prepared and then read with RouGe Software V1.11 from DLR (Braunschweig, Germany). All military flights, helicopters, departures, and parachutist transports were filtered. To validate the data visually, the datasets were converted to the Google Earth KML data format. Figure 14 illustrates the flown trajectories on 11 May 2019 that occurred under adverse weather conditions. One can recognize that multiple holdings were flown due to thunderstorms at and around the TMA area. Figure 15 shows the flown trajectories on 6 August 2019 under normal undisturbed weather conditions.

The FlightRadar24 data prepared as simulation scenarios contain the following information about callsign, aircraft type, weight category, position data plus time, speed, flight altitude, and heading.

The constructed simulation scenarios were each planned using an AMAN with no adverse weather as a baseline and weather forecasts from different forecast models for two different severe weather events. Flight paths, times, the number of clearances, and fuel consumption were used as key performance indicators (KPI) to compare these models. The considered traffic scenario is based on the dataset from 11 May 2019. Seven scenarios were planned using an AMAN and simulated:

- Without MET information (baseline);
- With MET data from the WRF-RUC, PhaSt, and RaNDeVIL models for the weather of 11 May 2019;
- With MET data from the WRF-RUC, PhaSt, and RaNDeVIL models for the weather of 6 August 2019.

Figure 14. Prepared FlightRadar24 datasets from 11 May 2019 with flights operated under adverse weather conditions.

Figure 15. Prepared FlightRadar24 datasets from 6 August 2019 with normal undisturbed weather conditions.

5.3.2. The Malpensa Airspace Implementation for the AMAN

For aircraft arrival planning and simulations, MXP airspace had to be implemented in DLR's AMAN and the traffic simulation environment (Figure 16). For this simulation, the airspace does not need all of the waypoints and sector boundaries of the real airspace; therefore, only waypoints relevant for arrivals have been inserted. Departure routes were not implemented, and this display thus looks less complex and simpler than usual radar images from MXP. For sequencing, all Standard Arrival Routes (STARs) with the related Flight Management System (FMS) waypoints were assigned as specified in the Aeronautical Information Publications (AIPs) of MXP [54]. The waypoints are connected with constraints regarding maximum and minimum altitudes, as well as maximum and minimum speeds. These data are needed for the 4D trajectory calculation since it is not only speed but also flight altitudes that have an influence on flight times until touchdown.

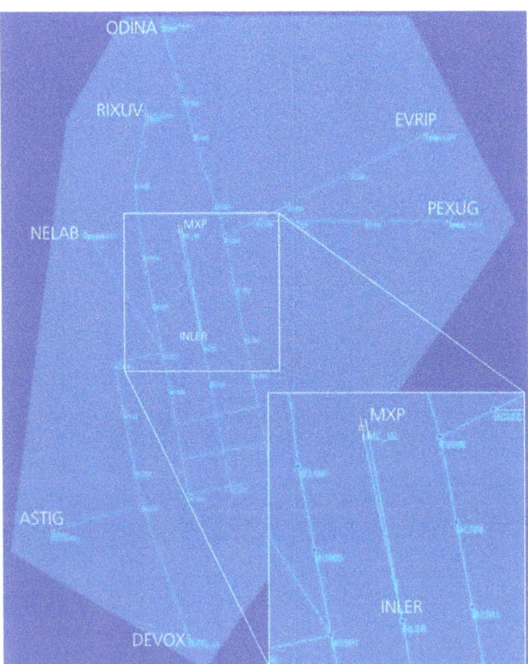

Figure 16. The Malpensa airport and airspace in the AMAN planning and simulation environment.

MXP has two runways in the south–north direction: 35L and 35R. Because of the closeness of the Alps, landings are exclusively northbound. With seven Entry Fixes or Metering Fixes, the TMA has an average number of entry points that, while not covering all points of the compass equally, are a good compromise with respect to the Alps, other nearby airports, and restricted military areas in the vicinity of the airport. The Path Stretching Area (PSA) to the south of the airport is formed by a trombone to the east and a double-trombone to the west, which allows for an even distribution of approaches across the two runways. According to the AIP, it is not until the Initial Approach Fix (IAF) with the name INLER on Final that a decision is made as to which runway aircraft will be guided to; however, analysis of radar data has shown that controllers guide aircraft to the appropriate centerline much earlier. Runway switches on centerline or final are rather the exception.

STARs are assigned for each aircraft based on the respective approach direction and, from the north, are also based on utilization of the western and eastern trombones. Aircraft from the north are routed accordingly via the ODINA and RIXUV fixes, aircraft from the east are routed via EVRIP and PEXUG, aircraft from the southeast are also routed via PEXUG, aircraft from the south are routed via MEBUR, aircraft from the southwest are routed via DEVOX, aircraft from the west are routed via ASTIG, and aircraft from the northwest are routed via NELAB. During the selected days for the validation, all routes were active and used by approaches.

6. Discussion of Approach Planning Results under Severe Weather Conditions

For validation of the planning and display software, the described scenarios were calculated and the results were then presented to an international group of ATCOs. They evaluated the individual scheduling and trajectory results of the AMAN 4D-CARMA, as well as the entire approach of the support system.

6.1. Results of Diversion Calculations and Sequence Planning

For validation of the new support functions implemented in DLR's AMAN, an evaluation was conducted with the AMAN that was linked to a traffic simulation and a questionnaire. The extended AMAN runs in 15 separate modules that communicate with each other via a Maria database. In addition, the DLR's radar display RadarVision is used, which can display aircraft movements as well as additional information on flight plans and the generated trajectories. In addition to the 2D route, this includes the planned altitudes and speed profiles and, during a simulation, the current air situation in relation to the planned 4D trajectory. For traffic simulation, the AMAN was coupled with DLR's ArrOS system. ArrOS enables precise movements along a trajectory and simulates cooperative controllers and pilots. A selection of screenshots from the simulation runs are used to illustrate the possibilities and results of the validation trials with 4D-CARMA. The following pictures are screenshots showing selected traffic and weather situations with different weather nowcast models.

Diversion route calculation has to consider different constraints in the TMA and the surrounding sectors. In the TMA, the routes are much more convergent, and the aircraft have to overfly more significant waypoints with specific constraints than in the adjacent sectors. As a result, the AMAN can plan larger-scale diversions in the sectors. In Figure 17, the AMAN plans an early diversion and guides the aircraft EJU34CD on a more southern route than the STAR intends.

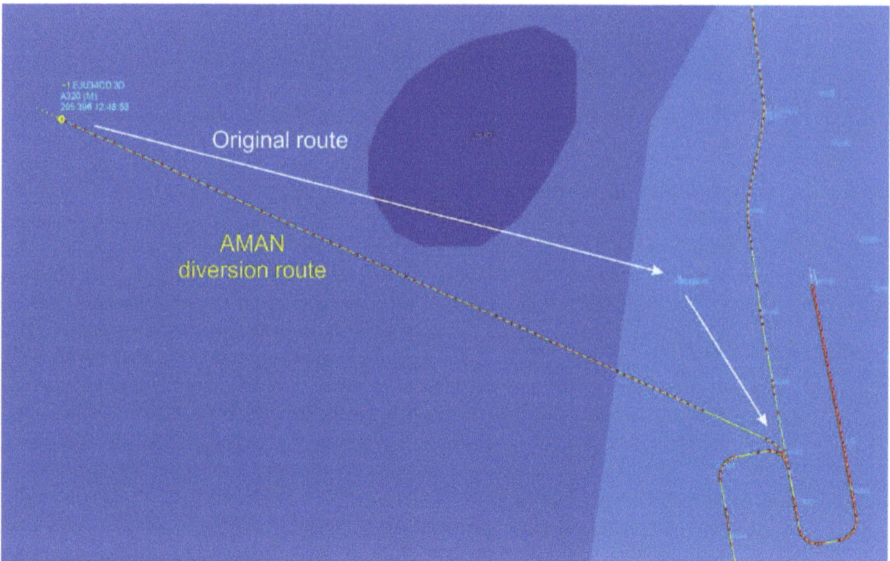

Figure 17. A large-scale diversion route plan for aircraft EJU34CD, currently No. 1 in the arrival sequence and approaching MXP from the west. The white arrows indicate the original STAR and the yellow–red dotted line represents the AMAN 4D trajectory. MXP traffic mix from 11 May 2019 with weather from 11 May 2019 and the WRF-RUC nowcast model.

For traffic visualization on radar screens, a dark image is preferred by some air traffic controllers. In Figure 18, a dark color scheme was selected with trajectories and a thunderstorm cell. In order to reduce the visual presence of the weather information, some controllers had suggested in the run-up to the SINOPTICA concept's development that convective zones should only be displayed as bordered areas [27]. RadarVision now allows controllers to choose between the two weather presentations of filled and bordered areas and also allows for switching between dark and light-colored traffic visualizations through keyboard input.

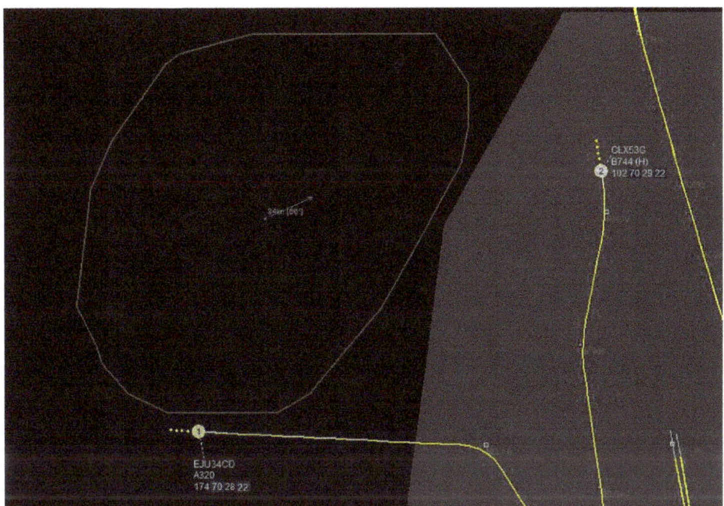

Figure 18. The dark color scheme with convective zones displayed as bordered areas. Malpensa traffic mix from 11 May 2019 with weather from 11 May 2019 and the WRF-RUC nowcast model.

A particular challenge was route planning that required diversions around several convective zones moving at different speeds in different directions. In Figure 19, an example of a diversion around four smaller convective areas is plotted for aircraft BEE156H arriving at Malpensa from the northwest. In weather situations of this type, the AMAN must decide independently whether it is more appropriate to fly through the loosely distributed convective zones or to fly completely around the entire affected area.

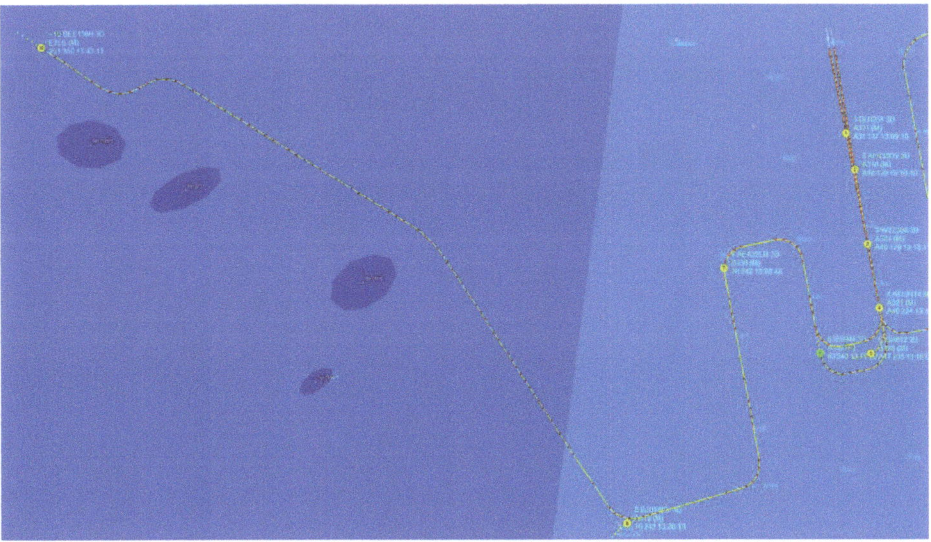

Figure 19. Route calculation for aircraft BEE156H approaching MXP from the northwest through a loose group of convective areas. If several convective cells occur as a group, the AMAN tries to guide aircraft around them as a whole. MXP traffic mix from 11 May 2019 with weather from 6 August 2019 and the RaNDeVIL nowcast model.

In Figure 20, a 4D trajectory was planned with a diversion right through a convective area. However, following the weather forecast, this cell shifted to the southeast in the course of the following minutes so that, when the aircraft arrived, the route precisely avoided this area. If convective cells appear as a loose group, the AMAN tries to find a straight-line route through them. It can happen that routes approach each other for up to a few miles but are continued differently due to a small spatial and temporal distance (Figure 21).

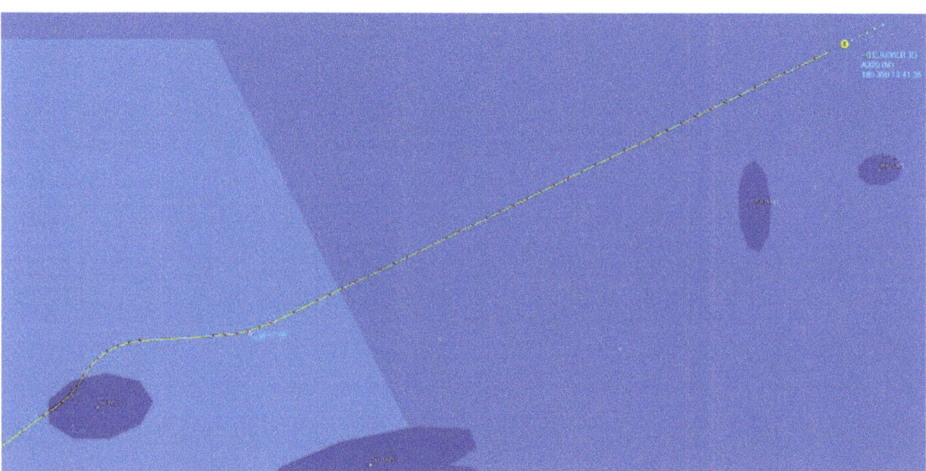

Figure 20. Route calculation for the aircraft EJU36LR approaching MXP from the northeast through a loose group of convective areas. Currently, this route is still blocked by a convective cell; however, based on the forecast, the AMAN knows that the route will be clear by the time the aircraft arrives. MXP traffic mix from 11 May 2019 with weather from 6 August 2019 and the RaNDeVIL nowcast model.

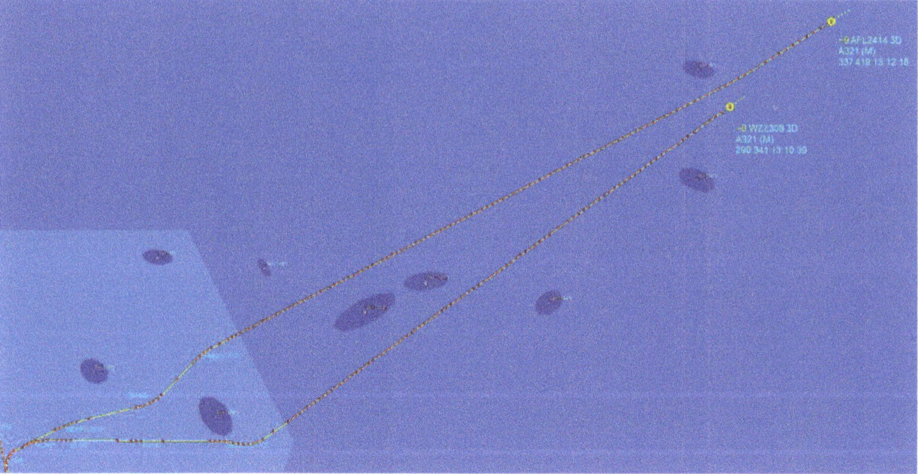

Figure 21. Two different route calculations for aircraft approaching Malpensa from the northeast through a loose group of convective areas. Even a small variance in temporally different initial conditions can lead to different optimal routing. Malpensa traffic mix from 11 May 2019 with weather from 6 August 2019 and the RaNDeVIL nowcast model.

All routes calculated and displayed in this paper are 4D trajectories that include altitude, speed, and time in addition to position. They are always calculated on the basis of EUROCONTROL's BADA 3.13 [36] and are therefore realistic within the scope of technical possibilities. They should also therefore be able to be safely flown by the respective types of aircraft. BADA data for the respective aircraft types are also used for sequence calculation on the finals and the runways so that the distances between the aircraft follow the separations corresponding to their weight classes. This is also valid for the 4D trajectories calculated as diversion routes around severe weather areas. As an example, an aircraft approaching from the north and its speed and altitude profile are presented in Figure 22. It can be seen from the altitude profile in the upper diagram that it is not possible for an aircraft to fly over or under a thunderstorm cell due to the descent that is usually already initiated at the distance of a few dozen miles from the airport.

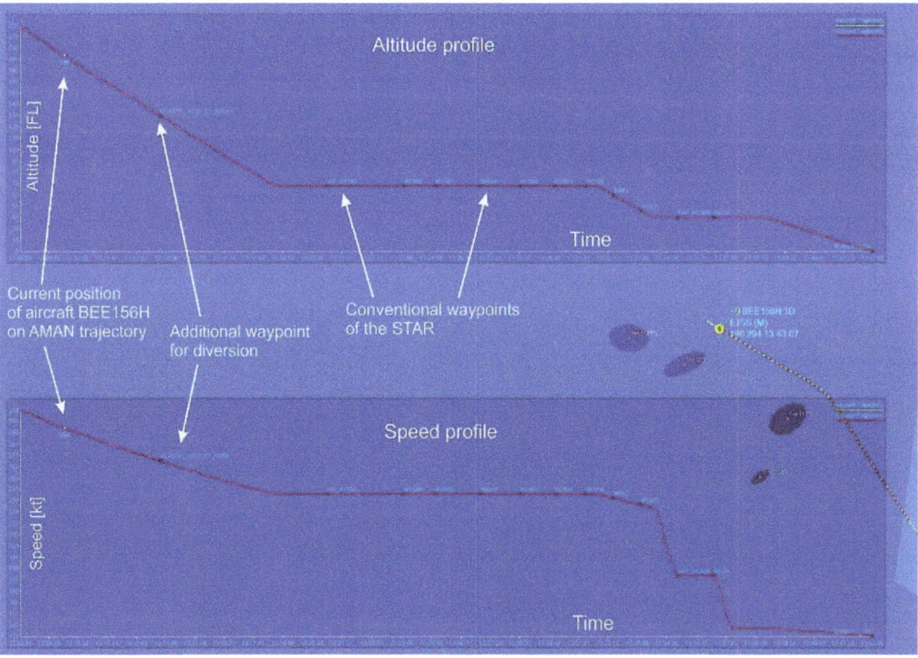

Figure 22. Visualization of an AMAN-calculated 4D diversion trajectory for aircraft BEE156H approaching MXP from the west and currently ninth in the arrival sequence. In the upper dark blue shaded diagram, the altitude profile is displayed, and in the lower one the speed profile is displayed. It can be seen that the aircraft is already descending as it flies around the thunderstorm cell. MXP traffic mix from 11 May 2019 with weather from 11 May 2019 and the PhaSt nowcast model.

Technically, 3D polyhedrons could be used for weather areas to specify the upper and lower boundary of convective cells; however, this does not bring any operational advantages in the context of approach planning. These visualizations indicate the strict dependence of AMAN-calculated 4D trajectories under the influence of local weather on the applied weather model and its assessment of meteorological severity for aviation.

6.2. Results of Controllers' Validations

In the validation process at the end of the project, five active and former ATCOs from Austria, Germany, and Poland received an introduction to the aims and the results of the functional extended AMAN with severe weather guidance support. Afterwards, an AMAN demonstration was held consisting of six videos comprising different AMAN-planned trajectories that changed depending on weather dynamics, traffic dynamics, and the position of the aircraft. Additionally, two different kinds of weather presentations (filled and bordered areas) on radar displays and 4D trajectories with predicted speed and altitude profiles were introduced.

During the validations, the main idea of combining meteorological data, air traffic approach procedures, and the topography of Malpensa in an AMAN for arrival support was presented. The new approach with the adjustment of routes at early stages through the use of forecast data and the organization of air traffic around dynamic weather areas was explained. Five new supporting functionalities with routing, sequencing, weather animation, aircraft tagging in the form of labels, and guidance advisories were introduced and demonstrated. After the introduction, the controllers viewed and analyzed the videos demonstrating AMAN planning and its impact on live traffic.

The controllers then filled out a questionnaire with a total of 17 main questions and sub-questions on the technical assessment. Similar to the user requirements survey, 14 questions offered a seven-point response option that ranged between "strongly agree" and "strongly disagree". Each question also had a comment field where additional comments and suggestions as well as explanations of the answers could be added. The controllers had no time constraints when completing the survey. The questionnaire consisted of 17 questions and sub-questions, 14 of which had graded response options and could be statistically analyzed:

1. In your opinion, can an AMAN be operationally improved by integrating adverse weather information?
    a. In the entire TMA?
    b. In arrival sectors?
    c. On anything else?
2. In your opinion, can an AMAN be operationally improved by integrating adverse weather guidance support?
    a. In the entire TMA?
    b. In arrival sectors?
    c. On anything else?
3. Would the concept proposed in SINOPTICA improve the situational awareness of the Flow Management Position (FMP) in adverse weather conditions?
4. Would the concept of the extended AMAN, proposed in SINOPTICA, lead to improved decision making?
    a. Would it lead to timelier planning?
    b. Would it anticipate decision making?
5. Would the SINOPTICA extended AMAN help to decide more appropriate ATFCM measures? Would it lead to improved Flow Management?
6. Would the proposed enhancements help reduce the negative effects of adverse weather?
7. Are the graphical displays proposed clear and easy to interpret?
8. Would you feel comfortable using the visual aids on the radar display?
9. All in all, do you find the proposed SINOPTICA concept for extended Arrival Management useful?

Analysis of the questionnaires showed that the controllers had professional experience ranging between 1 and 25 years, which thus covers a wide range of experience in the profession of air traffic control. It becomes clear that the newly developed support and visualization functions received clear overall approval from the controllers. Not all controllers were able to answer all the questions because their training and work experience did not give them sufficient knowledge of all the work positions where severe weather support could conceivably be provided (by their own appraisal). Particularly important for our controllers' support system evaluation were the comments, suggestions, and criticisms of the participants. For this reason, a generous comment field was added to each question, which was also used by most participants. These comments were intended to further develop the system so that it could also be operationally used in the future. In addition to a general level of approval, this kind of open user feedback is also important to ensure that further developments do not neglect the needs and requirements of air traffic controllers.

Evaluation of the responses showed the overall comprehensive satisfaction of the involved controllers with the extended system and its newly developed functionalities (Table 2). For the numbers in parentheses, the value 1 represents the answer "strongly disagree", the value 4 represents "neutral", and the value 7 represents "strongly agree", with the other values in between assigned accordingly. The ATCOs exhibited high agreement with the notion that an AMAN can be operationally improved by integrating adverse weather information (6.4/7) and adverse weather guidance support (6/7). They also showed a high level of agreement with the concept that an extended AMAN would lead to improved decision making (6/7). Furthermore, they reported their agreement with the statement suggesting that they would feel comfortable using the visual aids on the radar display (6/7). All in all, they found the proposed concept for extended Arrival Management very useful (6/7).

Table 2. Detailed overview of the results of the questionnaires with graded answer options.

	Question 1a	Question 1b	Question 1c	Question 2a	Question 2b	Question 2c	Question 3	Question 4a	Question 4b	Question 5	Question 6	Question 7	Question 8	Question 9
ATCO 1	7	6	4	7	5	4	-	7	5	-	6	5	6	6
ATCO 2	6	6	5	5	4	4	2	5	5	2	5	6	5	6
ATCO 3	6	6	4	6	6	4	6	5	6	4	7	6	7	6
ATCO 4	6	6	6	6	6	5	4	7	7	6	5	4	5	6
ATCO 5	7	6	-	5	4	-	4	6	4	6	6	6	6	6
Average agreement	6.4	6.0	4.75	5.8	5.0	4.25	4.0	6.0	5.4	4.5	5.8	5.4	5.8	6.0
	agree	agree	somewhat agree	agree	somewhat agree	neutral	neutral	agree	somewhat agree	somewhat agree	agree	somewhat agree	agree	agree

In summary, the evaluation of the validation with controllers showed that they would very much like to use a system that supports them in approach planning and flying around severe weather areas, both from a planning and visual point of view, as it would make their daily work easier and thus also safer. It is noteworthy that they placed more emphasis on weather information than on planning support.

7. Conclusions

In this paper, the extension of an approach controller support system for diversions around severe weather areas is presented that can use different meteorological nowcast models to automatically calculate approach routes and target times for arrival sequencing.

For this purpose, an AMAN was used in the H2020 project SINOPTICA as a planning system considering polygons enclosing severe weather areas as no-fly zones and the calculation of 4D trajectories based on aircraft type-specific parameters to avoid these areas. In the project, the AMAN considered not only standard approach routes, but also the constraints of waypoints; thus, realistic approach profiles were calculated that can be used for approach and touchdown sequencing. After a user requirements analysis with nine ATCOs, new functions were implemented so that approach controllers could receive guidance instructions from the AMAN for clearances in order to guide weather-affected aircraft along planned trajectories, and visual presentations were also provided on the primary display to help quickly and clearly identify aircraft concerned by a diversion. The AMAN and the newly developed diversion functionalities were tested with different extreme weather events that occurred in 2019 in northern Italy, which were modeled and reproduced with the PhaSt, WRF-RUC, and RaNDeVIL weather models.

During the first phase of the project, a user requirement survey was conducted in which different severe weather display types and support functionalities were evaluated by an international ATCO team with nine participants. As a result of the survey, two dynamic presentation variants were developed for the display of the current meteorological situation in the airspace, as well as its predicted development, and these variants are presented in detail in this paper. The first one is based on a linear extrapolation of the last meteorological measurements. It can be used if no dedicated weather forecasts are available. The second variant is based on a method similar to morphing in computer graphics. Here, the shapes of the actual and the predicted weather areas are decomposed into polygons and then merged in a flowing animation. In this way, an ATCO can track the development of severe weather areas at any time and can even adjust the organization of the airspace independent of planning support provided by a decision support system.

An evaluation with an international team of five active controllers showed that an AMAN is very helpful if there is a possibility of large-scale fly-around routing for the avoidance of severe weather areas. For this purpose, longer-term and highly precise forecasts that are precisely tailored to air traffic control requirements are essential. Additionally, the forecast model must correspond to the safety perception of ATCOs and pilots on site so that they can manage the traffic as efficiently and safely as possible. In summary, with the help of sophisticated nowcast models and an extended AMAN, SINOPTICA was able to show that it is possible to support ATCOs and pilots to guide air traffic safely and efficiently around severe weather areas in challenging meteorological situations, thus making planning more reliable and predictable for all stakeholders on the ground and in the air.

Author Contributions: Conceptualization, M.-M.T. and O.G.; methodology, M.-M.T., O.G., L.N., M.K. (Markus Kerschbaum) and O.O.; software, L.N., M.K. (Matthias Kleinert), O.O., K.M., H.E., N.G., V.M., M.M., M.L., E.R., L.E., T.R. and A.T.; validation, K.M., M.-M.T. and O.G.; formal analysis, O.G.; investigation, M.-M.T., O.G., M.K. (Matthias Kleinert), L.N. and O.O.; resources, H.E.; data curation, O.G. and A.T.; writing—original draft preparation, M.-M.T. and O.G.; writing—review and editing, A.P., R.B., O.G., V.M., M.M., M.L., S.F., R.C.T., L.E., M.C.L., R.B. and M.-M.T.; visualization, O.G., L.N., N.G. and M.-M.T.; supervision, A.P.; project administration, A.P.; funding acquisition, E.R., A.P., R.B., M.K. (Markus Kerschbaum) and M.-M.T. All authors have read and agreed to the published version of the manuscript.

Funding: This research was funded by H2020 SESAR, grant number No 892362.

Data Availability Statement: Restrictions apply to the availability of these data. Data was obtained from FlightRadar24.com and are available https://www.flightradar24.com/commercial-services/data-services. Further reports are available under "http://sinoptica-project.eu/index.php/articles/".

Conflicts of Interest: The authors declare no conflict of interest. The founding sponsor had no role in the design of the study, in the collection, analyses, or interpretation of data, in the writing of the manuscript, or in the decision to publish the results.

References

1. Ahlstrom, U. Work domain analysis for air traffic controller weather displays. *J. Saf. Res.* **2005**, *36*, 159–169. [CrossRef] [PubMed]
2. Masson-Delmotte, V.; Zhai, P.; Pirani, A.; Connors, S.L.; Péan, C.; Berger, S.; Caud, N.; Chen, Y.; Goldfarb, L.; Gomis, M.I.; et al. IPCC, 2021: Climate Change 2021: The Physical Science Basis. In *Contribution of Working Group I to the Sixth Assessment Report of the Intergovernmental Panel on Climate Change*; Cambridge University Press: Cambridge, UK, 2021.
3. Lee, D.S.; Fahey, D.W.; Skowron, A.; Allen, M.R.; Burkhardt, U.; Chen, Q.; Doherty, S.J.; Freeman, S.; Forster, P.M.; Fuglestvedt, J.; et al. The contribution of global aviation to anthropogenic climate forcing for 2000 to 2018. *Atmos. Environ.* **2021**, *244*, 117834. [CrossRef] [PubMed]
4. EUROCONTROL. *Performance Review Report—An Assessment of Air Traffic Management in Europe during the Calendar Year 2018 (PRR 2018). Performance Review Commission (PRC)*; European Organization for the Safety of Air Navigation (EUROCONTROL): Brussels, Belgium, 2019.
5. Dube, K.; Nhamo, G.; Chikodzi, D. COVID-19 pandemic and prospects for recovery of the global aviation industry. *J. Air Transp. Manag.* **2021**, *92*, 102–122. [CrossRef]
6. Gewerkschaft der Flugsicherung, E.V. Adverse Weather—New procedures help optimise air traffic management over the Alps in adverse weather conditions. *Der Flugleiter* **2018**, *6*, 49–50.

7. DFS Deutsche Flugsicherung GmbH. *Luftverkehr in Deutschland—Mobilitätsbericht 2020*; German Report of German Air Navigation Service Provider: Langen, Germany, 2021.
8. Steinheimer, M.; Kern, C.; Kerschbaum, M. Quantification of Weather Impact on Arrival Management. In Proceedings of the 13th USA/Europe Air Traffic Management Research and Development Seminar (ATM2019), Vienna, Austria, 17–21 June 2019.
9. Sueddeutsche.de. Flugchaos in London: Tausende Passagiere Sitzen Wegen Nebels Fest. Süddeutsche Zeitung, München, Germany, 21 November 2011, in German. Available online: https://www.sueddeutsche.de/reise/flugchaos-in-london-tausende-passagiere-sitzen-wegen-nebels-fest-1.1195637 (accessed on 15 February 2023).
10. Sueddeutsche.de. Flugverkehr: Winterchaos—Letzter Ausweg München. Süddeutsche Zeitung, München, Germany, 14 March 2011, in German. Available online: http://www.sueddeutsche.de/muenchen/muenchen/flugverkehr-winterchaos-letzter-ausweg-muenchen-1.1038829 (accessed on 15 February 2023).
11. Pejovic, T.; Noland, R.B.; Williams, V.; Toumi, R. A tentative analysis of the impacts of an airport closure. *J. Air Transp. Manag.* **2009**, *15*, 241–248. [CrossRef]
12. Polaschegg, M.; Kerschat, K.; Sawas, S. Advanced Secure Cockpit Connectivity Using the Thuraya Satellite Network. In Proceedings of the 10th Advanced Satellite Multimedia Systems Conference (ASMS) and 16th Signal Processing for Space Communications Workshop (SPSC), Graz, Austria, 20–21 October 2020.
13. Ohneiser, O.; Kleinert, M.; Muth, K.; Gluchshenko, O.; Ehr, H.; Groß, N.; Temme, M.-M. Bad Weather Highlighting: Advanced Visualization of Severe Weather and Support in Air Traffic Control Displays. In Proceedings of the 38th Digital Avionics Systems Conference (DASC), San Diego, CA, USA, 8–12 September 2019.
14. Tienes, C. Important Factors for a Pilot's Decision When Avoiding Severe Weather Conditions. Bachelor's Thesis, Rhein-Waal University of Applied Sciences, Kleve, Germany, 2018.
15. Matthews, M.P.; DeLaura, R. *Modeling Convective Weather Avoidance of Arrivals in Terminal Airspace. Massachusetts Institute of Technology, Lincoln Laboratory*; American Meteorological Society: Seattle, WA, USA, 2011.
16. Davison Reynolds, H.J.; DeLaura, R.; Venuti, J.C.; Wolfson, M. Uncertainty & Decision Making in Air Traffic Management. In Proceedings of the 2013 Aviation Technology, Integration, and Operations Conference, AIAA 2013-4345, Session: Uncertainty in ATM II, Los Angeles, CA, USA, 12–14 August 2013.
17. Ahlstrom, U.; Ohneiser, O.; Caddigan, E. Portable Weather Applications for General Aviation Pilots. *Hum. Factors* **2016**, *58*, 864–885. [CrossRef] [PubMed]
18. Temme, M.-M.; Tienes, C. Factors for Pilot's Decision Making Process to Avoid Severe Weather during Enroute and Approach. In Proceedings of the 37th Digital Systems Avionics Conference (DASC), London, UK, 23–27 September 2018.
19. Ahlstrom, U.; Caddigan, E.; Schulz, K.; Ohneiser, O.; Bastholm, R.; Dworsky, N. *The Effect of Weather State-Change Notifications on General Aviation Pilots' Behavior, Cognitive Engagement, and Weather Situation Awareness*; Technical Report DOT/FAA/TC-15/64; U.S. Department of Transportation, Federal Aviation Administration (FAA): Washington, DC, USA, 2015.
20. Ahlstrom, U. Weather display symbology affects pilot behavior and decision-making. *Int. J. Ind. Ergon.* **2015**, *50*, 73–96. [CrossRef]
21. Ahlstrom, U.; Della Rocco, P. *TRACON Controller Weather Information Needs: I. Literature Review*; Technical Report DOT/FAA/CT-TN03/18; U.S. Department of Transportation, Federal Aviation Administration (FAA): Washington, DC, USA, 2003.
22. Zhang, M.; Kong, X.; Liu, K.; Li, X. A Novel Rerouting Planning Model for the Terminal Arrival Routes under the Influence of Convective Weather. *J. Adv. Transp.* **2018**, *2018*, 7591932. [CrossRef]
23. Hayashi, M.; Isaacson, D.; Tang, H. Evaluation of a Dynamic Weather-Avoidance Rerouting Tool in Adjacent-Center Arrival Metering. In Proceedings of the 13th USA/Europe Air Traffic Management Research and Development Seminar, Vienna, Austria, 17–21 June 2019.
24. Evans, J.E.; Weber, M.E.; Moser, W.R., II. Integrating Advanced Weather Forecast Technologies into Air Traffic Management Decision Support. *Linc. Lab. J.* **2006**, *16*, 81–96.
25. Endsley, M.; Stein, E.S.; Sollenberger, R.L.; Nakata, A. *Situation Awareness in Air Traffic Control: Enhanced Displays for Advanced Operations*; DOT/FAA/CT-TN00/01; Federal Aviation Administration (FAA): Atlantic City, NJ, USA, 2000.
26. Parodi, A.; Mazzarella, V.; Milelli, M.; Lagasio, M.; Realini, E.; Federico, S.; Torcasio, R.C.; Kerschbaum, M.; Llasat, M.C.; Rigo, T.; et al. A nowcasting model for severe weather events at airport spatial scale: The case study of Milano Malpensa. In Proceedings of the 11th SESAR Innovation Days (SID), Online, 7–9 December 2021.
27. Temme, M.-M.; Gluchshenko, O.; Kerschbaum, M. *SINOPTICA Operation Concept Description and User Requirements for Adverse Weather Controller Support*; German Aerospace Center (DLR): Braunschweig, Germany, 2021.
28. Völckers, U. Arrival Planning and Sequencing with COMPAS-OP at the Frankfurt ATC-Center. In Proceedings of the 1990 American Control Conference, San Diego, CA, USA, 23–25 May 1990; pp. 496–501.
29. Gerling, W.; Seidel, D. Project 4-D Planner. In *Scientific Seminar 2002*; Institute of Flight Guidance: Braunschweig, Germany, 2002.
30. Harwood, K.; Sanford, B.D.; Lee, K.K. Developing ATC Automation in the Field: It Pays to Get Your Hands Dirty. *Air Traffic Control. Q. I* **1998**, *6*, 45–70. [CrossRef]
31. Helmke, H.; Ohneiser, O.; Buxbaum, J.; Kern, C. Increasing ATM Efficiency with Assistant Based Speech Recognition. In Proceedings of the 12th USA/Europe Air Traffic Management Research and Development Seminar (ATM2017), Seattle, DC, USA, 27–30 June 2017.
32. Besnard, X.; Guerin, E.; Clark, A.; Finke, M.; Easthope, G.; Azoulay, M.; Lacroix, A.; Zetsche, F.; Dieck, D. *xStream Demonstration Report—Arrival Management Extended to En-Route Airspace*; SESAR Joint Undertaking: Brussels, Belgium, 2019.

33. Dhief, I.; Wang, Z.; Liang, M.; Alam, S.; Schultz, M.; Delahaye, D. Predicting Aircraft Landing Time in Extended-TMA Using Machine Learning Methods. In Proceedings of the 9th International Conference for Research in Air Transportation (ICRAT), Tampa, FL, USA, 23–26 June 2020.
34. Seidel, D. Prädiktion von Anflugsequenzen mit Verfahren des maschinellen Lernens. (in German, Prediction of approach sequences using machine learning techniques). *Innov. Im Fokus* **2020**, *2*, 15–24.
35. Jun, L.Z.; Alam, S.; Dhief, I.; Schultz, M. Towards a greener Extended-Arrival Manager in air traffic control: A heuristic approach for dynamic speed control using machine-learned delay prediction model. *J. Air Transp. Manag.* **2022**, *103*, 102250. [CrossRef]
36. Nuic, A. *User Manual for the Base of Aircraft Data (BADA) Revision 3.13*; Eurocontrol Experimental Centre—European Organisation for the Safety of Air Navigation: Brétigny-sur-Orge, France, 2015.
37. Malkova, M. *Morphing of Geometrical Objects in Boundary Representation: The State of the Art and the Concept of Ph.D. Thesis*; Technical Report DCSE/TR-2010-02; University of West Bohemia in Pilsen: Pilsen, Czech Republic, 2010.
38. Gotsman, C.; Surazhsky, V. Guaranteed intersection-free polygon morphing. *Comput. Graph.* **2001**, *25*, 67–75. [CrossRef]
39. Guibas, L.; Hershberger, J.; Suri, S. Morphing Simple Polygons. *Discret. Comput. Geom* **2000**, *24*, 1. [CrossRef]
40. Malkova, M.; Parus, J.; Kolingerova; Benes, B. An intuitive polygon morphing. *Vis. Comput.* **2009**, *26*, 205–215. [CrossRef]
41. Moreira, J.; Dias, P.; Mesquita, P. Morphing techniques for creating and representing spatiotemporal data in GIS. International Environmental Modelling and Software Society (iEMSs). In Proceedings of the 7th International Congress on Environmental Modelling and Software, San Diego, CA, USA, 15–19 June 2014.
42. Liu, L.; Wang, G.; Zhang, B.; Guo, B.; Shum, H.-Y. Perceptually Based Approach for Planar Shape Morphing. In Proceedings of the 12th Conference on Computer Graphics and Applications, IEEE, Seoul, Republic of Korea, 6–8 October 2004; pp. 111–120.
43. Skamarock, W.C.; Klemp, J.B.; Dudhia, J.; Gill, D.O.; Barker, D.; Duda, M.G.; Powers, J.G. A Description of the Advanced Research WRF Version 3 (No. NCAR/TN-475+STR). *Univ. Corp. Atmos. Res.* **2008**. [CrossRef]
44. Lagasio, M.; Silvestro, F.; Campo, L.; Parodi, A. Predictive capability of a high-resolution hydrometeorological forecasting framework coupling WRF cycling 3dvar and Continuum. *J. Hydrometeorol.* **2019**, *20*, 1307–1337. [CrossRef]
45. Lagasio, M.; Campo, L.; Milelli, M.; Mazzarella, V.; Poletti, M.L.; Silvestro, F.; Ferraris, L.; Federico, S.; Puca, S.; Parodi, A. SWING, The Score-Weighted Improved NowcastinG Algorithm: Description and Application. *Water* **2022**, *14*, 2131. [CrossRef]
46. Mazzarella, V.; Milelli, M.; Lagasio, M.; Federico, S.; Torcasio, R.C.; Biondi, R.; Realini, E.; Llasat, M.C.; Rigo, T.; Esbrí, L.; et al. Is an NWP-Based Nowcasting System Suitable for Aviation Operations? *Remote Sens.* **2022**, *14*, 4440. [CrossRef]
47. Metta, S.; von Hardenberg, J.; Ferraris, L.; Rebora, N.; Provenzale, A. Precipitation nowcasting by a spectral-based nonlinear stochastic model. *J. Hydrometeorol.* **2009**, *10*, 1285–1297. [CrossRef]
48. Greene, D.R.; Clark, R.A. Vertically integrated liquid water—A new analysis tool. *Mon. Weather Rev.* **1972**, *100*, 548–552. [CrossRef]
49. Esbri, L.; Llasat, M.C.; Rigo, T.; Milelli, M.; Mazzarella, V.; Lagasio, M.; Parodi, A.; Temme, M.-M.; Gluchshenko, O.; Kerschbaum, M.; et al. Initial Results of the Project SINOPTICA (Satellite-Borne and IN-Situ Observations to Predict The Initiation of Convection for ATM). EMS Annual Meeting 2021. EMS2021-189. Available online: https://elib.dlr.de/143282/ (accessed on 15 February 2023).
50. Barker, D.M.; Huang, X.Y.; Liu, Z.; Aulігné, T.; Zhang, X.; Rugg, S.; Ajjaji, R.; Bourgeois, A.; Bray, J.; Chen, Y.; et al. The weather research and forecasting model's community variational/ensemble data assimilation system: WRFDA. *Bull. Am. Meteorol. Soc.* **2012**, *93*, 831–843. [CrossRef]
51. Barker, D.M.; Huang, W.; Guo, Y.-R.; Bourgeois, A.; Xiao, Q. A three-dimensional variational data assimilation system forMM5: Implementation and initial results. *Mon. Weather. Rev.* **2004**, *132*, 897–914. [CrossRef]
52. Fierro, A.O.; Mansell, E.R.; Ziegler, C.L.; MacGorman, D.R. Application of a lightning data assimilation technique in the WRF-ARW model at cloud-resolving scales for the tornado outbreak of 24 May 2011. *Mon. Weather Rev.* **2012**, *140*, 2609–2627. [CrossRef]
53. Parodi, A.; Mazzarella, V.; Milelli, M.; Lagasio, M.; Federico, S.; Torcasio, C.; Realini, E.; Temme, M.-M.; Gluchshenko, O.; Temme, A.; et al. Il progetto SINOPTICA: Come si può migliorare la gestione del traffico aereo durante gli eventi meteorologici severi? In Proceedings of the 4 Congresso Nazionale AISAM, Milano, Italy, 15–19 February 2022.
54. ENAV. *Aeronautical Information Publications LIMC Malpensa*; AD 2 LIMC 4; Società Nazionale per l'Assistenza al Volo (ENAV): Roma, Italia, 2020.

Article

Concepts for Increased Energy Dissipation in CFRP Composites Subjected to Impact Loading Conditions by Optimising Interlaminar Properties

Moritz Kuhtz [1,*], Jonas Richter [1], Jens Wiegand [2], Albert Langkamp [1], Andreas Hornig [1] and Maik Gude [1]

1 Institute of Lightweight Engineering and Polymer Technology, Technische Universität Dresden, Holbeinstraße 3, 01307 Dresden, Germany
2 COMPACT Composite Impact Engineering, 12 Gateway Mews, London N11 2UT, UK
* Correspondence: moritz.kuhtz@tu-dresden.de; Tel.: +49-351-4634-2089

Abstract: Carbon fibre-reinforced plastics (CFRP) are predestined for use in high-performance components due to their superior specific mechanical properties. In addition, these materials have the advantage that the material properties and in particular, the failure behaviour can be adjusted. Fibre-dominated failure modes are usually brittle and catastrophic. In contrast, delaminations successively absorb energy and retain in-plane structural integrity. Previous investigations have shown that interface modifications can be used to selectively adjust the interlaminar properties, which decisively influence the delamination behaviour and the associated failure behaviour of structures. However, a systematic analysis of the influences of the positioning and characteristics of the interface modifications on the structural failure behaviour is still missing. Based on existing experimental investigations on the energy dissipation of CFRP impact-loaded beams, the failure behaviour is described here with the help of numerical simulations. The structural failure behaviour and the energy dissipation are represented in a three-dimensional, parameterised finite element analysis (FEA) model. Furthermore, the parameterised models are used to maximise the energy absorption of the three-point bending test through three concepts of interface modification. The large number of model input parameters requires a metamodel-based description of the correlation between the positioning and characteristics of the interface modification and the energy dissipation. Within the scope of the present work, a procedure is therefore developed which enables an efficient design of interface-modified CFRP under impact loads.

Keywords: carbon fibre-reinforced plastics; energy absorption; impact; optimisation; simulation; three-point bending test

Citation: Kuhtz, M.; Richter, J.; Wiegand, J.; Langkamp, A.; Hornig, A.; Gude, M. Concepts for Increased Energy Dissipation in CFRP Composites Subjected to Impact Loading Conditions by Optimising Interlaminar Properties. *Aerospace* **2023**, *10*, 248. https://doi.org/10.3390/aerospace10030248

Academic Editors: Spiros Pantelakis, Andreas Strohmayer and Jordi Pons-Prats

Received: 30 January 2023
Revised: 16 February 2023
Accepted: 23 February 2023
Published: 3 March 2023

1. Introduction

With their highly specific mechanical properties, carbon fibre-reinforced plastics (CFRP) are suitable for use in high-performance components such as crash- and impact-loaded structures such as composite fan blades [1,2]. In addition, these materials have the advantage that the material properties and especially the failure behaviour can be adjusted. The focus is mostly on the adaption of the fibre orientation to the dominant stress in order to increase the stiffness and strength of the materials [3]. This approach often leads to a brittle failure behaviour of the corresponding structures so that on the one hand, the structural integrity is compromised, and on the other hand, the energy absorption capacity is not optimally utilised [4].

Besides the choice of fibre and matrix material as well as the textile architecture, there are essentially two approaches for increasing the energy absorption of fibre-reinforced plastics. Firstly, the fibre–matrix adhesion is specifically adjusted by influencing the boundary layer during manufacturing. Initial work on this goes back to [5], where the influence of the adhesion properties of an epoxy matrix to boron filaments and their influence on the fracture toughness of such composites is described. The work in [6] follows this approach and shows that the fracture toughness is further increased by alternating high and low fibre–matrix bonds along the fibres. Secondly, especially in multi-layered composites, suitable interlayers are inserted through the concept of controlled interlaminar bonding. In this methodology, primarily thermoplastic interlayers are inserted into layer-based thermoset composites. A good overview of methods that lead to an improvement of the interlaminar properties by interleaved layers on a thermoplastic basis, but also to a more brittle failure behaviour with lower energy absorption, is given in [7]. In contrast, methods with lower interlaminar properties generally result in significantly higher energy absorption with improved structural integrity and a moderate decrease in structural strength [8–10].

The mentioned studies are exclusively experimental, and their conclusions are based on purely empirical statements. The study presented here, however, introduces a simulation approach that can be used to identify, analyse, and target the phenomenology of the energy absorption mechanisms of impact-loaded composite structures. A methodology is presented that describes the interface modification of CFRP in the framework of finite element analysis (FEA) up to the structural scale. Thus, this approach can be used to extend the experimental database with virtual tests. Finally, it is shown how the interface modification approach can be adapted to an optimisation process [11,12]. Thus, it is possible to derive concepts with which the energy absorption behaviour of impact-loaded CFRP structures can be significantly increased.

2. Materials and Methods

Deformation, damage, and failure behaviour of interface-modified CFRP beams is investigated numerically based on the experimental data of [10]. An interfacial modification by means of perforated polytetrafluoroethylene (PTFE) foil is used (Figure 1a). Figure 1b shows the geometric dimensions of the perforated PTFE foil.

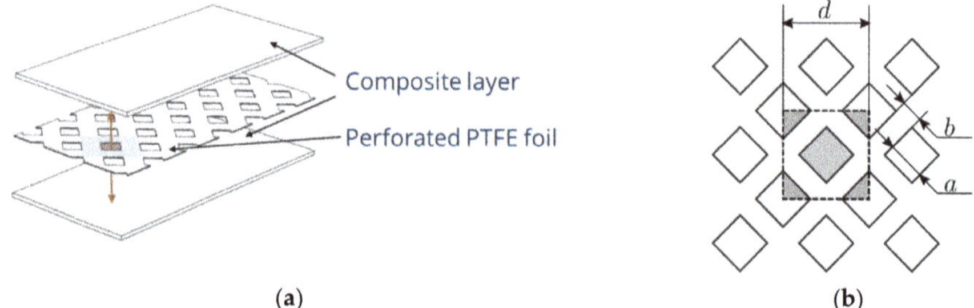

Figure 1. Modification concept. (**a**) Interface modification concept by inter-leaved perforated PTFE foils with quadratic holes; (**b**) geometry parameters of perforated PTFE foil: a: edge length of hole, b: distance between holes, d: edge length of unit cell [13].

As PTFE does not adhere to the composite material, the interface properties are influenced exclusively by the geometric (mesostructural) dimensions and can be adjusted by the so-called interlaminar contact area κ:

$$\kappa = \frac{2\,a^2}{d^2}. \tag{1}$$

In the experimental study, a constant value of 4 mm was chosen for the edge length of the perforation a [13]. Based of targeted interlaminar contact area κ, the distances between two perforations b and the length of the unit cell d are derived. The investigations are carried out using the example of a plain weave fabric composite based on a HexPly M49 200P prepreg semi-finished product from Hexcel. Consolidation is carried out in an autoclave process according to the manufacturer's specifications, resulting in an average fibre volume content of 55% [14]. Further details about the physical and mechanical properties are shown in Table A1 in Appendix A.

2.1. Influence of Interface Modification on In-Plane Properties

To determine the influence of the interlaminar interface modification on the in-plane properties of the composite, tensile tests according to DIN EN ISO 527-4 are performed. The rectangular specimens were 200 mm long, 25 mm wide, and 2 mm thick and were tested with a loading velocity of 0.5 mm/s. Figure 2a shows the fracture patterns of exemplary tensile specimens with the respective contact surface proportions κ. The fracture surfaces of specimens with high contact area fractions show smooth fracture surfaces, while specimens with low contact area fractions show more jagged fracture patterns. This behaviour is a consequence of different load redistribution processes during damage propagation. When a single layer of a specimen with high interlaminar properties fails, the load is redistributed to the other layers. These are subsequently overloaded in the immediate vicinity of the first point of damage and subsequently fail there. In contrast to test specimens with a high contact area, test specimens with a smaller contact area show a different failure mechanism. If a single layer in the laminate with low interlaminar properties fails, the crack does not run into the adjacent layers, but separates and decouples the individual layers from each other. This causes the adjacent individual layers to fail at their respective weak points, which do not necessarily have to be close to the position of the original initial failure.

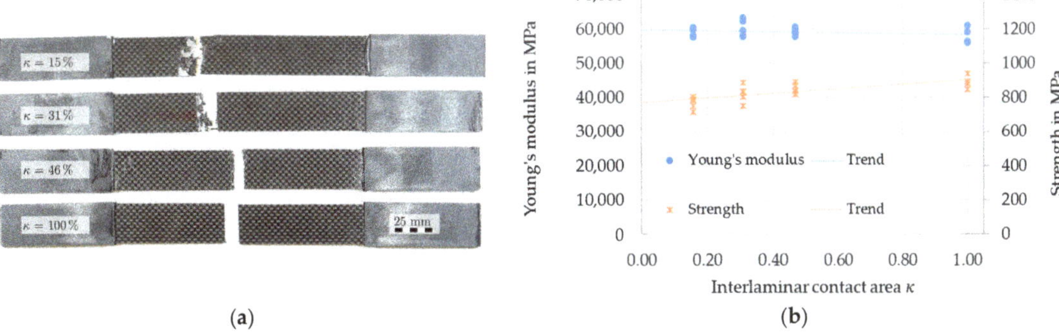

Figure 2. Failure behaviour of interface-modified textile-reinforced CFRP; (**a**) tensile specimens after tests and (**b**) corresponding mechanical in-plane properties in weft direction.

The Young's modulus remains almost constant with the increasing contact area κ. The strength, however, increases with increasing contact area κ, which is due to an influence on the local stress state as a result of the perforation as an interference point. Additionally, the load transfer between the individual layers is limited by the low interlaminar property. The investigations carried out here prove that the in-plane properties under tensile load are not significantly influenced by the selected type of interface modification.

The in-plane failure behaviour is described by an adaptation of Hashin's failure theory [15]. Here, no interaction between tensile σ_{11} and shear stresses σ_{12} and σ_{13} in tensile fibre mode is assumed:

$$\frac{\sigma_{11}}{xt} = 1, \text{ for } \sigma_{11} > 0. \tag{2}$$

The fibre compressive mode is modelled analogously:

$$\frac{|\sigma_{11}|}{xc} = 1, \text{ for } \sigma_{11} < 0, \tag{3}$$

where xt and xc are the tensile and compressive strengths in the fibre direction. Failure in the 2-direction is also modelled as fibre failure for woven fabrics due to symmetry.

Since out-of-plane failure is modelled with the cohesive element approach, the out-of-plane failure mode is supressed by setting strengths to very high values for the single-ply material model. The LS-DYNA material model 58 (*MAT_058) [16] is used for modelling. MAT_058 also enables the modelling of damage evolution using the Matzenmiller model [17]. The material's stiffness is gradually degraded until a residual level of stress (*slimt1, slimc1*) is reached. This is kept constant until the final failure strain (*fail1, fail2*) is reached. Since the material can still transfer loads in compression failure, higher values are assumed for *slimc1* and *fail2* than for the corresponding values for tensile failure *slimt1* and *fail1*. Figure 3 shows the modelled stress–strain curve in the fibre direction. Table 1 shows the key material parameters. The full LS-DYNA material cards are shown in Tables A2 and A3 in Appendix A.

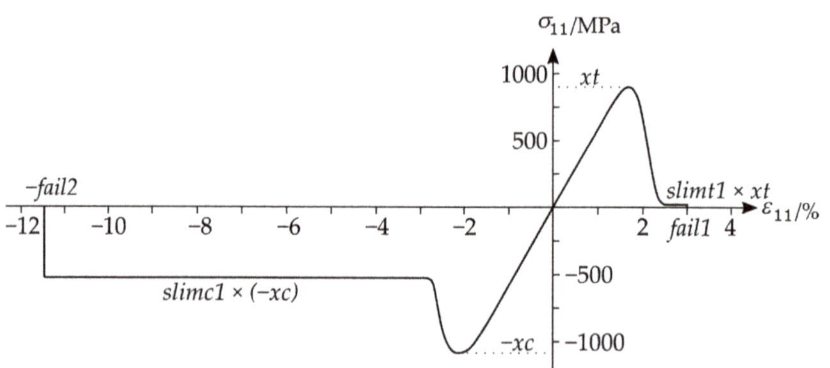

Figure 3. Investigated interface modification designs.

Table 1. Material properties used for three-point bending model.

Property	**Symbol**	**Quantity**	**Unit**
Elastic			
Young's modulus in fibre direction	*ea, eb*	53,300	MPa
Young's modulus in through-thickness direction	*ec*	10,000	MPa
Poisson ratio in-plane	*prba*	0.085	-
Poisson ratio out-of-plane	*prca, prcb*	0.0185	-
Shear modulus in-plane	*gab*	4099	MPa
Shear modulus out-of-plane	*gbc, gca*	1775	MPa
Strength			
Tensile strength in-plane	*xt, yt*	994	MPa
Compressive strength in-plane	*xc, yc*	1081	MPa
Shear strength in-plane	*sc*	100	MPa
Post failure			
Factor of minimum stress for fibre tension	*slimt1, slimt2*	0.03	-
Factor of minimum stress for fibre compression	*slimc1, slimc2*	0.581	-
Factor of minimum stress for in-plane shear	*slims*	0.950	-
Failure strains (element deletion)			
Tensile failure strain in fibre direction	*fail1, fail3*	0.030	-
Compressive failure strain in fibre direction	*fail2, fail4*	0.115	-
Shear failure strain in-plane	*fail5*	0.400	-

2.2. Influence of Interface Modification on Out-of-Plane Properties

The influence of the interlaminar contact area proportion on the delamination behaviour is investigated using five pre-cracked double cantilever beam (DCB) test specimens. The specimens have an interlaminar contact area κ of approx. 0.15, 0.30, 0.45, and 1.0, respectively, and are tested according to ISO15024. The results are presented in Figure 4. The relationship between through-thickness strength, the strain energy release rate, and κ is reasonably well-represented by a linear fit.

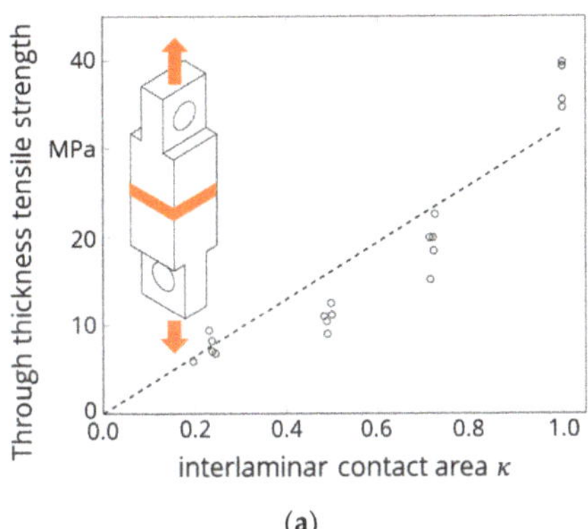

(a)

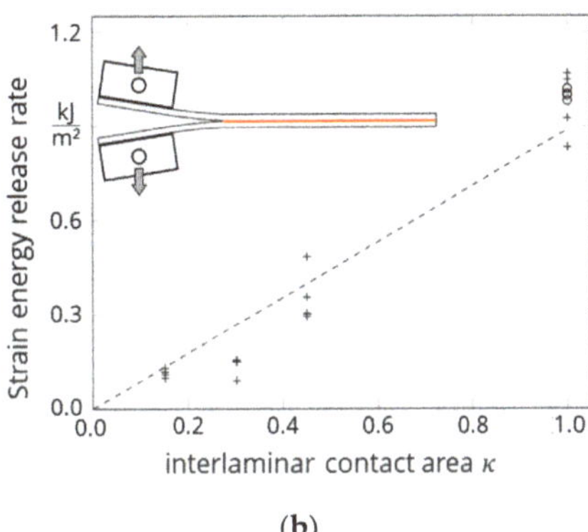

(b)

Figure 4. (a) Through-thickness tensile test results and (b) DCB test results for different interlaminar contact areas.

Cohesive zone approaches are used for modelling delamination between adjacent single plies. They allow an evaluation of the delamination initiation and a description of the delamination growth. LS-DYNA offers cohesive zone models as part of contact formulations or through special cohesive elements.

The accuracy of cohesive elements is higher in comparison to contact formulations [18]. Therefore, cohesive elements with a bilinear material model are used here for modelling the delamination behaviour of interface-modified multilayer composites.

Table 2 shows the key parameters for the cohesive model without interface modification (κ = 1.0). The full LS-DYNA material card is shown in Table A4 in Appendix A. The parameters of the strengths (t and s) and the critical energy release rates (gic and $giic$) are adjusted to account for the interface modification in the material model. Since PTFE does not adhere to the composite material, the interlaminar properties of the strengths and strain energy release rates are also zero at $\kappa = 0$ (no perforation, intact release foil). A linear relationship between the interlaminar contact area and strain energy release rate in Mode I applies to both the strengths and the characteristic values in the Mode II load case (Figure 4).

2.3. Three-Point Bending Test Simulation

The influence of the interface modification on the structural deformation and failure behaviour is investigated by modelling an impact-loaded three-point bending beam setup (Figure 5a). The supports and the load introduction are modelled with rigid shell elements. The indenter moves downwards at a constant velocity of 2 m/s for a total displacement of 20 mm. Each prepreg layer is modelled by two elements in the thickness direction using reduced integrated hexahedral elements (ELFORM 1). The element size is 1 mm in the longitudinal axis and 3.33 mm in the width direction, resulting in 5964 solid elements. In addition, potential delamination layers are distributed symmetrically to the centre plane corresponding to the experimental setup [10] with 1 µm thick cohesive elements (ELFOR 19), resulting in 1065 cohesive elements. A segment-based automatic surface-to-surface contact with a friction coefficient of 0.3 [19] is used to enable contact between the composite plies after the cohesive elements have failed.

Table 2. Cohesive properties.

Property	Symbol	Quantity	Unit
Elastic			
Normal stiffness	*en*	5×10^6	N/mm^3
Shear stiffness out-of-plane	*et*	1×10^6	N/mm^3
Strength			
Normal tensile strength	*t*	32	MPa
Shear strength out-of-plane	*s*	50	MPa
Post failure			
Normal strain energy release rate	*gic*	0.96	kJ/m^2
Shear strain energy release rate	*giic*	2.50	kJ/m^2

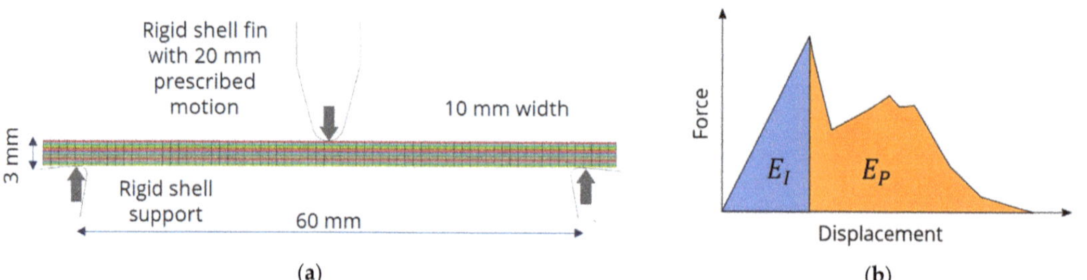

Figure 5. (**a**) Geometry of three-point bending test. (**b**) Force–displacement curve with according energies.

The test is evaluated using the force–displacement curve (Figure 5b). The curve's integral equals the total absorbed energy E_T. The integral up to the force maximum is referred to as initiation energy E_I and corresponds largely to the elastic energy stored in the test specimen. The remaining energy is referred to as propagation energy E_P and is largely related to damage and failure processes in the test specimen.

To improve the energy absorption capacity of impact-loaded CFRP structures, the total energy absorbed E_T should be maximised. In the case presented here, this can be achieved by maximising the propagation energy without significantly reducing the initiation energy. Furthermore, the structural integrity of such CFRP composites can be improved by the initiation and propagation of delaminations.

2.4. Optimisation Strategies

The desired increase of energy absorption and structural integrity of the impact-loaded CFRP beam is approached using three different interface designs (Figure 6). The interlaminar properties of all five interfaces are uniformly weakened by varying the interlaminar contact area in design A (Figure 6a). Design B establishes whether a layer-by-layer weakening of the interface leads to a further increase in energy absorption (Figure 6b). Finally, sectional interface modifications are used in the third concept to determine which positions are particularly suitable for weakening in design C (Figure 6c).

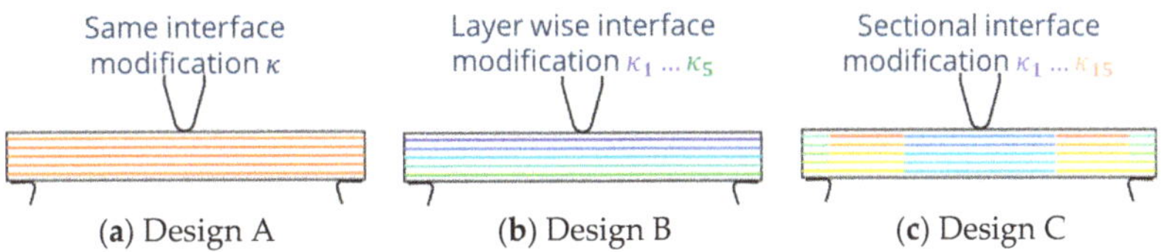

(**a**) Design A (**b**) Design B (**c**) Design C

Figure 6. Investigated interface modification designs.

The software LS-Opt is used to carry out the optimisation [20]. A metamodel-based optimisation of the resulting 15 model input parameters is performed using a sequential approach, which reduces the parameter space in each iteration (see Figure 7).

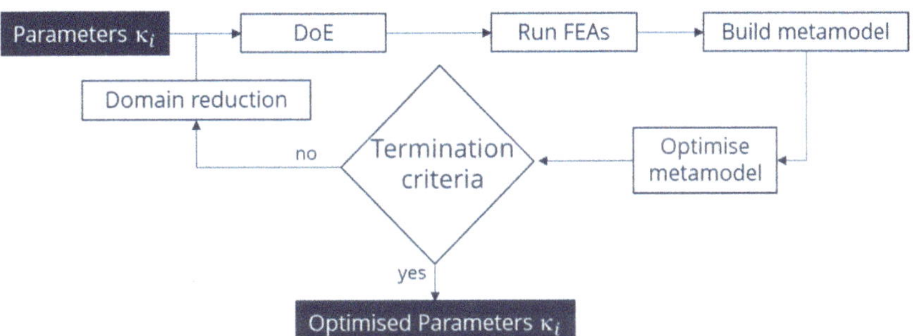

Figure 7. Framework of the applied metamodel-based optimisation.

A design of experiment (DoE) is created using Latin hypercube sampling. FE simulations are then generated for each parameter set. Based on these results, a feedforwarded neural network (FFN) metamodel is trained using standard LS-Opt settings. The FFN approximates the relationship between the interface parameters κ_i and the total energy absorption E_T. The input parameters are then optimised based on the predictions of the metamodel. The ASA (adaptive simulated annealing) algorithm is used for optimisation. The sequential optimisation is terminated when the parameter sets or the objective function differ less than 1% from the previous iteration. If this is not the case, new parameter sets are generated, whereby the parameter set domain is reduced by 20%.

3. Results

First, the FEA model is evaluated regarding its predictive quality for representing the complex deformation and failure behaviour. Then, the results of the optimisation for the three interface designs are presented.

3.1. Model Validation of the Reference Model

Figure 8 shows a comparison of the deformation and failure behaviour of impact-loaded CFRP beams with different interface modifications (design A) based on both experimental and numerical analyses. In the case of no interface modification $\kappa_i = 1$, the specimen fails brittle and central. This is well-predicted by the simulation. When the interfacial property is weakened ($\kappa_i = 0.3$), a delamination failure occurs as initial failure mode on only one side of the specimen with a subsequent fibre failure. This sequence is not predicted accurately by the simulation. The model predicts a symmetrical delamination failure on both sides of the specimen, which is due to the assumed perfect symmetry. Nevertheless, this FEA model can be used to investigate the influence of the interface design on the structural behaviour since the key failure modes are captured by the model.

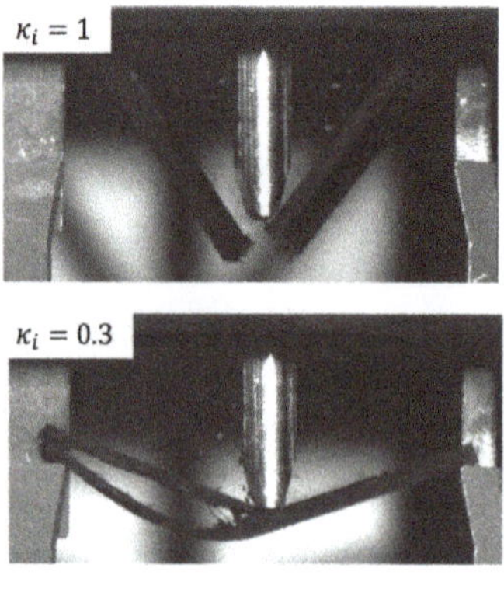

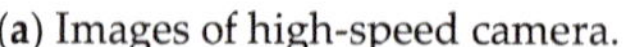

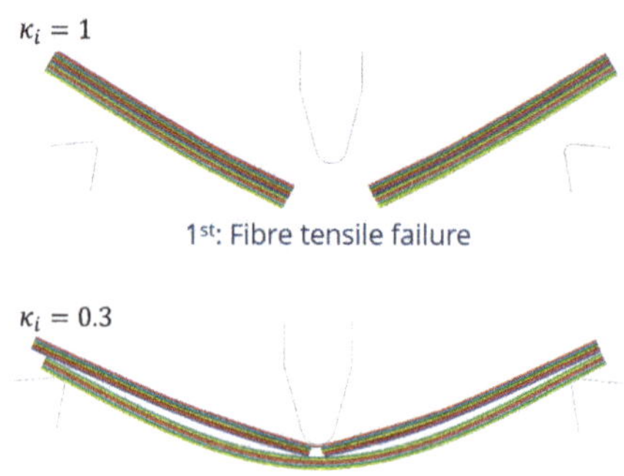

(**a**) Images of high-speed camera. (**b**) Corresponding simulation images.

Figure 8. (**a**) Images of high-speed camera during the three-point bending test for two different interface modification concepts [10] and (**b**) corresponding simulation images.

3.2. *Energy Absorption Capacity of Optimised Interface Modification Designs*

The following section presents the results of optimising the interface designs for improved energy absorption and structural integrity.

3.2.1. Design A (Same Layer-by-layer Interface Modification Concept—Figure 6a)

Figure 9 shows the energy absorption of the impact-loaded CFRP beam with different interlaminar contact areas κ. Figure 5b displays the various energies as functions of contact area. The total energy E_T is the sum of the initiation E_I and propagation energy E_P.

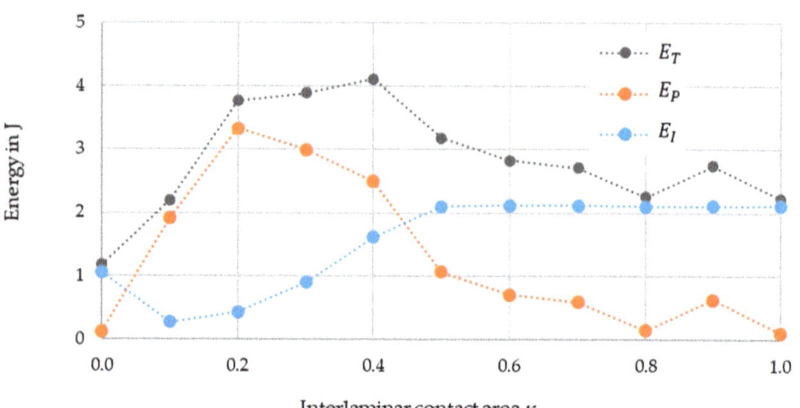

Figure 9. Simulated initial E_I and propagation energy E_P of the three-point bending test for different interface modifications κ_i.

The simulation with a contact area ($\kappa = 0.0$) shows a very compliant structural response, as the whole specimen is already delaminated in the simulation. It also shows no in-plane failure until the specimen slips through the support, and the force signal increases continuously from this point on. After the slippage, there is still some force transfer due to inertia and friction from the contact of the indenter with the specimen. However, these effects are neglectable; therefore, the energy absorption capacity of this interface modification is largely determined by the initiation energy.

From a κ of 0.1, E_I increases as the force rises more steeply due to a higher initial stiffness. Compared to the case $\kappa = 0.0$, E_I is lower for $\kappa = 0.1$. Although the maximum force is much higher, the displacement at maximum force is lower due to the much higher stiffness, resulting in lower E_I. Since the interlaminar properties of strengths (t and s) and critical energy release rates (gic and $giic$) are very low, extensive delamination occurs in all five interfaces from the time the maximum force is reached. As a result, E_P is significantly higher than E_I. In general, this indicates that an increasing contact area leads to an increased E_I due to the higher force maxima caused by increasing interlaminar strengths. However, at contact areas of 0.5 and above, the initial failure is no longer dominated by delamination failure, but by fibre failure. This is why the initiation energy remains almost constant from this interface modification onwards.

E_P decreases starting from $\kappa = 0.2$. Due to increasing interface strengths, not all five modified interfaces delaminate and dissipate energy accordingly. Reduced delamination occurs since the initial failure is determined by the in-plane failure. This trend intensifies above a κ of 0.5. E_P decreases with increasing interlaminar contact area and approaches zero. The small increase in the propagation energy at $\kappa = 0.9$ results from a subsequent force effect on the indenter due to a collision with the refracted beam arms and thus represents a numerical artefact.

In reference [10], the amounts for the interlaminar contact area are given as nominal values. For comparison with the simulations presented here, it is necessary to use the precise values for the specimens resulting from the manufacturing process. Table 3 contains these values, and Figure 10 compares the results from [10] with the simulation results based on the methods used here.

Table 3. Nominal values [10] and values resulting from the manufacturing process for interlaminar contact area.

	Interlaminar Contact Area κ			
Nominal [10]	0.20	0.40	0.60	1.00
Real	0.15	0.31	0.46	1.00

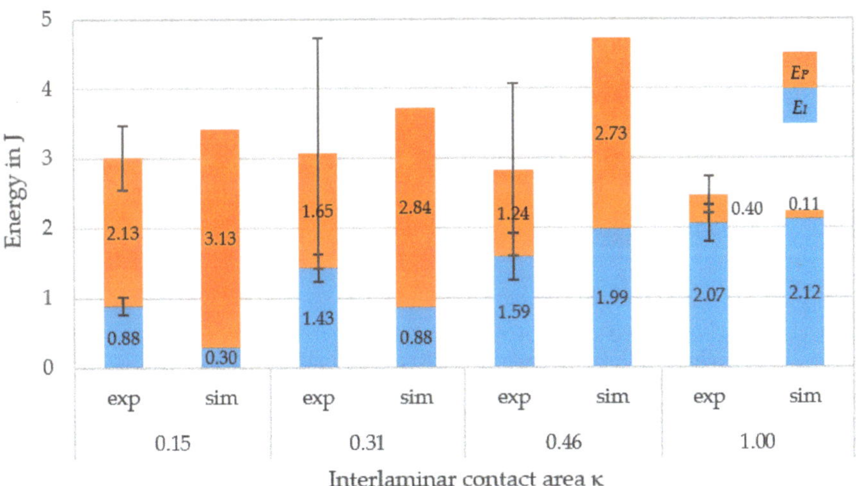

Figure 10. Experimental and numerical initial E_I and propagation energy E_P in three-point bending test for certain interface modifications κ.

Generally, the numerical model is capable of reproducing the complex deformation and failure behaviour of the impact-loaded CFRP beams. The simulations significantly underestimate the initial energy for a small contact area κ. The experimentally determined force–displacement curve is clearly nonlinear until the maximum force is reached. This rather gradual failure is explained by smaller delamination processes [10]. In contrast, the simulations predict a strongly linear force–displacement curve until a sudden large-scale delamination failure, which leads to an abrupt drop in force and the corresponding lower E_I. For the case of $\kappa = 0.46$, E_I is overestimated by the simulation as delamination failure is predicted at a higher force. The model of a fully interlaminar contact area predicts E_I with very good agreement with the experimental values.

All simulations overestimate E_P with the exception of the models with a κ of 1.0. This is caused by the symmetrical formation of delaminations. Nevertheless, E_T as the sum of E_i and E_P is predicted sufficiently well by the simulation models within the standard deviation of the experimental results, except for the results of $\kappa = 0.46$. Due to the moderate interface properties, the energy absorption in this case is higher due to overestimated delaminations than in the simulation with smaller κ.

The inserted foil increases the mass of the tested samples by 6–8% according to the chosen interlaminar contact area. This is negligible compared to the scattering of the energies. The mass is not varied in the simulation because the thickness and density of the cohesive elements are not varied with κ.

The numerically derived data are used to calibrate the FNN metamodel. The accuracy of the metamodel for Design A's concept is determined by the coefficient of determination R_A^2:

$$R_A^2 = \frac{\sum_{n=1}^{p}\left(\hat{E}_n(\kappa) - \overline{E}(\kappa)\right)^2}{\sum_{n=1}^{p}\left(E_n(\kappa) - \overline{E}(\kappa)\right)^2} = 0.919 \tag{4}$$

where p is the number of data sets, $\hat{E}_n$ is the predicted total energy, $\overline{E}$ is the mean of the analysed total energy, and E_n is analysed total energy with respect to the parameter set κ. The metamodel is then used to determine a parameter set κ in which E_T is maximised. Mathematically, the task can be formulated as follows:

$$\begin{cases} \max\limits_{\kappa} \hat{E}_n \\ 0 \leq \kappa \leq 1 \end{cases}, \tag{5}$$

where no further constraints for κ have to be fulfilled. An interlaminar contact area of 0.45 would result in the largest energy absorption of 4.9 J according to the metamodel predictions. This is an increase of about 120% compared to the reference model. Even if this value appears somewhat too optimistic due to the overestimated delamination energy, the simulation models and the applied metamodel-based optimisation confirm the results of the experimental investigations. The energy absorption can be significantly increased by the specific adjustment of the interlaminar properties and the resulting delamination behaviour.

3.2.2. Design B (Layer-Wise Interface Modification Concept—Figure 6b)

In the layer-wise modification concept, the interlaminar properties of all five interfaces are varied independently. This extends the parameter space to be investigated from one to five interlaminar contact areas κ_i with $i = 1, 2, \ldots, 5$. A total of 150 parameter sets p are simulated. For the investigated metamodel,

$$R_B^2 = \frac{\sum_{n=1}^{p}\left(\hat{E}_n(\kappa_i) - \overline{E}(\kappa_i)\right)^2}{\sum_{n=1}^{p}\left(E_n(\kappa_i) - \overline{E}(\kappa_i)\right)^2} = 0.862 \tag{6}$$

Again, the metamodel is then used to determine a parameter set κ_i in which E_T is the maximum:

$$\begin{cases} \max\limits_{\kappa_i} \hat{E}_n \\ 0 \leq \kappa_i \leq 1 \end{cases}. \tag{7}$$

where no further constraints for κ_i have to be fulfilled. Table 4 shows the optimal contact areas κ_i for the five layer-wise interface modifications. Overall, the values deviate slightly from the optimal value of κ for design A. Smaller values are obtained for the outer interfaces due to the low shear stress. The value of the energy absorption is only slightly higher than for the same interface modification, which is also due to the fact that the contact area does not change significantly.

Table 4. Optimal interlaminar contact areas κ_i for the layer-wise interface design.

Layer-Wise Interlaminar Contact Areas κ_i					
κ_1	κ_2	κ_3	κ_4	κ_5	E_T
0.32	0.51	0.44	0.44	0.30	5.18 J

3.2.3. Design C (Sectional Interface Modification—Figure 6c)

In the sectional interface modification, the five interfaces of the layer-wise concept are each divided into three sections, taking symmetry into account. For the creation of the metamodel, 205 parameter sets (p) for the 15 interlaminar contact areas κ_i with $i = 1, 2, \ldots, 15$ are analysed. The developed metamodel has an accuracy R^2_C of 0.689. The driving assumption of design C is the identification of the influence of specific locations for interface modifications on the global energy dissipation. Therefore, the influence of κ_i on the structural response of the three-point bending test is determined using a variance-based sensitivity analysis [21]. A global sensitivity analysis (GSA) index is determined for each κ_i, which describes the influence of the parameter on E_T. The sum of the indexes is normalised to 1.

Figure 11 shows the GSA index of the 15 interfaces. This shows that interface 3 has the greatest influence on energy absorption. Interfaces 2 and 4 are of secondary importance, and interfaces 1 and 5 are almost negligible. The inner interfaces seem to have a slightly greater influence on the structural behaviour than the middle interfaces, with the exception of interface 3. The inner interfaces in interface 1 are important because of the bulging effect, which describes delamination initiation due to local buckling of the upper plies under the compressive load. The outer interfaces have no significant influence on the structural behaviour.

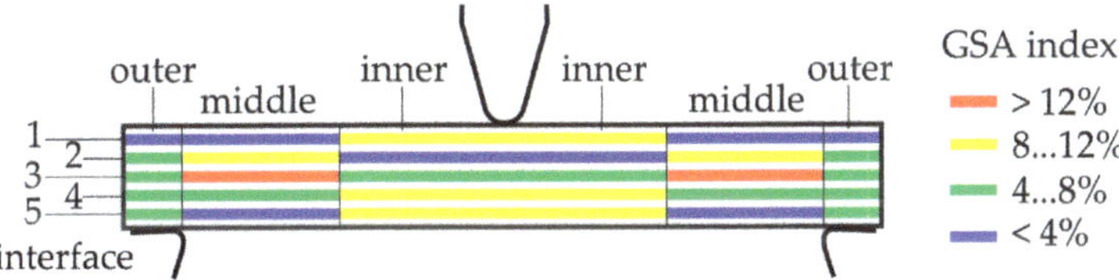

Figure 11. Influence of the sectional interface modification on the structural behaviour of the three-point-bended beam.

It can therefore be assumed that the out-of-plane regions subjected to shear stress are especially responsible for the delamination initiation and propagation because the three-point bending induces shear stress and it is not an edge effect. Otherwise, the outer interface design would have a greater GSA index.

4. Discussion and Conclusions

The delamination behaviour is promoted by selectively weakening the interlaminar properties of the impact-loaded CFRP beams via modified interfaces. This leads to an improved post-failure behaviour in terms of an increased energy absorption and thus improved structural integrity (Figure 10).

The used FEA modelling approach replicates the complex deformation and failure behaviour accurately and can model the trends of initiation and propagation energies for varying interlaminar contact areas κ. Future investigations should use improved discretisation in the delamination direction so that more cohesive elements are included in the delamination zone.

The sensitivity of delamination behaviour with respect to a deviation of the impact position should be investigated. It is possible that slightly offset impact positions in the experiments are responsible for the one-sided delamination formation. Additionally, stochastically distributed properties in the interfaces could be used to avoid the strict symmetry of failure observed in the simulations and thus improve the agreement between the simulation and experiments.

The failure behaviour of the single plies and delaminations can be described separately due to the negligible influence of the interface modification on the in-plane properties. The material model and the bilinear cohesive approaches describe the failure behaviour sufficiently accurately. The parameterisation of the cohesive properties based on the interlaminar contact area κ enables the adjustment of the interlaminar properties and thus the use within a DoE study. The optimisation framework employed here resulted in reasonable predictions for optimised interlaminar contact areas and is suitable for the three-point bending test investigated as part of this study.

Three distributions of the adapted interface modifications are investigated: design A—same layer-by-layer, design B—layer-wise, and design C—sectional interface modifications.

With the same layer-by-layer optimisation, a maximum possible energy absorption was determined at $\kappa = 0.45$. There, E_T is about 120% higher in comparison to the reference specimens without modification. The results for κ after the optimisation for the layer-wise interface modification are between 0.30 and 0.51 and thus deviate only slightly from the result of the layer-by-layer interface modification. The reason for this is that the energy absorption is particularly high when a maximum of delamination processes are initiated. The fact that the delaminations occur as a result of shear stress rather than as the consequence of an edge effect is revealed by the results of the sectional interface modification.

In summary, the results of the experimental work [10] are confirmed by the presented investigations. The observed effect of built-in interface defects was numerically confirmed. The adjusted interface modification significantly influences the energy absorption capacity of CFRP beams subjected to bending loads. Optimisation was successfully employed to maximise the total energy absorption.

Author Contributions: Conceptualization, M.K., A.L., A.H. and M.G.; methodology, M.K., A.L. and A.H.; software, M.K., J.R. and J.W.; validation, M.K., J.R., J.W., A.L. and A.H.; formal analysis, M.K., J.R. and J.W.; investigation, M.K., A.L., A.H. and M.G.; resources, J.W. and M.G.; data curation, M.K. and J.R.; writing—original draft preparation, M.K. and J.R.; writing—review and editing, J.W., A.L., A.H. and M.G.; visualization, M.K. and J.R.; supervision, J.W. and M.G.; project administration, M.G.; funding acquisition, M.G. All authors have read and agreed to the published version of the manuscript.

Funding: This research was funded by the German Research Foundation, DFG project number 407352905.

Data Availability Statement: The data presented in this study are available on reasonable request from the corresponding author.

Acknowledgments: The Article Processing Charge (APC) were funded by the joint publication funds of the TU Dresden, including Carl Gustav Carus Faculty of Medicine, and the SLUB Dresden as well as the Open Access Publication Funding of the DFG.

Conflicts of Interest: The authors declare no conflict of interest.

Appendix A

Table A1. HexPly M49 200P data sheet from supplier.

Prepreg Property		Quantity	Unit
Fibre		HS Carbon	
Tow		3K	
Weave		Plain	
Nominal cured ply thickness		0.234	mm
M49 resin content by weight		42	%
Mechanical property for 55% fibre volume content	Test method		
Tensile modulus in fibre direction	EN 2561	64,000	MPa
Tensile strength in fibre direction	EN 2561	900	MPa
Compressive strength in fibre direction	EN 2850B	725	MPa
Bending modulus in fibre direction	EN 2562	52,000	MPa
Bending strength in fibre direction	EN 2562	850	MPa
Interlaminar shear strength	EN 2563	63	MPa

Table A2. Used LS-DYNA material card *MAT_058 in unit system t-mm-s-N.

*MAT_LAMINATED_COMPOSITE_FABRIC_SOLID								
$	mid1	ro	ea	eb	(ec)	prba	tau1	gamma1
	58	1.47e-09	53300	53300	10000	0.085	42.4	0.007
$	gab	gbc	gca	slimt1	slimc1	slimt2	slimc2	slims
	4099	1175	1175	0.03	0.581	0.03	0.581	0.95
$	aopt	tsize	erods	soft	fs	epsf	epsr	tsmd
	0.0	0.0	0.0	0.0	−1.0	0.0	0.0	0.9
$	xp	yp	zp	a1	a2	a3	prca	prcb
	0.0	0.0	0.0	0.0	0.0	0.0	0.0185	0.0185
$	v1	v2	v3	d1	d2	d3	beta	lcdfail
	0.0	0.0	0.0	0.0	0.0	0.0	0.0	11
$	e11c	e11t	e22c	e22t	gms			
	0.0	0.0	0.0	0.0	0.345			
$	xc	xt	yc	yt	sc			
	1081	994	1081	994	100			
$	e33c	e33t	gm23	gm31				
	0.0	0.0	0.0	0.0				
$	zc	zt	sc23	sc31				
	10000	10000	10000	10000				
$	slimt3	slimc3	slims23	lsims31	tau2	gamma2	tau3	gamma3
	0.0	0.0	0.0	0.0	0.0	0.0	0.0	0.0
$	lcxc	lcxt	lcyc	lcyt	lcsc	lctau	lcgam	dt
	0	0	0	0	0	0	0	0.0
$	lce11c	lce11t	lce22c	lce22t	lcgms	lcefs		
	0	0	0	0	0			
$	lczc	lczt	lcsc23	lcsc31	lctau2	lcgam2	lctau3	lcgam3
	0	0	0	0	0	0	0	0
$	lce33c	lce33t	lcgms23	lcgms31				
	0	0	0	0				

$: comment lines.

Table A3. Used failure strains.

*DEFINE_CURVE								
$	lcid	sidr	sfa	sfo	offa	offo	dattyp	lcint
	11	0	1	1	0	0	0	0
$		a1		o1				
		1		0.030				
		2		0.115				
		3		0.030				
		4		0.115				
		5		0.400				
		6		1.0				
		7		1.0				
		8		1.0				

$: comment lines.

Table A4. Used LS-DYNA material card *MAT_138 for cohesive elements without interface modification (κ = 1.0) in unit system t-mm-s-N.

*MAT_COHESIVE_MIXED_MODE								
$	mid	ro	roflg	intfail	en	et	gic	giic
	138	1.56e-09	53300	1.0	5.0e+06	1.0e+06	0.96	2.50
$	xmu	t	s	und	utd	gamma		
	2.0	32.0	50.0	0.0	0.0	1.0		

$: comment lines.

References

1. Martynenko, V.; Hrytsenko, M.; Martynenko, G. Technique for Evaluating the Strength of Composite Blades. *J. Inst. Eng. India Ser. C* **2020**, *101*, 451–461. [CrossRef]
2. Zhou, Y.; Sun, Y.; Huang, T. Impact-Damage Equivalency for Twisted Composite Blades with Symmetrical Configurations. *Symmetry* **2019**, *11*, 1292. [CrossRef]
3. Kunze, E.; Galkin, S.; Böhm, R.; Gude, M.; Kärger, L. The Impact of Draping Effects on the Stiffness and Failure Behavior of Unidirectional Non-Crimp Fabric Fiber Reinforced Composites. *Materials* **2020**, *13*, 2959. [CrossRef]
4. Waimer, M.; Feser, T.; Schatrow PSchueler, D. Crash concepts for CFRP transport aircraft—Comparison of the traditional bend frame concept versus the developments in a tension absorbers concept. *Int. J. Crashworthiness* **2018**, *23*, 193–218. [CrossRef]
5. Marston, T.U.; Atkins, A.G.; Felbeck, D.K. Interfacial fracture energy and the toughness of composites. *J. Mater. Sci.* **1974**, *9*, 447–455. [CrossRef]
6. Atkins, A.G. Intermittent bonding for high toughness/high strength composites. *J. Mater. Sci.* **1975**, *10*, 819–832. [CrossRef]
7. Lobanov, M.V.; Gulyaev, A.I.; Babin, A.N. Improvement of the impact and crack resistance of epoxy thermosets and thermoset-based composites with the use of thermoplastics as modifiers. *Polym. Sci. Ser. B* **2016**, *58*, 1–12. [CrossRef]
8. Pegoretti, A.; Cristelli, I.; Migliaresi, C. Experimental optimization of the impact energy absorption of epoxy–carbon laminates through controlled delamination. *Compos. Sci. Technol.* **2008**, *68*, 2653–2662. [CrossRef]
9. Sorrentino, L.; Simeoli, G.; Iannace, S.; Russo, P. Mechanical performance optimization through interface strength gradation in PP/glass fibre reinforced composites. *Compos. Part B Eng.* **2015**, *76*, 201–208. [CrossRef]
10. Kuhtz, M.; Hornig, A.; Richter, J.; Gude, M. Increasing the structural energy dissipation of laminated fibre composite materials by delamination control. *Mater. Des.* **2018**, *156*, 93–102. [CrossRef]
11. das Neves Carneiro, G.; António, C.C. Reliability-based robust design optimization with the reliability index approach applied to composite laminate structures. *Compos. Struct.* **2019**, *209*, 844–855. [CrossRef]
12. Chen, Y.; Xu, C.; Wang, C.-H.; Bilek, M.M.M.; Cheng, X. An effective method to optimise plasma immersion ion implantation: Sensitivity analysis and design based on low-density polyethylene. *Plasma Process. Polym.* **2022**, *19*, 2100199. [CrossRef]
13. Kuhtz, M.; Hornig, A.; Gude, M.; Jäger, H. A method to control delaminations in composites for adjusted energy dissipation characteristics. *Mater. Design* **2017**, *123*, 103–111. [CrossRef]
14. Hexcel. HexPly M49 Epoxy Prepreg Data Sheet. Available online: https://www.hexcel.com/user_area/content_media/raw/HexPly_M49_eu_DataSheet.pdf (accessed on 4 March 2020).
15. Hashin, Z. Failure Criteria for Unidirectional Fiber Composites. *ASME J. Appl. Mech.* **1980**, *47*, 329–334. [CrossRef]
16. Livermore Software Technology (LST). LS-DYNA Keyword User's Manual II, Version R13. 2022. Available online: http://ftp.lstc.com/anonymous/outgoing/jday/manuals/LS-DYNA_Manual_Volume_II_R13.pdf (accessed on 30 January 2023).

17. Matzenmiller, A.; Lubliner, J.; Taylor, R.L. A constitutive model for anisotropic damage in fiber-composites. *Mech. Mater.* **1995**, *20*, 125–152. [CrossRef]
18. Dogan, F.; Hadavinia, H.; Donchev, T.; Bhonge, P. Delamination of impacted composite structures by cohesive zone interface elements and tiebreak contact. *Open Eng.* **2012**, *2*, 612–626. [CrossRef]
19. Van Paepegem, W.; De Geyter, K.; Vanhooymissen, P.; Degrieck, J. Effect of friction on the hysteresis loops from three-point bending fatigue tests of fibre-reinforced composites. *Compos. Struct.* **2006**, *72*, 212–217. [CrossRef]
20. Livermore Software Technology (LST). LS-OPT User's Manual, Version 7.0. 2020. Available online: https://www.lsoptsupport.com/documents/manuals/ls-opt/lsopt_70_manual.pdf (accessed on 30 January 2023).
21. Sobol′, I.M. Global sensitivity indices for nonlinear mathematical models and their Monte Carlo estimates. *Math. Comput. Simul.* **2001**, *55*, 271–280. [CrossRef]

 *aerospace*

Article

Kriging-Based Framework Applied to a Multi-Point, Multi-Objective Engine Air-Intake Duct Aerodynamic Optimization Problem

Przemysław S. Drężek [1,2,*], Sławomir Kubacki [2] and Jerzy Żółtak [1]

1 Łukasiewicz Research Network, Institute of Aviation, 02-256 Warsaw, Poland
2 Institute of Aeronautics and Applied Mechanics, Faculty of Power and Aeronautical Engineering, Warsaw University of Technology, 00-665 Warsaw, Poland
* Correspondence: przemyslaw.drezek@ilot.lukasiewicz.gov.pl

Abstract: The forecasted growth in dynamic global air fleet size in the coming decades, together with the need to introduce disruptive technologies supporting net-zero emission air transport, demands more efficient design and optimization workflows. This research focuses on developing an aerodynamic optimization framework suited for multi-objective studies of small aircraft engine air-intake ducts in multiple flight conditions. In addition to the refinement of the duct's performance criteria, the work aims to improve the economic efficiency of the process. The optimization scheme combines the advantages of Kriging-based Efficient Global Optimization (EGO) with the Radial Basis Functions (RBF)-based mesh morphing technique and the Chebyshev-type Achievement Scalarizing Function (ASF) for handling multiple objectives and design points. The proposed framework is applied to an aerodynamic optimization study of an I-31T aircraft turboprop engine intake system. The workflow successfully reduces the air-duct pressure losses and mitigates the flow distortion at the engine compressor's front face in three considered flight phases. The results prove the framework's potential for solving complex multi-point air-intake duct problems and the capacity of the ASF-based formulation to guide optimization toward the designer's preferred objective targets.

Keywords: optimization; multi-objective; multi-point; Kriging; metamodel; surrogate; intake; aerodynamics; CFD; mesh morphing; achievement scalarizing function

Citation: Drężek, P.S.; Kubacki, S.; Żółtak, J. Kriging-Based Framework Applied to a Multi-Point, Multi-Objective Engine Air-Intake Duct Aerodynamic Optimization Problem. *Aerospace* **2023**, *10*, 266. https://doi.org/10.3390/aerospace10030266

Academic Editors: Spiros Pantelakis, Andreas Strohmayer and Jordi Pons-Prats

Received: 28 January 2023
Revised: 5 March 2023
Accepted: 6 March 2023
Published: 9 March 2023

1. Introduction

From the time of the famous Wright brothers' first flight, global air traffic has grown continuously regardless of economic or political turmoil. In only the last two decades, it has more than doubled. Two major commercial aircraft suppliers, Boeing and Airbus, forecast further air traffic growth in the coming decades at a rate of nearly 4% annually, which will create a demand for global air fleet development of approximately 3% per year [1,2]. Such an expansion rate will result in doubling the number of passenger and freight airplanes by the end of the fourth decade of the 21st century.

The foreseen dynamic development coincides with a need to introduce new disruptive technologies supporting climate-neutral aviation. This target is imposed by the European Green Deal initiative [3], recently established by the European Commission. The ambitious goal of achieving net-zero emission air transport by 2050 requires the deployment of radical innovations within a short period. Such rapid anticipated progress forces aviation R&D entities to seek more streamlined design processes, which should be supported with advanced design tools and efficient optimization workflows. More efficient work schemes should be instituted simultaneously on an integrated aircraft level and for each subsystem and component.

This paper focuses on developing an optimization framework suited for airplane engine air-intake ducts. There is a long history of aerodynamic optimization studies on ducts of various shapes in the literature. Although surrogate-assisted optimization is a visible trend nowadays, conventional direct strategies are still present in recent investigations. For instance, Furlan et al. [4] applied a Genetic Algorithm (GA) [5] to optimize an S-shaped channel parameterized using Bézier curves. D'Ambros et al. [6] employed the Tabu Search algorithm supported by the Free-Form Deformation (FFD) technique to a multi-objective optimization problem of a generic S-duct. Zeng et al. [7] used the GA method to enhance the aerodynamic performance of an S-duct scoop inlet. Sharma and Baloni [8] solved a multi-objective optimization problem in a turbofan engine compressor transition S-duct using the Particle Swarm Optimization (PSO) technique [9].

All the abovementioned investigations were successful in improving the corresponding objectives; however, direct strategies come at a significant computational cost resulting from multiple objective function evaluations. This drawback occurs notably in aerodynamic problems for which objective values are determined using expensive Computational Fluid Dynamics (CFD) codes.

Using surrogate-based frameworks allows for a significant reduction in costly evaluations by approximating the objective functions with simple analytical representations. These substitutes are commonly referred to as metamodels [10,11] or surrogates [12–14]. In such a procedure, conventional optimization techniques (e.g., steepest descent [15], as well as various quasi-Newton and population-based methods) are used to search for a superior solution in an artificial response landscape. CFD solver calls are required predominantly to build a database necessary for the surrogate construction and, to some extent, for its subsequent improvement.

Surrogates are usually categorized by the class of mathematical functions used for their creation. Popular types supporting aerodynamic optimization problems are low-order polynomial Response Surface Models (RSM) [16], regression splines [17], and various Radial Basis Functions (RBF) [18]. Polynomial-based surrogates have the advantages of simplicity and ease of use, although they have limited capabilities to approximate complex objective functions. RBF-based metamodels, instead, can model functions of high curvature with reasonably higher fitting effort. More complex models, such as Artificial Neural Networks (ANN) [19], use a nonlinear regression process to fit the surrogate. Although they are very efficient in applications with numerous variables, the ANN training process might be computationally expensive as it requires a solution to a high-dimensional optimization problem.

A characteristic class of metamodels has the ability to consider a stochastic component in the function approximation to quantify the confidence of the surrogate predictions. Moreover, this property is extensively used to boost the global search by identifying areas in the objective space with high improvement potential. These models are predominantly based on the Gaussian Process (GP); among them, the Kriging surrogate [20–23] is the most widely exploited.

The abovementioned metamodels have received the most prominent attention in the field of aerodynamic shape optimization of various ducts and channels. Lu et al. [24] employed a third-order polynomial RSM to optimize an S-shaped compressor transition duct. The authors were successful in reducing the pressure losses along the channel, together with an improvement in outlet pressure and velocity distribution. In the problem of annular S-duct shape optimization, Immonen [25] used fourth- and fifth-order RSM to minimize energy loss and improve flow uniformity. The multi-objective study resulted in considerable refinement in both examined parameters.

An RBF-based metamodel served to approximate the objective function in the optimization study on a stealth aircraft diffusing S-duct performed by Gan et al. [26]. The problem of simultaneous maximization of the duct pressure recovery and minimization of total pressure distortion was solved by a GA performing a search in a surrogate-based space. A significant distortion coefficient improvement and slight pressure recovery factor improvement characterized the solution located by the optimizer. A similar GA-based approach was used by Donghai et al. [27]; however, the surrogate was constructed using

ANN. With such a method, the authors suppressed flow separation in a strutted annular S-duct, which resulted in a considerable reduction in total pressure loss.

Among contemporary literature sources on duct shape optimization, the use of Kriging has noticeable dominance, chiefly due to its ability to interpolate complex functions and inherent features supporting global optimization. Zerbinati et al. [28] used the so-called Multiple-Gradient Descent Algorithm to search for an optimum in an objective space approximated using the Kriging technique. The algorithm successfully reduced the pressure loss and velocity variance in an air-cooling S-duct. Verstraete et al. [29] compared the behavior of Kriging-based and ANN-based surrogates assisting a Differential Evolution (DE) algorithm [30] in the minimization of U-shaped cooling channel pressure loss. The authors reported a superior performance by the Kriging metamodel. The study was broadened by Verstraete and Li [31] to a multi-objective problem of pressure loss minimization and heat transfer maximization. Due to the previous research outcomes, Kriging solely assisted the DE algorithm. Despite having objectives of competing nature, the optimizer was successful in improving both measures. Koo et al. [32] executed a similar comparison of Kriging and ANN metamodels in a multi-objective optimization problem of a heat exchanger inlet duct. The study outcomes resulted in improved pressure loss and flow rate uniformity achieved by both metamodels, although preferences towards specific objectives were different for particular surrogates. Wang and Wang [33] employed a Kriging-assisted GA technique in a multi-objective optimization problem of a UAV S-shaped intake duct. The authors were successful in the combined improvement of the aerodynamic and electromagnetic performance of the S-duct diffuser.

The studies referenced in the previous paragraph used merely Kriging-based objective landscape approximation to search for the optimum. The full potential of Kriging, however, manifests itself in its ability to not only predict the objective value but also assess the uncertainty of this prediction. This property is extensively used to balance local and global searches by probing the objective space in regions with a high likelihood of improvement. The search strategy based on this unique Kriging feature constitutes the Efficient Global Optimization (EGO) proposed by Jones et al. [23]. This algorithm is proven to balance the exploration and exploitation properties efficiently and, as such, has been applied in numerous shape optimization studies.

Bea et al. [34] used the Kriging-based EGO strategy to successfully improve the pressure recovery factor in a diffusing S-duct. A similar objective measure was optimized by Dehghani et al. [35]. The authors employed EGO to enhance the performance of an axisymmetric diffuser channel. Marchlewski et al. [36] employed a similar search technique utilizing Kriging's uncertainty assessment in multi-objective optimization of U-shaped engine intake. Using Pareto front delimitation in an objective space, the authors reduced the duct pressure loss while maintaining the initial level of exiting flow uniformity. Drężek et al. [37] optimized analogous intake geometry using GA-assisted EGO. The two objectives were combined using the Achievement Scalarizing Function (ASF) [38]. The optimizer was successful in the simultaneous improvement of the duct's pressure loss and flow distortion.

Every aerodynamic optimization problem requires design variables representing geometry as the subject of improvement. Expressing generic models using independent parameters (variables) is usually referred to as parameterization, for which a variety of methods is possible. The choice of a specific technique may profoundly impact the optimization process's computational cost and the final result. The vast majority of literature sources referenced above use parameterization techniques that focus on modifying geometry definition, namely, direct engineering parameters [24,32], Bézier curves [4,8], B-splines [27–29,31], Non-Uniform Rational B-Splines (NURBS) [7,35], and Free-Form Deformation [6]. Such an approach requires a subsequent mesh regeneration at each reshaping step. The mesh morphing technique is an efficient alternative to reduce the overall computational cost of the process. This method requires only initial generation of a mesh, which is adjusted in the subsequent steps while preserving the original grid topology. Although some recent literature sources describe the combined use of surrogates and mesh morphing (applied to, e.g., fuselage–wing junction [39,40], generic car model [40], effusion cooling plate [41], and cooling channel rib [42]), only a few studies report such an application to intake duct optimization. To the best of our knowledge, these are [36,37].

Most intake duct optimization studies available in the literature consider only one operating condition, usually the nominal on-design point. Such an approach is commonly called Single-Point Optimization (SPO). As the conditions may change substantially at various mission stages, such a strategy may lead to a suboptimal performance at off-design points. Some studies (e.g., [33,43]) evaluate the off-design conditions in the post-optimization phase to secure the design against severe performance deterioration. A few literature sources report simultaneous optimization under multiple flight conditions—often referred to as Multi-Point Optimization (MPO). Brahmachary et al. [44] and Fujio and Ogawa [45] executed multi-point studies on axisymmetric scramjet intake. The authors employed a surrogate-assisted GA algorithm to improve multiple performance parameters via Pareto front evaluation. Chiang et al. [46] enhanced the shape of a Boundary-Layer Ingesting (BLI) engine S-duct intake. The performance objectives for cruise, descent, and climb conditions were combined using a simple weighted sum function. The foreseen importance of MPO in prospective design processes should lead to investigations into advanced scalarization methods allowing simultaneous involvement of multiple objectives and flight conditions while overcoming the well-known deficiencies of the weighted sum technique. Such studies are absent in the state-of-the-art literature.

This paper concentrates on designing an optimization scheme combining the advantages of the Kriging metamodel, mesh morphing technique, and advanced objectives scalarization method. The components are duly integrated to create a synergistic effect of improvements in the economic efficiency of the design process. To assess the usefulness of the proposed framework in practical engineering problems, the algorithm is applied to multi-point aerodynamic optimization of an I-31T turboprop aircraft's air intake. The procedure simultaneously reduces the air-duct pressure losses and mitigates the flow distortion at the engine compressor's front face while considering multiple flight conditions.

The novelty of this study is the use of the augmented Chebyshev-type ASF to combine multiple performance objectives under multiple flight conditions. Such a strategy integrated with the Kriging surrogate and the mesh morphing technique creates a synergistic effect of cost-efficient MPO.

2. Materials and Methods

2.1. Optimization Framework

The optimization framework reported in this study is founded on three pillars: Kriging surrogate for objective function approximation, mesh morphing technique for geometry parameterization, and ASF method for handling multiple objectives and conditions. The framework integrated from the abovementioned components is presented schematically in Figure 1.

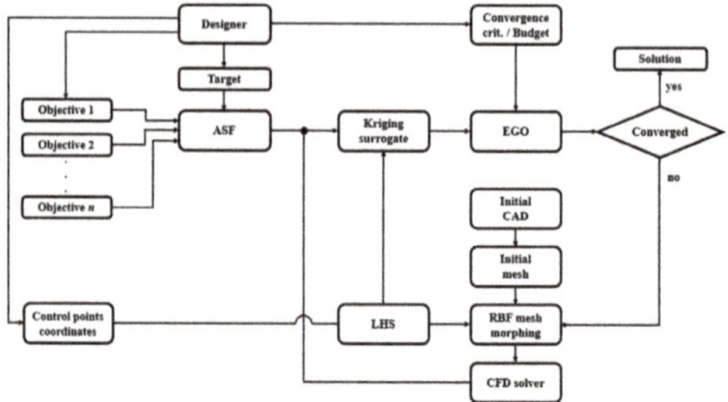

Figure 1. Schematic representation of the optimization framework.

2.1.1. Kriging Surrogate

The concept of Kriging originates from geostatistical modeling inspired by the work of a South African engineer named Daniel Krige [20], further formalized by Matheron [21], and eventually adapted for deterministic numerical simulations by Sacks et al. [22] and Jones et al. [23]. The Kriging-based surrogate is founded on the hypothesis that any black-box simulation response function $y = f(x)$ can be expressed as a realization of the Gaussian random process $Y(x)$.

The Kriging surrogate is composed of a global trend function $\mu(x)$ and centered GP $Z(x)$ with zero mean and non-zero covariance (Equation (1)).

$$Y(x) = \mu(x) + Z(x) \tag{1}$$

The trend function is a regression model that captures a general tendency in the observed data. It is expressed as a linear combination of n deterministic basis functions $\phi(x)$, based on regression of N response function evaluations ($n \leq N$):

$$\mu(x) = \sum_{k=0}^{n} \phi_k(x)\beta_k \tag{2}$$

Depending on the trend formulation, the Kriging metamodel can be classified into three categories: Simple Kriging (SK)—for which the trend is a known constant; Ordinary Kriging (OK)—for which the trend is constant but unknown; and Universal Kriging (UK)—for which the basis functions $\phi(x)$ are known and fixed, but the coefficients $\beta_k \in \mathbb{R}\backslash\{0\}$ are unknown and are estimated in the surrogate construction process. In this study, we use the most general, i.e., UK formulation, due to the anticipated high curvature of the objective function. The trend is defined as a p-dimensional full first-order polynomial (Equation (3)).

$$\mu(x) = \sum_{j=0}^{p} \beta_j x_j \tag{3}$$

The stochastic component $Z(x)$ in Equation (1) models a local deviation from the global trend to the true objective function. It is characterized by constant but unknown variance σ^2 and non-zero covariance (Equation (4)).

$$Cov(Z_i,\ Z_{i\prime}) = \sigma^2 R(x_i - x_{i\prime}, \boldsymbol{\psi}) \quad i, i\prime = 1, \ldots, N \tag{4}$$

Here, $R(x_i - x_{i\prime}, \boldsymbol{\psi})$ denotes the spatial correlation function between any two samples x_i and $x_{i\prime}$, and is often referred to as the covariance kernel. In multi-dimensional problems, the kernel turns into a tensor product of p one-dimensional correlation functions $r(x_i - x_{i\prime}, \psi)$:

$$R(x_i - x_{i\prime},\ \boldsymbol{\psi}) = \prod_{j=1}^{p} r\left((x_i - x_{i\prime})_j^{\delta_j},\ \psi_j\right) \tag{5}$$

The level of impact of j-th dimension correlations on the Kriging prediction is controlled by the covariance kernel hyperparameters grouped in the vector $\boldsymbol{\psi}$. The smoothness coefficient δ governs the differentiability of the metamodel surface. In the present study, we assume that the true objective function is smooth, which justifies using an infinitely differentiable Gaussian-type kernel ($\delta_j = 2\ \forall\ j = 1, \ldots, p$). In one dimension, the Gaussian kernel is formulated as:

$$r_j(x_i,\ x_{i\prime},\ \psi) = exp\left(-\frac{(x_i - x_{i\prime})_j^2}{2\psi_j^2}\right) \tag{6}$$

The values of unknown hyperparameters β, σ^2, ψ are estimated in the process called "fitting" the metamodel to the available data samples. This procedure is based on the maximization of a function $\mathcal{L}$, expressing the probability of predicting the evaluated objective values at sampled locations. This method is called Maximum Likelihood Estimation (MLE) and is a non-trivial optimization problem of minimizing a multi-modal likelihood function. For the purpose of this study, we used an algorithm relying on the quasi-Newton method [47], available in the R language package DiceKriging [48].

The Kriging predictor $\hat{Y}(x)$ at an arbitrary unobserved location x_0 is defined using random function evaluations at sampled positions $Y(x_i)$. The estimator holds the properties of the Best Linear Unbiased Predictor (BLUP), i.e.:

- Linearity—linear combination of random functions

$$\hat{Y}(x_0) = \sum_{i=1}^{N} \lambda_i(x_0) Y(x_i) \tag{7}$$

- Unbiasedness—absence of systematic bias

$$\mathbb{E}\left[\hat{Y}(x_0)\right] = \mathbb{E}[Y(x_0)] \tag{8}$$

- Minimal prediction variance—minimizing the mean squared error

$$MSE\left[\hat{Y}(x_0)\right] = \mathbb{E}\left[\left(\hat{Y}(x_0) - Y(x_0)^2\right)\right] \tag{9}$$

Solving the above system of equations for optimal weights $\boldsymbol{\lambda}^T \in \mathbb{R}^N$ allows computation of the GP predictor's mean and variance, which defines it entirely.

The Kriging surrogate requires input data from objective function evaluations at a set of discrete locations. The spatial distribution of these observations across the design space may profoundly influence the metamodel prediction quality [49]. Historically, various sampling strategies were developed for planning the empirical experiments, so the selection of an observation pattern is often called the Design of Experiment (DoE). The original DoE methods locate the majority of samples close to the design space boundaries and only a few observations in its interior [10]. Deterministic numerical simulations, however, are prone to systematic rather than random error. For this reason, space-filling strategies, distributing samples over the entire domain, are more convenient for such applications. Using classic DoE methods in computer experiments might be highly inefficient [22].

Although a variety of space-filling techniques well-suited for deterministic simulations is available, the Latin Hypercube Sampling (LHS) [50] strategy has attracted the most attention in recent years [49,51]. The LHS algorithm distributes the observations in a p-dimensional hypercube so that the samples' projections into $(p-1)$-dimensional space do not share the exact location along the axes [52]. In a two-dimensional simplification, this criterion would result in samples being distributed on a grid in which only one sample is located in each row and each column. The designer can arbitrarily select the number of observations, which makes the LHS method attractive for studies based on expensive aerodynamic simulations.

In this paper, we use the Optimal Latin Hypercube (OLH) technique, which improves LHS's space-filling properties using a columnwise-pairwise algorithm [53] to optimize sample distribution with respect to the S-optimality criterion [54]. The algorithm is executed using the lhs R package [55].

For the sake of computational efficiency of the optimization process, the initial dataset is limited to a number ensuring sufficient initial prediction quality to initiate the following adaptive sampling stage. The optimal number is unknown a priori, although some respected studies suggest approximation using the "$10p$" rule-of-thumb [23,56], where p is the number of design variables. Here, the design space has 12 dimensions, which would result in 120 initial samples; however, we follow the suggestion of a finite-decimal spacing between observations [23]. Ultimately, the DoE contains 126 elements with an inter-distance of $\frac{1}{(126-1)} = 0.008$.

The computational expense required to generate samples that construct the Kriging surrogate increases with the problem dimensionality, which restricts this method's applicability to a moderate number of design variables. The limiting problem size reported in the literature is not accurate (e.g., $p < 50$ in [11], $p < 20$ in [14]) as it is dependent on the available resources. Nevertheless, the Kriging technique is prone to the curse of dimensionality and, as such, is not suited for extensively large optimization problems.

The adaptive sampling, subsequent to the initial DoE, sequentially adds new observations to the dataset using current information about the predicted properties of the objective landscape. The process locates new samples with respect to the preselected infill criterion. For this purpose, we use the Expected Improvement (EI) function, proposed by Jones et al. [23], which takes advantage of the Kriging ability to estimate the prediction uncertainty.

The EI criterion uses the Kriging variance to assess the possibility of the objective function value being lower than the current best prediction y_{min} at any unobserved location. The addition of a new sample may bring an improvement $I(x)$ equal to $\max(y_{min} - y(x), 0)$. The value of $y(x)$ has yet to be discovered in unevaluated points, and it is modeled using the Kriging predictor $\hat{Y}(x)$. As a result, the improvement becomes a random variable for which an expected value constitutes the EI function (Equation (10)).

$$EI(x) = \mathbb{E}[I(x)] = \mathbb{E}\begin{Bmatrix} y_{min} - \hat{Y}(x) & if\ \hat{Y}(x) < y_{min} \\ 0 & otherwise \end{Bmatrix} \quad (10)$$

Integration by parts of Equation (10) gives the following analytical formulation of the EI:

$$EI(x) = \begin{Bmatrix} (y_{min} - \hat{m}(x))CDF\left(\frac{y_{min}-\hat{m}(x)}{\hat{s}(x)}\right) + \hat{s}(x)PDF\left(\frac{y_{min}-\hat{m}(x)}{\hat{s}(x)}\right) & if\ \hat{s}(x) > 0 \\ 0 & if\ \hat{s}(x) = 0 \end{Bmatrix} \quad (11)$$

Here, $\hat{m}(x)$ and $\hat{s}(x)$ are the Kriging prediction mean and variance, CDF is a cumulative distribution function, and PDF is a probability density function of a standard normal distribution.

Iterative introduction of new samples x^* at locations maximizing the EI function constitutes the EGO algorithm (Equation (12)).

$$x^* = \underset{x \in \mathfrak{X}}{\operatorname{argmax}} EI(x) \quad (12)$$

Maximization of the EI criterion might not be trivial because of its strong multimodality. This study uses for this purpose a derivative-supported GA algorithm genoud [57] available under the R package DiceOptim [48].

2.1.2. RBF-Based Mesh Morphing

Out of a variety of mesh morphing methods, this study employs the RBF-based technique. RBF are mathematical functions originating from a multivariate approximation of scattered data. Their value depends only on the distance from the function argument to the origin, called a control point. The RBF interpolates the value between control points according to the specific basis function properties while preserving it in the points' location.

A deliberate displacement of the control points introduces the desired deformation of the initial mesh. The motion is interpolated on the surrounding grid nodes' positions according to the characteristics of the selected RBF. The interpolation function $\mathfrak{h}(x) : \mathbb{R}^3 \to \mathbb{R}$, (Equation (13)) defines particular nodes' locations after deformation $x' = x + \mathfrak{h}(x)$.

$$\mathfrak{h}(\boldsymbol{x}) = \sum_{i=1}^{N_C} \gamma_i \varphi(\|\boldsymbol{x} - \boldsymbol{x}_{C_i}\|) + p(\boldsymbol{x}) \quad (13)$$

Here, $\boldsymbol{x} \in \mathbb{R}^3 : \boldsymbol{x} = (x, y, z)$ is an arbitrary spatial position, $\boldsymbol{x}_{C_i}$ is a known position of the i-th control point from a set of N_C elements, and γ_i is a corresponding weight coefficient describing the strength of the i-th interaction. The interpolation function comprises a linear combination of radial contributions described by RBF $\varphi(\cdot)$—depending solely on the Euclidean distance $\|x - x_{C_i}\|$—and a linear polynomial:

$$p(x) = \alpha_0 + \alpha_1 x + \alpha_2 y + \alpha_3 z \tag{14}$$

The introduction of the polynomial $p(\boldsymbol{x})$ ensures the uniqueness of the fit and allows for the affine motion.

From among several RBF suited for multivariate problems (see, e.g., [58–62]), we select the smooth Gaussian RBF to prioritize the preservation of the mesh quality (Equation (15)).

$$\varphi(\|x - x_{C_i}\|) = e^{-\|x - x_{C_i}\|^2/c^2} \tag{15}$$

The function's shape is tuned by a parameter c, whose value was selected for this study in a trial-and-error process as equal to 150.

The unknown values of weights γ_i (Equation (13)) and polynomial coefficients α_i (Equation (14)) are computed to satisfy the known displacement $\mathfrak{g}_i$ of each control point (Equation (16)).

$$\mathfrak{h}(x_{Ci}) = \mathfrak{g}_i \quad \text{for } i = 1, \ldots, N_C \tag{16}$$

The given displacements are stored in an $N_C \times 3$ matrix $\mathfrak{g}$. Moreover, additional orthogonality requirements (Equation (17)) are introduced to equate the number of equations with the number of degrees of freedom, increased by the presence of the polynomial $p(\boldsymbol{x})$.

$$\begin{aligned} \sum_{i=1}^{N_C} \gamma_i &= 0 \\ \sum_{i=1}^{N_C} \gamma_i x_{C_i} &= 0 \\ \sum_{i=1}^{N_C} \gamma_i y_{C_i} &= 0 \\ \sum_{i=1}^{N_C} \gamma_i z_{C_i} &= 0 \end{aligned} \tag{17}$$

A linear system of equations (Equation (18)) is constructed using the above conditions, which is subsequently solved for $N_C \times 3$ matrix γ of the interpolation function weights and 4×3 matrix $\boldsymbol{\alpha}$ of polynomial coefficients.

$$\begin{bmatrix} \boldsymbol{M} & \boldsymbol{P} \\ \mathcal{P}^T & 0 \end{bmatrix} \begin{bmatrix} \gamma \\ \boldsymbol{\alpha} \end{bmatrix} = \begin{bmatrix} \mathfrak{g} \\ 0 \end{bmatrix} \tag{18}$$

The $N_C \times N_C$ interpolation matrix $\boldsymbol{M}$ gathers RBF evaluations with respect to the distances between all considered control points (Equation (19)).

$$\boldsymbol{M} = \begin{bmatrix} \varphi(\|x_{C_1} - x_{C_1}\|) & \cdots & \varphi\left(\|x_{C_1} - x_{C_{N_C}}\|\right) \\ \vdots & \ddots & \vdots \\ \varphi\left(\|x_{C_{N_C}} - x_{C_1}\|\right) & \cdots & \varphi\left(\|x_{C_{N_C}} - x_{C_{N_C}}\|\right) \end{bmatrix} \tag{19}$$

The orthogonality conditions (see Equation (17)) result in $N_C \times 4$ matrix $\boldsymbol{\mathcal{P}}$ containing a unity column and all control points' positions (Equation (20)).

$$\mathcal{P} = \begin{bmatrix} 1 & x_{C_1} & y_{C_1} & z_{C_1} \\ \vdots & \vdots & \vdots & \vdots \\ 1 & x_{C_{N_C}} & y_{C_{N_C}} & z_{C_{N_C}} \end{bmatrix} \tag{20}$$

After solving the system above, the post-deformation spatial position $\boldsymbol{x\prime} = (x\prime, y\prime, z\prime)$ of an arbitrary location is computed as an incremental displacement (Equation (21)).

$$\begin{Bmatrix} x' = x + \mathfrak{h}^x(\boldsymbol{x}) \\ y' = y + \mathfrak{h}^y(\boldsymbol{x}) \\ z' = z + \mathfrak{h}^z(\boldsymbol{x}) \end{Bmatrix} \quad (21)$$

In the definition of the optimization problem considered in the present paper, the Cartesian components of control points' translation vectors serve as design variables.

This study implements the mesh morphing algorithms using the RBF module of PyGeM (Python Geometrical Morphing) [63], an open-source Python programming language library for parameterization and deformation of complex geometries.

A vital aspect of the functional mesh morphing process is maintaining the mesh quality. Such a problem usually comes down to controlling an appropriate quality metric. Selecting a measure essential for the specific flow solver's error-free operation is fundamental. From the various metrics available in the literature [64,65], we select the mesh element orthogonality angle diagnostic as appropriate for the employed fluid dynamics solver. This measure compares the angle between adjoining mesh cell faces to the ideal level (Equation (22)).

$$90^\circ - \cos^{-1}(\boldsymbol{d} \cdot \boldsymbol{n}) \quad (22)$$

The vector $\boldsymbol{d}$ connects two element centroids, and $\boldsymbol{n}$ is a face normal vector. The orthogonality metric positively correlates with the mesh quality as the grid element shape approaches the idealized form for high values. An area-weighted averaging over relevant faces gives a value for the specific control volume. Quality control involves monitoring the minimum of the metric among all mesh elements and comparing it to the value for the initial mesh as a reference.

2.1.3. Achievement Scalarizing Function

Multiple objectives and flight conditions considered in this study are combined using the augmented Chebyshev-type ASF [38]. This technique belongs to the a priori methods, for which the designer articulates their preferences about the relative importance of the objectives and conditions before the optimization process starts. Such an approach is particularly attractive for studies involving computationally expensive simulations, as only a fraction of the objective space needs to be resolved.

Combining preferences into one function, such that the problem can be solved with a single objective optimizer, is usually called scalarization. Consequently, the resulting function is referred to as a scalarizing function. Although the weighted sum method [66] is the most intuitive representant of this class, it has an inherent drawback in its inability to find solutions on a non-convex portion of a Pareto front (see Figure 2a). This issue is not present in methods minimizing a metric of distance to a reference point arbitrarily located in the objective space (see Figure 2b). The ASF selected for this study employs the Chebyshev metric as a measure of distance and the optimization target f^T, expressing the designer's aspirations about particular objectives, as a reference point. The scalarizing function in a multi-point formulation is defined as follows:

$$Y_{ASF}^{MP}(x) = max\left\{\lambda_{jk}\left(f_{jk}(x) - f_{jk}^T\right)\right\} + \varrho \sum_j \sum_k \lambda_{jk}\left(f_{jk}(x) - f_{jk}^T\right) \quad (23)$$

Above, subscripts j and k refer to particular design conditions and objectives, respectively. The augmentation coefficient ϱ (together with the following term) guarantees proper Pareto optimality and takes an arbitrarily small positive value, here 0.05.

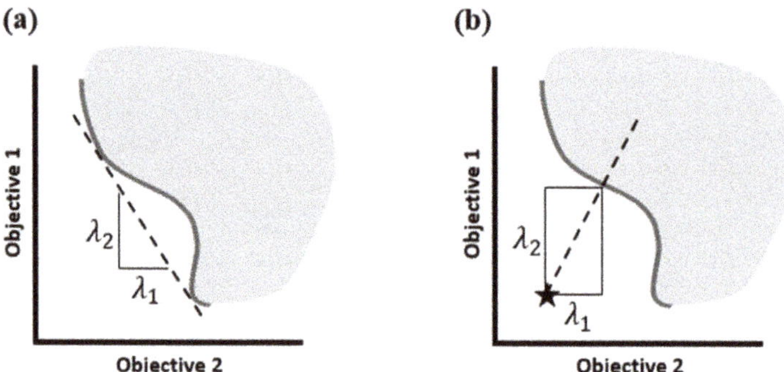

Figure 2. Non-convex Pareto front in different methods: (**a**) weighted sum, (**b**) reference point.

The coefficients λ_{jk} serve to normalize the objective values to a similar range (Equation (24)).

$$\lambda_{jk} = \frac{1}{f_{jk}^{N} - f_{jk}^{T}} \tag{24}$$

The f^N is called the *nadir* point and represents the upper bound of an objective value. The designer usually estimates its value based on knowledge about the landscape of the objective supported by their educated judgment.

Except for the advantages referred to above and the proven efficiency, the ASF is characterized by an intuitive concept of setting a target and aiming at advancing on it.

2.2. Optimization Problem

The surrogate-based framework is applied to the I-31T aircraft air-intake duct multi-point optimization problem. The I-31T flight demonstrator is a retrofitted make of the I-23 "Manager" airplane. The plane was designed in Łukasiewicz Research Network—Institute of Aviation [67] and modified under the EU R&D program ESPOSA (Efficient Systems and Propulsion for Small Aircraft) [68]. The redesign had in scope a replacement of the piston motor with the PBS TP100 turboprop engine [69–72]. Integration of the nominally pusher-purpose engine with the tractor-type airplane required the introduction of a tailored air delivery U-shaped duct. The duct was designed by Stalewski and Żółtak [73] using a parametric design strategy [74,75] relying on the NURBS-based in-house code PARADES [76]. This study takes the outcome of the abovementioned work as an initial point and aims to further improve the duct's performance characteristics in various flight conditions.

Figure 3 presents the initial air-intake duct geometry. The upstream portion of the channel is slightly S-shaped and transitions into a U-shaped duct with the flow direction. Further downstream, the U-duct interfaces with the engine compressor front face at the location marked as the Aerodynamic Interface Plane (AIP). The intake channel has a slightly converging character with a hydraulic diameter equal to 0.180 m at the duct's entrance and 0.147 m at the AIP. The dummy extension, of a length of two diameters, is an artificial portion introduced to secure the flow solver's stability. The duct's inlet and AIP shapes and positions are constrained, i.e., not subject to shape deformation.

The surface of the duct in Figure 3 shows the initial computational mesh. The nine points surrounding the geometry indicate the baseline location of the mesh morpher control points subject to displacement during optimization. They are located in regions where presumed deformation should have a significant impact on the duct performance. The horizontal and vertical translation vector components form a set of design variables. Three pairs of points are bounded so that each duplet shares the exact translation. This strategy guarantees the maintenance of duct symmetry. Such a formulation results in twelve design variables, shown in Figure 3, whose ranges are defined in Table 1. The ranges corresponding to control points close to the AIP are narrower to avoid excessive deformations in the vicinity of the constrained section. Such distorted geometries would have a low probability of generating superior solutions.

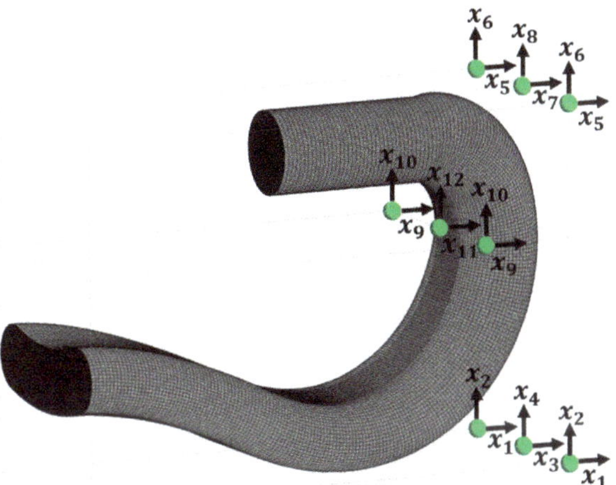

Figure 3. Initial mesh with an indication of flow stations and mesh morpher control points' locations.

Table 1. Ranges of design variables.

Design Variable	Range of Displacement
$x_1, \ldots, x_4$	±100 mm
$x_5,\ x_7,\ x_8$	±50 mm
x_6	±75 mm
$x_9, \ldots, x_{12}$	±15 mm

The amount of pressure loss along the air intake and the level of circumferential pressure distortion at the compressor entrance substantially influence the ultimate performance and operability of the airplane engine. These two objectives are subject to minimization in the present study:

$$\text{pressure loss coefficient}: \ f_1 \equiv dP = \frac{P_t^{IN} - P_t^{AIP}}{P_t^{IN}} \tag{25}$$

$$\text{distortion coefficient}: \ f_2 \equiv DC_{60} = \frac{P_t^{AIP} - \min\limits_{\theta \in [0,2\pi]} P_t^{AIP}\left(\theta - \frac{\pi}{6}, \theta + \frac{\pi}{6}\right)}{q^{AIP}} \tag{26}$$

In the above equations, P_t^{IN} and P_t^{AIP} denote total pressure values mass-averaged over the inlet and AIP cross-sections, respectively, and q^{AIP} indicates dynamic pressure mass-averaged across the AIP. The term $\min\limits_{\theta \in [0,2\pi]} P_t^{AIP}\left(\theta - \frac{\pi}{6}, \theta + \frac{\pi}{6}\right)$ refers to the lowest average pressure in the 60° sector, which is proven to have the most detrimental impact on the compressor performance [77]. This evaluation is realized by clocking a virtual sector around the AIP with $\Delta\theta = 10°$ angular intervals—as given schematically in Figure 4.

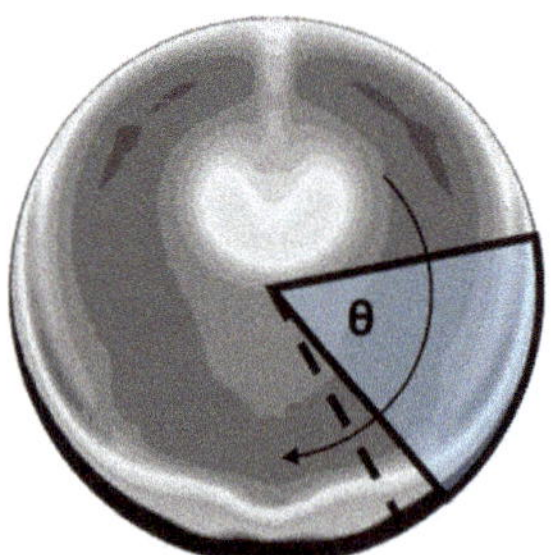

Figure 4. Illustration of $\min_{\theta \in [0,2\pi]} P_t^{AIP}(\theta - \frac{\pi}{6}, \theta + \frac{\pi}{6})$ evaluation process at the AIP. The blue segment represents the pressure averaging region.

The multi-point optimization problem considers three I-31T airplane flight conditions: DP1—nominal cruise, DP2—low-altitude climb, and DP3—high-altitude cruise. Details of the design points and related environmental properties are gathered in Table 2. The corresponding targets and *nadir* points are given in Table 3. The goal for the nominal cruise has a priority over the off-design conditions to reflect the contribution of this design point to the aggregated efficiency over the entire flight mission. Both performance objectives are considered to be of equal importance and share similar targets for particular design points. On the grounds of the designer's experience, the possible falloff of 200% for each objective gives a basis for the function's upper bound estimation.

Table 2. Details of operating conditions for considered design points.

Design Point	Altitude (m)	A/C Velocity (m/s)	Ambient Pressure (Pa)	Ambient Temperature (K)	Ambient Density (kg/m^3)	Engine Mass Flow Rate (kg/s)
DP1: Nominal cruise	3000	65	69,700	268.6	0.909	1.5
DP2: Low-altitude climb	100	46	100,129	287.5	1.213	1.9
DP3: High-altitude cruise	3700	71	64,089	264.1	0.845	1.3

Ambient conditions were evaluated with use of the U.S. Standard Atmosphere model [78].

Table 3. Target and nadir points for objectives under selected flight conditions.

Design Point	dP	DC_{60}
DP1: Nominal cruise	target: 10% nadir: 200%	target: 10% nadir: 200%
DP2: Low-altitude climb	target: 5% nadir: 200%	target: 5% nadir: 200%
DP3: High-altitude cruise	target: 5% nadir: 200%	target: 5% nadir: 200%

2.3. Evaluation of Objectives

2.3.1. Flow Solver Governing Equations

The values of the objective functions are derived from the flow field solution of the air-intake duct CFD simulations. For this purpose, we use a Reynolds-averaged Navier–Stokes (RANS) solver capable of modeling three-dimensional turbulent viscous flow to accurately predict flow features characterized by a secondary flow motion, strong pressure gradients, and occurrence of separation. Such flow features are highly expected in channels of high curvature [79]. The solver is implemented using a commercial CFD code: ANSYS CFX Release 18.0.

In this study, the code solves the following Favre-averaged Navier–Stokes conservation equations:

- The continuity equation:

$$\frac{\partial \overline{\rho}}{\partial t} + \frac{\partial}{\partial x_i}(\overline{\rho}\widetilde{u_i}) = 0 \tag{27}$$

- The momentum conservation equation:

$$\frac{\partial}{\partial t}(\overline{\rho}\widetilde{u_i}) + \frac{\partial}{\partial x_j}(\overline{\rho}\widetilde{u_i}\widetilde{u_j}) = \frac{\partial}{\partial x_j}(-p\delta_{ij} + \overline{t_{ij}} + \overline{\rho}\tau_{ij}) \tag{28}$$

In the above formulations, ρ, u_i, and p denote fluid density, velocity components, and static pressure, respectively. The tensor t_{ij} indicates the viscous shear stress component. The operators $\overline{(\cdot)}$ and $\widetilde{(\cdot)}$ stand for the Reynolds-averaged and Favre-averaged mean variables, respectively [80]. The presence of the turbulent (Reynolds) stress tensor $\overline{\rho}\tau_{ij} = -\overline{\rho u_i'' u_j''}$ requires additional closure for the system of RANS equations. By employing the Boussinesq hypothesis, which is the foundation of eddy viscosity-based turbulence models, the Reynolds stress tensor is approximated as:

$$\overline{\rho}\tau_{ij} = -\overline{\rho u_i'' u_j''} \cong \mu_t\left(\frac{\partial \widetilde{u_i}}{\partial x_j} + \frac{\partial \widetilde{u_j}}{\partial x_i} - \frac{2}{3}\frac{\partial \widetilde{u_k}}{\partial x_k}\delta_{ij}\right) - \frac{2}{3}\overline{\rho}k\delta_{ij} \tag{29}$$

where μ_t is the turbulent viscosity determined using the selected turbulence model. Here, we use the two-equation SST k-ω turbulence model by Menter [81,82]. This model computes μ_t from a solution of additional transport equations for two variables: the turbulent kinetic energy k and the specific dissipation rate ω (Equations (30) and (31)).

$$\frac{\partial \rho}{\partial t}(\overline{\rho}k) + \frac{\partial}{\partial x_j}(\overline{\rho}\widetilde{u_j}k) = \overline{\rho}\tau_{ij}\frac{\partial \widetilde{u_i}}{\partial x_j} - \beta^*\overline{\rho}k\omega + \frac{\partial}{\partial x_j}\left[(\mu + \sigma_k\mu_t)\frac{\partial k}{\partial x_j}\right] \tag{30}$$

$$\frac{\partial}{\partial t}(\overline{\rho}\omega) + \frac{\partial}{\partial x_j}(\overline{\rho}\widetilde{u_j}\omega) = \frac{\alpha}{\nu_t}\overline{\rho}\tau_{ij}\frac{\partial \widetilde{u_i}}{\partial x_j} - \beta\overline{\rho}\omega^2 + \frac{\partial}{\partial x_j}\left[(\mu + \sigma_\omega\mu_t)\frac{\partial \omega}{\partial x_j}\right] + 2(1 - F_1)\frac{\overline{\rho}\sigma_{\omega 2}}{\omega}\frac{\partial k}{\partial x_j}\frac{\partial \omega}{\partial x_j} \tag{31}$$

Details of the SST k-ω model formulation used in this paper and a specification of the closure coefficients are available in [83].

The system of transport equations is closed with the ideal gas law equation, solved for the density (Equation (32)).

$$\overline{p} = \overline{\rho}R\widetilde{T} \tag{32}$$

Here, R is the specific gas constant for air, and T is the fluid temperature. This study employs an isothermal model, for which the temperature is selected and kept constant according to the specifics of particular design points. Such an assumption is well grounded for a low Ma number flow regime and the absence of heat sources, which is accurate for the intake duct conditions.

2.3.2. Validation of Turbulence Modeling Technique

The selection of the SST k-ω turbulence model was validated prior to the optimization study. The model's predictive quality was compared with the eddy viscosity-based k-ε turbulence model and the non-linear k-ε Explicit Algebraic Reynolds Stress Model (k-ε EARSM). All three models were considered in two options: with and without the curvature correction (CC) term. The flow field predictions for the six abovementioned models were compared with the experimental data for a benchmark U-duct provided by Azzola et al. [84,85]. Details of the validation study are available in [37].

Figure 5 groups the study results for the longitudinal (U) and circumferential (V) velocity components normalized by the bulk velocity (Ub) at the entrance to the duct's curved portion. The k-ε turbulence model showed the most drastic underprediction in the core flow (r/D = 0) deceleration at the duct bend (see Figure 5a). Quite oppositely, the k-ε EARSM model failed to predict the circumferential flow motion at the station downstream of the curved portion. Overall, the study showed the best agreement between the flow simulations and the experimental reference data for the SST k-ω turbulence model. Moreover, no significant beneficial influence of the curvature correction term was detected. Following this reasoning, the SST k-ω model with deactivated CC term is employed in this work.

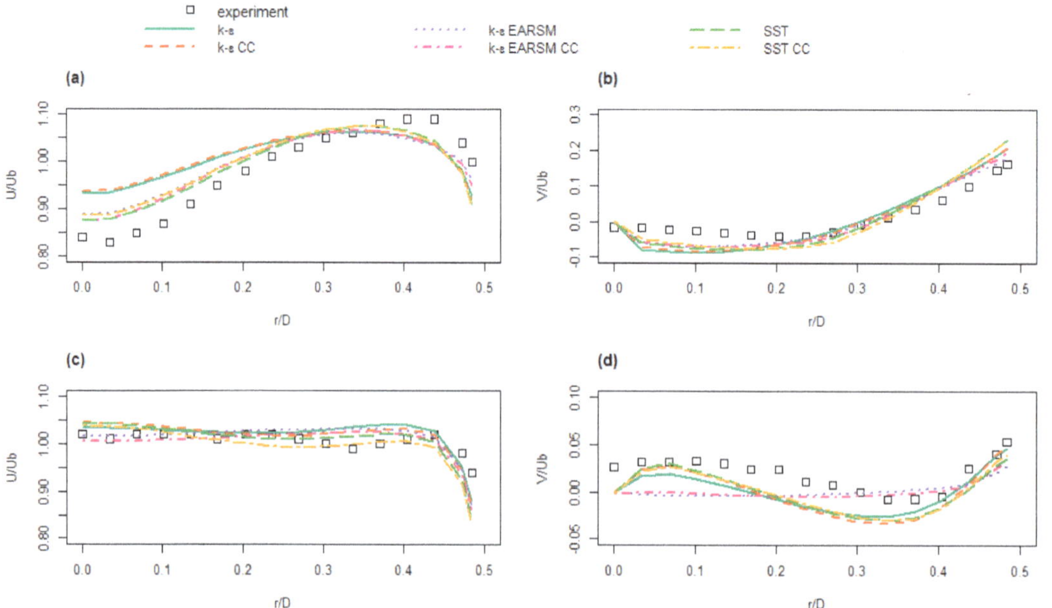

Figure 5. Prediction of normalized longitudinal (U) and circumferential (V) velocity components for various turbulence models in a 180° curved duct with a circular cross-section. Station: (**a**,**b**) at 90° bend; (**c**,**d**) two diameters downstream of the curved section. Adapted from [37].

2.3.3. Details of the Flow Field Modeling Approach

Steady Navier–Stokes equations are discretized by the element-based finite volume method, with a second-order upwind approximation of convective fluxes and a second-order central representation of diffusive fluxes. An additional limiter using the Barth and Jespersen principles [86] is activated to maintain the solution's boundedness. The solution to the equations is obtained using the coupled pressure-based algorithm iterated until the normalized residuals of the momentum and transport equations drop below 10^{-5}.

The boundary conditions mimic the mission characteristics given in Table 2. The duct's inlet flow conditions are defined, with the total pressure value calculated using Bernoulli's equation for specific flight speed and altitude. Furthermore, the inlet turbulence intensity is set to $Tu = 10\%$, and the eddy-to-molecular viscosity ratio equals $\nu_t/\nu = 100$. The amount of airflow accepted by the engine at specific design points is reflected through the mass flow rate condition at the domain outlet. The duct's walls obey the non-slip boundary condition.

The flow domain corresponding to the duct's interior is discretized using a structured hexahedral mesh. To secure a high-level resolution in the near-wall region accompanied by quality elements in the flow core, the mesh is built in the hybrid O-H grid topology (see Figure 6). On average, the non-dimensional wall distance is $y^+ \sim 1.7$ with extreme variation up to $y^+ \leq 3.5$. These values hold throughout the optimization process.

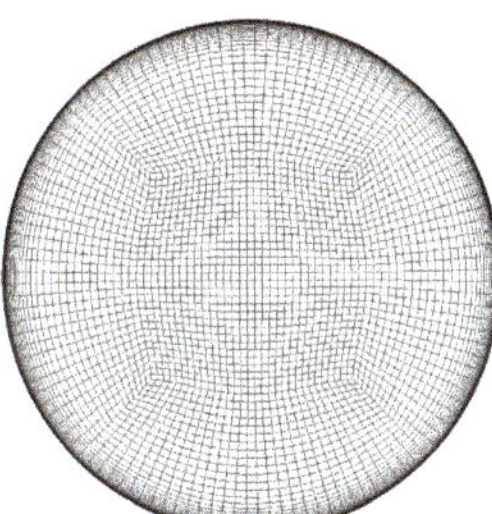

Figure 6. O-H grid at the duct's AIP.

The execution of the comparative study secures the optimization results' independence from the mesh size. Four different grid sizes are compared against the duct-wise pressure distribution. The mass-averaged total pressure values are normalized using the duct inlet pressure. The results are compared in Figure 7. The two finest grids (1,451,000 and 8,007,000 cells) show no difference in predictions for the vast majority of the duct length (~95%). A minor deviation is observable close to the duct's outlet. Hence, the mesh with 1,451,000 elements is used as it has sufficient prediction quality and protects the computational economy of the simulations.

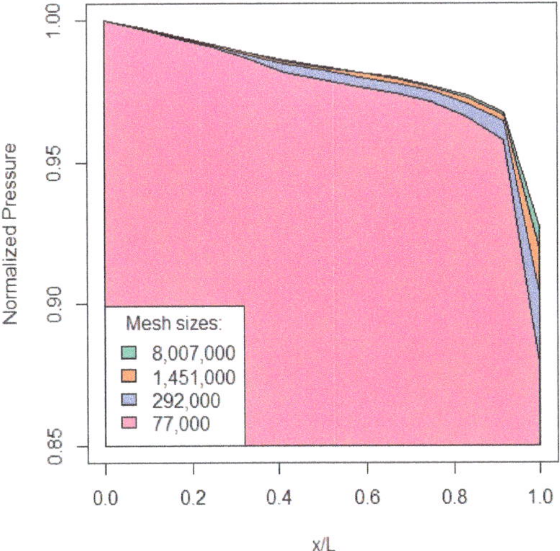

Figure 7. Results of the duct grid independence study. L = duct centerline length. Numerical values indicate the number of mesh elements.

3. Results and Discussion

3.1. Metamodel Fit Validation

Before the optimization began, the Kriging surrogate was assessed for its ability to generalize given data, i.e., to predict objective values at unobserved locations. For this purpose, we used the leave-one-out cross-validation (LOO-CV) procedure, described in detail in Appendix A.

The validation process was executed for the nominal ASF formulation and its mathematical transformations given in Table 4. This action anticipates potential deficiencies in the surrogate's prediction accuracy. In such a scenario, a deterministic mathematical function can be applied to the data set to deliberately impact its distribution and potentially improve the metamodel's fit quality.

Table 4. ASF transformation functions.

Transforming Function	Definition
Natural logarithm	$\ln y$
Square root	$\sqrt{y}$
Negative inverse square root	$\frac{-1}{\sqrt{y}}$
Negative inverse	$\frac{-1}{y}$
Cube root	$\sqrt[3]{y}$

The outcome of the LOO-CV is presented graphically using diagnostic plots in Figures 8–10. In Figure 8, values on the horizontal axis denote exact evaluations of each observation, while the vertical axis shows the surrogate prediction when the corresponding sample is omitted. The markers would be distributed perfectly on a diagonal in a utopian metamodel with zero-error predictions. A realizable high-performing surrogate would be characterized by samples located in close proximity to the imaginary line. Figure 8a shows the results for the nominal ASF. The samples approximately follow the diagonal and are reasonably distributed along the reference line. Only a few observations slightly depart from the ideal arrangement. The results for transformed data (see Figure 8b–f) present a similar pattern, with a slight tendency in some functions to cluster around the higher values (e.g., Figure 8d,e).

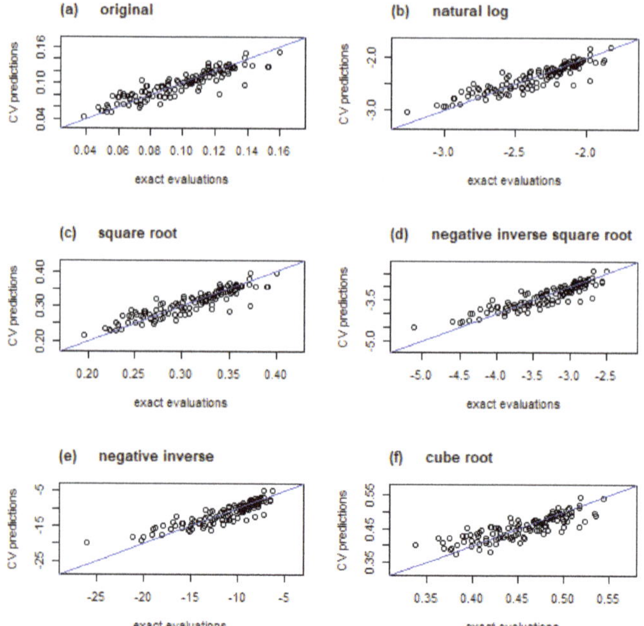

Figure 8. LOO−CV plots for various objective function transformations. Blue lines indicate locations of perfect predictions.

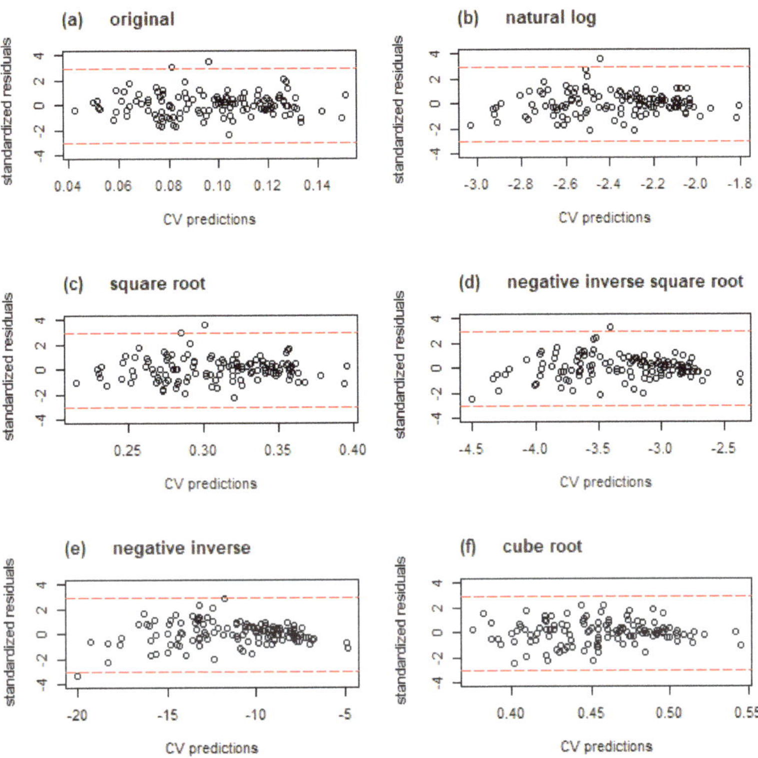

Figure 9. Plots of standardized cross-validated residuals for various objective function transformations. Red dashed lines indicate interval of +/− 3 standard deviations.

Figure 9 shows the distribution of the cross-validated residuals, which—as discussed in Appendix A—should be bounded in a ±3 interval. Inspection of the residuals for nominal ASF (see Figure 9a) reveals one observation that slightly exceeds the desired limits with a residual value of ~3.5. Except for that fault, the residuals are randomly distributed along the sample values with no apparent pattern. The abovementioned deficiency does not impose a rejection of the model, although it might adversely affect its prediction accuracy. Most of the considered data transformations (see Figure 9b–e) do not remove the outlier, although the negative inverse function reduces it to the threshold value. Quite oppositely, the cube root transformation, given in Figure 9f, reduces the error variance to a healthy range without affecting randomized distribution.

The quantile-quantile plot (Q-Q plot) in Figure 10 helps to examine the standardized residuals' normality and homoscedasticity. The assumption of normal distribution should be satisfied for the legitimate use of MLE in the surrogate's hyperparameter estimation. Moreover, following the Kriging principles, the residuals should be homoscedastic, i.e., the variance ought to be homogeneous across the design space [87]. The Q-Q plot draws the residuals' quantiles sorted in ascending order (vertical axis) against the quantiles from the theoretical Gaussian distribution (horizontal axis). For normally distributed residuals, markers on the plot would follow the diagonal. A random spread of points along this line manifests residuals' homoscedasticity.

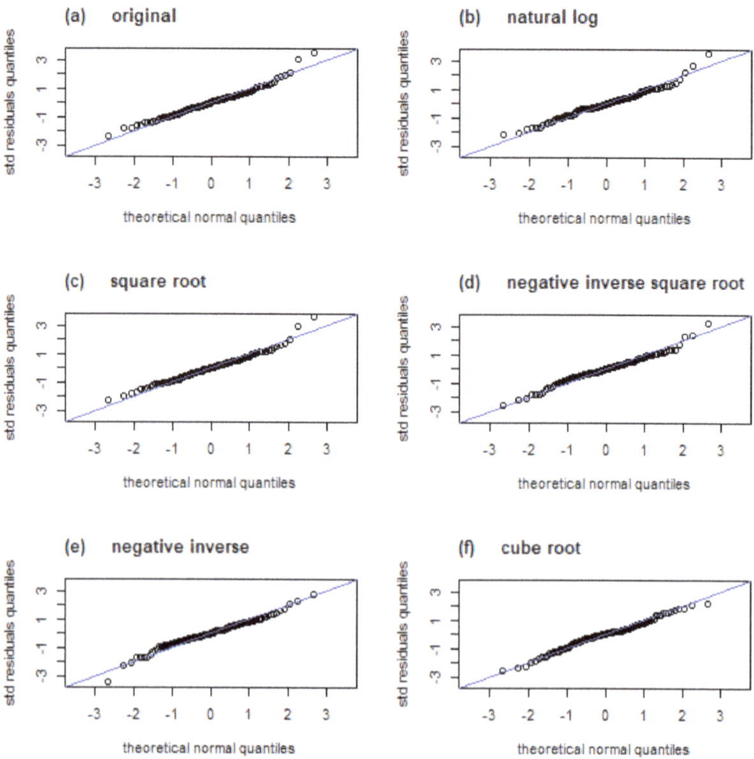

Figure 10. Q–Q plot for various objective function transformations. Blue lines indicate target normal distribution.

The Q-Q plot for the original ASF (see Figure 10a) reveals that the extreme positive residuals tend to deviate counter-clockwise relative to the reference line. Such a pattern indicates that more data are located on the range's right bound, i.e., the distribution is 'heavy-tailed'. Moreover, the lack of a similar pattern on the left tail suggests asymmetry in the distribution with a tendency to right-skewness. Inspection of the remaining Q-Q plots reveals that only two transformations, negative inverse square root (Figure 10d) and cube root (Figure 10f), normalize the distribution of the standardized residuals.

As the cube root transformation is superior in the normalization of residuals and maintains the error variance in a healthy span, we use it to transform the ASF in further studies. Moreover, this transformation is insensitive to the sign of input, which supports the robustness of the process. Although this is an optimistic scenario, the ASF formulation potentially produces negative values for solutions superior to the presumed target. Although less popular than other considered transformations, the cube root has been recognized in applications for long-tailed data [88].

3.2. Sensitivity Analysis

A sensitivity analysis was executed to assess how strong is the dependence of the surrogate's output (i.e., the objective function prediction) on variance in particular design variables. As a result, less impactful variables could be identified and removed from further consideration, thus reducing the computational cost of the optimization.

We used Functional Analysis of Variance (FANOVA), described in detail in Appendix B. This method assesses the contribution of particular inputs and interactions between them using the notion of Sobol' indices [89,90]. FANOVA is particularly beneficial in use with surrogates that might be non-linear and non-monotonic but are inexpensive in the output estimation.

Figure 11 displays graphically the outcome of the FANOVA analysis applied to the Kriging surrogate constructed on the DoE observations. The height of the bars represents the value of the total effect Sobol' indices, while black and gray bars' fractions denote the main effect and aggregated higher-order interactions, respectively.

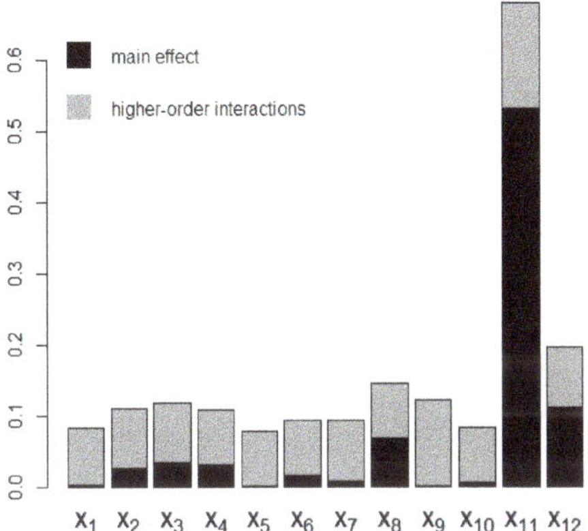

Figure 11. The effect of FANOVA analysis on model considering 12 variables. Bar height indicates the total effect Sobol' index value; black color represents the main effect index, and gray color denotes all higher-order interactions.

Undoubtedly, the design variable x_{11} has the most substantial influence on variance in the objective function. The second most impactful is the variable x_{12}. Both factors correspond to the spatial motion of the control point located in the duct's inner wall area (see Figure 3). Except for the high total index value, these variables are characterized by a strong direct effect, suggesting their crucial influence on the final optimization results.

Quite oppositely, the impact of variables $\{x_1, x_5, x_7, x_{10}\}$ is less significant. Their already low total effect value is related mainly to the higher-order interactions rather than to the direct impact. For this reason, these variables are removed from the design space, notably reducing its dimensions and increasing the process's cost-efficiency. Variable x_6 presents a similar level of total impact to x_7, although its main influence is slightly more prominent. Contrarily, the direct effect of x_9 is minor; however, strong higher-order interactions result in a meaningful value of the total effect index. It depends on the designer's intuition whether such variables should be further considered; here, we preserve them in the study.

3.3. *Assessment of Deformed Mesh Quality*

This section discusses the influence of shape deformation on the mesh quality using the metric defined in Section 2.1.2. Figure 12a displays the mesh orthogonality angle distribution plotted on a symmetry plane of selected exceptionally deformed duct geometry. Although a fall in the quality criterion is evident in the strongly warped regions, the bulk values are well above the warning limit of 20°.

Figure 12b shows a histogram of the minimum value of the orthogonality angle for all solutions produced in the course of optimization. The dashed line denotes the metric's reference value for the baseline geometry. The bulk of generated solutions maintain the quality criterion value close to the reference level. Some cases, however, report a reduction in the orthogonality angle, yet still without violating the warning limit. Finally, a few solutions show criterion values below 20°. Such distortion in the mesh elements might trigger a warning for the flow solver. After inspection, the cases were identified as excessively deformed and, as such, located far from the objective target. These instances did not harm the solver's stability and convergence; hence, they were accepted.

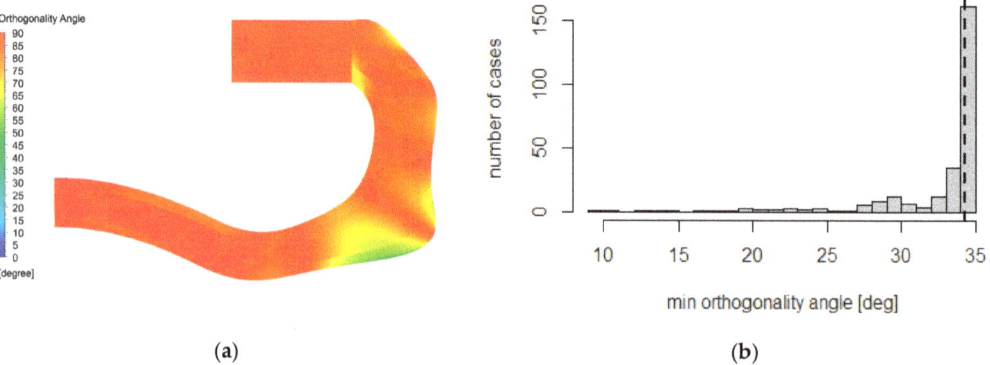

Figure 12. Assessment of deformed mesh quality: (**a**) mesh orthogonality angle distribution at the duct symmetry plane for a selected solution; (**b**) histogram of minimum value of mesh elements' orthogonality angle distribution in all optimization cases. The dashed line indicates the orthogonality angle value for the reference case.

3.4. Optimization Results

Figure 13 shows the contours of the baseline and optimized shape displayed on the duct's symmetry cross-section. The optimization affects the geometry mostly in the transition from the channel to the AIP. The sharp shape change from the inner wall to the compressor entrance is moderated in the final solution. Moreover, the optimizer reduced the curvature of the inner and outer walls; however, the effect on the concave side is more apparent.

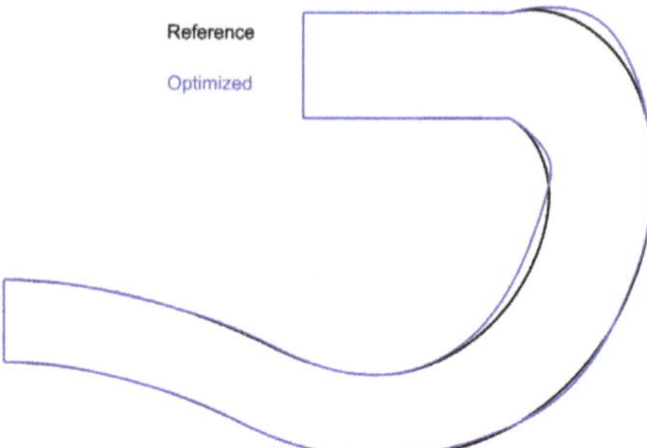

Figure 13. Contours of the duct's symmetry plane for reference and optimized shape.

Table 5 groups the optimized objective function values for all considered conditions and compares them with the reference levels. The results are presented on an absolute scale and as a percentage of improvement with respect to the baseline shape. Undoubtedly, the optimizer was successful in finding a solution that improves both objectives for all considered design points.

Table 5. Quantitative optimization results.

	Reference		Optimized Solution			
	dP	DC_{60}	dP		DC_{60}	
Design Point			Absolute	Relative	Absolute	Relative
DP1: Nominal cruise	0.00512	0.09640	0.00493	−3.71%	0.09190	−4.67%
DP2: Low-altitude climb	0.00459	0.09369	0.00443	−3.49%	0.08943	−4.55%
DP3: High-altitude cruise	0.00471	0.09679	0.00454	−3.61%	0.09252	−4.41%

The greatest improvement level in both objectives is attained for the nominal cruise, which corresponds with the predefined preferences in targets (see Table 3). Furthermore, comparable gains in the pressure loss and the distortion coefficient coincide with the equality in targets set for both objectives. However, a slight consistent preference towards refinement in the distortion coefficient is noticeable. Such an outcome confirms the ability of the ASF formulation to guide optimization toward the presumed target, although a margin has to be considered for potential moderate departures.

Figure 14 shows a cumulated best value of the ASF and the convergence criterion value along the EGO steps. A rapid convergence towards the optimized solutions is evident. The optimizer finds a case significantly improving the initial ASF value (marked by the dashed red line) already in the first iteration. Such superior performance results from a fine design space sampling and high-quality surrogate fit.

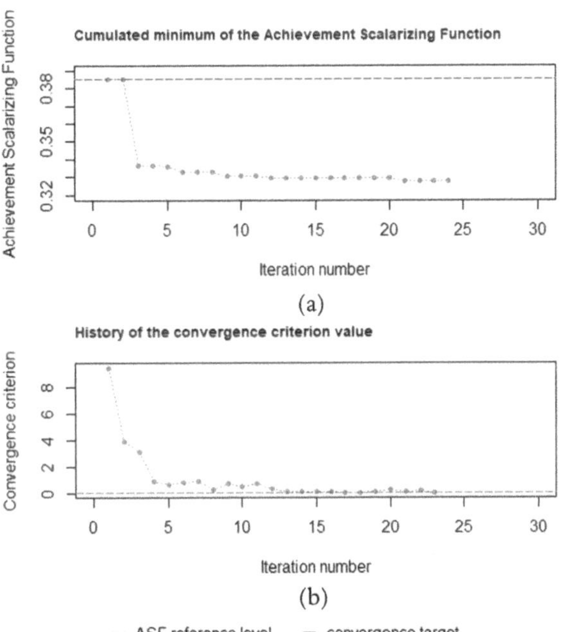

Figure 14. Cumulated minimum value of the ASF (a) and convergence criterion evolution (b) in the course of optimization.

The objective space with DoE and EGO evaluations is displayed in Figure 15 for the three considered design points. For all conditions, the instances from the optimization stage converge towards an imaginary line connecting the reference and target points. The vast majority of the Pareto front is not resolved, which is intentional and prioritizes the use of computational resources in the target's neighborhood. Such an observation shows the algorithm's ability to follow the direction defined by a predetermined target.

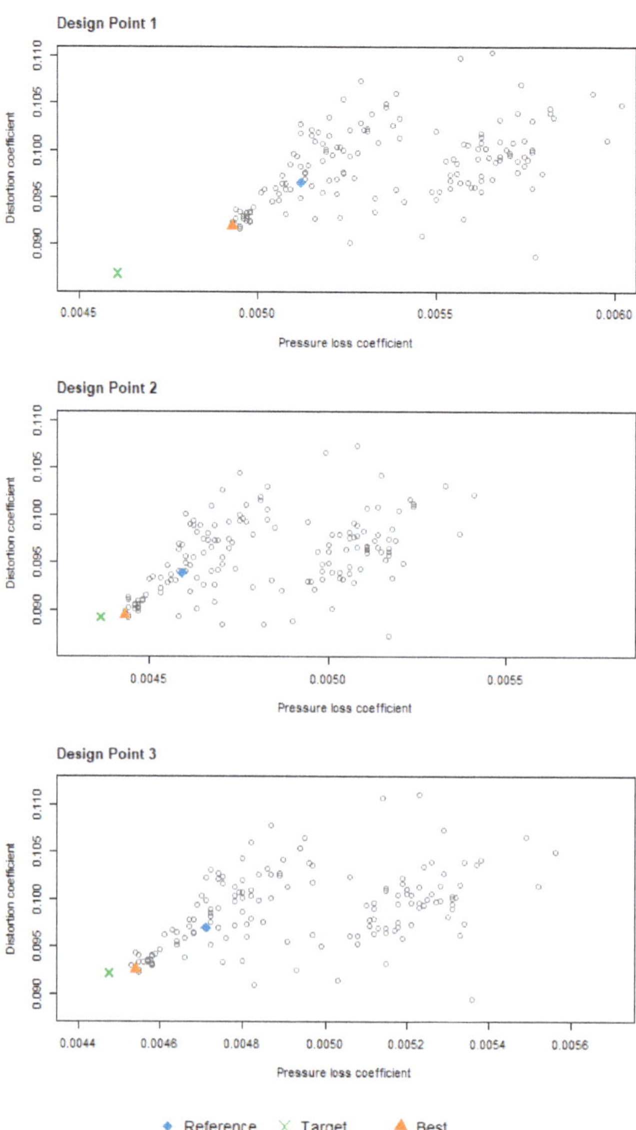

Figure 15. Optimization solutions in the objective function space for the three design points. Markers indicate optimization target (target), evaluation of initial geometry (reference), and best obtained solution (best).

Figure 16 gives insights into the improvements in the distortion coefficient. Figure 16a contrasts radially averaged total pressure values for the reference and optimized solutions. The pressure value is normalized against the average dynamic pressure over the AIP and plotted against the azimuthal position at the compressor face. For the optimized solution, an apparent reduction in the total pressure's lowest peak value manifests for all considered design points. The source of refinement is located in the 150–210° sector, corresponding with the transition area from the duct's inner wall to the AIP. Such an improvement translates directly to an enhancement in the distortion coefficient. A slight shift in the total pressure curves for the optimized solution toward higher values is an effect of reduction in the duct's pressure loss; however, it does not directly impact the distortion metric.

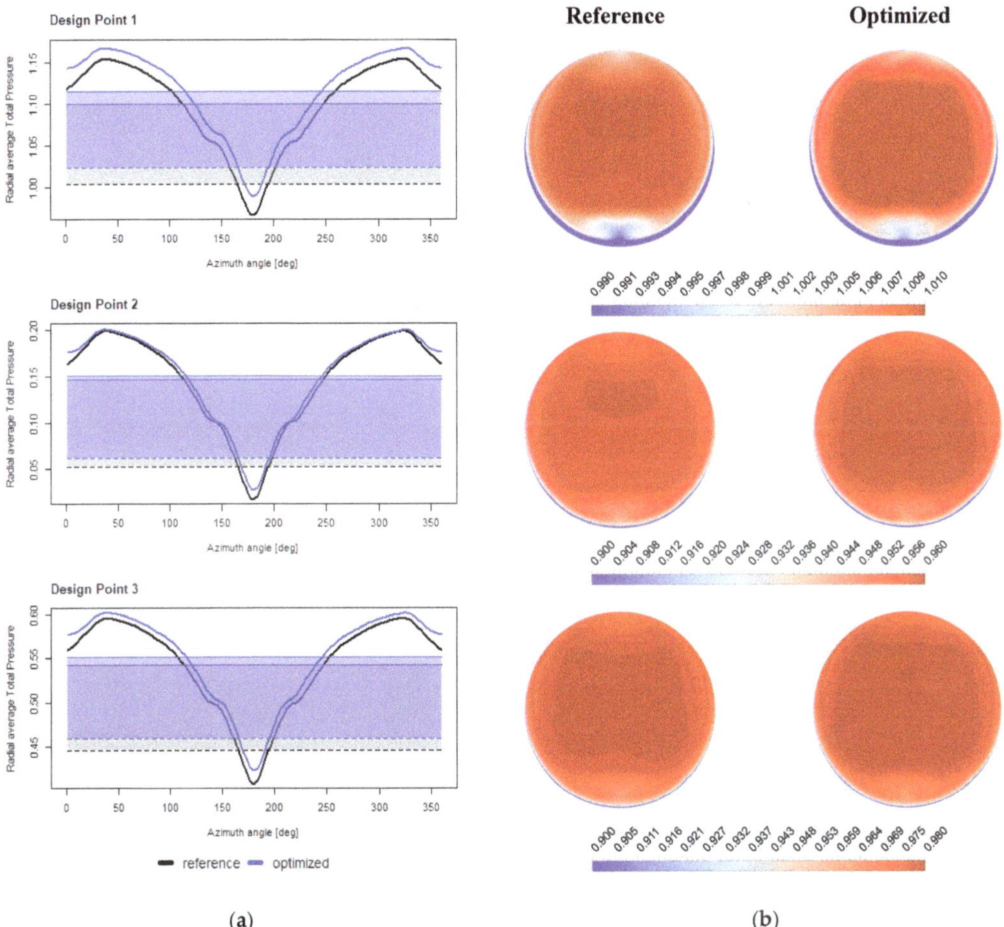

Figure 16. Details of the distortion coefficient improvements: (**a**) radially averaged total pressure distribution at AIP. Solid and dashed lines indicate total pressure levels averaged over the AIP and worst 60° sector, respectively; (**b**) total pressure at AIP normalized by the average dynamic head.

The distortion coefficient value can be visualized intuitively as an area between a plane-averaged total pressure (solid lines in Figure 16a) and total pressure averaged over the worst 60° sector (dashed lines in Figure 16a). The areas are marked with shaded fields with a color corresponding to the reference (gray) and optimized (blue) solutions. The area of the fields representing the improved solution is visibly smaller than the corresponding regions for the baseline case. Such an observation confirms the quantitative results given above in Table 5.

Figure 16b groups the maps of total pressure normalized by the average dynamic head and plotted at the AIP. The pressure field homogenization is visible for all conditions, although the most significant effect is noticeable for Design Point 1. This corresponds with the highest DC_{60} improvement observed in the quantitative data. The reduction in the low-pressure zone visible in the six o'clock position matches the improvement in the 150–210° sector discussed above. A slight improvement is observable in the twelve o'clock region corresponding to the convex wall transition to the AIP. This local enhancement can also be seen in Figure 16a for sector ~360 ± 30°. Even though such improvement does not influence the distortion metric, an increase in the pressure field uniformity at the compressor face is a positive effect.

Figure 17 shows details of the flow field with regard to the pressure loss coefficient improvements. Figure 17a displays the flow loss evolution along the duct through a local pressure loss coefficient contour. This metric is conceptually similar to the pressure loss objective; however, it quantifies the drop from the duct's inlet to each spatial location (Equation (33)).

$$dP^* = \frac{P_t^{IN} - P_t}{P_t^{IN}} \tag{33}$$

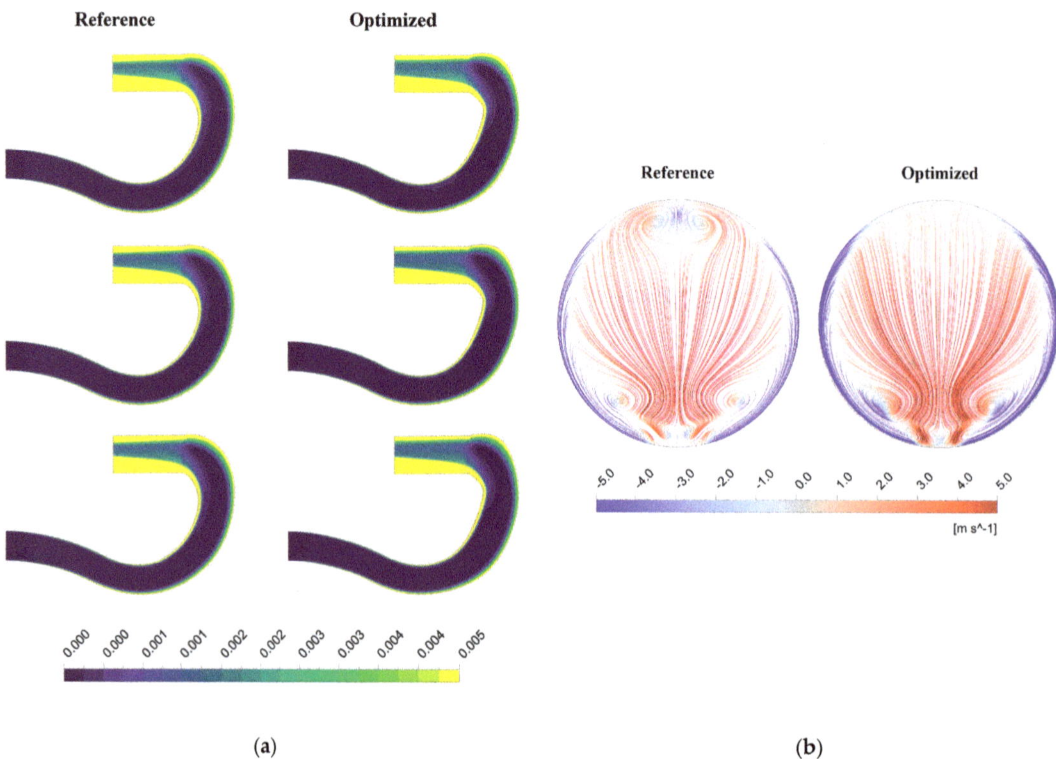

Figure 17. Details of the pressure loss coefficient improvements: (**a**) maps of local pressure loss coefficient; (**b**) flow streamlines colored by the vertical velocity component at cross−section plane downstream of the AIP.

The most apparent enhancement is the reduction in pressure loss in the transition region from the duct's convex wall to the AIP. Adjusting the shape of the wall's top sector and the above-discussed reduction in the low-pressure zone on the concave side transition removes a secondary flow motion downstream of the AIP. This improvement is visualized in Figure 17b, in which the vertical velocity component values color flow streamlines. The two counter-rotating vortices, visible at the twelve o'clock position in the baseline case, are evidently removed in the optimized solution.

The reduction in the concave wall curvature and expansion of the duct's cross-section results in a diffusing shape, bringing additional pressure recovery from the flow kinetic energy. This effect is subtle but still contributes to the duct's overall performance.

4. Conclusions

The proposed aerodynamic optimization framework, constructed from state-of-the-art components in advanced surrogates, mesh morphing, and distance-based scalarization, was applied to the multi-point optimization problem of an I-31T airplane air-intake duct. The study aimed to simultaneously improve the duct's pressure loss and flow distortion under three flight conditions: nominal cruise, low-altitude climbing, and high-altitude cruise.

The optimizer obtained both objective values superior to the reference configuration for all considered design points. The consistent level of improvements in the pressure loss and flow distortion confirms the capacity of the ASF-based formulation to guide optimization toward the presumed target. The study results prove the methodology's potential for optimizing complex multi-objective air-intake duct problems in multiple flight conditions while saving substantial computational resources.

Moreover, FANOVA-based sensitivity analysis was recognized as a valuable tool for assessing the importance of particular design variables. Application of this technique is particularly beneficial in Kriging-based frameworks where the use of surrogate predictions balances high computational costs related to the variance-based techniques.

This research, however, may be subject to potential limitations regarding the generalization of its results. Primarily, the number of design variables was relatively moderate, which favors the use of the Kriging surrogate. Although literature sources report a considerable margin for increased design space dimensions, studies on larger problems should be executed to prove the framework's efficiency. Moreover, the results revealed that all considered flight conditions were coherent in the direction of shape improvements. Although justified for this particular application, this scenario might only represent part of the class of problems. Future studies should cover scenarios with design points of contradicting performance requirements to generalize the results. Finally, the optimization was initiated from an already well-designed solution, which aimed to set an ambitious task for the optimizer but also resulted in a relatively low level of required deformations. Optimization studies demanding significant geometry adjustments may be subject to numerical errors resulting from an excessively distorted mesh. Such problems may require the implementation of a solution for handling erroneous data samples.

Optimization studies using the proposed framework could be expanded to more holistic multi-disciplinary problems. The most natural development would include solid body mechanics addressing the structural targets, although involvement of the intuitively more distant performance and economic objectives seems feasible. No fundamental reasons were identified that might inhibit the ASF-based formulation in handling such heterogeneous goals, although this would need to be proved in further studies.

Author Contributions: Conceptualization, P.S.D., S.K. and J.Ż.; methodology, P.S.D. and S.K.; software, P.S.D.; validation, P.S.D.; formal analysis, P.S.D.; investigation, P.S.D.; resources, P.S.D. and J.Ż.; data curation, P.S.D.; writing—original draft preparation, P.S.D.; writing—review and editing, P.S.D. and S.K.; visualization, P.S.D.; supervision, P.S.D., S.K. and J.Ż. All authors have read and agreed to the published version of the manuscript.

Funding: This work was co-funded by the Ministry of Education and Science as part of the "Implementation Doctorate first edition" program under the agreement no 3/DW/2017/01/1.

Data Availability Statement: Not applicable.

Acknowledgments: We would like to express our great appreciation to Karolina Dreżek for her support in the preparation of figures.

Conflicts of Interest: The authors declare no conflict of interest. The funders had no role in the design of the study; in the collection, analyses, or interpretation of data; in the writing of the manuscript; or in the decision to publish the results.

Appendix A. Leave-One-Out Cross-Validation

The leave-one-out cross-validation (LOO-CV) technique [91] sequentially omits one sample from the set of N observations, and the metamodel is constructed from the remaining $N - 1$ points. Afterward, the objective is predicted at the left-out sample's location, and the estimated value is compared with the known evaluation. The concept of LOO-CV is demonstrated schematically in Figure A1. The operation repeats N times, resulting in an assessment of prediction accuracy for the whole observation set.

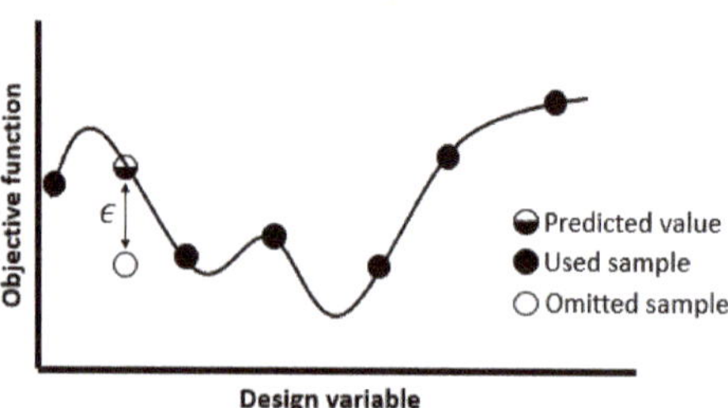

Figure A1. Schematic representation of LOO-CV concept for one-dimensional design space.

The measure of difference between the true objective value $y(\boldsymbol{x}_i)$ and the corresponding LOO-CV prediction $\hat{y}_{\mathfrak{D}-i}(\boldsymbol{x}_i)$ is termed a residual (Equation (A1)).

$$\epsilon_i = y(\boldsymbol{x}_i) - \hat{y}_{\mathfrak{D}-i}(\boldsymbol{x}_i) \tag{A1}$$

The residuals are standardized using the variance estimator $\widehat{s^2_{\mathfrak{D}-i}(\boldsymbol{x}_i)}$ available in the Kriging surrogate (Equation (A2)).

$$\epsilon_i^z = \frac{\epsilon_i}{\sqrt{\widehat{s^2_{\mathfrak{D}-i}}}} \tag{A2}$$

If the residuals approximately follow the Gaussian distribution, the objective evaluation locates within three standard deviations from the predicted Kriging mean with a confidence level of 99.7%. Thus, a surrogate with good prediction quality should be characterized by standardized residuals bounded in a ± 3 interval [23]. The probability of a residual being further away is less than 0.3%. Detecting such an outlier may suggest poor prediction quality in the vicinity of the corresponding observation.

Appendix B. Functional Analysis of Variance (FANOVA)

Consider that the design space $\mathfrak{X}$ forms a p-dimensional hypercube and $\boldsymbol{x} \in \mathfrak{X}$ is a vector of independent random variables normalized to a range $[0, 1]$. The surrogate model is described by a square-integrable function $Y = f(x)$ defined in $\mathfrak{X}$, which can be decomposed using the FANOVA representation [89]:

$$Y(\boldsymbol{x}) = f_0 + \sum_{i=1}^{p} f_i(x_i) + \sum_{1 \le i < j \le p} f_{ij}(x_i, x_j) + \cdots + f_{1,\ldots,p}(x_1, \ldots, x_p) \tag{A3}$$

In the above equation, the centered and orthogonal terms denote:

$$\text{mean value } f_0 = \mathbb{E}[Y(\boldsymbol{x})] \tag{A4}$$

$$\text{main effects } f_i(x_i) = \mathbb{E}[Y(\boldsymbol{x})|x_i] - f_0 \tag{A5}$$

$$\text{second-order interactions } f_{ij}(x_i, x_j) = \mathbb{E}[Y(\boldsymbol{x})|x_i, x_j] - f_0 - f_i(x_i) - f_j(x_j) \tag{A6}$$

The interactions of higher orders can be constructed accordingly as conditional expected values.

A similar technique serves for a decomposition of the model output's variance:

$$Var(Y(\boldsymbol{x})) = \sum_{i=1}^{p} Var(f_i(x_i)) + \sum_{1\leq i<j\leq p} Var\left(f_{ij}\left(x_i, x_j\right)\right) + \cdots + Var\left(f_{1,\ldots,p}\left(x_1, \ldots, x_p\right)\right) \tag{A7}$$

The individual contribution of variable x_i to the variance in the model's output is quantified by the *main effect Sobol index*:

$$S_i = \frac{Var(f_i(x_i))}{Var(Y(\boldsymbol{x}))} \tag{A8}$$

The influence of interaction between any two variables x_i and x_j is described by the second-order Sobol' index:

$$S_{ij} = \frac{Var\left(f_{ij}\left(x_i, x_j\right)\right)}{Var(Y(\boldsymbol{x}))} \tag{A9}$$

All higher-order interactions can be assessed by Sobol' indices, constructed following a similar concept of ratios between decomposed higher-order terms and the overall output variance. The values of main and higher-order indices for all variables sum to unity (Equation (A10)).

$$\sum_{i=1}^{p} S_i + \sum_{1\leq i<j\leq p} S_{ij} + \cdots + S_{1,\ldots,p} = 1 \tag{A10}$$

The aggregated contribution to the model's output variance of the i-th variable is measured by the total effect Sobol' index:

$$S_i^T = \frac{\mathbb{E}[Var(Y(x)|\boldsymbol{x}\backslash\{x_i\})]}{Var(Y(x))} \tag{A11}$$

In practice, only main and total indices are assessed for economic reasons. Knowing them allows for an estimation of the combined influence of all-order interactions.

The values of the Sobol' indices are to be computed numerically. For this purpose, this study employs the FAST algorithm [92] available in the R package sensitivity [93].

References

1. Airbus. Global Market Forecast 2022–2041. Available online: https://www.airbus.com/en/products-services/commercial-aircraft/market/global-market-forecast (accessed on 16 January 2023).
2. Boeing. Boeing Commercial Market Outlook 2022–2041. Available online: https://www.boeing.com/resources/boeingdotcom/market/assets/downloads/CMO_2022_Report_FINAL_v02.pdf (accessed on 16 January 2023).
3. European Commission Website. A European Green Deal-Striving to be the First Climate-Neutral Continent. Available online: https://ec.europa.eu/info/strategy/priorities-2019-2024/european-green-deal_en (accessed on 16 January 2023).
4. Furlan, F.; Chiereghin, N.; Kipouros, T.; Benini, E.; Savill, M. Computational design of S-Duct intakes for distributed propulsion. *Aircr. Eng. Aerosp. Technol.* **2014**, *86*, 473–777. [CrossRef]
5. Goldberg, D.E. *Genetic Algorithms in Search, Optimization, and Machine Learning*; Addison Wesley: Reading, UK, 1989.
6. D'Ambros, A.; Kipouros, T.; Zachos, P.; Savill, M.; Benini, E. Computational Design Optimization for S-Ducts. *Designs* **2018**, *2*, 36. [CrossRef]
7. Zeng, L.; Pan, D.; Ye, S.; Shao, X. A fast multiobjective optimization approach to S-duct scoop inlets design with both inflow and outflow. *Proc. Inst. Mech. Eng. Part G J. Aerosp. Eng.* **2019**, *233*, 3381–3394. [CrossRef]
8. Sharma, M.; Baloni, B.D. Design optimization of S-shaped compressor transition duct using particle swarm optimization algorithm. *SN Appl. Sci.* **2020**, *2*, 221. [CrossRef]
9. Kennedy, J. Particle Swarm Optimization. In *Encyclopedia of Machine Learning*; Sammut, C., Webb, G.I., Eds.; Springer: Boston, MA, USA, 2011; pp. 760–766. [CrossRef]
10. Wang, G.G.; Shan, S. Review of Metamodeling Techniques in Support of Engineering Design Optimization. *J. Mech. Des.* **2007**, *129*, 370–380. [CrossRef]
11. Simpson, T.W.; Poplinski, J.D.; Koch, P.N.; Allen, J.K. Metamodels for Computer-based Engineering Design: Survey and recommendations. *Eng. Comput.* **2001**, *17*, 129–150. [CrossRef]
12. Han, Z.-H.; Zhang, K.-S. Surrogate-Based Optimization. In *Real-World Applications of Genetic Algorithms*; Roeva, O., Ed.; InTech: London, UK, 2012; pp. 343–362.
13. Skinner, S.N. State-of-the-art in aerodynamic shape optimisation methods. *Appl. Soft Comput. J.* **2018**, *62*, 933–962. [CrossRef]
14. Forrester, A.I.J.; Keane, A.J. Recent advances in surrogate-based optimization. *Prog. Aerosp. Sci. 2009 45*, 50–79. [CrossRef]

15. Snyman, J.A.; Wilke, D.N. *Practical Mathematical Optimization–Basic Optimization Theory and Gradient-Based Algorithms*, 2nd ed.; Springer: Cham, Switzerland, 2018.
16. Khuri, A.I.; Mukhopadhyay, S. Response surface methodology. *WIREs Comp. Stat.* **2010**, *2*, 128–149. [CrossRef]
17. Friedman, J.H. Multivariate Adaptive Regression Splines. *Ann. Stat.* **1991**, *19*, 1–67. [CrossRef]
18. Regis, R.G.; Shoemaker, C.A. Constrained Global Optimization of Expensive Black Box Functions Using Radial Basis Functions. *J. Glob. Optim.* **2005**, *31*, 153–171. [CrossRef]
19. Haykin, S. *Neural Networks: A Comprehensive Foundation*, 3rd ed.; Pearson Prentice Hall: Upper Saddle River, NJ, USA, 2007.
20. Krige, D.G. A Statistical Approach to Some Basic Mine Valuation Problems on the Witwatersrand. *J. Chem. Metall. Soc. S. Min. Afr.* **1951**, *52*, 119–139. [CrossRef]
21. Matheron, G. Principles of geostatistics. *Econ. Geol.* **1963**, *58*, 1246–1266. [CrossRef]
22. Sacks, J.; Welch, W.J.; Mitchell, J.S.B.; Henry, P.W. Design and Experiments of Computer Experiments. *Stat. Sci.* **1989**, *4*, 409–423. [CrossRef]
23. Jones, D.R.; Schonlau, M.; Welch, W.J. Efficient Global Optimization of Expensive Black-Box Functions. *J. Glob. Optim.* **1998**, *13*, 455–492. [CrossRef]
24. Lu, H.; Zheng, X.; Li, Q. A combinatorial optimization design method applied to S-shaped compressor transition duct design. *Proc. Inst. Mech. Eng. Part G J. Aerosp. Eng.* **2014**, *228*, 1749–1758. [CrossRef]
25. Immonen, E. Shape optimization of annular S-ducts by CFD and high-order polynomial response surfaces. *Eng. Comput.* **2018**, *35*, 932–954. [CrossRef]
26. Gan, W.; Zhang, X. Design optimization of a three-dimensional diffusing S-duct using a modified SST turbulent model. *Aerosp. Sci. Technol.* **2017**, *63*, 63–72. [CrossRef]
27. Jin, D.; Liu, X.; Zhao, W.; Gui, X. Optimization of endwall contouring in axial compressor S-shaped ducts. *Chin. J. Aeronaut.* **2015**, *28*, 1076–1086. [CrossRef]
28. Zerbinati, A.; Désidéri, J.A.; Duvigneau, R. Application of metamodel-assisted Multiple-Gradient Descent Algorithm (MGDA) to air-cooling duct shape optimization. In Proceedings of the ECCOMAS—European Congress on Computational Methods in Applied Sciences and Engineering-2012, Vienna, Austria, 10–14 September 2012.
29. Verstraete, T.; Coletti, F.; Bulle, J.; Vanderwielen, T.; Arts, T. Optimization of a U-Bend for Minimal Pressure Loss in Internal Cooling Channels: Part I—Numerical Method. *ASME J. Turbomach.* **2013**, *135*, 051015. [CrossRef]
30. Storn, R.; Price, K. Differential Evolution—A Simple and Efficient Heuristic for Global Optimization over Continuous Spaces. *J. Glob. Optim.* **1997**, *11*, 341–359. [CrossRef]
31. Verstraete, T.; Li, J. Multi-Objective Optimization of a U-Bend for Minimal Pressure Loss and Maximal Heat Transfer Performance in Internal Cooling Channels. In Proceedings of the ASME Turbo Expo 2013: Turbine Technical Conference and Exposition Volume 3A: Heat Transfer, San Antonio, TX, USA, 3–7 June 2013. [CrossRef]
32. Koo, G.-W.; Lee, S.-M.; Kim, K.-Y. Shape optimization of inlet part of a printed circuit heat exchanger using surrogate modeling. *Appl. Therm. Eng.* **2014**, *72*, 90–96. [CrossRef]
33. Wang, B.; Wang, Q. Numerical Optimization of Electromagnetic Performance and Aerodynamic Performance for Subsonic S-Duct Intake. *Aerospace* **2022**, *9*, 665. [CrossRef]
34. Bae, H.; Park, S.; Kwon, J. Efficient global optimization for S-duct diffuser shape design. *Proc. Inst. Mech. Eng. Part G J. Aerosp. Eng.* **2013**, *227*, 1516–1532. [CrossRef]
35. Dehghani, M.; Ajam, H.; Farahat, S. Automated Diffuser Shape Optimization based on CFD Simulations and Surrogate Modeling. *J. Appl. Fluid Mech.* **2016**, *9*, 2527–2535. [CrossRef]
36. Marchlewski, K.; Łaniewski-Wołłk, Ł.; Kubacki, S. Aerodynamic Shape Optimization of a Gas Turbine Engine Air-Delivery Duct. *J. Aerosp. Eng.* **2020**, *33*, 04020042. [CrossRef]
37. Drężek, P.S.; Kubacki, S.; Żółtak, J. Multi-objective surrogate model-based optimization of a small aircraft engine air-intake duct. *Proc. Inst. Mech. Eng. Part G J. Aerosp. Eng.* **2022**, *236*, 2909–2921. [CrossRef]
38. Nikulin, Y.; Miettinen, K.; Mäkelä, M.M. A new achievement scalarizing function based on parameterization in multiobjective optimization. *OR Spectr.* **2012**, *34*, 69–87. [CrossRef]
39. Biancolini, M.E.; Costa, E.; Cella, U.; Groth, C.; Veble, G.; Andrejašič, M. Glider fuselage-wing junction optimization using CFD and RBF mesh morphing. *Aircr. Eng. Aerosp. Technol.* **2016**, *88*, 740–752. [CrossRef]
40. Kapsoulis, D.; Asouti, V.; Giannakoglou, K.; Porziani, S.; Costa, E.; Groth, C.; Cella, U.; Biancolini, M.V. Evolutionary aerodynamic shape optimization through the RBF4AERO platform. In *Proceedings of the VII European Congress on Computational Methods in Applied Sciences and Engineering (ECCOMAS Congress 2016) Crete, Greece, 5–10 June 2016*; Institute of Structural Analysis and Antiseismic Research, School of Civil Engineering, National Technical University of Athens (NTUA): Athens, Greece; pp. 4146–4155. [CrossRef]
41. Savastano, W.; Pranzitelli, A.; Andrews, G.E.; Biancolini, M.E.; Ingham, D.B.; Pourkashanian, M. Goal Driven Shape Optimisation for Conjugate Heat Transfer in an Effusion Cooling Plate. In *Proceedings of the ASME Turbo Expo 2015: Turbine Technical Conference and Exposition, Montreal, Quebec, Canada, 15–19 June 2015*; The American Society of Mechanical Engineers: New York, NY, USA, 2015; Volume 5A: Heat Transfer. [CrossRef]
42. Mastrippolito, F.; Aubert, S.; Ducros, F.; Buisson, M. RBF-based mesh morphing improvement using Schur complement ap-plied to rib shape optimization. *Int. J. Numer. Methods Heat Fluid Flow* **2020**, *30*, 4241–4257. [CrossRef]

43. Brahmachary, S.; Fujio, C.; Aksay, M.; Ogawa, H. Design optimization and off-design performance analysis of axisymmetric scramjet intakes for ascent flight. *Phys. Fluids* **2022**, *34*, 036109. [CrossRef]
44. Brahmachary, S.; Fujio, C.; Ogawa, H. Multi-point design optimization of a high-performance intake for scramjet-powered ascent flight. *Aerosp. Sci. Technol.* **2020**, *107*, 106362. [CrossRef]
45. Fujio, C.; Ogawa, H. Physical insights into multi-point global optimum design of scramjet intakes for ascent flight. *Acta Astronaut.* **2022**, *194*, 59–75. [CrossRef]
46. Chiang, C.; Koo, D.; Zingg, D.W. Aerodynamic Shape Optimization of an S-Duct Intake for a Boundary-Layer Ingesting Engine. *J. Aircr.* **2022**, *59*, 725–741. [CrossRef]
47. Park, J.-S.; Baek, J. Efficient computation of maximum likelihood estimators in a spatial linear model with power exponential covariogram. *Comput. Geosci.* **2001**, *27*, 1–7. [CrossRef]
48. Roustant, O.; Ginsbourger, D.; Deville, Y. DiceKriging, DiceOptim: Two R Packages for the Analysis of Computer Experiments by Kriging-Based Metamodeling and Optimization. *J. Stat. Softw.* **2012**, *51*, 1–55. [CrossRef]
49. Liu, H.; Ong, Y.-S.; Cai, J. A survey of adaptive sampling for global metamodeling in support of simulation-based complex engineering design. *Struct. Multidiscip. Optim.* **2018**, *57*, 393–416. [CrossRef]
50. McKay, M.D.; Beckman, R.J.; Conover, W.J. A Comparison of three methods for selecting values of input variables in the analysis of output from a computer code. *Technometrics* **1979**, *21*, 239–245. [CrossRef]
51. Viana, F.A.C. Things You Wanted to Know About the Latin Hypercube Design and Were Afraid to Ask. In Proceedings of the 10th World Congress on Structural and Multidisciplinary Optimization, Orlando, FL, USA, 19–24 May 2013.
52. Crombecq, K.; Couckuyt, I.; Gorissen, D.; Dhaene, T. Space-Filling Sequential Design Strategies for Adaptive Surrogate Modelling. In *The First International Conference on Soft Computing Technology in Civil, Structural and Environmental Engineering, Funchal Madeira, Portugal, 1–4 September 2009*; Civil-Comp Press: Stirlingshire, UK, 2009. [CrossRef]
53. Liefvendahl, M.; Stocki, R. A study on algorithms for optimization of Latin hypercubes. *J. Stat. Plan Inference* **2006**, *136*, 3231–3247. [CrossRef]
54. Santner, T.J.; Williams, B.J.; Notz, W.I. Space-Filling Designs for Computer Experiments. In *The Design and Analysis of Computer Experiments Springer Series in Statistics*; Springer: New York, NY, USA, 2018; pp. 145–200.
55. Carnell, R. lhs: Latin Hypercube Samples. R Package v. 1.0.1.2019. Available online: https://CRAN.R-project.org/package=lhs. (accessed on 11 January 2023).
56. Loeppky, J.L.; Sacks, J.; Welch, W.J. Choosing the Sample Size of a Computer Experiment: A Practical Guide. *Technometrics* **2009**, *51*, 366–376. [CrossRef]
57. Mebane, W.R.; Sekhon, J.S. Genetic optimization using derivatives: The rgenoud package for R. *J. Stat. Softw.* **2011**, *42*, 1–26. [CrossRef]
58. Duchon, J. Splines minimizing rotation-invariant semi-norms in Sobolev spaces. In *Constructive Theory of Functions of Several Variables Lecture Notes in Mathematics*; Schempp, W., Zeller, K., Eds.; Springer: Berlin/Heidelberg, Germany, 1977; pp. 85–100.
59. Sandwell, D.T. Biharmonic spline interpolation of GEOS-3 and SEASAT altimeter data. *Geophys. Res. Lett.* **1987**, *14*, 139–142. [CrossRef]
60. Buhmann, M.D. *Radial Basis Functions: Theory and Implementations*; Cambridge University Press: Cambridge, UK, 2003.
61. Wendland, H. *Scattered Data Approximation*; Cambridge University Press: Cambridge, UK, 2004.
62. Gneiting, T. Radial Positive Definite Functions Generated by Euclid's Hat. *J. Multivar. Anal.* **1999**, *69*, 88–119. [CrossRef]
63. Tezzele, M.; Demo, N.; Mola, A.; Rozza, G. PyGeM: Python Geometrical Morphing. *Softw. Impacts* **2021**, *7*, 100047. [CrossRef]
64. Knupp, P.M. Algebraic mesh quality metrics for unstructured initial meshes. *Finite Elem. Anal. Des.* **2003**, *39*, 217–241. [CrossRef]
65. Knupp, P.M. Remarks on mesh quality. In Proceedings of the 45th AIAA Aerospace Sciences Meeting and Exhibit, Reno, NV, USA, 8–11 January 2007. Also available as Sandia National Laboratories SAND2007-8128C.
66. Marler, R.T.; Arora, J.S. Survey of multi-objective optimization methods for engineering. *Struct. Multidiscip. Optim.* **2004**, *26*, 369–395. [CrossRef]
67. Baron, A. *Samolot Osobowy I-23 "Manager". Wybrane Problemy Badawcze*; Wydawnictwa Naukowe Sieci Badawczej Łukasiewicz–Instytutu Lotnictwa: Warszawa, Poand, 2012. (In Polish)
68. Efficient Systems and Propulsion for Small Aircraft Project. Available online: http://www.esposa-project.eu/ (accessed on 9 January 2023).
69. Guła, P.; Ulma, D.; Żurek, K.; Żurawski, R. Challenges of turboprop engine installation on small aircraft. *Aircr. Eng. Aerosp. Technol.* **2019**, *91*, 938–948. [CrossRef]
70. Iwaniuk, A.; Wiśniowski, W.; Żółtak, J. Multi-disciplinary optimisation approach for a light turboprop aircraft-engine integration and improvement. *Aircr. Eng. Aerosp. Technol.* **2016**, *88*, 348–355. [CrossRef]
71. Idzikowski, M.; Miksa, W. Flight Tests of Turboprop Engine with Reverse Air Intake System. *Trans. Aerosp. Res.* **2018**, *2018*, 26–36. [CrossRef]
72. Idzikowski, M.; Miksa, W. Ground and in-Fligh Testing of Cooling Efficiency of Turboprop Engine Compartment. *Trans. Aerosp. Res.* **2018**, *2018*, 17–25. [CrossRef]
73. Stalewski, W.; Żółtak, J. The preliminary design of the air-intake system and the nacelle in the small aircraft-engine integration process. *Aircr. Eng. Aerosp. Technol.* **2014**, *86*, 250–258. [CrossRef]

74. Stalewski, W.; Żółtak, J. Multi-objective and multidisciplinary optimization of wing for small aircraft. In Proceedings of the International Conference of the European Aerospace Societies Congress, Venice, Italy, 24–28 October 2011; pp. 1015–1024.
75. Stalewski, W.; Żółtak, J. Multicriteria Design and Optimisation of Helicopter Fuselage. In *Evolutionary and Deterministic Methods for Design, Optimization and Control with Application to Industrial and Societal Problems*; Poloni, C., Quaglarella, D., Periaux, J., Gauger, N., Giannakoglou, K., Eds.; CIRA: Capua, Italy, 2011; pp. 518–529.
76. Stalewski, W. PARADES—Program wspierający parametryczne projektowanie i optymalizację złożonych obiektów aerodynamicznych. In *Projektowanie i Optymalizacja Aerodynamiczna Wiropłatów*; Wydawnictwa Naukowe Instytutu Lotnictwa: Warszawa, Poland, 2017.
77. Reid, C. The Response of Axial Flow Compressors to Intake Flow Distortion. In *Proceedings of the ASME 1969 Gas Turbine Conference and Products Show, Cleveland, OH, USA, 9–13 March 1969*; The American Society of Mechanical Engineers: New York, NY, USA, 1969. [CrossRef]
78. United States Committee on Extension to the Standard Atmosphere, N.A.S.A.; United States Air Force, U.S. Standard Atmosphere, 1976. In *Report No.: NASA-TM-X-74335*; National Oceanic and Atmospheric Administration: Washington, DC, USA, 1976. Available online: https://ntrs.nasa.gov/citations/19770009539 (accessed on 14 January 2023).
79. Kalpakli Vester, A.; Örlü, R.; Alfredsson, P.H. Turbulent Flows in Curved Pipes: Recent Advances in Experiments and Simulations. *Appl. Mech. Rev.* **2016**, *68*, 050802. [CrossRef]
80. Wilcox, D.C. *Turbulence Modelling for CFD*, 3rd ed.; Dcw Industries, Inc.: Los Angeles, CA, USA, 2006.
81. Menter, F. Zonal Two Equation Kw Turbulence Models for Aerodynamic Flows. In Proceedings of the 23rd Fluid Dynamics, Plasmadynamics, and Lasers Conference, Orlando, FL, USA, 6–9 July 1993. [CrossRef]
82. Menter, F.R. Two-equation eddy-viscosity turbulence models for engineering applications. *AIAA J.* **1994**, *32*, 1598–1605. [CrossRef]
83. Menter, F.R.; Kuntz, M.; Langtry, R. Ten Years of Industrial Experience with the SST Turbulence Model. In *Turbulence Heat and Mass Transfer 4*; Hanjalic, K., Nagano, Y., Tummers, M., Eds.; Begell House: Danbury, UK, 2003.
84. Azzola, J.; Humphrey, J.A.C.; Iacovides, H.; Launder, B.E. Developing Turbulent Flow in a U-Bend of Circular Cross-Section: Measurement and Computation. *J. Fluids Eng.* **1986**, *108*, 214–221. [CrossRef]
85. Azzola, J.; Humphrey, J.A.C. Developing Turbulent Flow in a 180° Curved Pipe and Its Downstream Tangent. Lawrence Berkeley National Laboratory; Report No.: LBL-17681. 1984. Available online: https://escholarship.org/uc/item/1fg887fq (accessed on 11 January 2023).
86. Barth, T.; Jespersen, D. The design and application of upwind schemes on unstructured meshes. In Proceedings of the 27th Aerospace Sciences Meeting, Reno, NV, USA, 9–12 January 1989; p. 366. [CrossRef]
87. Congdon, C.; Martin, J. On Using Standard Residuals as a Metric of Kriging Model Quality. In Proceedings of the 48th AIAA/ASME/ASCE/AHS/ASC Structures, Structural Dynamics, and Materials Conference, Honolulu, HI, USA, 23–27 April 2007. [CrossRef]
88. Miles, B.W.J.; Stokes, C.R.; Vieli, A.; Cox, N.J. Rapid, climate-driven changes in outlet glaciers on the Pacific coast of East Antarctica. *Nature* **2013**, *500*, 563–566. [CrossRef]
89. Sobol', I.M. Global sensitivity indices for nonlinear mathematical models and their Monte Carlo estimates. *Math. Comput Simul.* **2001**, *55*, 271–280. [CrossRef]
90. Sobol', I.M. Sensitivity estimates for nonlinear mathematical models. *Math. Model. Comput. Exp.* **1993**, *1*, 407–414.
91. Martin, J.D.; Simpson, T.W. Use of Kriging Models to Approximate Deterministic Computer Models. *AIAA J.* **2005**, *43*, 853–863. [CrossRef]
92. Saltelli, A.; Tarantola, S.; Chan, K.P.S. A Quantitative Model-Independent Method for Global Sensitivity Analysis of Model Output. *Technometrics* **1999**, *41*, 39–56. [CrossRef]
93. Iooss, B.; Janon, A.; Pujol, G.; Boumhaout, K.; Da Veiga, S.; Delage, T.; Monari, F.; Oomen, R.; Ramos, B.; Sarazin, G.; et al. Sensitivity: Global Sensitivity Analysis of Model Outputs. 2018. R Package v. 1.15.2. Available online: https://cran.r-project.org/package=sensitivity (accessed on 11 January 2023).

Article

Structured Expert Judgment Elicitation in Conceptual Aircraft Design

Vladislav T. Todorov [1,*], Dmitry Rakov [2] and Andreas Bardenhagen [1]

1 Aircraft Design and Aerostructures, Institute of Aeronautics and Astronautics, Technische Universität Berlin, 10587 Berlin, Germany
2 Blagonravov Mechanical Engineering Research Institute (IMASH), Russian Academy of Sciences, 101990 Moscow, Russia
* Correspondence: vladislav.t.todorov@tu-berlin.de

Abstract: Disruptive technologies and novel aircraft generations represent a potential approach to address the ambitious emission reduction goals in aviation. However, the introduction of innovative concepts is a time-consuming process, which might not necessarily yield an optimal design for a given flight mission and within the defined time frame. In order to address the need for a structured and more exhaustive search for novel concept generations, the Advanced Morphological Approach (AMA) and its further enhancement was introduced earlier. It implies the decomposition of design problems into functional attributes and appropriate technological alternatives. Subsequently, these are evaluated and combined into solutions, which are then projected onto a solution space. The current paper focuses on the technology evaluation step by deriving and integrating structured expert judgment elicitation (SEJE) techniques into conceptual aircraft design with the AMA. For this purpose, the first aim of the work is to justify the developed method by giving an overview and discussing the most prominent SEJE methods and their applications in aerospace. Then, the derived SEJE concept is described and applied in the form of an expert workshop on the use case of wing morphing architecture. As a result, a solution space of wing morphing architecture configurations is generated and analyzed. The workshop conduction and the expert feedback serve as valuable findings for both the further AMA enhancement and similar research.

Keywords: conceptual design in aerospace; Advanced Morphological Approach; structured expert judgment elicitation

Citation: Todorov, V.T.; Rakov, D.; Bardenhagen, A. Structured Expert Judgment Elicitation in Conceptual Aircraft Design. *Aerospace* **2023**, *10*, 287. https://doi.org/10.3390/aerospace10030287

Academic Editors: Spiros Pantelakis, Andreas Strohmayer and Jordi Pons-Prats

Received: 3 February 2023
Revised: 4 March 2023
Accepted: 7 March 2023
Published: 14 March 2023

1. Introduction

The conceptual design of novel aerospace vehicles is usually based on the assessment of initially suggested configurations. Such an approach often implies the consideration of a limited number of initial ideas resulting from earlier designer experience or collective multidisciplinary brainstorming [1]. When designing new aircraft generations however, one should acknowledge that the optimal solution integrating certain disruptive technologies may be left out of scope [2]. Furthermore, one should consider that an early concept fixation reduces the room for later design adjustments, also associated with additional costs [1,3]. However, the certain choice of an optimal concept can be assured by parametric optimization and a detailed analysis of each alternative. This approach is not feasible for novel technologies lacking test data.

The mentioned challenges indicate the necessity for (a) the systematic generation of a larger number of concepts and (b) an efficient and robust approach to assess and compare alternative configurations of complex systems such as aerospace vehicles without relying on quantitative data, especially when such is not available (yet).

These are key focal aspects of the Advanced Morphological Approach (AMA) by Rakov and Bardenhagen [3]. It aims to structure a complex design problem in an intuitive way and generate an exhaustive and consistent solution space [3]. In order to replace lacking test data of innovative components, the method intents to lean on the professional opinion of dedicated experts, who would be required to assess the technological alternatives. These evaluations will be used as a scientific basis for the qualitative evaluations and the generation of a wider solution space. The resulting exhaustive solution space allows the consideration of solutions possibly let out of scope during conventional idea generation. By clustering the solutions, the designer is able to identify sub-groups of similar, especially advantageous configurations. These optimal sub-spaces could be then defined as the search boundaries within parametric optimization with Multidisciplinary Aircraft Optimization (MDAO). In other words, the AMA aids to find the optimal design sub-spaces for further investigation with MDAO.

In order to define the AMA as a structured and robust method for conceptual design in aerospace (and complex engineering products in general), the implemented techniques should be carefully studied and justified. For this reason, the enhancement of the initial AMA represents a multi-stage project, visualized in Figure 1. The first stage was dedicated to the definition of the objectives and benefits of the AMA, the overview of other work using the MA in Aerospace, as well as the AMA positioning in the scientific context, presented in Reference [2]. Then, the problem structuring, uncertainty modeling and the main data flow was establishes in the next stage, shown in Reference [4]. The current project stage (the third stage in the figure) focuses on the use of expert opinions as a scientific basis for the qualitative assessment of innovative technologies lacking test data. This context is covered by the current paper which aims to integrate structured expert judgment elicitation (SEJE) methods in pre-conceptual aircraft design and its application in the form of an expert workshop on a wing morphing use case. A deeper solution space analysis, integration of technology interaction aspects, verification, and validation of the "full-scale" methodology (fourth and fifth stage) remain subjects of future work.

In this context, it is first necessary to give a brief overview of the AMA method and define the concrete objectives of the current paper.

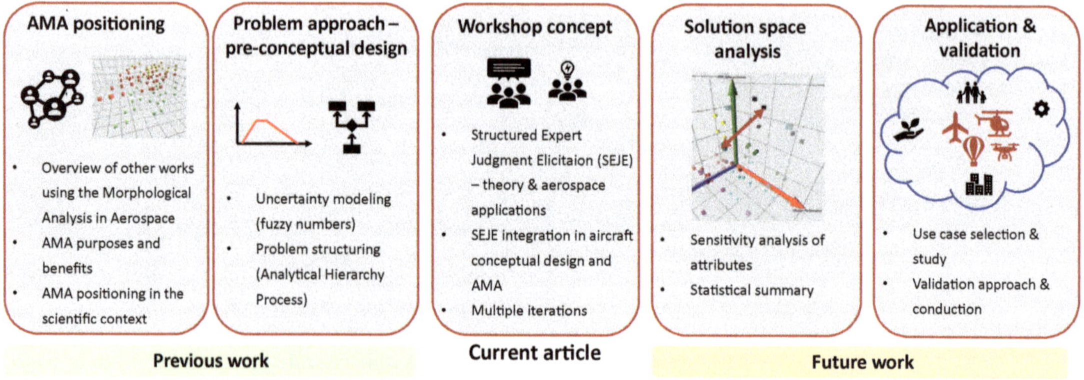

Figure 1. Main AMA enhancement stages.

1.1. Advanced Morphological Approach

The AMA is based on the classical Morphological Analysis (MA) by Fritz Zwicky [5] dating to the middle of the twentieth century. The initial MA introduced product decomposition into functional and/or characteristic attributes, which can be each fulfilled by corresponding sets of given implementation alternatives (denoted as "options"), systematized in a Morphological matrix (MM). Such problem structuring allows to obtain an exhaustive solution space, containing all possible option combinations (denoted as "solutions").

The AMA by Rakov and Bardenhagen [3] is based on the MA and its application in the search for promising technical systems by Rakov [6] from the late twentieth century. The main steps of the AMA combined with a brief presentation of the workshop data flow are schematically shown in Figure 2. The method extends the MA by assigning evaluations to each option according to pre-defined criteria on a qualitative scale from 0 to 9 [3]. The generated solutions combine the scores of the selected options by adding their separated criteria evaluations, which can also be weighted. Subsequently, one can visualize the resulting solutions based on their summed criteria scores, allowing to compare their criterion-specific and multi-criteria performance [3]. The lower left part of Figure 2 exhibits the main data processing steps after obtaining individual pairwise technology comparisons during an expert workshop and their integration into the main AMA methodology.

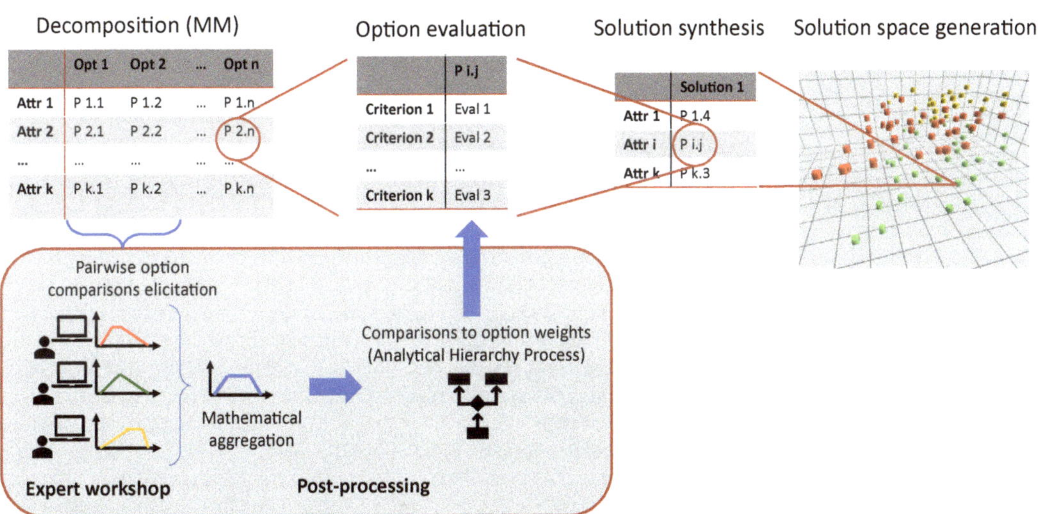

Figure 2. Main steps of the enhanced AMA including the workshop. Attr.—attribute; Opt.—option; Eval.—evaluation.

One of the main problems of morphological analysis is to reduce the dimensionality of the problem and to reduce the number of solutions to be solved [7]. Often, the reduction of the morphological set of solutions by the analysis of incompatible options is used. In some cases, graph theory and genetic algorithms are used [8,9]. In the case of AMA, clustering is used to aid the identification of similar solutions in vast solution spaces.

The MA has been applied in conceptual aircraft design in different forms and contexts, discussed and compared with the AMA in detail in Reference [2]. For example, the Technology Identification, Evaluation and Selection method by Mavris and Kirby [10] uses the MM for impact estimation of technologies and the search for their optimal combination for improved resource allocation. Although direct system simulation is also avoided, the method uses "physics-based analytical models" [10] which are applicable for a more extensive configuration modeling/representation up to a preliminary configuration design. In another work, Ölvander et al. [11] implement conceptual design of sub-systems by describing the MM options with quantitative data and mathematical models. In contrast with these MA applications, the AMA uses a qualitative approach to compare non-existent technologies lacking experimental data or/and are hard to model.

As shown in Figure 1, the AMA has underwent further development by implementing uncertainty modeling with fuzzy numbers, problem-specific hierarchical problem definition by means of the Fuzzy Analytical Hierarchy Process (FAHP) as well as multi-criteria decision-making (MCDM) methods (see Reference [4]). Based on this approach, a methodology for organized expert panels shall be studied, which would evaluate the MM options and serve as a scientific ground for the generated solution space. A first iteration of the methodology testing has already been conducted within a first workshop on the design use case of a search and rescue aircraft (SAR), described in detail in Reference [12]. As a result, a solution space containing 54 configurations was generated which yielded a multi-criteria optimum implementing hybrid aerodynamic/aerostatic lift generation, fully hydrogen-based non-distributed propulsion and wing morphing. The first workshop implemented a set of initial features such as solely individual evaluations, mathematical aggregation and the use of fuzzy numbers for uncertainty modeling. It served as a starting point for the development of a full scale concept, aimed in the second workshop iteration.

In this context, the second workshop extended the functionality of the workshop and its post-processing to a next level and was applied on the use case of wing morphing architectures. The added major improvements include the behavioral aggregation of evaluations in the form of group discussions focused on aircraft design aspects, the possibility for the experts to edit their evaluations, and the weighting of the participants' assessments based on their expertise. The current article presents the methodology and results of the second workshop, as well as the aspects needed for the integration of SEJE methods into aircraft conceptual design with the AMA.

1.2. Current Objectives

Among the current challenges of the AMA remains the scientific acquisition of qualitative technology evaluations from domain experts, in order to appropriately assess innovative technologies still lacking deterministic test data. Hence, the specific adaptation and integration of SEJE methods in conceptual aircraft design and its application represents the paramount objectives of the current paper. This introduces the necessity to justify the developed methodology and selected/adapted SEJE techniques. For this purpose, a brief overview of workshop types and established SEJE methods will be given and analyzed by positioning the present work within the global SEJE context found in the literature. Next, the current state of the developed workshop methodology will be described and justified.

The second part of the paper is dedicated to the structure, conduction and outcome of the second AMA workshop. Its first aim is to obtain and analyze a generated solution space containing different implementations of wing morphing technologies. Another equally important target is the thorough analysis of the workshop conduction, collection of valuable participant feedback, and the derivation of improvement proposals for next potential iterations.

Hence, the following global research questions have been defined for the current project stage:

- Can qualitative expert knowledge be suitable for global conceptual design problems in aerospace?
- How to make subjective expert opinions a scientific basis for conceptual design tasks?

2. Structured Expert Judgment Elicitation

One of the AMA's focal points is the conceptual design integrating non-existent or prominent innovative technologies, often lacking deterministic test or performance data. This represents a challenge for the objective, transparent and scientific comparison of the generated aircraft solutions. In such cases, one usually seeks the professional opinion of domain experts [13,14]. Regardless of their expertise level, however, such assessments would always exhibit a certain level of subjectivity and uncertainty simply due to their human nature [15]. This is the motivation behind the research questions defined in Section 1.2.

For such purposes, researchers usually apply structured expert judgment elicitation (SEJE) methods. These define aspects such as elicitation type (remote or in person), questionnaire design, knowledge aggregation possibilities, uncertainty handling, etc. The current section will start by giving a definition of a SEJE method and presenting the most prominent approaches in this domain. Subsequently, the current research will be placed into the literature context and positioned in reference to similar works. Finally, the concepts applied within this paper will be selected and justified.

2.1. Literature Overview

The concepts of expert judgment elicitation originate from the fields of psychology, decision analysis and knowledge acquisition [14]. The literature suggests the following definitions for expert judgments and their elicitation:

- "data given by an expert in response to a technical problem." [14]
- "we consider each elicited datum to be a snapshot that can be compared to a snapshot of the expert's state of knowledge at the time of the elicitation." [14]
- "Judgements are inferences or evaluations that go beyond objective statements of fact, data, or the convention of a discipline. ... a judgement that requires a special expertise is defined as "expert judgement"." [16]
- "... elicitation, which may be defined as the facilitation of the quantitative expression of subjective judgement, whether about matters of fact or matters of value." [17]
- "Elicitation is a broad field and in particular includes asking experts for specific information (facts, data, sources, requirements, etc.) or for expert judgements about things (preferences, utilities, probabilities, estimates, etc.). In the case of asking for specific pieces of information, it is clear that we are asking for expert knowledge in the sense of eliciting information that the expert knows." [18]

In this context, one seeks the judgments from qualified professionals (experts) in their respective domains. According to the definitions, the elicitation process is not simply asking for the required information. It also implies the extraction of the required subjective knowledge, the filtration of biases and its possibly precise representation in a quantitative form—be it the evaluation of concrete physical (one might dare say "tangible") parameters or simply placing it on a qualitative scale.

One should note the various terminologies used throughout the literature to denote structured elicitation of expert opinions such as structured expert judgments, structured expert knowledge, expert knowledge elicitation or elicitation protocol. These in general stand for the principle of aiming to scientifically elicit expert knowledge. In the current paper, the abbreviation SEJE (structured expert judgment elicitation) will be used for this purpose. Similarly, the widely used notation of the participating experts as decision-makers (or DMs) is also adopted in this work.

The main advantage of formalized SEJE methods is the transparency and the possibility to review the process [14], therefore contributing to the scientific significance of the results. In this context, Cooke [19,20] advocates the achievement of rational consensus by adhering to the principles of Scrutability/Accountability (reproducibility of methods and results), Empirical Control (empirical quality control for quantitative evaluations), Neutrality (application of methods aiming to reflect true expert opinions) and Fairness (no prejudgment of the experts). However, it is necessary to underline that these requirements have been outlined during the definition of the Classical Model, which serves as a method for quantitative elicitation.

SEJE approaches have been developed starting from at least the 1950–60s or earlier [16,21]. In some cases, these are formalized into structured elicitation protocols and guidelines by/for authorities and companies [21], e.g., by the European Food Safety Authority [18] or the procedures guide in the field of nuclear science and technology [22]. Further applications can be found in the domains of human health, natural hazards, environmental protection [20], and provision of public services [17]. Some aeronautical disciplines have also profited from SEJE methods, which will be discussed in the next subsection.

Available SEJE methods establish the main components of a structured elicitation and present various possibilities to implement these. In order to identify and derive the appropriate SEJE process for the integration into the AMA methodology, the most prominent SEJE methods are briefly presented, summarized and compared—namely the

Classical Method by Cooke [19], the The Sheffield Elicitation Framework [23], the Delphi method [24] and the IDEA protocol [25].

2.1.1. The Classical Model

The Classical Model (CM) was developed by Cooke [19] and represents an established method for the elicitation of quantitative values. Instead of qualitative approximations, the experts are asked to roughly estimate a value of a given parameter on a continuous scale, e.g., "From a fleet of 100 new aircraft engines how many will fail before 1000 h of operations?" [20]. This approach advocates uncertainty quantification in the form of subjective probability distributions by eliciting from the experts distribution percentiles—typically at 5%, 50% and 95% [20,26]. These allow the fitting of a "minimally informative non-parametric distribution" [26] for the estimation.

In order to cope with the empirical control requirement defined by Cooke [19], the method involves the attribution of an individual weight to each expert, called calibration. This is executed by preparing two sets of questions—the so-called seed and target questions, incorporating the same types of elicitation heuristics. The seed evaluations involve quantitative inquiries with known correct answers and are used to "assess formally and auditably" [20] the expert's deviation from the precise solution in a domain, denoted as a calibration score. However, even a perfect calibration does not stand for an informative decision in the form of a narrow probability distribution around the precise value. For this purpose, the information score is introduced which reflects the precision of the answer. Then, the correction factors or weights for each expert represent a combination of the calibration and the information score. The target questions refer to the information of interest. Hence, in order to obtain elicited data with highest possible precision, the experts' assessments are mapped with their corresponding weights.

The CM advocates mathematical aggregation of the varying expert evaluations, e.g., in the form of a Cumulative Distribution Function (see Reference [20] for more detail).

2.1.2. The Sheffield Elicitation Framework

The Sheffield elicitation framework (SHELF) has established itself as another notable SEJE methodology [23]. Opposed to the mathematical aggregation used in the CM, SHELF strongly relies on behavioral approaches to combine knowledge from different experts. The process is executed via gathering an expert panel moderated by a dedicated facilitator. After discussing the evaluation subject(s), the DMs give their individual assessments which are then fitted into an aggregated distribution. In a next step, the individual and group evaluations are shown to the panel and discussed, while the experts have the opportunity to edit their previous judgments. The discussions and reevaluations are repeated until a consensus on the aggregated distribution(s) is reached.

It becomes obvious that the method was initially developed to elicit knowledge on a single variable, however extensions for multivariate assessments have been introduced as well [23].

Although SHELF incorporates mathematical aggregation, the result still depends on the common acceptance of the combined distribution, which leans on behavioral aggregation. Accordingly, one expects an increased influence of group dynamics and the corresponding group biases.

2.1.3. Delphi Method

Some authors mark the Delphi method as one of the first formal elicitation approaches of structured expert judgments [16] dating back to the 1950–60s. By applying a structured questionnaire, the experts are asked remotely, individually and anonymously/confidentially for their responses. After that, their answers are collated [16,24] (while removing the 25% upper and 25% lower responses in some cases [16]). In a second round, selected questions are sent back to the DMs for them to review and eventually correct their initial evaluations. This is repeated until the expert group reaches consensus [16,24].

The outlined advantages of the Delphi technique include the avoidance of group biases common for personal meetings as well as the participation of larger expert panel through the remote inquiries [24]. The main drawbacks pointed out are the time-consuming character [24], possible questionnaire ambiguity and the possible alteration of the responses before the next evaluation round [16].

2.1.4. IDEA Protocol

Similarly to the SHELF method, the IDEA protocol is attributed to the mixed approaches using both mathematical and behavioral aggregation of expert knowledge [27]. This is done by structuring the process in four main steps, forming the IDEA acronym [25,27]:

1. Investigate —clarification of the questions and entering of individual estimations;
2. Discuss—the answers of the experts in relation to the rest are shown and serve as a basis for a group discussion, aiming to discover further meanings, reasoning and dimensions to the question;
3. Estimate—the DMs may give a second/final private answer as a correction to their initial one;
4. Aggregate—final results are obtained through mathematical aggregation of the experts' estimations.

The original formulation of the IDEA protocol as defined in References [25,27] aims at eliciting numerical quantities or probability by implementing sets of questions.

2.1.5. Summary of SEJE Components

In their guide on SEJE, Meyer and Booker [14] give an overview of methodology components and how to select among their available implementation options. The source emphasizes on different elicitation situations, response modes, dispersion measures, types of aggregation and documentation methods. Their summary is given in a form of a MM which could be used for the design of new SEJE processes, shown in Table 1. It combines the elements from [14] as well as other aspects from the literature mentioned earlier.

Table 1. Identified SEJE components and their implementation options in the form of a morphological matrix.

Component	Option 1	Option 2	Option 3	Option 4	Option 5	Option 6
Elicitation format/ situation	remote	individual interviews	interactive group	individual evaluations combined with group discussions		
Variable character	quantitative	qualitative				
Response modes	probability estimate	probability distribution	continuous scales	pairwise comparisons	ranks	Bayesian updating
Dispersion measures	ranges	percentiles	standard deviations	fuzzy numbers		
Aggregation method	mathematical	behavioral	mixed			
Expert weighting/ calibration	yes	no				

Based on the descriptions in the previous subsection, Table 2 summarizes the implementations of the SEJE components by the prominent methods found in the literature. It is necessary to underline that these methods have been shown according their original definition. Many of these have been further extended throughout the years and numerous modifications can be found in the literature—e.g., by adding expert calibration or extending the elicitation to multiple variables [16,23,28].

Table 2. Summary of established SEJE methods.

Component	Classical Model	SHELF	Delphi	IDEA
Elicitation format	remote	personal	personal or remote	personal or remote
Aggregation method	mathematical	behavioral and mathematical	mathematical	behavioral and mathematical
Expert weighting/calibration	yes	no	no	no
Variable type(s)	quantitative	quantitative	quantitative	quantitative

2.1.6. Bias as a Source of Uncertainty

One of the main reasons for the thorough structuring of SEJE approaches is the difficulty to obtain data reflecting expert knowledge and experience as exactly as possible. This is mostly due to the presence of multiple types of bias [14,15], often resulting in systematic errors when a person is asked to give objective scientific judgment based on knowledge and experience. The literature knows multiple descriptions and overviews of bias for the purposes of knowledge elicitation [14,15]. Instead of giving another review, the current work will use the bias definitions and guidance in order to construct a tailored SEJE method for the AMA aiming to minimize uncertainties.

Meyer and Booker [14] define two views on bias—motivational and cognitive. Bias can be considered motivational when the elicitation task aims at reflecting the expert's opinion as precisely as possible. This is for example the case when one's purpose is to understand the DM's way of thinking or problem solving. In such cases, ambiguous definitions or faulty elicitation processes could alter the expert's point of view and therefore be the source of motivational bias. The main types of motivational biases are social pressure (in group dynamics), misinterpretation (influence of sub-optimal elicitation methodology or questionnaires), misrepresentation (flawed modeling of expert knowledge) and wishful thinking (influence of one's involvement in the subject of inquiry) [14].

Meanwhile, the cognitive bias view should be preferred for the purpose of likelihood estimation or a mathematically/statistically correct quantification of given parameters [14]. In such cases, it is linked to the cognitive shortcuts people use to process information. Examples of cognitive bias are inconsistency (the inability to yield identical results to the same problem throughout time), anchoring (resolving a problem under the influence of a first impression), availability (vastly relying on a easier retrievable from memory event), and underestimation of uncertainty [14].

To the knowledge of the current article's authors, the majority of developed SEJE methods (originally) aim at eliciting probability distributions for physical values. In this context, an observation has been made that the literature pays more attention to the cognitive view of bias [15,20].

Although the mentioned motivational and cognitive biases might appear complimentary to each other, Meyer and Brooke [14] advise to take a single view of uncertainty depending on the purpose of the project. They bring forward the following justification. On the one hand, the reduction of motivational bias aims to help the data reflect the knowledge of the expert as well as possible by adjusting the methodology accordingly. On the other hand, the cognitive view states the inability of the expert to represent their opinion in an exact mathematical manner and tries to guide the DMs to express themselves in a correct statistical way [14].

2.1.7. Elicitation Format

The main elicitation formats are the remote inquiry or individual interviews opposed to group interactions, which bring their own advantages and drawbacks in view of the different bias types these invoke.

The remote or separate elicitation from the experts implies no interaction among them, which excludes biases originating from group dynamics such as social pressure, hierarchy influence or personality traits [14]. Such an approach could be strengthened by expert calibration, weighting and mathematical aggregation (e.g., in the CM) to contribute to the empirical control and method transparency.

Although the researcher should acknowledge the presence of group think biases within interactive groups, personal meetings of expert panels has a spectrum of advantages as well [14]. In particular, it can contribute to creative thinking by combining different expertise and generate a wider variety of ideas. Furthermore, data with bigger precision could be acquired through interactions.

2.1.8. Qualitative versus Quantitative Variables

As previously mentioned, the majority of studied sources on SEJE aim the elicitation of physical parameters describing a system or a phenomenon. Such elicitation variables can be labeled as quantitative, since there theoretically exists a precise correct value, which should be approximated by the experts.

However, as the purposes of the AMA have shown, a SEJE might be required under the following circumstances:

1. A global assessment of a system or product in the early stages of conceptual design;
2. Evaluation according to qualitative criteria in a vague linguistic form such as "mission performance" or "system complexity" which summarize multiple characteristics;
3. The necessity to consider innovative concepts or technologies in a current study;
4. The lack of test data or statistical estimations on such technologies.

In such cases, the expert would have insufficient experience or knowledge on nonexistent concepts. Therefore the attempts to quantify system parameters would not only be challenging but would also lead to a significant increase of epistemic uncertainty.

The alternative is to use qualitative scale for the assessment regarding non-deterministic criteria. Usually, it reflects a linguistic evaluation grades such as "very good", "good" or "bad" in a categorical form. For the purposes of quantification and further data processing, these statements have been transferred to a continuous numerical scale from 1 to 9 within the initial AMA as defined in [29]. One of its main drawbacks is obviously the ill-defined character of the numerical definitions for the assessment of technological alternatives—should a certain technology be evaluated with 7 ("very good") or 6 (between "good" and "very good")? This is found ambiguous especially by representatives of technical fields used to precise quantities (which has been observed during both conducted workshops within the current project so far). Further disadvantage is the lack of a reference for the qualitative values, at least for the extremas of the scale. This issue is addressed within the Analytical Hierarchy Process (AHP) by Saaty [30], where pairwise comparisons should be evaluated on the scale from 1 (equal) to 9 (absolutely superior) to denote the superiority of a certain alternative over another one.

At this point, one should recall the SEJE definition in the beginning of the current subsection, denoting such a process as a quantification of subjective expert knowledge. Although such scales are considered "qualitative" in the current work, these still represent a way to quantify the DM's experience and knowledge in a certain data format. Hence, this can also be defined as a form of elicitation.

2.1.9. SEJE Integration with Multi-Criteria Decision-Making

The aim of the SEJE methods is to ensure a transparent and scientific elicitation of expert knowledge by offering sometimes multiple steps of aggregation, reevaluation or discussion. However, the thorough conduction of such routines might be applicable for the elicitation of a relatively small amount of variables simply due to time constraints and limitations related to participant engagement and their attention span. This was experienced during the first [12] and second workshop of the current project.

However, one might need to select among a significant amount of alternative scenarios or technologies. Furthermore, the selection might be defined as the outcome of a multi-criteria assessment, which is the case for the AMA. Hence, a certain structure is required both for the elicitation process and the subsequent data processing in order to obtain the option comparison results in the form of comparisons, ranks, etc. The algorithms which represent a framework to transform expert inputs into final option ranking or assessments by accounting for multiple criteria simultaneously are denoted as Multi-Criteria Decision-Making (MCDM) methods [31].

A selection and the possible integration of a MCDM approach into the AMA process is given in Reference [4]. The source also justifies the choice of FAHP for the problem structuring. Once a MCDM algorithm/framework has been selected, it is necessary to choose or adapt an appropriate SEJE approach.

This raises the question of integration of MCDM and SEJE methodologies which is subject to the following challenges:

- Time- and energy-efficient elicitation of a larger number of variables;
- Ensure compatible format of the elicited variables from the SEJE and the input variables into the MCDM;
- Ensure transparent value transformations within the methods;
- Minimize the additional bias or noise of the uncertain data caused by the methods.

2.1.10. Purposes for SEJE Applications

Regardless of the application domain, an extended research on SEJE has yielded a spectrum of terms using participative approaches such as "technology assessment", "scenario workshops", "future workshops" or "stakeholder workshop". Therefore, it is first necessary to distinguish among these notations in order to position the AMA workshops in this scientific context.

Technology Assessment (TA) has the general purpose of identifying promising innovations and evaluating the socio-technical impact of new technologies or improvements [32]. Ultimately, this could aid the political decision-making on administrative or company level. Such assessments might involve cost-benefit and risk analysis, as well as the potential relationships to markets and society [32]. In particular, some sources lay focus on the technology integration in policy making and public acceptance [32,33], as well as the "co-evolution of technology and society" [32]. For this purpose, stakeholder workshops are conducted which allow the gathering of representatives from concerned circles such as companies, society, political and research institutions [34]. In a structured process, these attempt to form visions of the innovative technology interaction with the public and its administrative regulation by aiming to resolve existing concerns. This vision is denoted as a "scenario", which encompasses a possible implementation of the novelty and its impact on a spectrum of dimensions such as various society levels, the environment and public/consumer acceptance [34]. There exist numerous types of workshops and conferences aiming to define multiple scenarios, select the most promising ones and/or detail their implementation. Prominent examples are the scenario workshop, future workshop, consensus conference, future search or search conference [35]. The current article will not describe these in detail. Instead, it is purposeful to introduce the process of creativity use for the creation and definition of scenarios (or visions, futures, etc.).

Vidal [35] summarizes that creativity encompasses two types of thinking: divergent, which allows to see multiple perspective of a certain situation, and convergent thinking which helps to "continue to question until satisfaction is reached" [35]. Methods for creative problem solving incorporate switching between these two thinking approaches. In this context, both the scenario workshops [34] and the future workshops [36] include steps which encourage divergent thinking to generate new ideas and convergent thinking to select among these the most appropriate one(s) and to improve their level of detail. To date, the conducted workshops applying the enhanced AMA have used only the convergent thinking by asking the experts to compare technological alternatives.

2.2. SEJE Applications in the Aerospace Domain

The uncertain character of the conceptual design phase of aerospace vehicles has encouraged researchers to seek expert opinions on upcoming innovations. An extensive research and development of SEJE techniques was conducted at NASA (National Aeronautics and Space Administration) to estimate parameters of weight, sizing and operations support for a launch vehicle [37]. In particular, probabilistic values were elicited by creating appropriate expert calibration [38] and aggregation [39] techniques in combination with a specially designed questionnaire [13].

Additionally, the use of scenarios in the aircraft design process has been discussed by Strohmayer [40]. The author argues that based on market analysis, one could derive requirements and identify technologies for new aircraft by organizing and evaluating these in the form of consistent scenarios.

Authorities in the aeronautical domain have also shown interest in expert judgments. The Federal Aviation Administration (FAA, United States Department of Transportation) assumed the use of SEJE for the development of a risk assessment tool for the electrical wire interconnect system [41]. This idea has been further studied and refined by Peng et al. [42] by checking the validation of expert opinions and estimating the agreement within the panel.

Despite of the available statistics and databases on accidents in aviation, Badanik et al. [43] used expert judgments as a possible aid for airlines to estimate accident probabilities. For this purpose, the CM was applied to analyze the answers of airline pilots to assess the probabilities of some IATA (International Air Transport Association) accident types for occurring Flight Data Monitoring events on multiple aircraft.

2.3. Justification of the Developed Workshop Concept

In order to position the current research and justify the developed SEJE methodology, it is first necessary to summarize the objectives of the AMA and the elicitation as follows:

- Idea generation for conceptual aerospace design;
- Consideration of innovative technologies lacking historical data;
- Evaluation of the technological options by experts;
- Benefiting from the expertise and creativity of the domain professionals.

Therefore, one can categorize the AMA workshops as a variation of TA. However, depending on the selected use case and the attending participants, the research conducted so far does not necessarily focus on the social impact of new concepts, but rather on an optimal selection of technologies to use in vehicle designs (e.g., during the first and second workshops).

The consideration of novel or non-existent technologies with scarce or no historical performance data implies the challenging elicitation of deterministic physical parameters of the configurations. In addition, the limited experience of the professionals with such components will further increase the epistemic uncertainty during the quantitative elicitation of system parameters. Hence, a qualitative expression of the performance evaluations of the options has been defined. In combination with the FAHP by Buckley and Saaty, the experts are required to enter pairwise comparisons of the technological options in the form of trapezoidal fuzzy numbers on the scale from 1 (both technologies are equal) to 9 (technology A is absolutely superior to technology B)—see Reference [4] for more details.

By using the qualitative evaluation character, the current work leans on the assumption that intuitive elicitation of expert knowledge and experience are a reliable scientific basis for concept derivation. In this context, the motivational view on bias is selected, as defined by Meyer and Booker [14]. The reason for that is the dominant importance of clear problem and methodology definition leading to a cleaner elicitation of expert knowledge rather than the accurate statistical estimation of precise physical parameters (as in the case of the cognitive view on bias).

In order to engage the full potential of the experts' knowledge and creativity, the method would benefit from both mathematical and behavioral aggregation. For this purpose, a modification of the IDEA protocol is derived to fit the contextual evaluation of

MM options. A detailed description of the entire methodology and its implementation can be found in the following section on the second AMA workshop.

3. Second Workshop Methodology

For the first development iteration of the enhanced AMA, a first workshop was conducted and described in Reference [12], where the conceptional design of a search and rescue aircraft was studied. The current article discusses the conduction and results of a second workshop which aimed to consider the lessons learned and test new ideas. The following subsections will reveal major aspects related to the workshop, namely use case definition, developed methodology, questionnaire design, data post-processing, and results analysis.

3.1. Airfoil Morphing and Integration—Use Case Definition and Morphological Matrix

In this context, the conceptual design of a subsystem was conducted for the purpose of selection and integration of wing morphing technologies. Firstly, this is a representative use case to demonstrate the current methodology state on innovative technologies. Secondly, the subject reflects the demand for future work on morphing wings technologies as stated at an earlier project stage in Reference [3].

The use case focuses on two main aspects: the geometric shape modification and the morphing technology to implement it. These were defined in a MM as three attributes with their corresponding implementation possibilities (Figure 3). The morphing mode selected in the current work represents the modification of the cross-section wing geometry, denoted here as "airfoil morphing". The selected responses (first attribute in the MM) are the deformation solely of the trailing edge, of both the trailing and leading edge or the deformation of the entire airfoil geometry. The next questions that arises is the positioning of the morphable airfoil sections on the wing. Therefore, the second attribute allows to select possible wing sections for morphing, namely an area near the wingtips, near the wing root or the whole wing. For the third attribute, a research yielded the following technologies studied for the purpose of airfoil morphing:

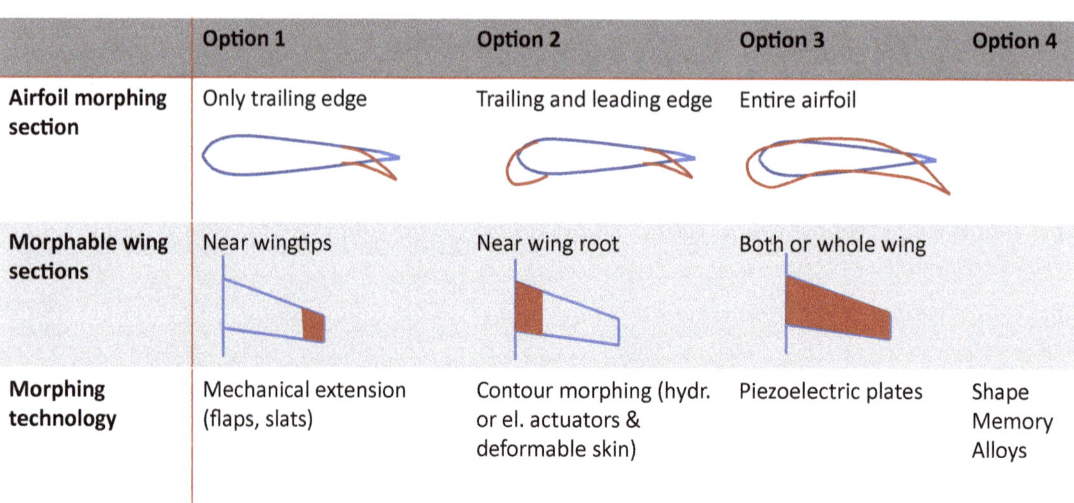

Figure 3. Morphological matrix for the conceptual design of wing morphing architectures.

- Mechanical extension—these are mechanical devices which extend the trailing and/or leading edge of the airfoil [44]. In this context, solely the concepts of conventional flaps and slats are considered, serving as an existing reference option.

- Contour morphing—the deformation of the airfoil contour via hydraulic or electric actuators as presented in References [45].
- Piezoelectric macro-fiber composite plates—such plates exhibit the ability to deform when submitted to electric current and vice versa [46].
- Shape memory alloys—these smart materials deform under certain loading and temperature conditions [47].

Such a definition of the MM by no means aims to represent an exhaustive outline of all possible morphing technologies or their integration, neither to support a complete design process. As previously mentioned, the main target of the second workshop is to test the enhanced AMA methodology and the evaluation of perspective technologies with scarce test data.

3.2. Evaluation Criteria

In order to compare the alternative morphing technologies and their integration options, a set of relevant evaluation criteria should be chosen. By considering the main advantages and challenges for such architectures, the following criteria have been selected:

- Flight performance
- Required energy
- System complexity

Improvements in flight performance represent the primary reason for the consideration of morphing wing structures in the first place. Evaluations according to this criterion should reflect the difference in aerodynamic qualities, flight stability, integration of innovative flight controls, etc. However, a trade-off should be made with the energy required for different morphing modes and morphable wing areas, making it a second criterion. Last but not least, the MM options should be compared according to their system complexity and weight in order to round-up the subject-relevant main global aspects.

For the purpose of easier results interpretation, the criteria will not be weighted for the final solution evaluation.

3.3. Uncertainty Modeling

In order to model the uncertainties incorporated into the experts' subjective answers, the concept of ordinary fuzzy number is used for each technology comparison. The benefits of this approach and the justification for its use within the AMA are described in more detail in References [4,12].

3.4. Problem Hierarchy Structure

In order to capture the multi-dimensional character of the design problem, the workshop evaluations are structured by applying the FAHP suggested Buckley [48], which builds on the classical AHP by Saaty [30]. Reference [4] explains the use of the approach based on fuzzy comparisons within the AMA and the definition of necessary system hierarchies based on the problem statement.

The hierarchy used for the current workshop is exhibited in Figure 4. The organization of options and criteria in levels has been executed analogically to the problem structuring in the first workshop (see Reference [12]). The options from the same attributes are compared regarding each criterion of the level above. The generic right part of the diagram showcases the positioning of options from different attributes into separate hierarchy branches due to their incomparability.

After obtaining all option comparisons from the DMs the algorithm yields the global weights of each option according the criteria. Naturally, the comparisons are elicited only for options from the same attribute.

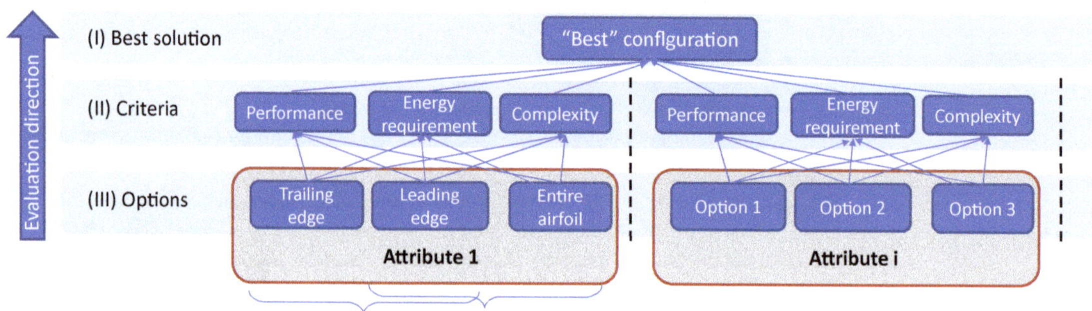

Figure 4. Hierarchical representation of the problem definition by using the AHP.

3.5. Workshop Structure

The previous workshop aimed to test only a single step of a SEJE methodology, namely the individual expert evaluations by leaving DMs interactions and behavioral knowledge aggregation out of scope. The current workshop implements both features by adapting the IDEA protocol for the pairwise comparisons of MM options. Further focus has been laid on contextual consistency of the tasks (the discussion of attribute options are followed immediately after the DMs evaluated these) and on the optimization of workshop duration and expert concentration.

The workshop was introduced with a presentation outlining the objectives of the study, the problem statement and instructions on the interactive elicitation. The evaluations were entered on a software platform specially developed for SEJE workshops within the AMA. As an upgrade of the version used for the first workshop [12], it also integrated a carefully designed User Interface (UI) in the front-end, and a back-end, which ensured the smooth execution of the background operations and the evaluation storage in a database. The architecture allowed the interactive input of the DMs' evaluations according to a specially developed questionnaire design, presented in the next subsection.

The main steps of the IDEA protocol are repeated for each attribute row of the MM. In the Investigate part, the experts individually evaluate the pairwise option comparisons according to all criteria. Subsequently, a moderated discussion round is conducted which aims to share ideas among the participants. The purpose is to broaden their horizon and point their attention at forgotten or maybe unknown aspects which might influence the evaluation. If this is the case, the DMs have the possibility to edit their previous evaluations. After finishing the evaluations for a single attribute row, the same procedure is conducted for the next one.

The discussion rounds consisted of the following steps:

1. Mind mapping—for each criterion, the sub-criteria relative for the current attribute are derived;
2. "Best" options mapping—based on a group consensus, an option from this attribute with the best performance is assigned for each sub-criterion;
3. Visual overview—individual identification of option dominance for the sub-criteria;
4. Re-evaluation—based on this mapping, the DMs might consider editing their initial evaluations.

The result of such mapping from the second workshop is described in Section 4.3.2. Such an approach benefits from divergent thinking and allows the group to gather more aspects in order to increase the objectivity of an evaluation. One should stress that a consensus is required only for the definition of a most suitable option for each sub-criterion, roughly based on majority agreement. Beyond this point, each expert is left to decide for themselves whether the mapping of the options is convincing enough for them to edit their evaluations.

3.6. Questionnaire Design

The questionnaire consisted of two main parts: the professional background questionnaire and a technology assessment section.

3.6.1. Professional Background Questionnaire

The background questionnaire aims to elicit the participants' level of expertise in the relevant domains based on their own perception. This approach was inspired by the questionnaire design from the NASA's SEJE method mentioned in Section 2.2 and described in Reference [38]. However, the source implies asking for the DM's age and expertise in the form of integers from 1 to 5 [38] and combining these into a coefficient. Instead, the current questionnaire requires self-assessment of knowledge or experience in the domains as fuzzy numbers on the scale from 0 to 9, reflecting the progressive increase in theoretical knowledge and practical hands-on experience. An example is exhibited in Figure 5. Taking into account the MM structure and the represented domain expertise in the panel, the expertise of the DMs has been elicited in the disciplines aircraft design, aerostructures, aeroengines, flight mechanics and aerodynamics.

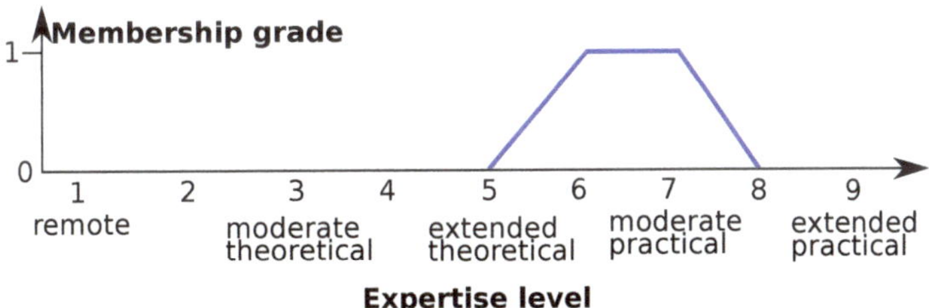

Figure 5. An elicitation example of the professional background questionnaire with a trapezoidal fuzzy number.

3.6.2. Technology Assessment Questionnaire

The technology assessment questionnaire design builds upon the one used in the first workshop by considering the lessons learned in Reference [12]. The same diagram for the input of fuzzy estimates is used. Since the experts are asked to evaluate pairwise comparisons, the positive values of the qualitative x-axis express the superiority of one technological option over another and the negative—its inferiority. The fuzzy evaluations are then placed by the DMs according to their perception.

The feedback after the first workshop included the abundance of labels and text around the diagram as well as the tiresome and confusing switching between evaluations of the same comparisons but regarding different criteria [12]. For this purpose, the labeling has been simplified and a single answer page included diagrams for the same option comparisons according to all available criteria. This allows an easier visual perception of multiple questions simultaneously and a strict division of the page into two vertical halves, each corresponding to a technological option, while the diagrams (or the "rows" of the page) are assigned to the criteria.

In order to further increase simplicity, the terms "superiority" (expressed with positive values) and "inferiority" (expressed with negative values) of one option over another have been omitted. Instead, the positive side of the x-axis was denoted to be in favor of option A and the negative—in favor of option B.

Furthermore, the first workshop required the participants to enter evaluations for reciprocal comparisons as well (e.g., separate questions "compare technology A referred to technology B" and "compare technology B referred to technology A") in order to encourage the DMs to correct their answers [12]. This was regarded by the participants as a rather useless and time-consuming feature, leading to its removal in the current questionnaire. The new version requires an input on the comparisons of each option pair only once.

3.7. Expert Weighting

Based on the background questionnaire explained in the previous subsection, an approach to assign weights to the expert evaluations has been derived, which is summarized in Figure 6. Each expert leans on their own perception to enter their expertise in the domains aircraft design, aerostructures, aeroengines, flight mechanics and aerodynamics. The DM's abstract "total" relevant expertise is then calculated by simply adding or weighting the domain self-evaluation. Subsequently, the technology comparisons made by different experts are aggregated by applying the weighted geometric mean. The weights are calculated by normalizing the expertise of each participant with the sum of the expertise of all experts.

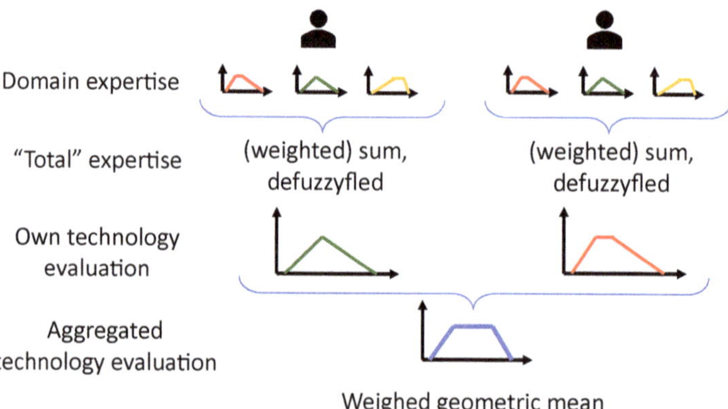

Figure 6. Schematic overview of expert weighting based on their self-determined expertise in relevant domains.

3.8. Expert Anonymity

The level of participant anonymity during and after the elicitation is an aspect concerning not only the experts themselves. It also serves as a potential source of social pressure biases able to influence the evaluations and therefore the final results [14]. Such biases can occur during the group discussions as well as when the experts enter their evaluations. In the first case, the group think bias might be present, representing the alteration of one's own opinion or action in order to be aligned with common positions [14]. The same effect could be observed when hierarchical structures are present in the group. In order to reduce these effects, the current workshop implements individual evaluations which are not shown to the panel. However, even such a setting hides risks of revealing the participants identity to the researcher during the data evaluation or/and to society, if the evaluations are associated with their person later on. This is connected to the impression management bias, which stands for the concern with the reaction of people not present [14].

For this reason, the workshop concept ensures the anonymity of the participants to the researchers and to the public. This is done by requiring them to log in to the questionnaire with a random identification number chosen by them. In the further data processing, the evaluations are associated solely with the identification number which cannot be related to any person.

3.9. Post-Processing of the Raw Results

The raw output from the workshop yields fuzzy comparisons of the MM options from each expert. These should be adapted to comply with the input format of the AMA process in order to analyze the final solution space. The conducted steps for that purpose are depicted in Figure 7 and explained in the following:

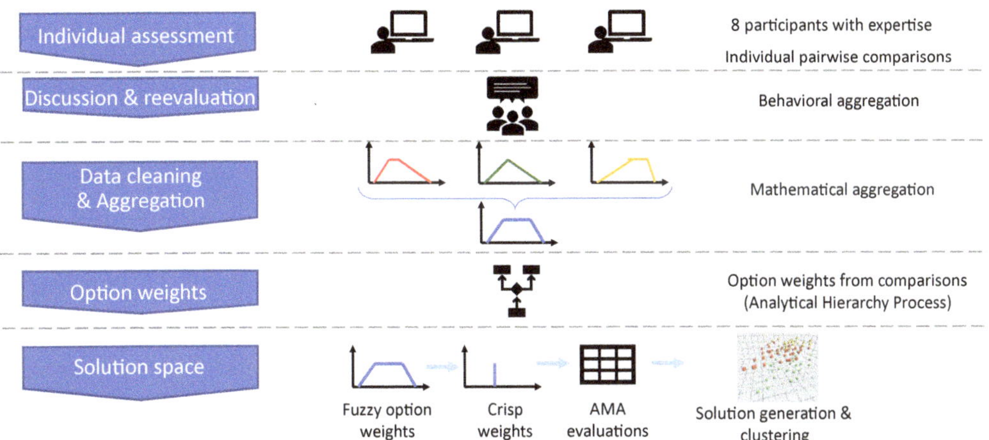

Figure 7. Schematic steps of the data post-processing.

1. **Workshop conduction**
   The workshop stages represented the first three aspects of the IDEA protocol—Investigate (individual assessment), Discuss and Estimate (possible reevaluation).
2. **Data cleaning**
   As with the first workshop, there was a set of typical errors and inconsistencies observed in the raw user inputs, a more detailed overview of which can be found in Reference [12]. One should stress that these did not represent meaningless data noise but rather comprehensible deviations which could be corrected in a logical manner. These were dealt with by applying the previously developed automatic workflow which cleaned the inconsistencies.
3. **Mathematical aggregation**
   In the current use case, one aims to generate a single solution space based on the evaluations of all experts. Hence, it is necessary to combine their input. During the workshop, behavioral knowledge aggregation took place in the form of group discussions. The post-processing involves mathematical aggregation of the DMs' opinions. For this purpose, the fuzzy comparisons of the same options according to the same criteria but from different experts are combined with a geometric mean.
4. **Option weights**
   The aggregated fuzzy comparisons are then used as input to the FAHP algorithm by Buckley [48] in order to obtain the fuzzy option weights regarding to the criteria of the level above.
5. **Defuzzification**
   At this stage, the AMA accepts only crisp (real) numbers for option evaluations. For this reason, the fuzzy option weights are defuzzified by using the center of gravity method analogically to the first workshop [12].
6. **Solution space generation**
   The AMA then combines the option weights into solutions and generates the solution space, which is analyzed in more detail.

3.10. Software Implementation Aspects

Own specialized software has been developed for the following purposes:

- Workshop conduction platform—back-end and front-end with a user interface for evaluation input;
- Mathematical aggregation;
- Implementation and application of the FAHP;
- AMA software—definition of MM and criteria, solution space generation and clustering.

Major sub-tasks have been developed and validated with available examples, e.g., the implementation of the FAHP. Single smaller sub-tasks have been accomplished by using the following external packages or libraries:

- Scikit-learn [49]—an implementation of the K-Means clustering algorithm for the solution space analysis in the Python programming language;
- Helix Toolkit [50]—certain objects for the three-dimensional visualization of the solution space;
- Plotly.NET [51]—plotting of the interactive fuzzy evaluations within the workshop software platform;
- Scikit-Fuzzy [52]—implementation of single operations with fuzzy numbers, based on Reference [53]. However, the majority of fuzzy functions and definitions have been developed autonomously.

4. Results of the Second Workshop

Based on the option evaluations obtained from the workshop, the AMA software generated a solution space (Figure 8) with the following parameters:

- Three diagram dimensions, corresponding to the criteria flight performance (red), system complexity (blue) and needed energy (green);
- Size—36 generated solutions resulting from the exhaustive combinations of the MM options. No inconsistent options were observed;
- Clustering—application of the K-Means clustering method.

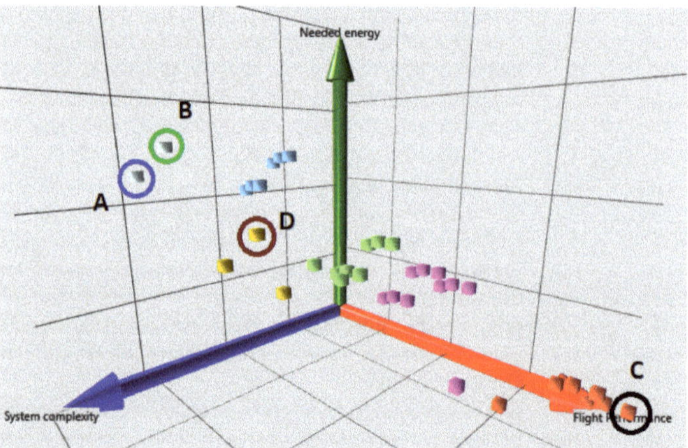

Figure 8. The generated solution space, which is zoomed for a better visibility of all solutions. The colors represent different clusters—see Table 3. The marked solutions can be interpreted as follows: A—local maximum according to system complexity; B—local maximum according to needed energy; C—local maximum according to flight performance; D—global optimum/compromise considering all criteria.

Table 3. Clusters overview.

Cluster ID & Color	Sol. Count	Max. Norm. Sol. Score	Avg. Norm. Score	Score Std. Dev.	Rel. Hamming Distance
1	3	1.08	1.04	0.40	0.64
2	7	1.02	0.96	0.37	0.69
3	8	1.07	1.03	0.21	0.96
4	6	1.05	1.02	0.23	0.67
5	10	1.03	0.97	0.31	1.02
6	2	1.06	1.05	0.12	0.48

Similarly to the result presentation of the first workshop [12], the further position of a configuration along a certain axis implies its improvement according to the criterion. For example, further position along the system complexity axis stands for better/less system complexity.

The following subsections will be dedicated to the different visualizations of the solution space within the AMA software and its analysis.

4.1. *Influence of the Expert Weights*

Regarding the influence of expert weighting, two solution spaces have been generated—one applying geometric mean aggregation with equal expert weights and another one using the DMs' expertise as weights (see Section 3.7). Multiple weighting possibilities have been experimented with when obtaining the total expertise of each DM—e.g., no weighting of the expertise domains, their equal weights, as well as the selection of specific (most relevant) domains for the corresponding attribute. However, the analysis of the aggregation results and the obtained solutions yields a negligible difference in the final option evaluations and thus almost identical solution spaces. Two possible causes for such outcome might be:

- Insufficient variation of the elicited expertise among the participants or
- Undetected methodology flaws.

The first aspect might be due to the homogeneity of the professional background of most participants coming mostly from the domains of aircraft design and aerostructures.

4.2. *Cluster Overview*

4.2.1. Optimal Amount of Clusters

The K-Means clustering algorithm uses the number of clusters as an input parameter. The optimal amount of six clusters was obtained beforehand by using the "elbow method" [54] as in the first workshop [12]. The generated clusters are visualized in Figure 8 and described in Table 3.

4.2.2. Cluster Metrics

The metrics used to describe the clusters in Table 3 have been introduced in Reference [12]. These are defined as follows:

- Max. norm. sol. score—"the maximum value of the cluster solution scores referred to the average solution score in the entire solution space. Indicates the score of the "best" solution within the cluster compared to the whole solution space." [12]
- Score std. dev.—"the standard deviation of the total solution scores within the cluster. Indicates the numerical compactness of the cluster based on the total scores." [12]
- Rel. Hamming distance—"the average Hamming distance within the cluster relative to the average Hamming distance of the whole solution space. Represents the qualitative cluster compactness, based on the variation of the selected technological options." [12]

According to Table 3, no cluster yields a solution which is a definitive global optimum considering all three criteria simultaneously—the maximal solution scores in each cluster vary within a 5% interval referred to the average solution space score. Meanwhile, the average cluster scores are distributed around the global solution space average, exhibiting normalized scores between 0.96 and 1.05.

Nevertheless, if a compromise among all three criteria is sought, one recognizes the solution marked with the letter D. It has a normalized score of merely 8% above the solution space average, which is still the maximal solution score in the whole space. This morphing configuration implements morphing of the entire airfoil with mechanical components solely near the wingtips.

4.2.3. Local Maxima

The local maxima are defined as solutions indicating maximal scores regarding at least one criterion (along the respective diagram axis). These are marked with the letters A, B and C in Figure 8 and described in Table 4. The local maxima according to the needed energy (B) and system complexity (A) criteria are located close to each other in the solution space and integrate classical mechanical trailing edge morphing near the wing root or the wing tips. Opposed to these, solution C shows the maximum value for flight performance and comes with contour morphing of the leading and trailing edge on the entire wing.

Table 4. Selected technologies for the local maxima configurations.

Attribute	Selected Options for A/B (Less System Complexity & Needed Energy)	Selected Options for C (Best Flight Performance)	Selected Options for D (Global Optimum)
Airfoil morphing area	Only trailing edge	Leading & trailing edge	Entire airfoil
Morphing area on wing	Near wing root/Near wing tips	Entire wing	Near wingtips
Morphing technology	Mechanical extension	Contour morphing	Mechanical extension

4.2.4. Trend Analysis

With the AMA software, it is possible to filter the solution coloring according to the applied technological options for a given attribute. Figure 9 showcases the distribution of the options for the morphable wing section attribute. It points out the definitive advantage of entire wing morphing regarding flight performance. Simultaneously, morphing only at wing root appears to contribute to slightly less system complexity than morphing at the wing tip. The influence of these two options while fixing the other attributes results in pure solution translation along the axes, which, however, is not very significant compared to the size of the solution space. This speaks for small solution score sensitivity against these options.

Furthermore, the distribution of different morphing technologies is observed in Figure 10. Mechanical extension of the airfoil with flaps and/or slats showcases less system complexity compared to the other mechanisms. Similarly to the previous visualization, compact groups of solutions can be observed, which incorporate different morphing technologies with the other attributes fixed. Again, this testifies the reduced influence of the contour morphing, the memory alloys and the piezoelectric plates referred to the solution space size.

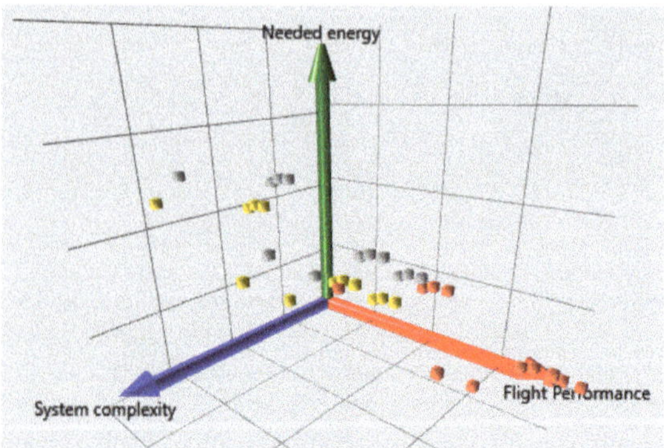

Figure 9. Option distribution for the morphable wing section attribute. Gray—near wingtips; yellow—near wing root; red—both or entire wing. The solution space is zoomed for a better visibility of all solutions.

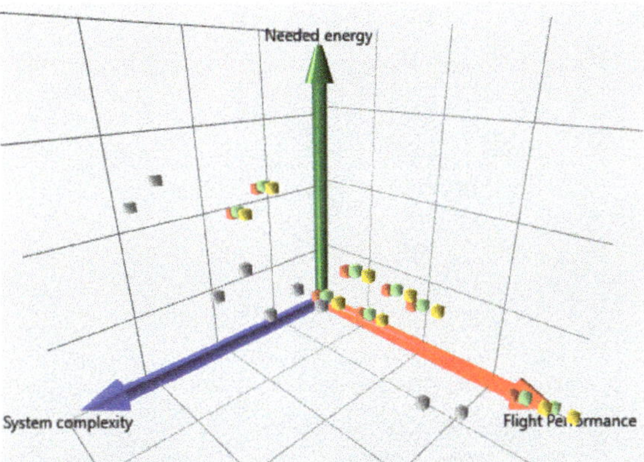

Figure 10. Option distribution for the morphing technology attribute. Gray—mechanical extensions (flaps, slats); yellow—contour morphing; green—memory alloys; red—piezoelectric plates. The solution space is zoomed for a better visibility of all solutions.

4.3. Participant Feedback

Since the current article describes another iteration of the method development, the feedback of the participating experts represents an important result combined with the solution space.

4.3.1. Overall Impression

Concerning the overall workshop structure and conduction, the DMs reported an improvement compared to the first workshop. This is reflected in a less time-consuming questionnaire and the presence of group discussions. The enhanced and more compact questionnaire layout has also been remarked, which contributed to an easier visual perception and reduced effort during the evaluations.

4.3.2. Group Discussions

The outcomes of group discussions can be summarized in the mapping of best options described in Section 3.6. The example shown in Figure 11 shows the option mapping for the morphing wing section attribute and the system complexity criterion. The first question that the participants answer together is: which sub-criteria or aspects associated with system complexity are related to or influenced by different morphable wing sections? After outlining the sub-criteria, the experts discuss which options of this attribute exhibit best performance according to the sub-criteria. These promising options are then noted in "Best option" column. Based on the dominance level of the different options obtained during the discussion, the experts might consider editing their initial evaluations.

Regarding one of the new methodology elements, namely the group discussions and the behavioral aggregations, one could observe both positive feedback and wish for further improvement. Fruitful conversations and idea sharing were an important part of the behavioral aggregation. The derivation of sub-criteria and the associated best options contributed to largely structured conversations. The initial idea behind the visual mapping of best options was to introduce new aspects (reduce the "availability" bias) and probably motivate the experts to correct some initial evaluations. However, this purpose was not entirely supported by the majority of the participants, so the question arose whether such a discussion was necessary in the first place. Nevertheless, two participants (25%) stated they had indeed wished to correct their initial evaluations after the discussion.

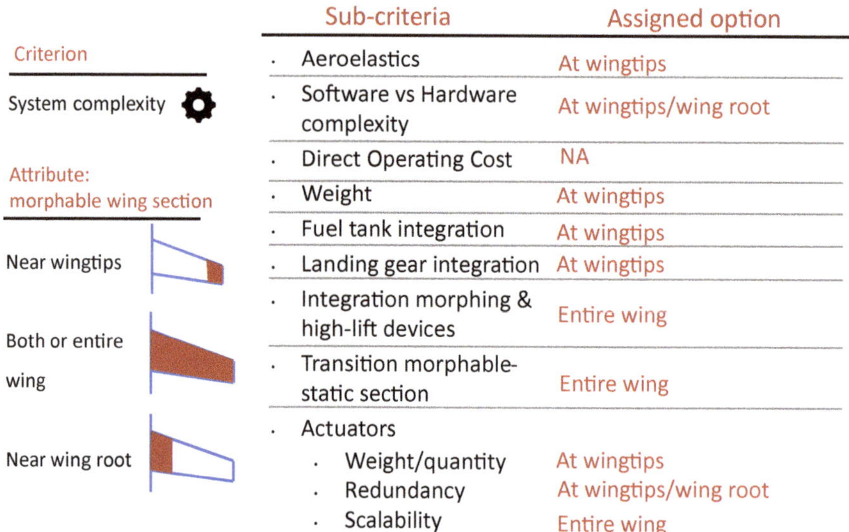

Figure 11. Option mapping of the morphable wing section attribute to the derived sub-criteria of the system complexity criterion. This mapping resulted from one of the group discussions during the second workshop.

4.3.3. Technology and Terminology Definition

As a considerable drawback was mentioned the ambiguity of some criteria and option definitions. Furthermore, the lack of knowledge on specific technologies such as the morphing mechanisms served as a source of epistemic uncertainty, which clearly reflected in the evaluations. For example, the energy input criterion was a subject of clarification discussions. The struggle was between the absolute energy needed for morphing in a special case and the ratio of required energy and the energy savings due to the new drag-efficient wing.

5. Discussion

As mentioned in the discussion of the first workshop [12], a full-scale data-based validation of this AMA stage is hardly possible for such studies, considering the lacking experimental data on the morphing attributes. For this purpose, the relative positioning of the option implementation is qualitatively verified. For instance, the definitive flight performance advantage of an entire morphable wing over separate morphable sections confirms the logical expectation. Similarly, mechanical extensions such as flaps and slats exhibit slightly less system complexity than the other morphing mechanisms included in the MM. However, as stated in the previous section, the difference among solutions involving certain options is seen to be relatively weak, which is reflected in the small solution translations along the axes. This might speak for two things: (a) prominent epistemic uncertainty (lack of knowledge on these options) or/and (b) overly general definition of the options preventing the experts from focusing on the most relevant aspects. Based on these observations, one could declare a partial reliability of the results.

The expert feedback also plays a key role as a partial method validation. In this context, the aspects put forward by the participants not only indicate more space for improvement possibilities. Together with a global workshop conduction analysis, these also point out further dimensions of SEJE integration with the AMA design process. Such is, for example, the purpose of group discussions. On the one hand, a significant number of profound professional arguments have been heard when deciding on the best options for the sub-criteria. On the other hand, the purpose of "re-convincing" the expert to possibly correct their evaluation was rejected by the majority of the DMs. Its influence on the results has been assessed as low by the researchers as well. This brings forward the following questions:

- Which tasks in conceptual aircraft design are more suitable for individual assessments and which would benefit more from a group effort?
- Which tasks in conceptual design would benefit more from convergent or divergent thinking?
- How should the discussions be structured?
- How should the influence of technologies on criteria be structured and integrated into a collective expert assessment?

Furthermore, based on the results and the participant feedback, the following improvement proposals have been drawn:

- **Use case handouts.** The lack of knowledge on specific topics such as morphing mechanisms have shown the necessity of preliminary preparation of the experts. In this context, it would be useful to prepare a brief summary of the study case, which would be sent to the experts prior to the workshop.
- **Avoid ambiguous or overly-general formulations of technologies and criteria.**
- **Expert selection from heterogeneous professional backgrounds.** The first two workshops involved participants mainly from the aircraft design and aerostructures domains. This was reflected in the almost identical solution spaces resulting from the weighted and non-weighted aggregation of the evaluations.
- **Group discussions improvement.** In order to fully benefit from such expert panels, it would be senseless to reject behavioral aggregation and ideas sharing in group discussions. Instead, a clearer purpose and a better structure of the exchange should be developed.

6. Conclusions

The present paper describes the definition and application of the technology evaluation stage of the AMA design process based on expert workshops. In particular, the derivation and integration of SEJE techniques in the conceptual aircraft design with the AMA. The following research objectives have been fulfilled: (a) justification for the development of a SEJE approach adapted to the AMA needs and (b) the extraction of use case results and expert feedback for further method development from a second conducted workshop.

In order to justify the derived methodology, an extensive literature review on SEJE has been conducted first. As a result, the most prominent SEJE methods (the Classical Method, the SHELF framework and the IDEA protocol) were summarized, compared and used for the identification of common SEJE components. Further important aspects for the preparation and conduction of such elicitation have been researched and commented, namely the role of bias, the remote and personal formats as well as qualitative and quantitative elicitation variables. Additionally, the integration of these aspects and the combination of SEJE methods with MCDM frameworks were considered. The importance of such approaches is underlined by SEJE application examples in aerospace. This analysis was used for the development of a SEJE methodology adopted specifically for conceptual aircraft design within the AMA design framework and its positioning in the scientific context.

The second part of the paper is dedicated to the description and application of this methodology in the form of a second AMA workshop. Its design took into account the lessons learned from the first workshop (see Reference [12]). The consideration of novel technologies was implemented by selecting the design of wing morphing concepts as a use case for the workshop. In particular, it studied the influence of morphing technologies and their integration in different airfoil and wing sections. During the workshop, these were compared by the expert panel regarding the criteria flight performance, system complexity, and energy input by entering their fuzzy numbers in the specially developed software platform. The expert weighting was based on their self-designated expertise and did not show any significant influence on the results. This might be due either to the homogeneity of the panel's professional background or to an undiscovered methodological flaw. The workshop consisted of two main steps: individual evaluation of the options and group discussions. The latter aimed to combine ideas and help the experts correct their evaluations, thus reducing bias and increasing results objectivity.

After applying the FAHP, the AMA software was used to generate the solution space. On the one hand, it confirmed some expected trends, e.g., the flight performance advantage of an entirely morphable wing referred to only partial morphing. On the other hand, it revealed the unexpectedly low difference of some option scores, mostly among the morphing mechanisms. The reason for this might be the very similar evaluations given by the experts as a result of lacking knowledge on this very specific topic (epistemic uncertainty).

Along with the various visual results presentation and their analysis, a vital outcome of the work was the participants' feedback. It serves not only as a partial validation source of the methodology, but also as a valuable guideline for its further development.

Based on these results, improvement proposals for future work on the project have been drawn. These refer mostly to ambiguity in the definition of technology options and criteria. The need for additional use case structuring and definition was pointed out along with the necessity for preliminary preparation on specialized topics. Furthermore, the full capacity of expert panel discussions should be used by restructuring these and defining clearer purpose and deliverables.

Although based on a thorough literature study on SEJE methods and the corresponding biases, a deeper look into technology assessments and scenario workshops for a better process structuring is required. Future studies would benefit not only from the lessons learned through the current work. Potential workshops would also profit from widening their purpose, namely by extending the pure technical assessment of technologies to their integration with the environment and society.

The findings are not only an integral part of the AMA enhancement process, but can also serve as a solid basis for technology evaluation via SEJE methods focused mostly on technical qualities of the solutions, which is rarely found in the existing literature. This work represents a "stand-alone" methodological component for similar studies in the aerospace domain and beyond.

The major elicitation components that were added to the methodology take the extended workshop concept to a new level. It exhibits not only the development of a full-scale SEJE method, but most importantly its integration into the early stages of conceptual aircraft design. This has been achieved by the appropriate definition of attributes, options, and criteria, as well as by orienting the elicitation towards the problematic of aircraft architecture design. In particular, the evaluations required references to specific aspects from design and operational point of view. Additionally, the group discussions aimed to broaden the experts' horizon on a certain technology group by gathering different views from various engineering disciplines in order to obtain results with increased objectivity.

In a global perspective, the AMA, the qualitative evaluation approach, and the corresponding expert workshops are seen as a step prior to the application of "classical" MDAO algorithms. The flexible definition of the MM and the workshops implies the qualitative multi-criteria comparison of options from possibly different aircraft classes (e.g., aerostatic lift generation against fixed wings), otherwise hardly feasible within a single MDAO framework. The resulting exhaustive solution space allows the consideration of solutions possibly let out of scope during conventional idea generation.

Author Contributions: Conceptualization, V.T.T., D.R. and A.B.; methodology, V.T.T. and D.R.; software, V.T.T.; validation, V.T.T. and A.B.; formal analysis, V.T.T.; investigation, V.T.T.; resources, D.R. and A.B.; data curation, V.T.T.; writing—original draft preparation, V.T.T. and D.R.; writing—review and editing, D.R. and A.B.; visualization, V.T.T.; supervision, D.R. and A.B.; project administration, A.B.; funding acquisition, A.B. All authors have read and agreed to the published version of the manuscript.

Funding: This research was funded by Deutsche Forschungsgemeinschaft (DFG, German Research Foundation) grant number 443831887.

Institutional Review Board Statement: Not applicable.

Data Availability Statement: Data can be made available on demand.

Conflicts of Interest: The authors declare no conflict of interest.

Abbreviations

The following abbreviations are used in this manuscript:

AMA	Advanced Morphological Approach
SEJE	Structured expert judgment elicitation
MA	Morphological Analysis (classical)
MM	Morphological matrix
FAHP	Fuzzy Analytical Hierarchy Process
AHP	Analytical Hierarchy Process
SAR	search and rescue
CM	Classical Model
SHELF	Sheffield Elicitation Framework
IDEA	Investigate, Discuss, Estimate, Aggregate
DM	decision-maker
MCDM	Multi-Criteria Decision-Making
TA	Technology Assessment
SW	scenario workshop
NASA	National Aeronautics and Space Administration
FAA	Federal Aviation Administration
IATA	International Air Transport Association
UI	user interface
MDAO	Multidisciplinary Aircraft Optimization

References

1. Bernstein, J.I. Design Methods in the Aerospace Industry: Looking for Evidence of Set-Based Practices. Master's Thesis, Massachusetts Institute of Technology, Cambridge, MA, USA, 1998.
2. Todorov, V.T.; Rakov, D.; Bardenhagen, A. Creation of innovative concepts in Aerospace based on the Morphological Approach. *IOP Conf. Ser. Mater. Sci. Eng.* **2022**, *1226*, 012029. [CrossRef]
3. Bardenhagen, A.; Rakov, D. *Analysis and Synthesis of Aircraft Configurations during Conceptual Design Using an Advanced Morphological Approach*; Deutsche Gesellschaft für Luft- und Raumfahrt—Lilienthal-Oberth e.V.: Bonn, Germany, 2019. [CrossRef]
4. Todorov, V.T.; Rakov, D.; Bardenhagen, A. Enhancement Opportunities for Conceptual Design in Aerospace Based on the Advanced Morphological Approach. *Aerospace* **2022**, *9*, 78. [CrossRef]
5. Zwicky, F. *Discovery, Invention, Research—Through the Morphological Approach*; The Macmillan Company: Toronto, ON, Canada, 1969.
6. Rakov, D. Morphological synthesis method of search for promising technical system. *IEEE Aerosp. Electron. Syst. Mag.* **1996**, *11*, 3–8. [CrossRef]
7. Ritchey, T. General morphological analysis as a basic scientific modelling method. *Technol. Forecast. Soc. Chang.* **2018**, *126*, 81–91. [CrossRef]
8. Frank, C.P.; Marlier, R.A.; Pinon-Fischer, O.J.; Mavris, D.N. Evolutionary multi-objective multi-architecture design space exploration methodology. *Optim. Eng.* **2018**, *19*, 359–381. [CrossRef]
9. Villeneuve, F. A Method for Concept and Technology Exploration of Aerospace Architectures. Ph.D. Thesis, Georgia Institute of Technology, Atlanta, GA, USA, 2007.
10. Mavris, D.N.; Kirby, M.R. *Technology Identification, Evaluation, and Selection for Commercial Transport Aircraft*; Georgia Tech Library: Atlanta, GA, USA, 1999.
11. Ölvander, J.; Lundén, B.; Gavel, H. A computerized optimization framework for the morphological matrix applied to aircraft conceptual design. *Comput.-Aided Des.* **2009**, *41*, 187–196. [CrossRef]
12. Todorov, V.T.; Rakov, D.; Bardenhagen, A. *Improvement and Testing of the Advanced Morphological Approach in the Domain of Conceptual Aircraft Design*; Deutsche Gesellschaft für Luft- und Raumfahrt—Lilienthal-Oberth e.V.: Bonn, Germany, 2022. [CrossRef]
13. Monroe, R.W. A Synthesized Methodology for Eliciting Expert Judgment for Addressing Uncertainty in Decision Analysis. Ph.D. Thesis, Old Dominion University Libraries, Norfolk, VA, USA, 1997. [CrossRef]
14. Meyer, M.; Booker, J. *Eliciting and Analyzing Expert Judgment: A Practical Guide*; Technical Report NUREG/CR-5424; LA-11667-MS; OSTI.GOV: Oak Ridge, TN, USA, 1990. [CrossRef]
15. Tversky, A.; Kahneman, D. Judgment under Uncertainty: Heuristics and Biases. *Science* **1974**, *185*, 1124–1131. [CrossRef]
16. Simola, K.; Mengolini, A.; Bolado-Lavin, R. *Formal Expert Judgement: An Overview*; Technical Report EUR 21772 EN; Office for Official Publications of the European Communities: Luxembourg, 2005.

17. Dias, L.C.; Morton, A.; Quigley, J. (Eds.) Elicitation: State of the Art and Science. In *Elicitation*; International Series in Operations Research & Management Science; Springer International Publishing: Cham, Switzerland, 2018; Volume 261, pp. 1–14. [CrossRef]
18. European Food Safety Authority. Guidance on Expert Knowledge Elicitation in Food and Feed Safety Risk Assessment. *EFSA J.* **2014**, *12*, 3734. [CrossRef]
19. Cooke, R.M. *Experts in Uncertainty: Opinion and Subjective Probability in Science*; Environmental Ethics and Science Policy Series; Oxford University Press: New York, NY, USA, 1991.
20. Quigley, J.; Colson, A.; Aspinall, W.; Cooke, R.M. Elicitation in the Classical Model. In *Elicitation*; Dias, L.C., Morton, A., Quigley, J., Eds.; International Series in Operations Research & Management Science; Springer International Publishing: Cham, Switzerland, 2018; Volume 261, pp. 15–36. [CrossRef]
21. French, S.; Hanea, A.M.; Bedford, T.; Nane, G.F. Introduction and Overview of Structured Expert Judgement. In *Expert Judgement in Risk and Decision Analysis*; Hanea, A.M., Nane, G.F., Bedford, T., French, S., Eds.; International Series in Operations Research & Management Science; Springer International Publishing: Cham, Switzerland, 2021; Volume 293, pp. 1–16. [CrossRef]
22. Cooke, R.M.; Goossens, L.J.H. *Procedures Guide for Structured Expert Judgment: Final Report*; Number 18820 in EUR Nuclear Science and Technology; Office for Official Publications of the European Communities: Luxembourg, 2000.
23. Gosling, J.P. SHELF: The Sheffield Elicitation Framework. In *Elicitation*; Dias, L.C., Morton, A., Quigley, J., Eds.; International Series in Operations Research & Management Science; Springer International Publishing: Cham, Switzerland, 2018; Volume 261, pp. 61–93. [CrossRef]
24. Williams, P.L.; Webb, C. The Delphi technique: A methodological discussion. *J. Adv. Nurs.* **1994**, *19*, 180–186. [CrossRef]
25. Hemming, V.; Burgman, M.A.; Hanea, A.M.; McBride, M.F.; Wintle, B.C. A practical guide to structured expert elicitation using the IDEA protocol. *Methods Ecol. Evol.* **2018**, *9*, 169–180. [CrossRef]
26. Hanea, A.M.; Nane, G.F. An In-Depth Perspective on the Classical Model. In *Expert Judgement in Risk and Decision Analysis*; Hanea, A.M., Nane, G.F., Bedford, T., French, S., Eds.; International Series in Operations Research & Management Science; Springer International Publishing: Cham, Switzerland, 2021; Volume 293, pp. 225–256. [CrossRef]
27. Hanea, A.M.; Burgman, M.; Hemming, V. IDEA for Uncertainty Quantification. In *Elicitation*; Dias, L.C., Morton, A., Quigley, J., Eds.; International Series in Operations Research & Management Science; Springer International Publishing: Cham, Switzerland, 2018; Volume 261, pp. 95–117. [CrossRef]
28. Hanea, A.M.; Nane, G.F.; Bedford, T.; French, S. *Expert Judgement in Risk and Decision Analysis*; International Series in Operations Research & Management Science; Springer Nature Switzerland AG: Cham, Switzerland, 2021; Volume 293.
29. Bardenhagen, A.; Rakov, D. Advanced Morphological Approach in Aerospace Design During Conceptual Stage. *Facta Univ. Ser. Mech. Eng.* **2019**, *17*, 321–332. [CrossRef]
30. Saaty, T. *Decision-Making. Analytic Hierarchy Process*; Radio i Svyaz: Moscow, Russia, 1993.
31. Mota, P.; Campos, A.R.; Neves-Silva, R. First Look at MCDM: Choosing a Decision Method. *Adv. Smart Syst. Res.* **2013**, *3*, 25–30.
32. Rip, A. Technology Assessment. In *International Encyclopedia of the Social & Behavioral Sciences*; Elsevier: Enschede, The Netherlands, 2015; pp. 125–128. [CrossRef]
33. Grin, J.; Graaf, H.v.d.; Hoppe, R. *Technology Assessment through Interaction: A Guide*, 1st ed.; Number 57 in Working Document; Rathenau Institute: The Hague, The Netherlands, 1997.
34. Quist, J.N.; Vergragt, P.J. System innovations towards sustainability using stakeholder workshops and scenarios. In Proceedings of the 3rd POSTI international conference, London, UK, 1–3 December 2000; pp. 1–18.
35. Vidal, R.V.V. *Creative and Participative Problem Solving—The Art and the Science*; DTU: Albertslund, Denmark, 2006.
36. Vidal, R.V.V. The Future Workshop: Democratic problem solving. *Econ. Anal. Work. Pap.* **2006**, *5*, 21.
37. Unal, R.; Keating, C.; Conway, B.; Chytka, T. *Development of an Expert Judgement Elicitation Methodology Using Calibration and Aggregation for Risk Analysis in Conceptual Vehicle Design*; Technical Report 20040016143; NASA Langley Research Center, Engineering Management Department Old Dominion University: Norfolk, VA, USA, 2004.
38. Conway, B.A. Calibrating Expert Assessments of Advanced Aerospace Technology Adoption Impact. Ph.D. Thesis, Old Dominion University Libraries, Norfolk, VA, USA, 2003. [CrossRef]
39. Chytka, T.M. Development of an Aggregation Methodology for Risk Analysis in Aerospace Conceptual Vehicle Design. Ph.D. Thesis, Old Dominion University Libraries, Norfolk, VA, USA, 2003. [CrossRef]
40. Strohmayer, A. Improving Aircraft Design Robustness with Scenario Methods. *Acta Polytech.* **2001**, *41*, 4–5. [CrossRef] [PubMed]
41. Linzey, W.G. *Development of an Electrical Wire Interconnect System Risk Assessment Tool*; Technical Report DOT/FAA/AR-TN06/17; Federal Aviation Administration, U.S. Department of Transportation: Springfield, VA, USA, 2006.
42. Peng, W.; Zan, M.; Yi, T. Application of Expert Judgment Method in the Aircraft Wiring Risk Assessment. *Procedia Eng.* **2011**, *17*, 440–445. [CrossRef]
43. Badanik, B.; Janossy, M.; Dijkstra, A. The Use of Expert Judgement Methods for Deriving Accident Probabilities in Aviation. *Promet-Traffic Transp.* **2021**, *33*, 205–216. [CrossRef]
44. Raymer, D.P. *Aircraft Design: A Conceptual Approach*; AIAA Education Series; American Institute of Aeronautics and Astronautics, Inc.: Reston, VA, USA, 2018.
45. Sinapius, M.; Monner, H.P.; Kintscher, M.; Riemenschneider, J. DLR's Morphing Wing Activities within the European Network. *Procedia IUTAM* **2014**, *10*, 416–426. [CrossRef]

46. Atique, M.S.A. Energy Harvesting & Wing Morphing Design Using Piezoelectric Macro Fiber Composites. Ph.D. Thesis, University of North Dakota, Grand Forks, ND, USA, 2022.
47. Costanza, G.; Tata, M.E. Shape Memory Alloys for Aerospace, Recent Developments, and New Applications: A Short Review. *Materials* **2020**, *13*, 1856. [CrossRef] [PubMed]
48. Buckley, J. Fuzzy hierarchical analysis. *Fuzzy Sets Syst.* **1985**, *17*, 233–247. [CrossRef]
49. Pedregosa, F.; Varoquaux, G.; Gramfort, A.; Michel, V.; Thirion, B.; Grisel, O.; Blondel, M.; Prettenhofer, P.; Weiss, R.; Dubourg, V.; et al. Scikit-learn: Machine Learning in Python. *J. Mach. Learn. Res.* **2011**, *12*, 2825–2830.
50. Helix Toolkit Contributors. Helix Toolkit 2022. Available online: https://github.com/helix-toolkit/helix-toolkit (accessed on 3 February 2023).
51. Schneider, K.; Venn, B.; Mühlhaus, T. Plotly.NET: A fully featured charting library for .NET programming languages. *F1000Research* **2022**, *11*, 1094. [CrossRef]
52. Warner, J.; Sexauer, J.; Unnikrishnan, A.; Castelão, G.; Pontes, F.A.; Uelwer, T.; Batista, F. JDWarner/Scikit-Fuzzy: Scikit-Fuzzy Version 0.4.2. Available online: https://scikit-fuzzy.github.io/scikit-fuzzy/ (accessed on 3 February 2023).
53. Dong, W.M.; Shah, H.C.; Wongt, F.S. Fuzzy computations in risk and decision analysis. *Civ. Eng. Syst.* **1985**, 2, 201–208. [CrossRef]
54. Syakur, M.A.; Khotimah, B.K.; Rochman, E.M.S.; Satoto, B.D. Integration K-Means Clustering Method and Elbow Method For Identification of The Best Customer Profile Cluster. *IOP Conf. Ser. Mater. Sci. Eng.* **2018**, *336*, 012017. [CrossRef]

Article

A Novel Process to Produce Ti Parts from Powder Metallurgy with Advanced Properties for Aeronautical Applications

Tamas Miko [1,*], Daniel Petho [1], Greta Gergely [1], Dionysios Markatos [2] and Zoltan Gacsi [1]

1 Institute of Physical Metallurgy, Metalforming and Nanotechnology, University of Miskolc, 3515 Miskolc, Hungary
2 Laboratory of Technology & Strength of Materials (LTSM), Department of Mechanical Engineering and Aeronautics, University of Patras, 26504 Patras, Greece
* Correspondence: femmiko@uni-miskolc.hu

Abstract: Titanium and its alloys have excellent corrosion resistance, heat, and fatigue tolerance, and their strength-to-weight ratio is one of the highest among metals. This combination of properties makes them ideal for aerospace applications; however, high manufacturing costs hinder their widespread use compared to other metals such as aluminum alloys and steels. Powder metallurgy (PM) is a greener and more cost and energy-efficient method for the production of near-net-shape parts compared to traditional ingot metallurgy, especially for titanium parts. In addition, it allows us to synthesize special microstructures, which result in outstanding mechanical properties without the need for alloying elements. The most commonly used Ti alloy is the Ti6Al4V grade 5. This workhorse alloy ensures outstanding mechanical properties, demonstrating a strength which is at least twice that of commercially pure titanium (CP-Ti) grade 2 and comparable to the strength of hardened stainless steels. In the present research, different mixtures of both milled and unmilled Cp-Ti grade 2 powder were utilized using the PM method, aiming to synthesize samples with high mechanical properties comparable to those of high-strength alloys such as Ti6Al4V. The results showed that the fine nanoparticles significantly enhanced the strength of the material, while in several cases the material exceeded the values of the Ti6Al4V alloy. The produced sample exhibited a maximum compressive yield strength (1492 MPa), contained 10 wt.% of fine (milled) particles (average particle size: 3 μm) and was sintered at 900 °C for one hour.

Keywords: titanium; powder metallurgy; aerospace; dual scale microstructure

Citation: Miko, T.; Petho, D.; Gergely, G.; Markatos, D.; Gacsi, Z. A Novel Process to Produce Ti Parts from Powder Metallurgy with Advanced Properties for Aeronautical Applications. *Aerospace* **2023**, *10*, 332. https://doi.org/10.3390/aerospace10040332

Academic Editors: Spiros Pantelakis, Andreas Strohmayer and Jordi Pons-Prats

Received: 28 February 2023
Revised: 16 March 2023
Accepted: 21 March 2023
Published: 27 March 2023

1. Introduction

Titanium and its alloys have a significant role in the aerospace industry. Demonstrating a density of 4.5 g/cm^3, titanium alloys weigh about half as much as steel or Ni-based super alloys, resulting in superior mechanical properties to the latter, while their exceptional corrosion resistance makes them excellent candidates for use in the aviation and aerospace sectors [1]. Currently, this metal and its alloys account for 14% of the total weight of modern aircrafts [2,3]. For non-structural applications in which corrosion resistance, good formability and low weight are the main requirements (e.g., welded pipes and ducts, bolts, seat rails, water supply systems for galleys and sanitary, etc.), Cp-Ti is generally used. However, there are several application areas wherein high strength is important as well. Airframe joints, engine parts (e.g., fan blades, fan case, shaft, compressor) and landing gears require the use of high strength Ti alloys [4,5]. For these purposes, the Ti6Al4V is the most widely used alloy. Titanium-based alloys have the highest tensile strength/density ratio among the metals [6]. Unlike aluminum alloys, Titanium can preserve its strength at elevated (up to 600 °C) and cryogenic temperatures as well [7], which is favorable considering the temperature conditions affecting aircraft. Moreover, the ongoing focus on achieving closed-loop circularity in the aviation sector for the accomplishment of sustainability objectives [8] can be supported by using recycled titanium alloys. Titanium alloys can be remelted and reprocessed into high purity ingots with lower energy requirements compared to virgin ones. In this context, specialized forging plants that recycle titanium from the aerospace sector are already functioning [9].

Yet, the biggest obstacle to the wider use of titanium is its price. Generally, the cost of the traditional production method is higher compared to other metals, such as steel or aluminum [10]. In addition, cold forming of thin titanium alloys is challenging due to their high yield strength (YS) and their significant strain hardening effect. Furthermore, they present low thermal conductivity, which increases the heat of the tools during machining, while their low Young's modulus causes a significant spring back during traditional processing [11]. Figure 1 shows the price per unit volume of some traditionally produced alloys used in the aerospace industry at different thicknesses [12]. Obviously, the Ti6Al4V alloy has the highest price compared to the unalloyed Ti grade 2, 17-4PH stainless steel and 7075 aluminum alloy. In the case of 7075 Al and Ti grade 2, the price does not increase significantly with decreasing thickness, due to their good formability. However, the specific prices of the 17-4PH stainless steel and the Ti6Al4V increase significantly with the thickness decrease. This is due to the resulting increase in hardness and strength. The latter can make the material more resistant to deformation during the cold forming, which in turn may require more energy to shape it. Additionally, thinner materials may require more precise and specialized equipment and processes to be shaped properly, which can also increase costs. Therefore, cold forming of these alloys is extremely difficult and energy demanding, especially for thinner sheets, something which increases processing costs significantly.

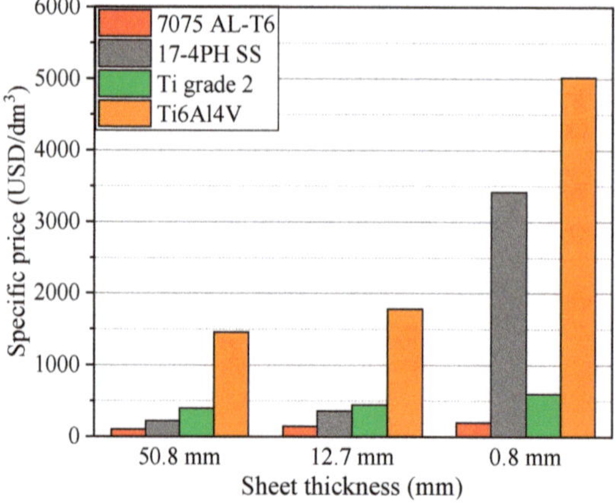

Figure 1. Specific price (USD/dm^3) of different metal sheets (304 × 304 mm) at different thickness.

Furthermore, due to the low yield rate of the traditional Ti metallurgy, 82% of the initial Ti becomes scrap, e.g., in the case of F-22 fighter jet Ti parts [13]. In the above context, reduction of scrap material and consequent manufacturing costs, as well as increases in the mechanical properties of Cp-Ti and titanium alloys, represent the two main challenges of the research community [14]. Within this framework, the powder metallurgical (PM) approach can significantly increase the yield rate and thus reduce manufacturing costs, as it is suitable for the production of ready-to-use parts and components [10]. Among the PM processes, the traditional press and sinter process is the lowest cost method for converting metal powder into a near net shape part [15]. This method involves the cold pressing of the prepared powder into the desired shape in a die, then creating metallic bonding between the cold-welded powder particles via sintering at temperatures below the melting temperature. The productivity of this process is high, but the size and complexity of the cold pressed parts are limited due to technological reasons [16]; therefore, fan blades, cockpit window frames or hydraulic pipes cannot be produced this way. Cold isostatic pressing (CIP) and hot isostatic pressing (HIP) can be alternative solutions to the press and sinter technique [17]; however, productivity is greatly decreased with these solutions.

However, the use of press and sinter technology has its limitations. The produced parts must be relatively small, as the required press force increases as the surface area increases. In addition, they must have an axisymmetric simple shape to be manufacturable with cold pressing tools. For example, fasteners (nots, screws, clevis pins, washers) can be produced as a finished or semi-finished product this way. These products are traditionally made from rolled sheets and bars by chipping or punching, which results in a lot of scrap material. PM appears as a suitable method for the production of these small parts as it does not generate a considerable amount of scrap. This manufacturing process is particularly economical for expensive materials such as Ti and its alloys. From the observed high costs of small parts produced through conventional methods [18], it becomes clear that the PM technology could emerge as a potential cost-efficient method for the production of small parts such as thin washers, as such parts can be produced without generating any scrap material, thus saving on processing costs and energy.

The properties of PM products are significantly influenced by the size, morphology, and hardness of the initial powders. Duriagina et al. has investigated the effect of the size and morphology of the VT20 Ti alloy powders on the mechanical properties of coatings. The coatings deposited from the $-160 + 40$ μm fraction showed an optimum ratio of strength and plasticity. Nonspherical VT20 titanium alloy powders are characterized by a finer structure than the coatings produced from spherical powders [19].

Although powders with spherical morphology are ideal for the additive manufacturing (AM) processes such as metal injection molding (MIM), they are not favorable for cold pressing due to the weak compressive bonds between the pressed powder particles. Powders with sponge-like particles ensure the highest green (i.e., not yet sintered part) strength [15] by promoting higher plastic deformation during cold compaction and better interlocking behavior compared to spherical powders [20]. The morphology of the Cp-Ti sponge results from the reduction of the Ti ore using the Kroll, Hunter or Armstrong method, which is the first main step of commercial titanium's production route [10]. The size of the coral-like spongy particles can be decreased to specified ranges by using a crushing or milling technique. This size reduction can easily be carried out after hydrogenating the powder. The resulting TiH_2 can be easily crushed to different particle sizes, ranging from 45 μm to 300 μm. After this relatively inexpensive hydrogenation-dehydrogenation process (HDH), we obtain a powder with irregularly shaped particles with high purity [21]. Currently, the spherical Ti6Al4V powder is also produced from alloyed Kroll sponge; it is formed into a wrought product which is then reduced to powder by an atomization process (e.g., the plasma rotating electrode process (PREP)). The resulting spherical powder's cost is approximately 15–30 times of the cost of sponge Cp-Ti powder [14]. This feedstock ensures much higher sintered strength, but the cost could be higher than the production cost of the same wrought alloy. In the case of PM parts, the effect of porosity has an important role. In order to increase the relative density close to its theoretical maximum, most studies used high sintering temperatures and long sintering times, but in these cases, coarsening is a significant issue [22]. The applied value of the cold pressing has also an important role in the green density [23]. Decreasing the porosity from 35% to 5%, the strength (YS, UTS) and the elongation to fracture value increases by about 5 to 10 times, respectively [24]. The difference between the green density of the samples made of Ti6Al4V and the Cp-Ti is ~20% [20]. However, if we compare the tensile properties of the Cp-Ti and Ti6Al4V (Table 1), it can be stated that the strength (YS, UTS) of the sintered/wrought Ti6Al4V are double, while the elongation is half of the Cp-Ti's values [25].

Table 1. Tensile properties of Cp-Ti and Ti alloys.

Type	Production Method	Specification	Tensile YS (MPa)	UTS (MPa)	Tensile Strain (%)
CP-Ti (grade 2)	Metal Injection Molding (MIM)	ASTM F2989 MIM 2	360	420	17
	Wrought (forging stock)	EN 3451:2017	290	390	20
	Wrought (forging stock)	ISO 5832-2:1999	275	345	20
Ti6Al4V	Metal Injection Molding (MIM)	ASTM F2885 grade 5	680	780	10
	Additive Layer Manufacturing (ALM)	ASTM F2924	825	895	10
	Wrought (sheet or bar)	ISO 5832-3:2016	780	860	10
Ti-10V-2FE-3Al	Wrought hot rolled bars	EN 4685:2011	1110	1240	4

From the above table (Table 1), it is clear that the alloying elements play an important role on the strength of the alloys. There are several studies which have investigated the role of the different alloying elements on the strength of the titanium. By adding 42 wt.% Nb, the tensile yield strength increases to 675 MPa [26], and by adding 7 wt.% Fe, the tensile strength increases to 916 MPa [27]. The strength can even be increased more by adding ceramic reinforcement to Ti. Jeong et al. mixed pure titanium powder with TiB_2 and B_4C powder, then produced bulk samples using the press and sinter approach. The compressive yield strength of the in situ processed composites was higher than that of the Ti6Al4V alloy at ambient temperature. The highest compressive yield strength obtained was 1400 MPa [28]. The oxygen content has also an important role on the strength of the Ti. Chen et al. increased the oxygen content of the Cp-Ti up to 0.8 wt.%. using different PM methods. The highest tensile YS was measured around 900 MPa [20].

Besides the alloying and reinforcing elements, the role of the grain structure and the grain size are crucial as well, and this will be the focus of the present work. Nano-grained (NG) and ultrafine-grained (UFG) metals and alloys show significantly higher strength compared to coarse-grained metals [29]. However, with the increased strength, the deformability and the room temperature ductility significantly decrease as well, limiting the applicability. To tackle the latter issue, dual scale grain size can be considered. The big advantage of a dual scale grain size is that the coarse grains retain the toughness of the material, while the fine grains improve its strength [30]. This microstructure can be achieved by PM method. Sun et al. prepared non-milled and cryo-milled Ti6Al4V alloy by plasma-activated sintering. The highest hardness measured was 470 HV, and the highest compressive YS was 1706 MPa [31]. Li et al. made a Ti-Bi bimodal alloy by using high-energy ball-milled Ti-Bi and spark plasma sintering, achieving 1080 MPa tensile YS [32]. Attar et al. made in situ titanium–titanium boride composites using the mixture of fine TiB_2 and coarse CP-Ti powder. The solidification was carried out with the selective laser melting (SLM) method, achieving 1400 MPa compressive YS [33].

Based on the cited literature, the economically feasible production of Ti parts with satisfactory mechanical properties can still be further improved. In this study, Cp-Ti was considered the initial material, with the aim being to improve its performance using a dual-scale microstructure which was produced by mixing fine and coarse titanium grade-2 type powders through the PM process. The goal of the study was to produce a material with increased mechanical properties, to be considered as a potential substitute to commercially produced Ti alloys widely used in the aviation sector, such as the Ti6Al4V (grade5) alloy. To achieve the above, the cavities of Cp-Ti sponge powders were filled with fine nanoscale-milled Cp-Ti powder during the mixing process, which resulted in a special feedstock for the applied cold press and sinter process. The density, hardness, yield strength, compressive strength and strain were systematically investigated on the sintered parts. The results showed that the fine nanoparticles of the dual-scale grain have significantly enhanced the strength of the material, while in several cases, the corresponding strength exceeded the values of the Ti6Al4V alloy. The novelty of the work lies in the simplicity and inexpensiveness of the production route.

2. Experimental

The Ti powder used is a 99.4% purity Cp-Ti (grade 2) with nominal average particle size of 150 µm, produced by Alfa Aesar (product number: 10383), Kandel, Germany. The received (coded as "initial") powder was used as coarse-grained Ti, and the fine-grained Ti was produced from the same powder by use of high-energy ball milling. The milling was carried out with a Pulverisette 5 high-energy planetary ball mill (Fritsch, Idar-Oberstein Germany) at room temperature for 20 h. In order to prevent excessive cold welding, alcohol was used as a process control agent (PCA). The milling was performed in Ar atmosphere using a 250 mL hardened steel vial and hardened steel balls of 8 mm diameter. The ball-to-powder (BPR) weight ratio and rotational speed were 30:1 and 200 rpm, respectively. The mixing of the initial and milled powders was also carried out with the same mill, but the milling parameters were changed to 10:1 BPR, 30 min and 100 rpm. Ethanol was not used in this case. During mixing, 2; 4; 6; 8 and 10 wt.% milled Ti was added to the initial Ti powder.

Instron 5982 (Norwood, MA, USA) equipment was used for the consolidation of the powders to green parts at ambient temperature. The mixed powders were placed in a hardened steel die between hardened steel punches to produce cylindrical specimens with a diameter of 8 mm and 8 mm height. Graphite 33 spray (Iffezheim, Germany) was used as lubricant on the die walls to decrease the friction, and during the cold uniaxial pressing, 1.6 GPa pressure was applied. The sintering of the green samples was carried out in an electric furnace (SF 16 type Three-Zone Split Tube Furnace, Norwood, MA, USA) in Ar 6.0 atmosphere (Siad Hungary, Miskolc, Hungary). Four different sintering temperatures were considered: 800, 850, 900 and 950 °C. The heating and the cooling rate were 100 °C/min and 1000 °C/min, respectively and the holding time was 60 min.

The morphology of the powders, the green samples and the sintered samples was investigated by Scanning electron microscopy (SEM) using a C. Zeiss EVO MA 10. and a Zeiss Crossbeam 540 LA-FIB SEM (Zeiss, Oberkochen, Germany). For the determination of APS values, an analysis of the SEM pictures was conducted; five SEM pictures representing different regions of the initial as well as the milled powder were analyzed, and the particle size (n = 100) was measured through the use of Image J 1.47 software [34].

The density of the samples in both green and sintered state was determined by measuring the weight and the dimensions of the samples. The microhardness (HV0.025) of the initial and milled powders was measured by an Instron Wilson Tukon 2100 B (Norwood, MA, USA). The hardness of the green and sintered samples was measured with the Brinell method (HB2.5/62.5) on a Wolpert UH930 machine (Illinois Tool Works Inc. Shanghai, China). Compression tests were carried out in order to determine yield, compression strength and the compression strain. These tests were conducted with Instron 5982 type (Norwood, MA, USA) universal material testing equipment, using a crosshead speed of 3 mm/min.

3. Results and Discussion

3.1. Morphological Analysis

The initial powder has a sponge-like structure with a high specific surface and lots of cavities (Figure 2a). Based on the SEM measurements, the average particle size (APS) of the initial powders was about 180 µm. After ball milling for 20 h, the particles' size and morphology changed to very fine and flake-like with 3 µm APS (Figure 3b). By conducting a wet chemical analysis (ICP-OES Varian 720 ES, Palo Alto, CA, USA) (Table 2), the chemical composition of the initial powder was assessed and compared to the milled powders. Table 2 shows the results. It can be concluded that the composition of the initial and the milled powder slightly differs due to contamination derived from the milling equipment. The amount of contaminant elements has increased due to the long-term high-energy milling, which was, however, an inevitable phenomenon. If we consider that the maximum milled content of the produced sample is 10 wt.%, this impurity overall is not significant (max. 0.5 wt.%). Moreover, due to the applied rapid and low-energy mixing, the contamination level is not further increased.

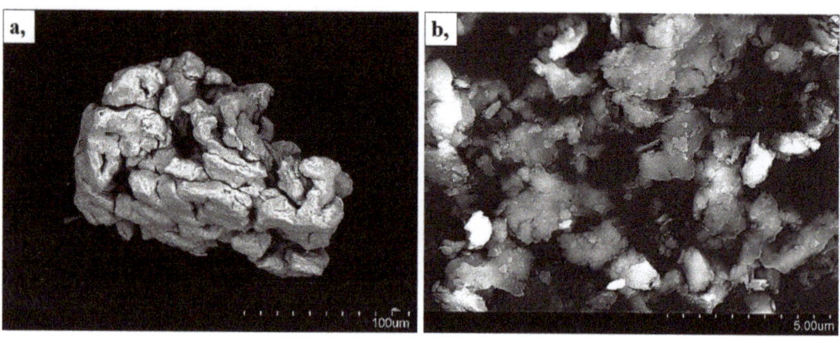

Figure 2. (**a**,**b**) Show the morphology of the initial and the milled Ti powders, respectively.

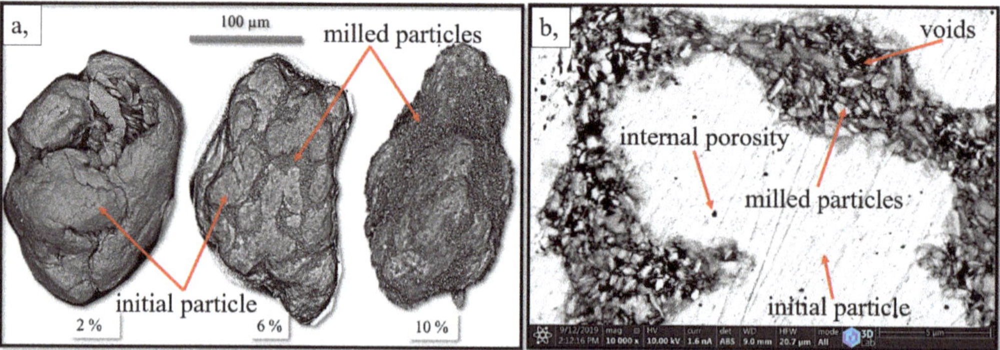

Figure 3. (**a**) Morphology of the typical particles after mixing the initial powder with 2 wt.%, 6 wt.% and 10 wt.% milled powder; (**b**) typical SEM images of the green sample which contain 10 wt.% milled powder.

Table 2. Chemical composition of initial and milled Ti powder (wt.%).

Elements	Al	Cr	Cu	Fe	Mn	Ni	Si	Ti
Initial powder	0.0615	0.0015	0.0027	0.0408	0.0003	0.0032	0.365	99.525
Milled powder	0.9701	0.04624	0.0050	1.22	<0.0037	0.0058	1.97	94.473

Due to the intensive plastic deformation caused by high-energy ball milling, not only the particle size, but also the crystallite size, as determined by X-ray diffraction (XRD), was decreased from 180 nm to 4 nm. This value was determined in previous research [35].

The microhardness of the milled powder shows a four-fold increase compared to the initial powder, namely from 200 to 800 HV0.025. During the mixing, these hard and fine particles were trapped within the soft and big initial particles, as can be seen in Figure 4a, wherein different amounts of milled powder content have been considered (2, 6 and 10 wt.%).

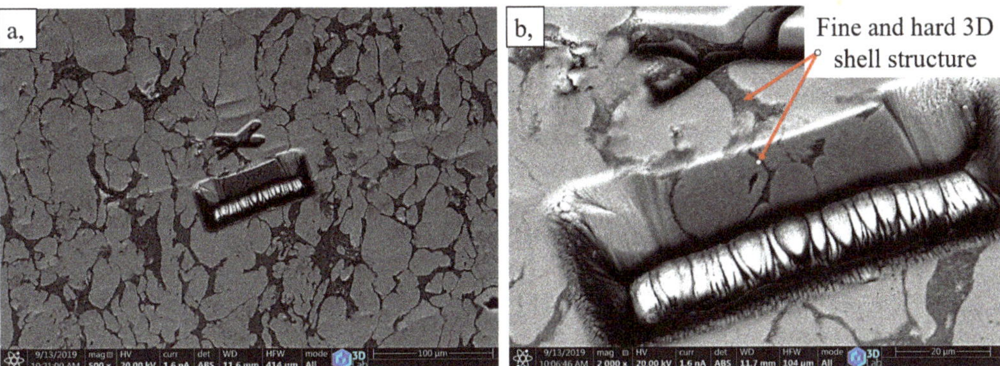

Figure 4. (**a**,**b**) SEM images of the samples containing 10 wt.% milled powder after sintering at 850 °C, 1 h at different magnifications.

Based on the SEM images (Figure 3a), it is evident that during the mixing process, the fine, milled powder fills out the pores of the large Ti spongy particles first; afterwards, the surface becomes covered with the fine Ti particles. This structure is visible in the cross-section of the green sample (Figure 3b). Some pores can be seen on Figure 3b, which are mainly closed porosity inside the coarse unmilled particles or voids inside the fine grain shell structure. Figure 4 shows SEM images of a sintered sample with 10 wt.% of ball-milled powders at different magnifications. This sample was sintered at 850 °C for 1 h. The images display both at low and high magnification that the dual-scale structure was not altered during the applied sintering process. The fine milled particles formed a 3D shell structure around the coarse unmilled grains. The thickness of this boundary layer is not constant. It varies between ~1 and ~20 µm. This is a consequence of the different sized voids of the spongy initial powder. However, the APS value of the initial powders was 180 µm, while the size of the coarse grains covered with the fine shell layer was less than ~100 µm in the case of the sintered sample. This means that during the mixing process, the fine particles managed to fill in not just the surface cavities but also the internal ones. The investigated surface has two incisions made by plasma ablation. The aim of these incisions was to proves that this dual-scale structure is truly a 3D structure.

3.2. Density Analysis

Figure 5 shows the effect of the added milled fine powder on the relative density of different green and sintered samples. Based on the density measurements, the milled powder has a notable effect on the properties of the samples. Due to the applied high compacting force (1600 MPa), the relative density of the green sample, which does not contain any milled titanium, was determined to be around ~97.5%. Due to the powder production, the ~2.5% porosity was typically closed porosity. The high green density is beneficial to obtaining high sintered density and high mechanical properties. Robertson et al. applied the same load (1600 MPa) during the cold pressing of Ti6Al4V powder, and the measured green density was ~95% [36]. Raynova et al. has investigated the effect of the green density (65.6% to 96.5%) on the sintered density in the case of HDH Cp-Ti powder. After sintering, the density of the sample with the highest green density remained the highest (~97%) [24]. In our research, the results showed the same tendency. The density of the green sample, which does not contain any milled titanium, practically does not change during the sintering temperatures. As it can be seen in Figure 5, the green density values were decreased by increasing the fraction of the milled powder. This is due to the fact that porosities take place between the hard particles of the milled fraction (Figure 3b) which add to the micro porosities inside the powder particles. Moreover, the increase in sintering temperature resulted in an exponential increase in density. This is due to the fact that the specific interface increases with fine grains, which increases the driving force of sintering [37].

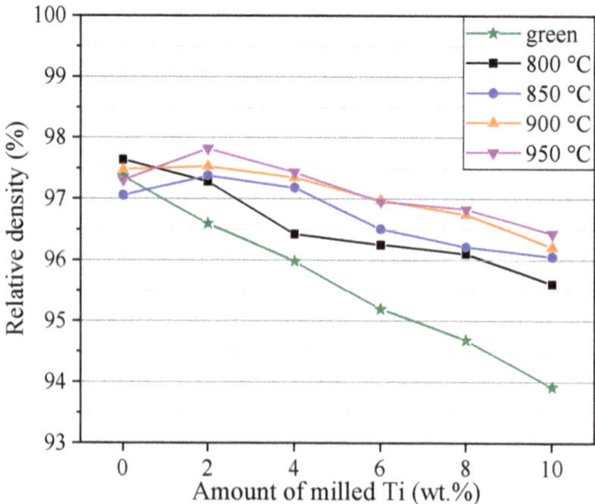

Figure 5. Relative density of the samples containing different amount of milled Ti.

Despite the applied low sintering temperature and short sintering time, the sintered density of the produced samples was found to be similar or better, compared with those observed in the literature (Table 3). As the values show, the applied processing parameters have a significant effect on the density.

Table 3. Density values of different titanium and Ti alloy sintered at different conditions.

Powder Type	D50 (μm)	Green Density (%)	Sintered Density (%)	Sintering Temperature (°C)	Sintering Time (min)
Current research Ti sponge+6% milled Ti	180	95.2	97	900	60
Ti irregular [38]	30	60	95.8	1250	120
Ti sphere [39]	23	67	89.6	1000	120
Ti sphere [37]	74	52	63.5	850	60
Ti irregular [40]	18	68	82.5	1250	60
Ti6Al4V sphere [41]	35	65	97.5	1400	120
Ti6Al4V irregular [42]	39	69	93.5	1300	120

3.3. Hardness Results

The hardness of the green samples was not influenced by the amount of milled Ti content, simply due to the fact that these samples do not present an adequate integrity in the green state (Figure 6). The hardness of all sintered samples, which had no milled powder ranges, was between 180 and 210 HB. This hardness value corresponds to the standard hardness of the Cp-Ti grade 2 (200 HB). However, the amount of milled powder had a significant effect on the hardness. The hardness increased with an increase of the milled Ti content. The highest hardness was obtained at a sintering temperature of 850 °C. Higher temperatures resulted in a slight decrease in hardness. This can be explained by the recrystallization or coarsening process, which occurs above 850 °C, resulting in grain growth and a decrease in hardness or strength [43]. The sample that contained 10 wt.% milled powder showed the highest hardness (350 HB) sintered at 850 °C. This value is equal to the hardness of the Ti6Al4V [10].

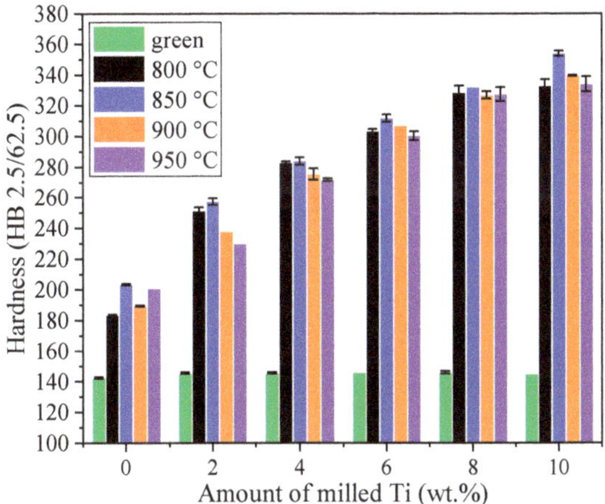

Figure 6. Hardness of the samples with different amounts of milled Ti.

3.4. Compression Test Results

The yield strength is probably the most important mechanical property of the material. This value determines the maximal mechanical loads that the material can withstand without deformation. This value is also influenced by the direction of the load. There are some studies which applied both the tensile and compression test on the sintered Ti samples. Schulze et al. determined the YS of SLM-produced Ti-42Nb alloy on tensile and compression specimen. The YS was 674 MPa upon tensile test and 831 MPa upon compression test, respectively [26]. Raghavendra et al. compared the tensile and compressive properties of additively manufactured (SLM) Ti6Al4V porous samples. The ultimate tensile strength was lower than compressive strength by an average value of 100 MPa [44]. We have to account for the fact that the PM samples have a significant number of defects, e.g., porosities. Generally, the sensitivity of the tensile test properties on the presence of defects is much higher than in the case of a compression test. Figure 7 shows the measured engineering stress–engineering strain curves of samples sintered at 800 °C, determined by compression test.

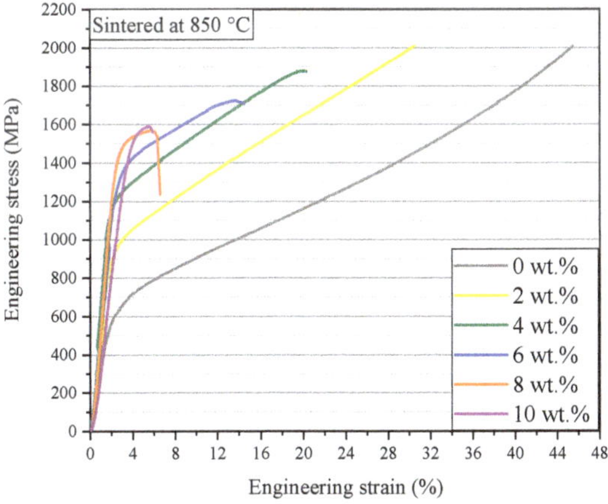

Figure 7. Stress–strain curves of the samples sintered at 850 °C with different milled Ti content.

Strength values (YS, compressive strength) determined on the measured curves show the same tendency as the hardness results. The samples without milled Ti content had an average YS (500–640 MPa). On the other hand, for the samples with an increased milled

Ti content, the YS increased proportionally (Figure 8). The highest strength values in most cases belong to the samples sintered at 850 °C. Regarding the strain values of the compression tests, it can be concluded that increasing the milled fraction decreases the deformability of the material.

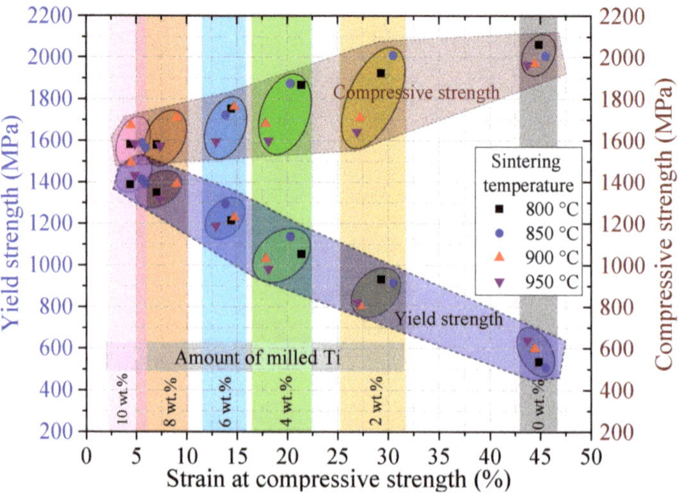

Figure 8. Strength and strain values of the measured samples.

The samples with 8 wt.% milled Ti presented a reduced formability around 5%, meaning that the material is too brittle for engineering applications. Comparing the measured YS values with data from the literature, it can be stated that the sample that has 6 wt.% milled powder shows a higher YS than the wrought Ti6Al4V (Table 4).

Table 4. Mechanical properties of titanium alloys fabricated by various processing technologies. (compressive YS; ultimate compressive strength, plastic strain to maximum compressive stress).

Material	Processing Method	Yield Strength (MPa)	Compressive Strength (MPa)	Strain at Maximum Stress (%)	Source
Ti-6% milled Ti (850 °C)	Press and Sinter	1295	1724	14	Current Research
CP-Ti	ECAP	700	900	35	[43]
CP-Ti	SLM	560	1136	51	[45]
Ti6Al4V	Annealed	1000	1300	10	[46]
Ti6Al4V	Wrought	1200	1400	20	[47]

4. Summary

In this study, different mixtures of coarse (180 µm APS) and fine (3 µm APS) Cp-Ti 2 powders were used for the production of a pure Ti material. The fine powder was milled from the initial coarse powder by use of high-energy planetary ball milling.

During the mixing process, the finely milled flakes were trapped and filled the pores of the coarse spongy particles and covered their surface. Cylindrical samples were produced from the different mixtures via uniaxial cold pressing. In order to minimize the porosity of the green samples, the various feedstock mixtures (0–10 wt.% fine powder) were cold-pressed at an extremely high uniaxial pressure (1.6 GPa).

The relative density of the cold-pressed samples varied between 97% and 94% depending on the proportion of fine particles. These density values are particularly high compared to literature data (52–69%), which is due to the good deformability of the rough and soft titanium sponge, which accounts for 90–98 wt.% of the produced samples.

This high green density allows the use of low-temperature and rapid sintering (800–950 °C, 60 min in 6.0 argon atmosphere), as long-distance atomic movement is not required during the densification process. The relative density of the sintered samples varied between 95.6% and 97.8%, which is comparable with corresponding results from the literature (93.5 to 97.5%).

Due to the applied sintering conditions, the coarsening of the initial dual-scale structure has not occurred, and the comparable hardness (max. 350 HB) and compressive properties (max. 1492 MPa YS) of the Ti6Al4V alloy were measured. The production parameters of the optimal strength–strain version of the alloy were 6 wt.% amount of milled Ti and 850 °C sintering temperature. It is possible that the inhomogeneous microstructure with large and fine features could contribute to differences between the compression and tensile test results. The presence of these features could cause variations in stress distribution and lead to localized deformation or failure during testing. Additionally, the presence of large features could affect the sample geometry and cause stress concentrations, which could also influence the test results. This constitutes a matter for further research.

In summary, a quasi-pure Ti material was produced with comparable to the Ti6Al4V alloy properties by use of a production method which is cheaper, more energy efficient, and consequently more environmentally friendly than conventional techniques. The latter factor is due to the low temperature and the rapid sintering applied. Moreover, the produced Ti parts are more easily recyclable due to the absence of alloying elements. The latter advantages make our material a potential candidate for consideration in aerospace applications requiring high strength, as the above characteristics contribute to the goals and aims of sustainable and circular aviation [48,49].

A complete LCA and LCC analysis of the production route could be the subject of further research in order to quantify the exact environmental and cost gains of the whole process compared to conventional production processes. Moreover, further research will be needed, including tensile and fatigue tests, to address the complete mechanical characterization of novel materials. The measuring of the effect of contamination by, for example, iron or oxygen should be part of a future research. Finally, further optimization of the production process will probably boost and lead to more uniform properties.

Author Contributions: Conceptualization, T.M., Z.G. and D.M.; methodology, T.M.; software, T.M.; validation, T.M. and D.P.; formal analysis, D.P. and G.G.; investigation, T.M.; resources, Z.G. and G.G.; data curation, T.M.; writing—original draft preparation, T.M., D.M. and D.P.; writing—review and editing, T.M., D.M., D.P. and Z.G.; visualization, T.M. and Z.G.; supervision, Z.G.; project administration, G.G. All authors have read and agreed to the published version of the manuscript.

Funding: The research was supported by the UMA^3 project, funded by the European Union's Horizon 2020 research and innovation program under grant agreement No. 952463.

Conflicts of Interest: The authors declare no conflict of interest.

References

1. Kulkarni, K.G.; Bhattacharya, N. Use of titanium and its alloy in aerospace and aircraft industries. *Int. J. Creat. Res. Thoughts* **2020**, *8*, 1383–1396.
2. Gholizadeh, S. Impact behaviours and Non-Destructive Testing (NDT) methods in Carbon Fiber Composites in Aerospace Industry: A Review. *Authorea Prepr.* **2022**. [CrossRef]
3. Liu, S.; Shin, Y.C. Additive manufacturing of Ti6Al4V alloy: A review. *Mater. Des.* **2019**, *164*, 107552. [CrossRef]
4. Williams, J.C.; Boyer, R.R. Opportunities and Issues in the Application of Titanium Alloys for Aerospace Components. *Metals* **2020**, *10*, 705. [CrossRef]
5. Inagaki, I.; Takechi, T.; Ariyasu, Y.S. Application and Features of Titanium for the Aerospace Industry. *Nippon. Steel Sumitomo Met. Tech. Rep.* **2014**, *106*, 22–27.
6. Niinomi, M.; Kagami, K. Recent topics of titanium research and development in Japan. In Proceedings of the 13th World Conference on Titanium, San Diego, CA, USA, 16–20 August 2016.
7. Zang, M.C.; Niu, H.Z.; Yu, J.S.; Zhang, H.R.; Zhang, T.B.; Zhang, D.L. Cryogenic tensile properties and deformation behavior of a fine-grained near alpha titanium alloy with an equiaxed microstructure. *Mater. Sci. Eng.* **2022**, *840*, 142952. [CrossRef]

8. Markatos, D.N.; Pantelakis, S.G. Assessment of the Impact of Material Selection on Aviation Sustainability, from a Circular Economy Perspective. *Aerospace* **2022**, *9*, 52. [CrossRef]
9. Available online: https://www.aubertduval.com/wp-media/uploads/2022/07/Titane_brochure_202207.pdf (accessed on 10 February 2023).
10. Frohes, F.H. *Titanium Physical Metallurgy, Processing, and Applications*; ASM International: Almere, The Netherlands, 2015.
11. Oleksik, V.; Trzepiecinski, T.; Szpunar, M.; Chodola, L.; Ficek, D.; Szczesny, I. Single-Point Incremental Forming of Titanium and Titanium Alloy Sheets. *Materials* **2021**, *14*, 6372. [CrossRef]
12. OnlineMetals.com. Available online: https://www.onlinemetals.com (accessed on 5 February 2023).
13. Fang, Z.Z.; Paramore, J.D.; Sun, P.; Chandran, K.S.R.; Zhang, Y.; Xia, Y.; Cao, F.; Koopman, M.; Free, M. Powder metallurgy of titanium—Past, present, and future. *Int. Mater. Rev.* **2017**, *63*, 407–459. [CrossRef]
14. Zwitter, T.M.; Nash, P.; Xu, X.; Johnson, C. Energy Efficient Press and Sinter of Titanium Powder for Low-Cost Components in Vehicle Applications Final Scientific Report. 2011. Available online: https://www.osti.gov/servlets/purl/1020890 (accessed on 10 February 2023).
15. Henriques, V.A.R. Titanium production for aerospace applications. *J. Aerosp. Technol. Manag.* **2009**, *1*, 7–17. [CrossRef]
16. Abkowitz, S.M.; Abkowitz, S.; Fisher, H. Breakthrough claimed for titanium PM. *Met. Powder Rep.* **2011**, *66*, 16–21. [CrossRef]
17. Araci, K.; Mangabhai, D.; Akhtar, K. Production of titanium by the Armstrong Process. In *Titanium Powder Metallurgy, Science, Technology and Applications*; Elsevier: Amsterdam, The Netherlands, 2015; pp. 149–162.
18. Allied Titanium, Inc. Available online: https://www.alliedtitanium.com/ (accessed on 5 February 2023).
19. Duriagina, Z.A.; Lemishka, I.A.; Trostianchyn, A.M.; Kulyk, V.V.; Shvachki, S.G.; Tepla, T.L.; Pleshakov, E.I.; Kosbasyuk, T.M. The effect of morphology and particle-size distribution of VT20 Titanium alloy powders on the mechanical properties of deposited coatings. *Powder Metall. Met. Ceram.* **2019**, *57*, 697–702. [CrossRef]
20. Chen, B.; Shen, J.; Ye, x.; Umeda, J.; Kondoh, K. Advanced mechanical properties of powder metallurgy commercially pure titanium with a high oxygen concentration. *J. Mater. Res.* **2017**, *32*, 19. [CrossRef]
21. Dong, S.; Ma, G.; Lei, P.; Cheng, T.; Savvakin, D.; Ivasishin, O. Comparative study on the densification process of different titanium powders. *Adv. Powder Technol.* **2021**, *32*, 2300–2310. [CrossRef]
22. Xu, X.; Zheng, Y.; Liang, H.; Yang, M.; Zhao, Y.; Zhou, W. Influence of sintering parameters on the grain growth and mechanical properties of Ti(C,N)-based cermets prepared via mechanical activation and in-situ carbothermal reduction. *Mater. Today Commun.* **2022**, *33*, 104667. [CrossRef]
23. Wang, H.; Fang, Z.Z.; Sun, P. A critical review of mechanical properties of powder metallurgy titanium. *Int. J. Powder Metall.* **2010**, *46*, 45–57.
24. Raynova, S.; Collas, Y.; Yang, F.; Bolzoni, L. Advancement in the Pressureless Sintering of CP Titanium Using High-Frequency Induction Heating. *Metall. Mater. Trans. A* **2019**, *50*, 4732–4742. [CrossRef]
25. Sidambe, A.T. Biocompatibility of Advanced Manufactured Titanium Implants A Review. *Materials* **2014**, *7*, 8168–8188. [CrossRef]
26. Schulze, C.; Weinmann, M.; Schweigel, C.; Keßler, O.; Bader, R. Mechanical Properties of a Newly Additive Manufactured Implant Material Based on Ti-42Nb. *Materials* **2018**, *11*, 124. [CrossRef]
27. Chen, B.Y.; Hwang, K.S.; Ng, K.L. Effect of cooling process on the phase formation and mechanical properties of sintered Ti–Fe alloys. *Mater. Sci. Eng. A* **2011**, *528*, 4556–4563. [CrossRef]
28. Jeong, H.W.; Kim, S.J.; Hyun, Y.T.; Lee, Y.T. Densification and Compressive Strength of In-situ Processed Ti/TiB Composites by Powder Metallurgy. *Met. Mater. Int.* **2002**, *8*, 25–35. [CrossRef]
29. Tellkamp, V.L.; Melmed, A.; Lavernia, E.J. Mechanical Behavior and Microstructure of a Thermally Stable Bulk Nanostructured Al Alloy. *Metall. Mater. Trans. A* **2001**, *32*, 2335–2343. [CrossRef]
30. Okulov, I.V.; Kühn, U.; Marr, T.; Freudenberger, J.; Schultz, L.; Oertel, C.G.; Skrotzki, W.; Eckert, J. Deformation and fracture behavior of composite structured Ti-Nb-Al-Co(-Ni) alloys. *Appl. Phys. Lett.* **2015**, *104*, 071905. [CrossRef]
31. Sun, Y.; Luo, G.; Zhang, J.; Chen, J.; Wang, G.; Shen, Q.; Zhang, L. Microstructure, mechanical properties and reinforcement mechanism of dual-scale TC4 titanium alloy prepared by cryomilling and plasma activated sintering. *Mater. Sci. Eng. A* **2018**, *736*, 120–129. [CrossRef]
32. Li, Z.; Dong, A.; Xing, H.; Xu, H.; Du, D.; Zhang, T.; She, H.; Wang, D.; Zhu, G.; Sun, B. Microstructure and mechanical properties of bimodal Ti-Bi alloys fabricated by mechanical alloying and spark plasma sintering for biomedical applications. *Mater. Charact.* **2020**, *161*, 110134. [CrossRef]
33. Attar, H.; Ehtemam-Haghighi, S.; Kent, D.; Dargusch, M.S. Recent developments and opportunities in additive manufacturing of titanium-based matrix composites: A review. *Int. J. Mach. Tools Manuf.* **2018**, *133*, 85–102. [CrossRef]
34. Rasband, W.S. *ImageJ*; U.S. National Institutes of Health: Bethesda, MD, USA, 1997–2018. Available online: https://imagej.nih.gov/ij/ (accessed on 5 February 2023).
35. Angel, D.A.; Miko, T.; Kristaly, F.; Benke, M.; Gacsi, Z. Development of TiB and nanocrystalline Ti-reinforced novel hybrid Ti nanocomposite produced by powder metallurgy. *J. Mater. Sci.* **2022**, *57*, 4130–4144. [CrossRef]
36. Robertson, I.M.; Schaffer, G.B. Review of densification of titanium based powder systems in press and sinter processing. *Powder Metall.* **2010**, *53*, 146–162. [CrossRef]
37. German, R.M. Titanium sintering science: A review of atomic events during densification. *Int. J. Refract. Met. Hard Mater.* **2020**, *89*, 105214. [CrossRef]

38. Cai, Y.X.; Chang, Q.; Ding, Y. Research of injection molding titanium parts. *Powder Metall. Tech.* **2005**, *23*, 449–455.
39. Ieki, T.; Katoh, K.; Matsumoto, A.; Masui, T.; Andoh, K. Tensile properties of sintered Ti compacts by metal injection Molding process. *J. Jap. Soc. Powder Powder Metall.* **1997**, *44*, 448–452. [CrossRef]
40. Panigrahi, B.B.; Godkhindi, M.M.; Das, K.; Mukunda, P.G. Ramakrishnan, P. Sintering kinetics of micrometric titanium powder. *Mater. Sci. Eng.* **2005**, *A396*, 255–262. [CrossRef]
41. Ferri, O.M.; Ebel, T.; Bormann, R. The influence of small boron addition on the microstructure and mechanical properties of Ti-6Al-4V fabricated by metal injection moulding. *Adv. Eng. Mater.* **2011**, *13*, 436–447. [CrossRef]
42. Bolzoni, L.; Ruiz-Navas, E.M.; Gordo, E. Influence of sintering parameters on the properties of powder metallurgy Ti-3Al-2.5V alloy. *Mater. Charact.* **2013**, *84*, 48–58. [CrossRef]
43. Long, F.W.; Jiang, Q.W.; Xiao, L.; Li, X.W. Compressive Deformation Behaviors of Coarse- and Ultrafine-Grained Pure Titanium at Different Temperatures: A Comparative Study. *Mater. Trans.* **2011**, *52*, 1617–1622. [CrossRef]
44. Raghavendra, S.; Molinari, A.; Fontanari, V.; Luchin, V.; Zappini, G.; Benedetti, M.; Johansson, F.; Klarin, J. Tensile and compression properties of variously arranged porous Ti-6Al-4V additively manufactured structures via SLM. *Procedia Struct. Integr.* **2018**, *13*, 149–154. [CrossRef]
45. Attar, H.; Bönisch, M.; Calin, M.; Zhang, L.C.; Scudino, S.; Eckert, J. Selective laser melting of in situ titanium–titanium boride composites: Processing, microstructure and mechanical properties. *Acta Mater.* **2014**, *76*, 13–22. [CrossRef]
46. Srinivasan, K.; Venugopal, P. Compression Testing of Ti-6Al-4V in the Temperature Range of 303–873 K. *Mater. Manuf. Process.* **2008**, *23*, 342–346. [CrossRef]
47. Fadida, R.; Rittel, D.; Shirizly, A. Dynamic Mechanical Behavior of Additively Manufactured Ti6Al4V with Controlled Voids. *J. Appl. Mech.* **2015**, *82*, 041004. [CrossRef]
48. European Commission. *Flightpath 2050, Europe's Vision for Aviation*; Report of the High Level Group on Aviation Research, Di-rectorate-General for Research and Innovation, Directorate General for Mobility and Transport; European Commission: Brussels, Belgium, 2011; p. 28.
49. Communication from the Commission to the European Parliament, the European Council, the European Economic and Social Committee and the Committee of the Regions the European Green Deal com/2019/640 Final. Available online: https://eur-lex.europa.eu/legal-content/EN/TXT/?uri=COM%3A2019%3A640%3AFIN (accessed on 13 February 2023).

Article

Design and Preliminary Performance Assessment of a PHM System for Electromechanical Flight Control Actuators

Antonio Carlo Bertolino, Andrea De Martin *, Giovanni Jacazio and Massimo Sorli

Politecnico di Torino, Department of Mechanical and Aerospace Engineering, 10129 Torino, Italy
* Correspondence: andrea.demartin@polito.it

Abstract: The evolution toward "more electric" aircraft has seen a decisive push in the last decade due to growing environmental concerns and the development of new market segments (flying taxis). Such a push has involved both the propulsion components and the aircraft systems, with the latter seeing a progressive trend in replacing traditional solutions based on hydraulic power with electrical or electromechanical devices. Flight Control Systems (FCSs) are one of the aircraft systems affected the most since the adoption of Electromechanical Actuators (EMAs) would provide several advantages over traditional electrohydraulic or mechanical solutions, but their application is still limited due to their sensitivity to certain single points of failure that can lead to mechanical jams. The development of an effective and reliable Prognostics and Health Management (PHM) system for EMAs could help in mitigating the risk of a sudden critical failure by properly recognizing and tracking the ongoing fault and anticipating its evolution, thus boosting the acceptance of EMAs as the primary flight-control actuators in commercial aircraft. The paper is focused on the results of the preliminary activities performed within the CleanSky 2/Astib research program, dedicated to the definition of the iron bird of a new regional-transport aircraft able to provide some prognostic capabilities and act as a technological demonstrator for new PHM strategies for EMAs employed in-flight control systems. The paper is organized as follows. At first, a proper introduction to the research program is provided, along with a brief description of the employed approach. Hence the simulation models adopted for the study are presented and used to build synthetic databases to inform the definition of the PHM algorithm. The prognostic framework is then presented, and a preliminary assessment of its expected performance is discussed.

Keywords: PHM; EMA; flight control actuators; prognostics; iron bird; actuators

Citation: Bertolino, A.C.; De Martin, A.; Jacazio, G.; Sorli, M. Design and Preliminary Performance Assessment of a PHM System for Electromechanical Flight Control Actuators. *Aerospace* **2023**, *10*, 335. https://doi.org/10.3390/aerospace10040335

Academic Editors: Spiros Pantelakis, Andreas Strohmayer and Jordi Pons-Prats

Received: 27 February 2023
Revised: 23 March 2023
Accepted: 24 March 2023
Published: 28 March 2023

1. Introduction

Prognostics and Health Management (PHM) is a relatively new, multidisciplinary research field aimed at the definition of routines capable of predicting the time of failure (ToF) of a defective system or component based upon a set number of signals (or "features") extracted from the system itself. The capability to anticipate the failure occurrence and to estimate the remaining useful life (RUL) of a system or a component would provide a set number of valuable advantages. If completely realized, it would provide important strategic information pertaining to the opportunity to perform maintenance operations, the available time window to successfully replace the faulty component, and eventually to provide advice or an automatic reconfiguration of the defective system to compensate for the effects of the degradation or to extend the RUL [1]. Although application-agnostic in nature, PHM is of particular interest for aerospace applications, where the occurrence of unanticipated failures causes the disruption of aircraft availability, which is a costly and potentially dangerous situation in both commercial and military aviation. As such, the benefits of PHM are not limited to the optimization of the maintenance policy and a reduction of its costs but have significant ramifications over the maintenance logistics (e.g., spare parts, personnel, and dedicated facilities), business choices (spare aircraft number) and eventually strategic decisions (finish the mission or return to base). In this context, onboard actuation systems are one of the most critical aircraft systems and one of the major causes of disruption of aircraft availability. The vast majority of currently in-service aircraft are equipped with electrohydraulic or electrohydrostatic actuators. However, the growing push toward the design of "more-electric" aircraft has encouraged several research activities aimed at the design and certification of electromechanical solutions. Compared with hydraulic technology, Electromechanical Actuators (EMAs) completely avoid the environmental and cost issues associated with the use of aggressive hydraulic fluid, provide significant advantages in terms of reliability and system layout design, and offer a combination of weights competitive with the hydraulic counterpart, especially for low power requirements. Despite these advantages, EMAs are seldom used in flight-control systems and are mostly limited to UAVs or nonsafety critical controls due to their susceptibility to single points of failure, which can cause potentially catastrophic events like the jamming of the aerodynamic surface. Although these issues must be solved by design or through changes in the flight control architecture, the definition of a reliable PHM system would potentially help mitigate the probability of jamming, thus pushing the adoption of EMA technology. Although the literature on the definition of health monitoring schemes for electromechanical actuators and their most important components is fairly extensive [2–5], little can be found on the subject of the implementation and performance of such PHM logics on a real operating platform. There are two main reasons for this: first, the number of flying aircraft equipped with electromechanical actuation systems is still extremely low; and second, it is often impossible to gain access to and record the signals generated by the EMA sensors due to hardware constraints and data propriety restrictions. The lack of in-flight data represents a significant stopgap toward the definition of the PHM algorithm for flight-control systems since it becomes more difficult to fully represent the features uncertainty due to the widely varying operating conditions (temperature, loads, command pattern). To overcome these issues and reproduce as closely as possible the scenario in which the flight controls usually operate, researchers have resorted to highly complex simulation environments [6] and challenging experimental procedures, such as the in-flight test bench [7]. The aim of the ASTIB project (Development of Advanced Systems Technologies and hardware/software for the flight simulator and iron bird demonstrators for regional aircraft) is to bridge the gap between preliminary analysis and full-scale implementation through the design and realization of a technological demonstrator, or iron bird, for a new regional aircraft with fully electrical flight controls and PHM functionalities [8]. In the literature, very few iron birds have been presented, mainly aimed at studying electrohydraulic solutions. Li et al. [9] presented the development of an iron bird for an already in-service regional jet aircraft aimed at validating its troubleshooting plans, making some design improvements, and understanding the root of certain observed failures on the hydraulic nose landing gear system. A similar approach was followed by Spangenberg et al. [10], which developed a hardware-in-the-loop iron bird to perform certification tests during the early stage of aircraft design without constructing prototypes, focusing on electrohydraulic actuators for primary flight-control surfaces. Blasi et al. [11] analysed the control architecture of a modular iron bird oriented to the test of small/medium UAV's electromechanical actuators subjected to realistic in-flight load conditions obtained by a real-time flight simulator. In [12], an iron bird of an F-18 research aircraft was constructed to execute in-flight tests to perform system integration tests, verification, validation, and failure mode analyses for electromechanical aileron actuators. The novelty of the iron bird presented in this paper resides in its implementation: in fact, it has a hybrid layout in which a half-wing is composed of real equipment, while the other half is completely simulated with a hardware-in-the-loop scheme. Furthermore, it not only represents a commercial technological demonstrator but also embraces innovative research goals, being the first iron bird also aimed at the definition of PHM algorithms. This paper deals with the activities pertaining to the development of such PHM systems and their implementation within the iron bird structure and is organized as follows. First, the research objectives, perimeter, and workflow are detailed, framing the research activities and highlighting the expected goals and known limitations of the proposed work. Then, the simulation activities necessary for the definition of the PHM framework are detailed and justified. Such activities include the definition of a high-fidelity dynamic model of the system under analysis needed to study the effects of the most common failure modes on the actuators signals and a real-time declination of such models to quickly generate data. The PHM system is then introduced and tested against several synthetic datasets to provide a preliminary assessment of its expected performances.

2. Research Workflow

The main objective of the "CleanSky2/Astib" consortium was the definition of an iron bird for a regional transport aircraft with fully electrical flight-control systems acting as a technological demonstrator for several new technologies, including PHM functionalities monitoring the health status of the flight-control actuators.

The iron bird followed the scheme provided in Figure 1 [8] and was roughly divided into 2 major sections, namely a "real" half-wing and a "simulated" half-wing. The electromechanical actuators developed during the program were seated within the "real" half-wing and coupled with a set of electrohydraulic actuators controlled in force. The "simulated" half-wing was instead realized through a real-time representation of the very same actuators present in the "real" half-wing (Actuators Simulation Module), while an Engineering Test Station allowed changing the parameters of the simulation and introducing faults or failures within the "simulated" actuators. Both the "real" and the "simulated" half-wings interacted with the Flight Control Computers (FCC), which received input from a real cockpit. A real-time simulation of the aircraft dynamics (Flight Mechanics Simulation Module) dialogued with the FCCs and computed the vehicle attitude considering the behavior of the flight-control system, of the simulated throttle, and of the simulated atmospheric conditions [13]. The choice to divide the FCS into "real" and "simulated" halves resided in the need to simulate the effects that ongoing faults or degradation may have at the component, system, and aircraft level without damaging the iron bird and while reducing the need for spare equipment. The Health Management System Module (HMSM) finally collected the data coming from both the "real" and the "simulated" actuators and ran the PHM routines. The PHM routines were developed following the scheme provided in Figure 2. The case study was first analyzed, assessing which were the constraints (only component-level signals available, system-level signals, and so forth), which were the sensors that could be employed to monitor the health status of the actuators, and which were the most critical failure modes. This information drove the definition of a first high-fidelity model of the system. Such a model was built to represent the component dynamics in detail and was exploited to assess which signals could be useful for PHM, and which features could be reasonably extracted to perform diagnostics/prognostics tasks with the help of an experimental setup [14]. In parallel, a streamlined version of the high-fidelity model, capable of representing the effects of ongoing degradations over the system dynamics and suitable for real-time deployment, was prepared. The goal of such a model was twofold. On the one hand, it was prepared to be deployed within the Actuators Simulation Module ("simulated" half-wing) of the iron bird. On the other hand, it could be used offline to quickly produce the datasets required to test the PHM routines before their installation on the iron bird, thus allowing the code to be debugged off-site and providing a preliminary performance assessment prior to the iron bird commissioning.

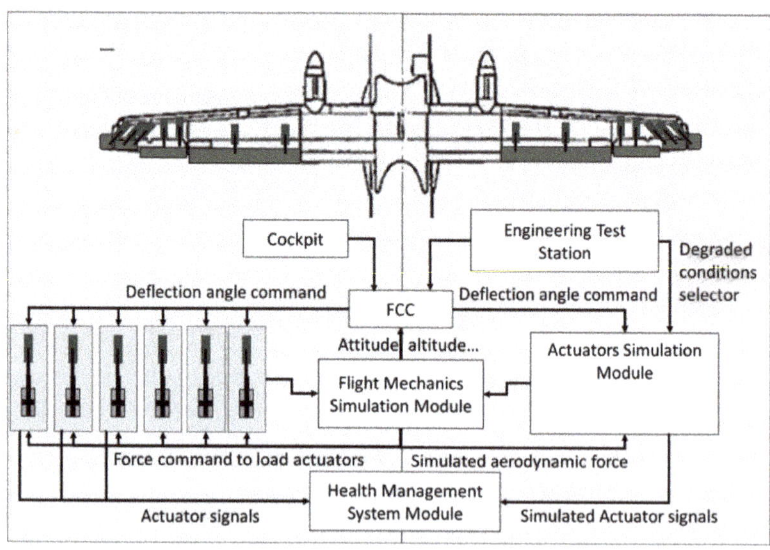

Figure 1. Iron-bird schematics.

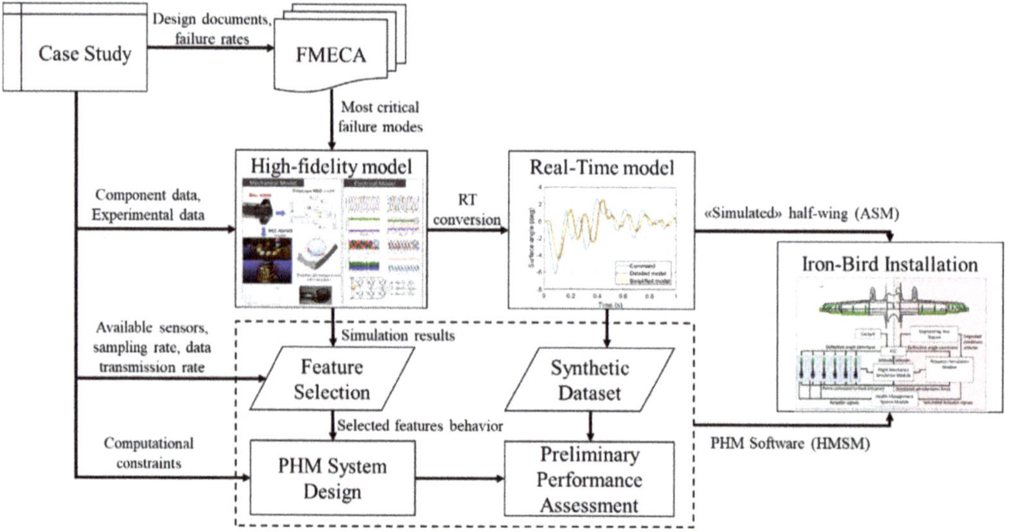

Figure 2. Research program workflow.

3. Preliminary Activities and Simulation Environment

Prior to the definition of the PHM system, a number of preliminary activities were performed to better assess the perimeter of the research program, define the architecture and the type of the simulation models to be employed (e.g., physics-based, black-box, surrogate), and how to translate such simulation models to make them suitable for deployment in the real-time environment of the iron bird. This section is intended as a summary of such preliminary activities, highlighting the most significant considerations for each passage and referring to the literature for additional details.

3.1. Approach and Design Goals

Before presenting the case study and proceeding to the description of the simulation models prepared for the project, it is necessary to detail the scope, objective, and limitations of the research activities. The main research goals pursued with the definition of the PHM system was to provide a proof of concept of a prognostic framework, demonstrating its feasibility within the constraints of a civil aviation application in terms of signals management and computational constraints, while assessing the maturity level of the algorithms available in the literature. As such, it is underlined that the objective of the research activities was not to provide a high-TRL solution, such as a PHM framework ready to be deployed in the operational scenario. In particular, the experimental activities supporting the definition of the high-fidelity models and the real-time models did not involve the study of real actuators under degraded health conditions. Such constraints could be justified considering that the actuators mounted on the iron bird were prototypes produced in very limited numbers, meaning that no spare actuators could be brought to failure during the experimental activities. Faulty conditions were studied through simulations only, through the degradation models available in the literature. It is worth mentioning that even if mutual interaction between different degradations is potentially available, in the context of the present work, only 1 fault at a time was considered for the definition of the PHM framework. Although the results obtained through this approach are expected to be realistic, it is worth noting that the complete validation of the PHM system falls outside the perimeter of the project.

3.2. Case Study and FMECA Analysis

The flight-control system considered for analysis is based on a set of new direct-drive Electromechanical Actuators (EMA) responsible for the movement of a few secondary flight-control surfaces (active winglet and wingtip). As depicted in Figure 3, each actuator was driven by a Brushless-AC (BLAC) electric motor and controlled through 3 nested control loops, each monitoring the motor currents, its angular frequency, and the actuator linear position. A ball screw integral with the motor shaft was used to transform the rotary motion imposed by the BLAC into the sliding motion of the end user. Spherical joints connected the actuator to the aerodynamic surface on 1 end and to the airframe on the other. Each actuator came with a number of sensors used for control purposes; phase currents were measured, while a resolver was mounted on the motor shaft to monitor its position, dictate the phases commutation, and infer the angular speed. An LVDT was instead integral with the ball screw and used to precisely measure the position of the end-user, providing this information to the associated control loop. Phase voltages were also acquired. Such signals were then considered available for the development of the PHM system and the basis on which to evaluate the possible features and health indexes.

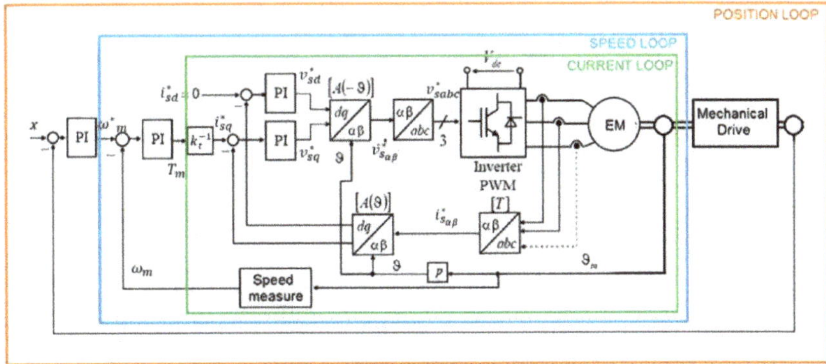

Figure 3. Architecture of the case study.

Once the available sensors were identified, guidelines provided in [1,15] suggested that an FMECA analysis tailored for PHM be performed, with the objective of assessing which failure modes needed to be prioritized when designing the PHM system. The most significant failure modes for each component were studied and their causes, symptoms, and failure effects were detailed, while a composite score was computed based on the fault's frequency of occurrence, the severity of effects, expected observability, and replaceability of the component. The results were hence ranked, and a priority list was derived, stating which failure modes were more interesting or more probably observable by a Health Monitoring framework. Such operations are described in more detail in [16] and allowed to select the following failure modes. For the electrical motor, 2 failure modes were selected: the occurrence of turn-to-turn shorts in the motor windings and the degradation of the motor's permanent magnets [17,18]. The reasoning behind this choice is that the first was the most probable failure mode involving short circuits [16], while the second was critical from a severity perspective since it involved the loss of the actuator damping capability, which prevented flutter in the case of a sudden loss of power. Two additional failure modes pertaining to the motor were addressed for fault detection only, due to their causes and dynamics: the occurrence of static eccentricity in the Brushless-AC motor and the occurrence of MOSFET Base-Drive Open circuit conditions within 1 leg of the inverter supplying the actuator. The first is related to mistakes during the assembly and is tracked to avoid the triggering of a false alarm on the "real" half-wing of the iron bird, while the second is often indicated as one of the most probable failure modes involving PWM-driven inverters [19,20]. Switching the attention to the mechanical drive, 3 failure modes were assessed as preferential. The first was the effect of lubricant degradation, and the consequent efficiency losses within the ball screw [21]. The other 2 were the occurrence of backlash due to wear within the mechanical transmission and within the spherical rod end connecting the actuator to the aerodynamic surface. The lubricant degradation was chosen since it was expected to be a frequent and inevitable occurrence over prolonged usage, while the progressive increase of backlash was selected due to its expected frequency of occurrence and severity of the possible effects, especially when considering the rod end [22]. All of the selected failure modes were expected to be observable or detectable.

The turn-to-turn short failure mode is covered in the literature [3,16], and several studies are available on the topic of magnet degradation and its detection [18,23]. Similar considerations can be taken for the selected failure mode for the MOSFET [20], the occurrence of static eccentricity [24], and the issues associated with lubricant degradation and backlash opening [25,26].

3.3. *High-Fidelity Modeling*

As previously stated, the creation of highly detailed mathematical models was paramount to studying the effect and evolution of selected failure modes on the different components of the electromechanical actuator without involving the real actuators' damage. The high-fidelity models are expected to represent the real behavior of each component under investigation under both nominal and degraded conditions. In such a way, the model assumed the role of a virtual test bench on which to inject artificial flaws to study their effects on the available measurable signals.

The mathematical models were created following a complete white box approach, where the dynamics of every phenomenon were derived and dictated by physical laws and equations.

3.3.1. Electric Drive

The dynamic model of the electric drive was made of 3 main subsystems: the Electronic Control Unit (ECU), the Electronic Power Unit (EPU), and the brushless electric model. The model of the Electronic Power Unit made use of a functional description of the PWM-modulated inverter. It received the commands from the ECU, which was responsible for the motor current control in the d-q-0 axis and modulated the electric power exchanged with the motor. The d-q axis control featured PI regulators receiving as input the current command and the filtered current feedback subject to Park's transformation. The output of the controllers was then transformed back to the 3-phase system and used inside a PWM modulator based on a triangular bipolar wave carrier that generated the digital control signal for each of the 3 commutation poles, thus modulating the DC-link tension and generating the 3 motor phase voltages as per [27]. The motor model was derived from that adopted in [16]. As depicted in Figure 4, it was based on the representation of the 3-phase dynamics through a lumped parameters representation, where each electrical parameter is a variable as a function of the windings' temperature, itself computed at each time step as a function of the thermal power generated through the Joule effect and commutation losses and the external temperature according to a simplified first-dynamic order model of the motor's thermal dynamics. The introduction of faults within the model was achieved according to the literature models. In particular, the occurrence of a turn-to-turn short was described according to [2], where the fault process caused the progressive decrease of 1 phase resistance, auto-inductance, and mutual inductance.

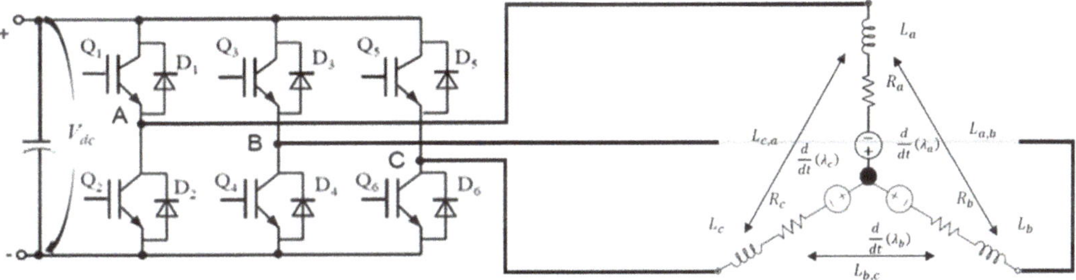

Figure 4. Schematics of the Electric-Drive model.

Such a failure mode was mainly caused by the progressive degradation of the insulating material separating the coils from each other. Such a degradation process was mainly driven by thermal issues and was modeled according to a modified Arrhenius Law as proposed by [28]. As evidenced in Figure 5, its occurrence caused asymmetric behavior between the 3-phase currents and increased a common node current which is not present under nominal health conditions. Similar considerations can be performed for the other considered failure modes. The progressive degradation of the motor permanent magnets was simulated by progressively reducing its magnetic flux term, while the occurrence of eccentricity was represented by changing the flux term according to the expected gap size variation as a function of the rotor angular position.

3.3.2. Mechanical Transmission

For what concerns the mechanical part of the EMA, particular attention was paid to the mechanical transmission between the rotatory motion of the electric motor and the linear displacement required to move the aerodynamic surface, i.e., the ball screw. It was mainly composed of a screw shaft, 1 or 2 nuts, and a set of spheres that allowed the mechanical efficiency of this component to reach extremely high values due to the replacement of sliding with rolling friction. To accurately describe its behavior and performance, a 3-dimensional modeling approach was necessary. In fact, in order to transmit the motion, the spheres rolled between the screw shaft and the nut's grooves along a helical path. Moreover, a recirculating insert was present to maintain them within the nut body. As a result, the motion of each sphere was governed by an intrinsically 3D contact pattern.

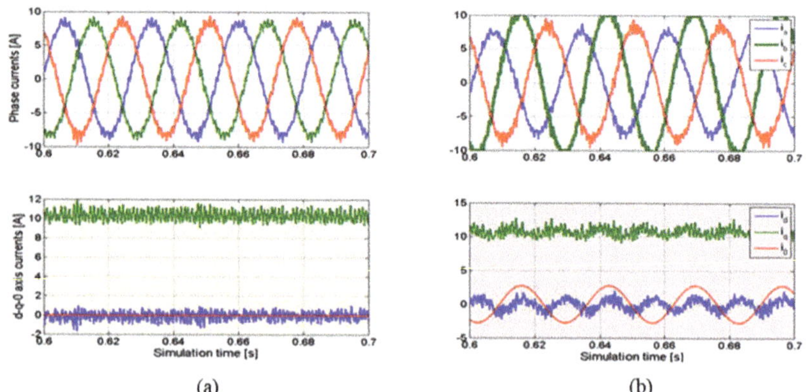

Figure 5. Effect of an occurring turn-to-turn short in a BLAC drive. (**a**) Healthy condition. (**b**) Developing short.

The developed dynamic model, whose main screen is shown in Figure 6, considers all these peculiarities of the mechanism through a multibody approach, integrated with a highly detailed description of the contact conditions. In fact, a generic gothic arch profile of the raceways was considered; hence each sphere could enter in contact with each groove in a maximum of 2 points simultaneously, 1 for each circular side of the profile. The contact force and footprint extent were calculated at each time step by means of a penalty method with variable contact stiffness, dependent also on the contact angle. The contact parameters were calculated according to the approximated Hertzian theory [29]. The latter allows sensibly reducing the computational time with a precise closed form solution and without the need to run implicit iterative calculi at each integration time step. The normal contact force was obtained by the extent of the contact constraint violation and was composed by an elastic and a dissipative term, dependent on the approaching speed, related to energy dissipation that occurs during the deformation of the material.

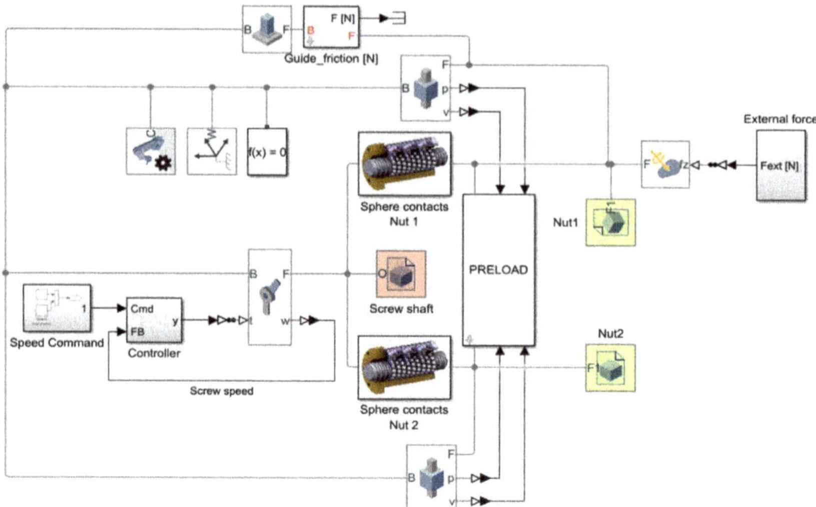

Figure 6. Main screen of the Simscape Multibody high-fidelity model of the ball screw component.

To avoid unrealistic discontinuities in the contact force, the elastic component was also used as a saturation value for the damping part.

From the evaluation of the relative sliding speed in each contact point and of its direction, the friction forces were calculated. Although sliding friction was replaced with rolling friction, a little amount of slippage always occurs due to the elastic deformations of the bodies in the contact area and to the kinematics of the mechanism itself. Therefore, ball screws are usually lubricated with grease to create a proper separation of the mating bodies to avoid wear on the rolling surfaces and fatigue damage. The developed high-fidelity model took into account the presence of grease lubrication, considering the effect of its base oil, which is the main actor in the surface separation. The lubricating media rheology was considered by means of Roeland's model for the viscosity dependence on shear stress and temperature, and of the Eyring model, which describes the nonlinear dependence of the shear stress from the shear rate in the lubricant film [30]. It depends on the film height that, following the Grubin approach [31] for the sake of simplicity, was obtained by the solution of the Nijembanning approximated formulation [32].

Generally, as in rolling bearings, elastohydrodynamic lubrication regime takes place in the contact regions; therefore hydrodynamic rolling resistance forces, rolling hysteresis torques, lubricant pressure components, and microslip losses were taken into account according to [33], as well as the spinning friction torque. Finally, for low entrainment speeds and/or high normal loads, the 2 surfaces had direct contacts: the model considered the variability of the lubrication regime, calculating the sliding friction force as a weighted sum of the dry and full film lubrication forces based on the Tallian parameter [34]. When the mating surfaces of the spheres and the grooves are sufficiently separated, almost no wear phenomena occur, while in the mixed and boundary lubrication regimes, wear becomes not negligible. The grooves' wear is one of the most important issues in ball screws and is detrimental to their positioning accuracy as it causes preload loss, increased vibration, and, eventually, complete backlash [35]. Indeed, it was included as one of the ball screw degradation models through the well-known Archard equation, calculating the volume worn in each sphere/groove contact point at each time step and summing all the contributions to obtain a uniform mean wear of each raceway. The Archard coefficient was assumed to be dependent on the lubrication regime, a function of the Tallian parameter [36].

The probability of contact of the mating surfaces increases with the component's usage. In fact, grease aging involves a degradation of the lubricity property of the base oil in time, i.e., viscosity reduction, meaning a decrease of the load-bearing capacity and a thinner film thickness, with increased direct contact probability and wear. Furthermore, particularly at high speeds, film thinning is also caused by lubricant starvation, which was considered following the approach presented in [37,38]. The adopted lubricant aging model calculated the reduction of viscosity as a function of the entropy generated within the lubricant due to internal mechanical shearing [21,39].

Figure 7 depicts the results of a study executed with the ball screw high-fidelity model on a ball screw with a 5 mm lead and a nominal diameter of 16 mm, directly connected to the electric motor and the grease, lubricated and subjected to different external loads levels while operating at various rotational speeds. Lubricant starvation, wear of the grooves, and lubricant aging were considered. The analyses were aimed at investigating the effect of the latter phenomenon on the macroscopic behavior of the component. It can be seen from the plot that higher external forces, combined with low speeds, lead to higher wear rates and, ultimately, to a shorter operative life. In the case of excessive wear, backlash can occur, as shown in Figure 8, where the effects of an excessive backlash are depicted for what concerns the speed and position. The left graph depicts the linear position of the nut and the equivalent linear position of the screw obtained considering the ideal transmission ratio, i.e., the 2 lines should overlap. The position of the screw shaft, being speed controlled, follows a sinusoidal trend; however, because of the presence of backlash, the nut fails to replicate its positioning and stops close to the motion reverse points. The internal spheres lose contact and engage again when the relative position reaches the backlash gap. This can be more easily observed in Figure 8b, which shows the equivalent rotational speed of the screw and the nut. At motion reversal points, due to the friction on the linear guides and the presence of backlash, the nut speed settles at 0 and remains steady until the backlash is closed; at this time, the screw's speed also oscillates because of the new engagement.

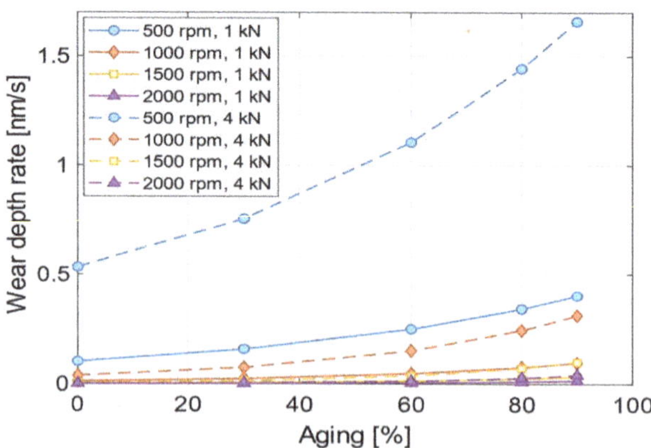

Figure 7. Wear rate versus aging level for different speeds and loads for a ball screw with 5 mm lead and 16 mm nominal diameter (© 2022 IEEE. Reprinted, with permission, from [21]).

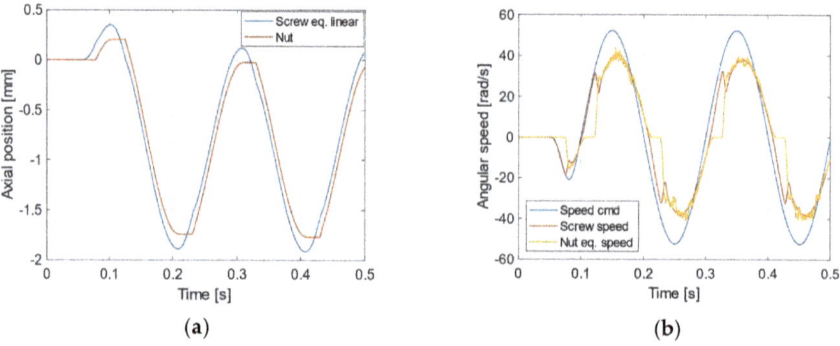

Figure 8. Simulation results of a ball screw with a high backlash of 0.3 mm. (**a**) Comparison between the nut position and the equivalent linear position of the screw shaft; (**b**) Comparison between the equivalent angular speeds of the screw shaft and the nut (© 2020 ASME. Reprinted, with permission, from [33]).

3.3.3. Rod-End

Another important element of mechanical transmission that has been an object of study is the spherical rod-end, which connects the actuator to the airframe and the aerodynamic surface, performing a critical safety function at the aircraft level. Rod-ends are subjected to different failure modes, such as crack openings, lubricant degradation, surface denting, and so forth. However, the most common failure is represented by the wear of the internal surface of the spherical joint, with the consequent increase of backlash, which, if outside the acceptable range, could result in a significant accuracy reduction in surface positioning. Furthermore, a significant rise in friction forces within the spherical joint, due to lubricant degradation and improper lubrication regimes, can lead to excessive bending stress on the actuator. This component has been studied in conjunction with a high-fidelity model of a complete actuator to also investigate the effect on its expected performance [40]. To simplify the model definition and prepare for more complex models, some simplifying assumptions were made. This involved reducing the 3D spherical joint geometry to a 2D cylindrical approximation, ignoring misalignment and lubricant leakages. The rod-end model considered 3 different operating conditions based on the relative position between the shaft and bush. As the clearance decreased, the system transitioned from hydrodynamic lubrication to mixed and then to direct metal-on-metal contact. As for the ball screws, grease lubrication was considered and, therefore, the lubricant media was represented by its base oil. Conducting a PHM study on aerospace rod-ends requires defining models that are both computationally efficient and accurately reflect the component's physics; for this reason, a finite difference scheme has been selected as a resolution method of the simplified Reynolds equation. To streamline the simulation and avoid the need for iterative solvers, the Half-Sommerfeld solution was chosen to describe cavitation regions, while the viscosity was computed at each FDM node, dependent on the temperature and local pressure according to the Vogel-Barus equation [31]. Depending on the operative conditions, the model was able to represent the variation of the lubrication regime from full film to mixed and boundary lubrication, monitoring the local Tallian parameter and weighting the results from the hydrodynamic and direct contact models.

Figure 9 illustrates the effect of the rod-end backlash on the final surface positioning when the control surface is moved by 2 parallel servo-actuators according to the active-standby control strategy. It can be observed in Figure 9a how the backlash in the rod-end due to excessive wear causes important variation in the deflection of the control surface, which can lead to increased surface vibration, damage, and controllability issues. In particular, it can be seen from Figure 9a that the equivalent surface displacement considering surface kinematics in the presence of backlash (green line) differs from that with no backlash (red line) by approximatively 2 mm, which corresponds to half the width of the imposed backlash for the current simulation for the fully deflected surface condition; while the surface position varies by about 0.6 mm for null commands of the active actuator, mainly due to the presence of external aerodynamic loads, internal damping, and the attachment with the second stand-by actuator.

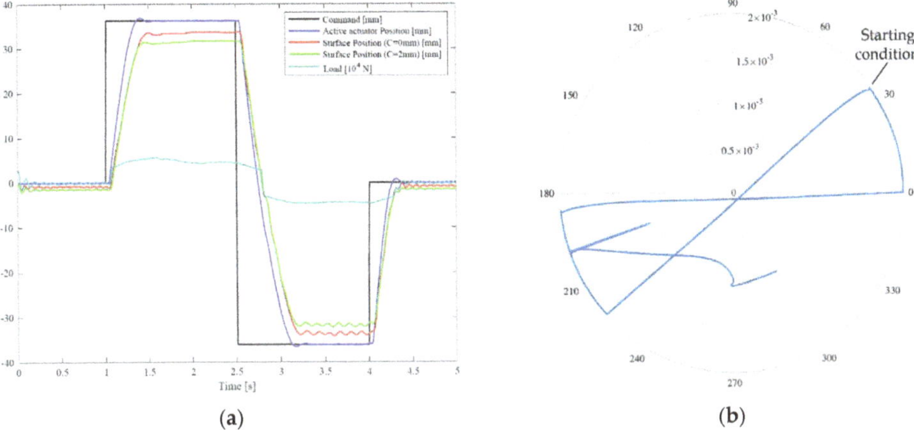

Figure 9. Simulation results of a rod-end with a high backlash of 2 mm connecting 2 servo-actuators to the control surface: (**a**) Pin trajectory within the rod-end housing during the simulation; (**b**) Comparison between the surface position obtained with (green) and without (red) backlash in the rod-end (© 2021 ASME. Reprinted, with permission, from [40]).

Figure 9b depicts the motion of the internal sphere within the rod-end housing during the simulation and shows how it impacts the bushing's race, contributing to damage progression and rod-end degradation. The model gives in the output the normal and frictional contact forces and the wear information, which are useful insights to understand the component's behavior under various working conditions, helping in the definition of a suitable PHM framework.

3.4. Real-Time Modeling

The abovementioned high-fidelity models are accurate, metaphysical, and extremely detailed, based on the physical descriptions of the involved phenomena. However, the high level of detail counterposes the computational constraints required to run PHM routines. In fact, such models are very computationally intensive and are not suitable for prolonged simulations of a complete servosystem. With the aim of streamlining the computational time to make them available to generate a large dataset of nominal and degraded system responses and to be used in real-time simulations on the iron bird, several reduction techniques were applied to the models to obtain a simplified EMA model capable of reproducing the main features required by the iron bird to enable the simulation of a complete flight, i.e., the position of the actuator and control surface. The various considered degradations were then inserted into the model with equivalent simplified models; they aimed to simply reproduce the effects of each fault progression on the relative signals which were seen as being affected by such degradations during the offline phase, according to the detailed results obtained with high-fidelity models. Because of the practical implementation of the iron bird and the performance of its hardware components, the real-time (RT) model must comply with the time constraint of a maximum integration frequency of 12 kHz.

Hence, only the more important dynamics were considered in the RT model to balance the need for realism with the computational effort. It is worth highlighting that not only 1, but multiple EMA models should be run simultaneously to enable the correct functionality of the iron bird; therefore, the actual model complexity must remain as low as possible to avoid CPU overloads. Furthermore, jitter usually occurs in the RT hardware, so a certain amount of security margin must be considered.

Following these constraints, the RT model was assembled, as shown in Figure 10. It was composed of several subsystems, each representing an interconnected subcomponent reflecting the physical arrangement. The main blocks represent the Power Drive Unit, controller, electric motor, mechanical transmission, and control surface. A number of simplifying assumptions were made to speed up the simulations. The 3-phase electric motor was modeled with the equivalent d-q axes representation, exploiting the Clarke-Park transformation. Also, the PDU was represented by a 3-leg inverter with conduction losses dependent on the temperature. The ball screw and rod-ends were contained in the mechanical transmission subsystem, describing the motion transformation from rotational to linear considering mechanical efficiency, stiffness, and friction. The latter was modeled with the fast and simple Karnopp model to maintain the computational requirements as low as possible. The surface's block enclosed the description of the variable lever arm which allowed the linear displacement of the actuator to create a controlled rotation of the aerodynamic surface. The system was then basically reduced to a 2-body lumped mass model: 1 representative of the control surface and the other of the entire mechanical drive from the rotor to the translating part of the ball screw. The rod-end connection stiffness was considered in the connection point between the 2 lumped masses. All the sensors were represented with dynamic transfer functions and, together with the controller states, were sampled at the various control loops' frequencies. A simple single-body thermal model was used to estimate the global temperature change of the EMA during the simulation.

The RT model was created according to the following principles:

- Replace expressions with high computational cost with approximated low-order forms,
- Minimize the number of integrators and state variables,
- Reduce or avoid repeated calculi or Boolean checks,
- Optimize the C++ executable creation options for RT target, and
- Substitute continuous-time integrators with discrete-time state variable.

As stated in the previous paragraphs, 7 faults were considered, 5 of which described a progression in time, and 2 (i.e., motor's static eccentricity and PDU's MOSFET base drive open circuit) were imposed a priori before the simulation started. The introduction of the selected failure modes within the real-time models was achieved in 2 separate ways, depending on the model modifications required to adapt the high-fidelity model to the real-time framework requirement. The 1st type was represented by those failure modes whose progression could be successfully described by the reduced-order model, such as the magnet degradation, the occurrence of rotor eccentricity, and the degradation modes affecting the mechanical transmission. In such cases, the very same models adopted for the high-fidelity model were ported within the real-time simulation with minimal adjustments. For other failure modes, including the turn-to-turn short and the MOSFET base drive open circuit, a different approach was needed; the physics-based representation of such degradation process required a 3-phase dynamic model of the motor, while the real-time representation worked on a streamlined d-q axis description.

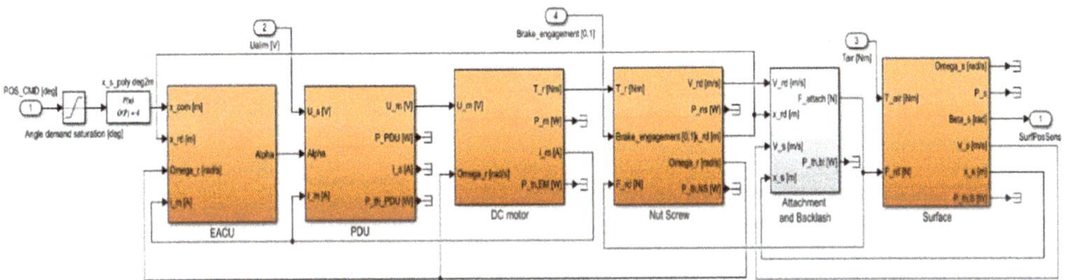

Figure 10. Main screen of the RT model.

Starting with the latter case, the approach was to map the effects that the degradation mode had on the quantities represented in the real-time model through the high-fidelity environment, then proceeding to modify the signals of the real-time model according to such maps. To better explain this approach, please consider the case of the turn-to-turn short and the scheme provided in Figure 11. The real-time model described the nominal health behavior of the motor according to a d-q axis description, computing the expected windings temperature T_w, according to a streamlined thermal model of the motor. Such temperature was then used to feed the turn-to-turn short model, which computed the expected variations on each of the 3-phase currents according to the fault mapping.

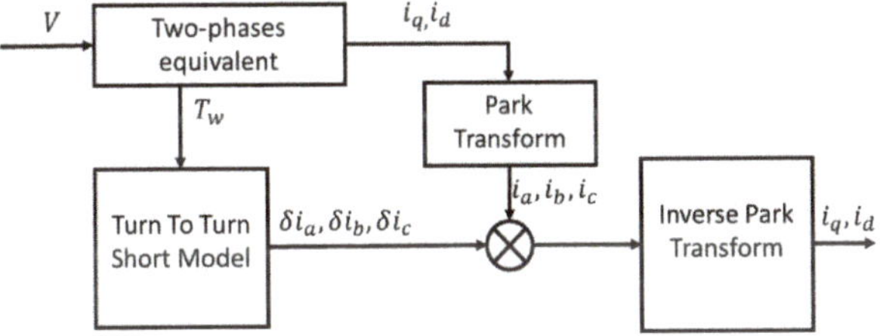

Figure 11. Implementation of the turn-to-turn short in the real-time model.

Such variations were then applied to the 3-phase currents ($i_a, i_b,$ and i_c) derived from the direct, quadrature currents computed by the real-time model. Once the variations were applied, the updated d-q axis currents were obtained through the inverse Park transformation.

The faults of the mechanical transmission included the wear of ball screw and rod-end and the lubricant degradation. The latter, as seen, also caused an increment in the wear rate, but, as previously stated, mutual excitations of different failure modes were ignored in this work for being outside the original scope of the research project. Wear was considered by means of the Archard equation. The effect of wear is a backlash increase that was modeled by means of a double-sided elasto-backlash with hysteretic damping based on the relative position of the 2 bodies. Since only 2 bodies represented the mechanical dynamics of the system, the backlash models for both the ball screw and the rod-end were condensed into a single backlash located in the rod/surface joint. A smart approach was implemented to differentiate the 2 situations, consisting in sending the actual rod position back to the controller if no ball screw backlash was considered, while conveying the linear position on the surface's side of the backlash if the degradation was introduced. Despite this, the backlash of the rod-end is normally taken into consideration in the model. As a result, the effective backlash seen by the RT model, concentrated in the rod/surface connection, was the sum of the 2 backlashes in series. The effect of lubricant degradation was obtained with a mechanical efficiency reduction in the RT model.

Figure 12 shows the results obtained with the RT model when a surface position command is imposed to the EMA with aerodynamic loads dependent on the surface deflection. To highlight the effect of 1 of the selected degradations, 2 simulations with low and high levels of backlash in the rod-end are shown; for the second case, the surface position exhibits flutter oscillations for little commands due to the excessive backlash under the action of the external loads. The discrepancies between the detailed and simplified models emphasized during the highly dynamic phases of the simulation, i.e., the first part, while tending toward 0 for steady conditions, assumes an RMS value of the absolute error between the results of the RT model and the detailed model, shown in Figure 12, of 0.23 deg, which is a reasonably small value and an acceptable compromise when considering the computational constraint imposed by the hardware implementation. The presence of the backlash model does not introduce substantial variations to the RMS error.

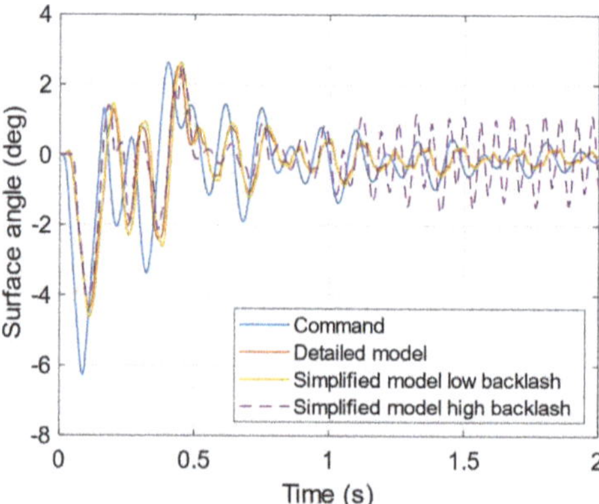

Figure 12. Comparison between detailed model and RT model with low and high levels of backlash.

4. PHM Framework Design

Once the simulation environment, both in its high-fidelity and real-time declination, is defined, it is possible to use the data generated through repeated simulation cycles to inform the definition of the PHM system, study the expected behavior of the actuators under degraded health conditions, and assess which signals to use for health monitoring. This section details the approach followed in the definition of the PHM system, starting from the definition of the operational scenario.

4.1. PHM Framework Design-Preliminary Operations

The design of the PHM framework followed the procedures suggested by [1], where the high-fidelity model is first used to study the effects of the injected failure modes on the system performances, inform the feature selection process, and then lead to the choice of the fault diagnosis and failure prognosis routines.

The feature selection process was performed through the analysis of the simulation results, leading to the definition of a pool of possible feature candidates. Such feature candidates were then ranked according to metrics such as correlation with the failure mode behavior, signal-to-noise ratio, and precision. During this first stage of the design process, it was paramount to properly characterize the uncertainty surrounding the simulation results, including the effects that external disturbances, such as the external temperature, load, and command patterns, have on the system performance and on the possible feature candidates to obtain a statistically representative database of the possible operating conditions.

For the case study under analysis, this translates into the definition of a complete operative scenario, following the scheme provided in Figure 13. A sequence of position commands and aerodynamic load, representative of the expected operating cycle of the flight-control actuators under analysis, was provided by the industrial partners of the project. The operating cycle can be represented as a pair of (t, x_{set}) and (t, F_{ext}) time series where t is the time vector, x_{set} is the position command, and F_{ext} is the external force. Before each simulation, the sequence was warped and modified through the following equation

$$\begin{cases} t = t\,\sigma_t \\ x_{set} = x_{set}\sigma_x + \sigma_{x0} \\ F_{ext} = F_{set}\sigma_F + \sigma_{F0} \end{cases} \tag{1}$$

where σ_t is a random number drawn from a normal distribution with mean 1 and a standard deviation equal to 0.1. σ_x and σ_F are again drawn from a normal distribution with mean 1 and a standard deviation of 0.3, while σ_{x0} and σ_{F0} are chosen from a normal distribution with a 0 mean and standard deviation equal to 10% of the actuator stroke and of the nominal aerodynamic load, respectively.

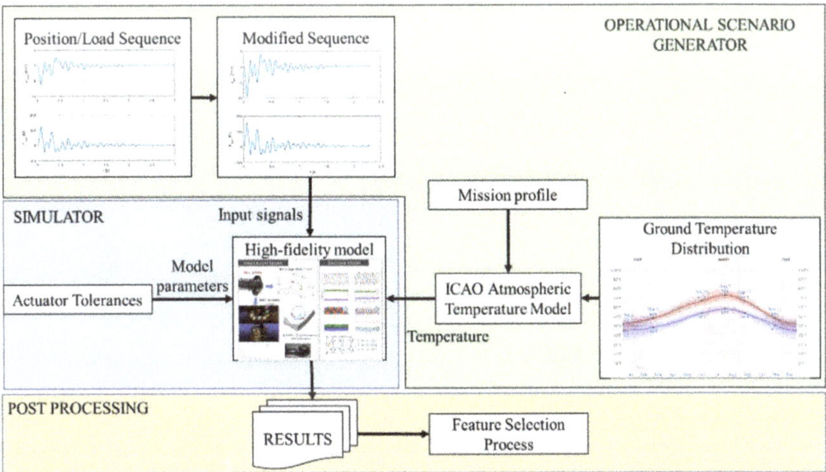

Figure 13. Data generation process to support PHM design.

Ten possible aircraft were then selected by imposing small perturbations, compatible with the expected production tolerances, on the actuator parameters. Of these aircraft, 4 are supposed to operate in temperate climate conditions, 3 in cold conditions, and 3 in hot conditions. Three reference location were then chosen: Turin for temperate conditions, Abu Dhabi for hot conditions, and Vancouver for cold conditions, and historical temperature distribution for each were saved. For each simulated flight, the temperature varied between the ground value, randomly drawn from the reference location datasets, and −40 °C, which is the expected temperature during cruise for the considered aircraft type. A total of 100 flights for each aircraft were simulated under nominal health conditions. Degradations were then injected and their progression was simulated through accelerated fault-to-failure model. Results were then analyzed to search for the most significant effects of each failure mode and provide the basis for the feature selection process. Among a pool of more than 50 candidates, 1 preferential feature for failure mode was chosen according to correlation and signal-to-noise ratio scores. A brief overview of the considered failure modes, along the signals required to compute the associated feature, is reported in Table 1. Between these failure modes, due to their interplay, 4 were analyzed in more detail: the occurrence of turn-to-turn short, magnet degradation efficiency loss, and wear-induced backlash in the mechanical transmission. This choice was justified by looking at the other failure modes. The occurrence of MOSFET Base Drive Open Circuit was expected to be a fast-evolving failure mode, thus, was treated only for fault detection and isolation. Similarly, the occurrence of a static eccentricity within the motor was traced to misalignments or mounting errors persistent in time and not to slowly evolving degradation. The wear in the spherical joint was finally tracked through a feature that was not affected by the occurrence of the other failure modes. Figure 14 depicts the correlation of the chosen features for the 4, possibly interplaying, failure modes against the progression of each considered degradation. As anticipated in Section 3, the main symptom of the occurrence of a turn-to-turn short in the electric motor windings was the formation of a growing asymmetric behavior between the 3-phase currents. The feature F_{TTS} exhibited high correlation marks as expected, while the other 3 selected features were less affected. A very loose correlation occurred with the feature F_{MTEL} due to the reduction of the motor efficiency due to the ongoing short circuit.

Table 1. Selected features.

Component	Failure Mode	Feature Symbol	Signals	Feature
Motor	Turn-to-turn short	F_{TTS}	Phase currents Phase voltage	Variance of the common node current over the average phase voltage
Motor	Magnets degradation	F_{DMD}	Phase currents	Distance between the mean root square of the phase currents from an expected baseline
Motor	Static eccentricity	F_{SE}	Phase currents Motor shaft position	Periodic disturbances over the phase currents signals
EPU	MOSFET Base Drive Open Circuit	F_{BDO}	Phase voltage	Phase voltage standard deviation
Mechanical transmission	Efficiency loss, lubricant aging	F_{MTEL}	Phase currents Expected aerodynamic load based on deflection angle	Efficiency estimate at still actuator
Mechanical transmission	Wear induced backlash	F_{MTWEAR}	Rotor shaft position LVDT measurement	Difference between LVDT output and shaft position
Spherical joints	Wear-induced backlash	F_{REWEAR}	LVDT measurement from 2 actuators on the same aerodynamic tab	Difference between LVDT outputs

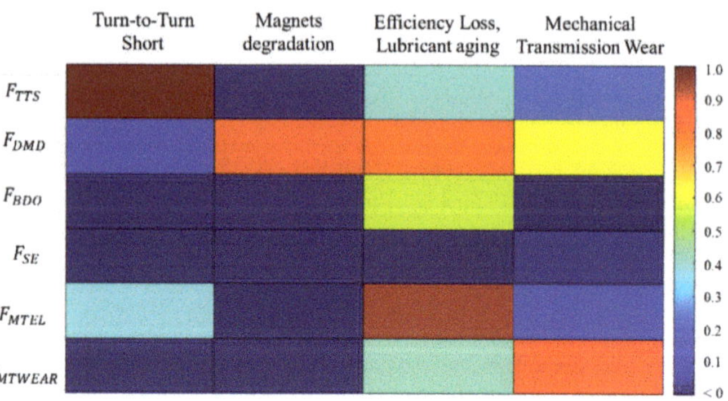

Figure 14. Feature correlation against several failure modes propagation.

The occurrence of a distributed magnet degradation is associated with good correlation indexes for the associated feature F_{DMD} only. The occurrence of efficiency loss in the mechanical transmission can instead affect more features, showing the highest correlation value with F_{MTEL}, while also exhibiting a high correlation with F_{DMD} due to an increase in the absorbed current. This result is, of course, suboptimal but does not represent a critical issue: F_{MTEL} shows good correlation only for the efficiency-loss case and is not significantly correlated with the occurrence of magnet degradation. The opening of an increasing backlash in the mechanical transmission is also highly correlated with its own associated feature (F_{MTWEAR}). The feature F_{DMD} is also affected due to the current spikes caused by the impacts which originate whenever the freeplay is recovered. Its correlation is, however, lower than that associated with F_{MTWEAR}, while the correlation of F_{MTWEAR} with the occurrence of magnet degradation is negligible. As such, the selected feature set is expected to be suitable for fault diagnosis and failure prognosis.

4.2. PHM Framework Design-Framework and Algorithms Choice

The PHM framework was designed considering the need to keep the computational effort required to perform the diagnostics/prognostics tasks at a minimum to comply with the requirement of being installed on board the iron bird. As such, the scheme provided in Figure 15 was adopted, where the data coming from the iron bird are at first checked to verify their coherence, used to evaluate the features at any time, and then sent to the fault-detection routine.

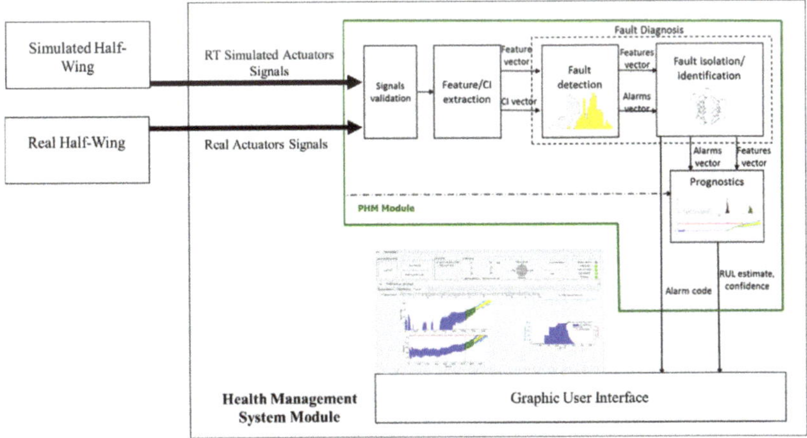

Figure 15. Architecture of the proposed PHM Framework.

Before being analyzed, the features were at first downsampled (80 Hz, against the 800 Hz of the iron bird signals) to reduce the computational effort and then preprocessed by an exponential filter to improve their signal-to-noise ratio. The fault-detection algorithm was based upon a simple, purely data-driven logic, comparing at each timestep the running feature distributions against the baselines obtained for health conditions. The remaining parts of the PHM algorithms were switched off. If the running distribution of 1 (or more) features deviated excessively from their own baselines, an alarm was triggered and the remaining routines of the PHM frameworks were activated. Such a solution has the benefits of keeping the most computationally expensive routines dormant until they are needed, hence, reducing the computational burden. An example of the fault-detection algorithm behavior is highlighted in Figure 16, where the running distribution of the feature overcomes a separation threshold, triggering the corresponding alarm of a fault declaration confidence equal to 0.98. In this instance the fault-detection algorithm observes a deviation of the feature associated with the occurrence of a turn-to-turn short in the motor windings, prompting a fault declaration once the confidence threshold is met.

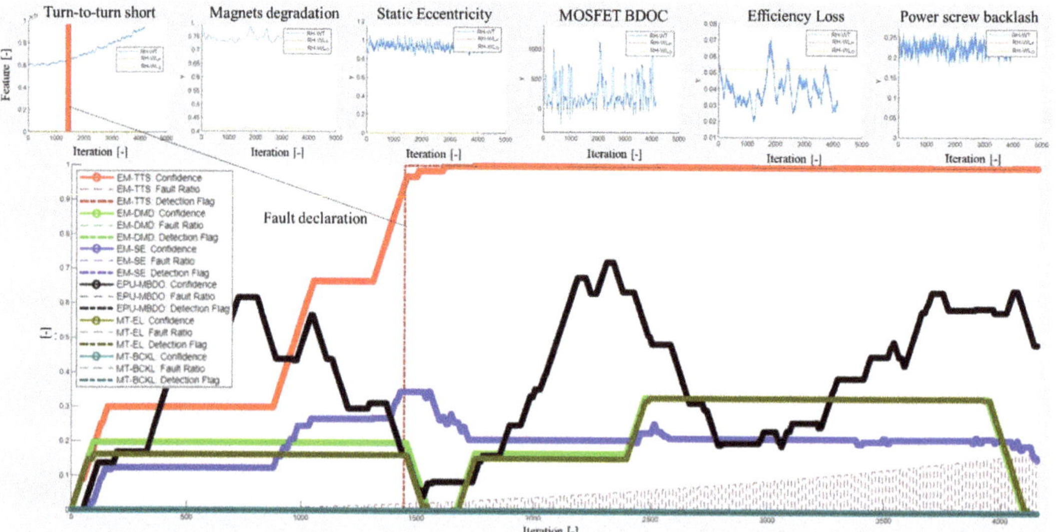

Figure 16. Output of the fault-detection algorithm in response to fault occurrence in the winglet actuator.

If a fault is detected, the fault classification algorithm is called to assess which failure mode is occurring (or is being simulated) within the flight-control actuator. Given the preliminary nature of the study, only the case where 1 degradation may occur at any given time was considered. As such, a simple Linear Support Vector Machine, operating over the features vector, was considered. The LSVM was trained with a randomly chosen subset representative of 70% of the data generated through the high-fidelity model, while the remaining 30% was used for verification purposes.

Results of the training process are aimed at a total accuracy rate higher than 95%. Once a fault was detected and classified, the information was sent to the prognostic algorithm.

The prognostic routine was based on the particle-filtering structure, depicted in Figure 17, to infer the fault severity and forecast the degradation growth [41]. The particle filter scheme tracks the fault progression by iterating at each time stamp 2 consequential steps, the first being the "prediction" stage and the latter being the "filtering" stage. The prediction step combines the knowledge of the previous state estimate $p(x_t|y_{t-1})$ with a process model $p(x_{0:t-1}|y_{1:t-1})$ to generate the a priori estimate of the state probability density functions for the next time instant,

$$p(x_{0:t}|y_{1:t-1}) = \int p(x_t|y_{t-1})p(x_{0:t-1}|y_{1:t-1})dx_{0:t-1} \tag{2}$$

This expression usually cannot be analytically solved; the Sequential Monte Carlo algorithms can be used in combination with efficient sampling strategies for such purpose [42]. Particle filtering approximates the state probability density function through samples or "particles" characterized by discrete probability masses, or "weights", as,

$$p(x_t|y_{1:t}) \approx \widetilde{w}_t\left(x^i_{0:t}\right)\delta\left(x_{0:t} - x^i_{0:t}\right)dx_{0:t-1} \tag{3}$$

where $x^i_{0:t}$ represents the state trajectory; thus, the fault severity while $y_{1:t}$ is the measurements up to time t. During the "filtering" stage, a resampling scheme is employed to update the state estimates by updating the particle weights. One of the most common versions of this algorithm, the Sequential Importance Re-sampling (SIR) particle filter [43], updates the state weights using the likelihood of y_t as:

$$w_t = w_{t-1}p(y_t|x_t) \tag{4}$$

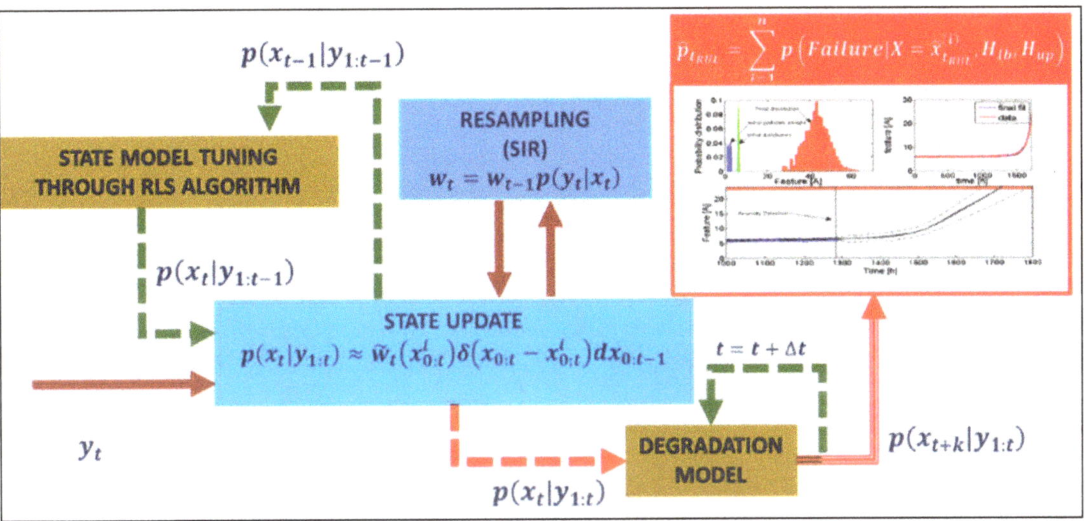

Figure 17. Scheme of the employed particle-filtering algorithm.

Although widely adopted within the PHM community due to its simplicity and relatively low computational requirements, it is worth mentioning that this resampling scheme has limitations in the description of the distribution tails, and that more advanced resampling schemes, aimed at obtaining and improving representation of the distribution tails, have been proposed [44]. However, given the low-TLR nature of the study, the SIR scheme was deemed sufficient. The algorithm performs long-term predictions of the fault evolution in time by iterating the "prediction" stage. Evaluating the particle trajectories, it is then possible to estimate the probability of failure by comparing their behavior against a hazard zone, defined via a probability density function with lower and upper bounds for the domain of the random variable [45]. As shown in Figure 17, this approach is based on particle-filtering schemes in which tunable degradation models are adopted as process models. These models are then used both to estimate the current a priori state of the system, $p(x_t|y_{1:t-1})$, and to perform the iterative steps necessary to achieve the prognosis $p(x_{t+k}|y_{1:t})$. Auto-tuned models are required to describe and follow changes in the degradation process and to describe the process and measurement noise. For the case study under consideration, the particle filter employs a nonlinear process model $y_t = f(x_t) + \nu$, obtained considering the dependency of the selected features on degradation growth. The process noise ν is then obtained through a kernel function mirroring the feature distribution around the fitted model. The state model $x_{t+1} = f(x_t, t) + \sigma$ is a time-dependent nonlinear model whose parameters are automatically tuned through a Recursive Least Square [46] algorithm operating over the state estimates provided by the particle filter itself. The measure of noise σ is also estimated at each time stamp computing the state estimate variance with respect to the noiseless output of the state model itself. The particle filter code also employs the noise compensation techniques described in [47] and self-adjusts the process noise by comparing the particle distribution at a given time instant (usually a few time steps ahead of the prediction time) of a previous long-time prediction against the particle distribution obtained for the same time instant from the prediction/filtering loop. The output of the prognostic algorithm is represented as the RUL distribution for each prediction, coupled with the evaluation of the risk of failure as detailed in [45]. An example of the results achievable with such a framework is provided in Figure 18, where its response against a simulated fault, the occurrence of accelerated degradation of the motor's permanent magnets, is depicted for a single prediction step highlighting the trajectories of the long-term prognosis on both the estimated fault size (hidden state) and feature. Figure 19, instead, provides an example of the RUL distributions obtained for different prediction steps for the very same failure mode. It can be noticed that the RUL uncertainty estimate tends to decrease along with the degradation process. However, when the degradation severity has grown close to the failure declaration threshold, the prediction uncertainty tends to increase again. Such behavior is expected since the closer the degradation process approaches the failure status, the more it becomes susceptible to small variations of the physical variable responsible for its progression, thus, leading to increased uncertainty over its future expected behavior.

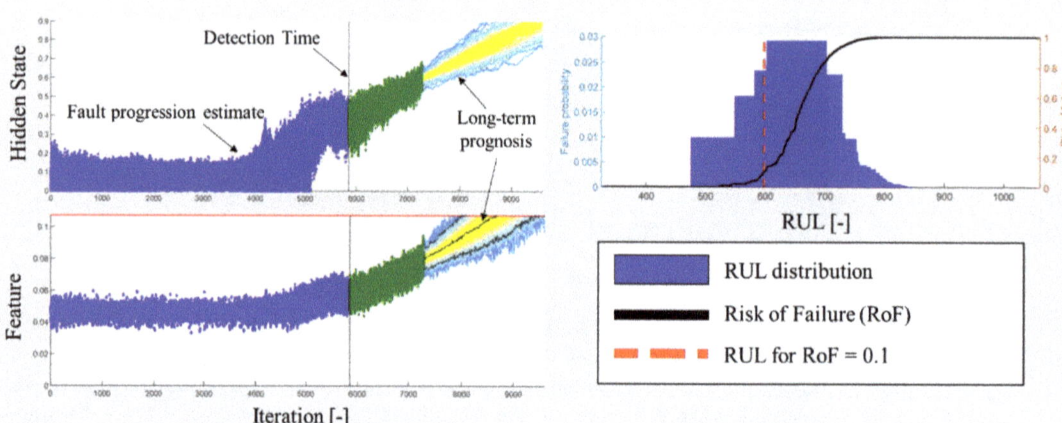

Figure 18. Example of the particle filter framework output.

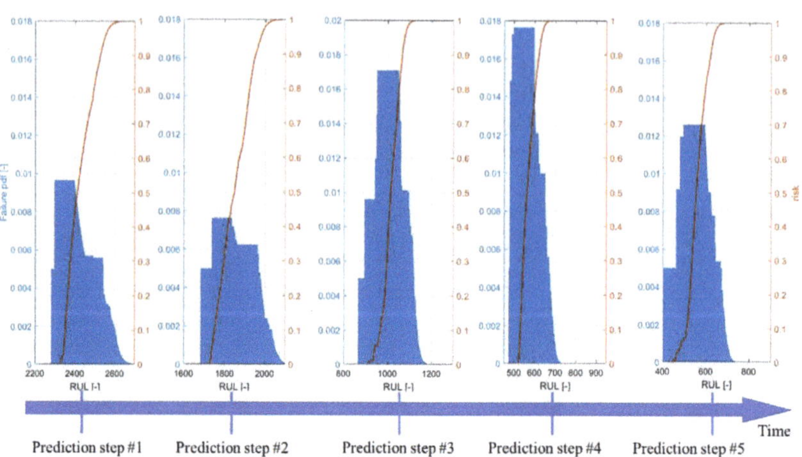

Figure 19. Example of RUL estimate for 5 consequent prediction steps for one of the degradation patterns associated with the demagnetization of the motor permanent magnets.

4.3. PHM Framework Design-Implementation and Verification Process

The proposed PHM framework was implemented within a Matlab GUI running as a standalone executable file within the HMSM module of the iron bird. The application to an integrated technological demonstrator, of which PHM is a part but not the only objective, means that some limitations were introduced. For the application at hand, it was chosen to replicate a system operating a periodical download of the actuators signals to be analyzed. For each simulated flight (or series of simulated flights), the Health Management System Module receives the signals from both the "simulated" half-wing and the "real" half-wing through the optical ring operating with a data transmission rate of 800 Hz. This posed a significant issue for data analysis, since all the electric current signals associated with a motor angular frequency higher than such transmission rate divided by 10 times the number of the magnetic pole couples of the Brushless-AC were not properly reconstructed. As such, it was necessary to limit the data analysis for low-speed conditions only. Before testing the behavior of the PHM algorithm on board the iron bird, it was necessary to verify its behavior and expected performances offline to avoid the need to debug or rethink parts of the framework once that system was installed. Such a verification process was achieved through the approach proposed in Figure 20, thus stressing the PHM routines with datasets generated through the real-time version of the model of the actuator, which are offline versions of the simulation models employed on the iron bird for the "simulated" half-wing. The use of the RT models was chosen because of 2 main reasons. The first was the significant cut in computational effort required to generate data. The second was the need to remain as close as possible to the iron bird configuration, to test possible issues with data exchange, labels, computational effort, and memory leaks. The test was performed again considering 10 different aircraft—different from those used during the design phase—subjected to several degradation patterns while using the same scenario generator described in Section 4.1. Data were then sent to the PHM GUI prepared for iron bird deployment. Data were analyzed, subjected through the fault-detection and classification routine, and then sent to the prognostic algorithm, which performed 10 RUL predictions at equally spaced time instants ranging from the time-at-detection (the time instant corresponding to the fault detection) to the end-of-life of the component (failure conditions).

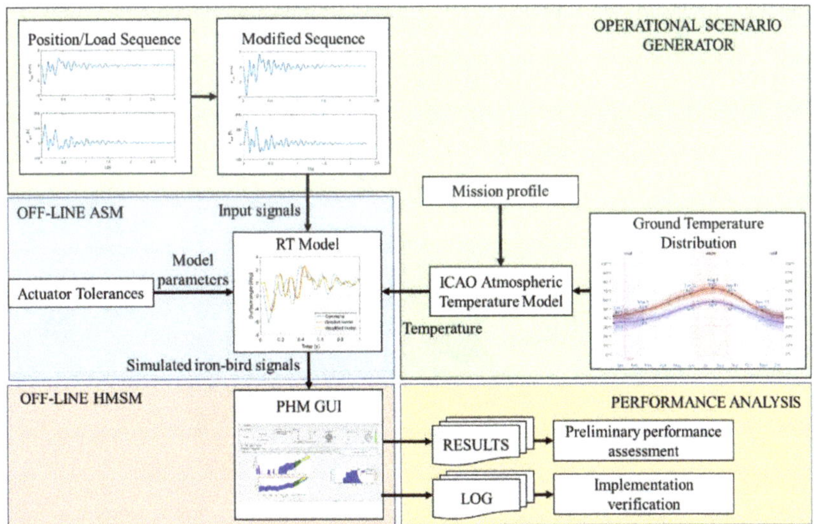

Figure 20. Procedure for the preliminary performance assessment.

Results were then analyzed by checking the percentage of false alarm and misclassification rates and the convergence of the RUL estimate to the ground truth provided by the simulations.

5. Results

Results of the preliminary performance assessment procedure are summarized in this section, divided between the verification of the implementation targets, in terms of computational effort, execution time of each routine, and the evaluation of the performance of the fault detection, fault classification, and failure prognosis routines against the simulated degradation histories.

5.1. Implementation Targets

As stated in Section 4, the proposed PHM system works off-line with respect to the iron bird, analyzing data periodically; thus, no specific targets are set in terms of execution time of the algorithms. However, since the iron bird is meant to act as a technological demonstrator, it is useful to check whether the proposed algorithms are suitable for real-time applications. For this purpose, the PHM system employed in the GUI is embedded with a tool monitoring the computational performances of the PHM routines. Monitored activities include the execution time of the fault-diagnosis algorithm and of the failure-prognosis routine with its subroutine. Test conditions include the analysis of signals coming from three identical electromechanical actuators, two operating on a morphing winglet and one on the wingtip surface, for a total of 24 signals, sampled at 800 Hz, and 19 features, and downsampled at 80 Hz. The analysis was performed on an Intel Core i9-9880H CPU running at 2.30 GHz with 16 GB of DDR4 RAM, which was compatible with the performances of the HMSM and was conducted considering 200 degradation patterns, each comprehensive of a number ranging between 150 and 400 position/load sequences, which correspond to 25 degradation patterns for the seven considered faults and 25 simulation cycles performed in healthy conditions. The fault-detection algorithm completed the analysis of one data batch corresponding to advancement of a single time step in 11.3 ms. Such a result is compatible with the features sampling (80 Hz), hence suggesting that the feature extraction and fault-detection scheme is suitable for real-time applications. The same simulation cycles were also used to assess the computational performance of the long-term prognosis algorithm. As depicted in Figure 21, a single cycle of prediction and subsequent state updates of the particle-filtering routine requires 7.4 ms on the employed test machine, with the RLS algorithms responsible for the tuning of the degradation model accounting for the 2.25% of such a number. Such results were obtained for a particle-filtering scheme operating with 200 particles and are expected to scale almost linearly, increasing the number of particles.

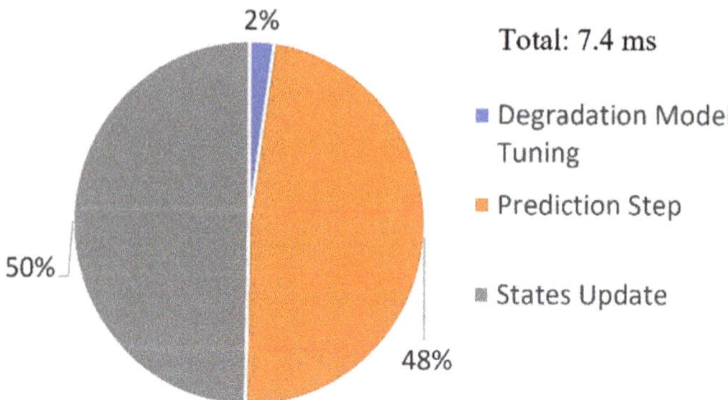

Figure 21. Elapsed time breakdown—average execution time of a particle-filtering step.

5.2. *Preliminary Evaluation of Prognostic Performances*

The same simulated degradation patterns adopted for the evaluation of the execution time of the PHM algorithms were also used to evaluate the expected performances in terms of fault-detection metrics, fault classification (or misclassification) rates, convergence of the long-term prognosis, and expected accuracy of the RUL estimate. Results are provided below and divided between those pertaining to the fault diagnosis and those pertaining to the failure prognosis.

5.2.1. Fault Diagnosis

The fault-diagnosis routine is at first tested against the database of 25 simulation cycles representative of health conditions to check for the eventual occurrence of false alarms. As anticipated, such a database is built considering the expected distribution of actuator parameters, such as geometrical quantities, physical properties, and according to production tolerances. Similarly, temperature variations, load, and command histories are also randomly drawn from probability distributions representative of the expected operative conditions. Under such a hypothesis, the fault-detection algorithm provided no false alarms. The fault diagnosis was then employed to detect and classify the faults affecting the simulated dataset, achieving the results depicted in Figure 22, where the classification rate is expressed in percentage. Under the assumption of just one failure mode occurring at any time, consistent with the objectives of the technological demonstrator, the fault-diagnosis algorithm provides acceptable results, with misclassification occurring only between the efficiency loss and magnet degradation failure modes. In particular, 4% of the cases associated with the occurrence of magnet degradation were incorrectly classified as efficiency losses within the mechanical transmission, while the opposite situation occurred for 12% of the simulated efficiency losses patterns. No misclassifications were observed for the other failure modes, although this result is probably skewed by the absence of possible concurrent degradation modes.

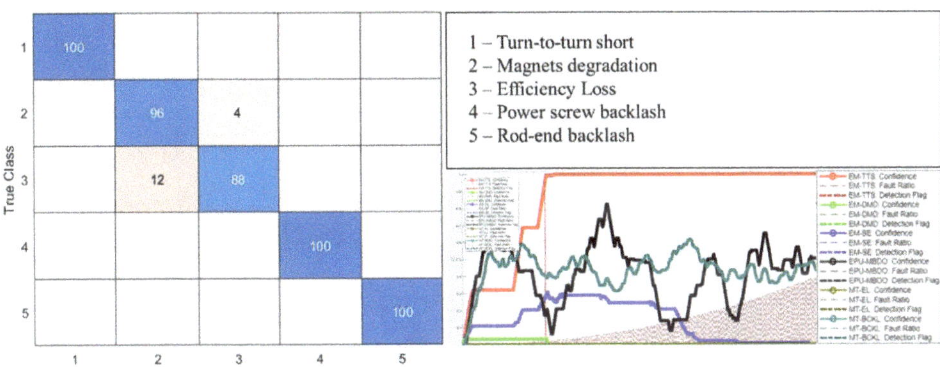

Figure 22. Confusion matrix for fault classification and example of the fault diagnosis algorithm for the turn-to-turn short failure mode.

5.2.2. Failure Prognosis

The failure prognosis behavior is studied through the algorithm response to the 175 simulated degradation patterns previously described and analyzed to check their convergence to the simulated ground truth and its expected performances in terms of nondimensional Prognostic Horizon and mean accuracy metrics [48]. Figure 23 depicts the prediction trajectories associated with the most probable RUL estimate for each prediction step of each degradation pattern, comparing their behavior against the simulated ground truth. It can be observed that all the considered failure modes converge toward the ground-truth solution, although with different patterns and different performances.

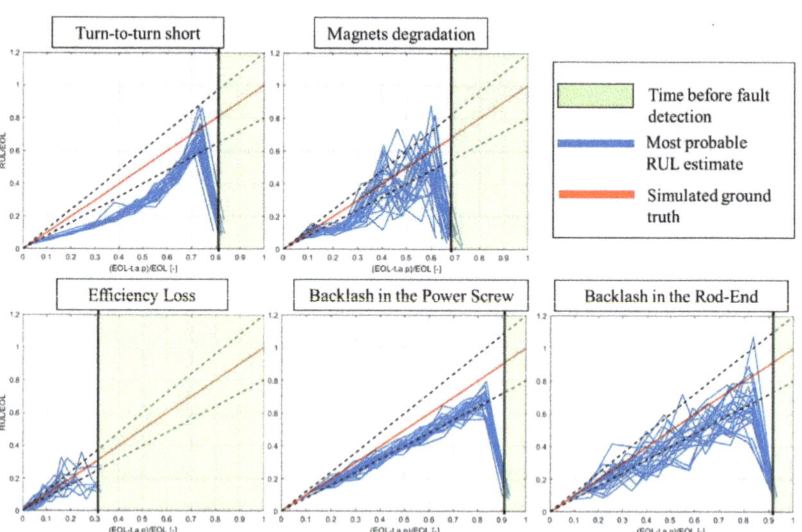

Figure 23. Elapsed time breakdown—average execution time of a particle-filtering step.

In particular, the RUL associated with a turn-to-turn short is often underestimated with respect to the ground truth. Although suboptimal, such a result is still positive, since it would cause, at worst, a slight anticipation of the maintenance action. On the other hand, the remaining investigated failure modes all fare well, quickly converging within an uncertainty cone equal to ±20% of the ground-truth RUL value.

The first prediction step, that is the one performed right after the fault detection, is affected by significant uncertainty and provides the less accurate results. Such behavior is expected, since the initial parameters of the degradation models employed within the particle-filtering routines are first-trial values for the first prediction step, with the tuning process requiring more data to converge to the ongoing degradation process. Results are also analyzed according to two traditional metrics, the Prognostic Horizon and the average accuracy [48], and reported in Table 2. The Prognostic Horizon is defined as the remaining useful life for which the most probable RUL estimate falls, and remains, within a certain accuracy threshold. Since the results pertain forcefully to accelerated degradation and not to a naturally evolving one, such a metric is provided as a nondimensional value over the simulated lifetime of the equipment for each considered failure mode. All the results are the average obtained for the considered degradation patterns.

Table 2. Expected performance metrics.

Component	Failure Mode	Prognostic Horizon [−]	Relative Accuracy [%]
Motor	Turn-to-turn short	0.11	64.3%
Motor	Magnet degradation	0.55	75.3%
Mechanical transmission	Efficiency loss, lubricant aging	0.15	78.2%
Mechanical transmission	Wear-induced backlash	0.82	81.2%
Spherical joints	Wear-induced backlash	0.65	79.3%

Overall, results are encouraging for long-lasting degradations, where high values of PH and good accuracy levels are observed. The low scores associated with the turn-to-turn failure mode denounce a lower level of accuracy associated with the prediction, but as already underlined, such a result is due to an underestimation of the RUL, which is less critical than an overestimation. Similarly, results for the efficiency loss are suboptimal in terms of the Prognostic Horizon. Such degradation is, however, a slowly evolving process requiring hundreds of flight hours; thus, the results still provide sufficient time to issue an early warning and plan the due maintenance accordingly.

6. Conclusions

The CleanSky 2/Astib research program was launched with the objective of building an iron bird for a new regional transport platform acting as a demonstrator for new technologies such as fully electrical flight-control systems and the definition of a PHM system. This paper detailed the activities performed to support the definition of such a PHM system, starting from a high-fidelity representation of the involved subsystems, then transitioning to a real-time representation to support both the implementation of such models within the iron bird and the rapid generation of a simulated dataset to study the behavior of the case study in both nominal and degraded health conditions. A PHM framework is proposed based on a simple data-driven routine to perform fault detection, an SVM for fault classification, and a particle-filtering routine with auto-tuning, and time-variant degradation models. Such a system was then stressed with several simulated degradation patterns to test the algorithm's readiness to be deployed on the iron bird and its expected performances. Results are encouraging, as the algorithm is suitable for installation on board the technological demonstrator and its expected performances are coherent with the target of the research project. Future work will include the verification of the algorithm behavior in response to flight cycles simulated on the iron bird, which will allow stressing the PHM framework with a widely varying set of maneuvers and include in the analysis the effect of pilot-induced compensations on the fault evolution in time. Outside the scope of the research project, but of significant interest for further research, is the study of the effect of different concurring degradations and their effect on the PHM routines.

Author Contributions: Conceptualization, A.C.B., A.D.M., G.J. and M.S.; methodology, A.C.B., A.D.M., G.J. and M.S.; software, A.C.B. and A.D.M.; validation, A.C.B. and A.D.M.; formal analysis, A.C.B. and A.D.M.; investigation, A.C.B. and A.D.M.; resources, A.C.B. and A.D.M.; data curation, A.C.B. and A.D.M.; writing—original draft preparation, A.C.B. and A.D.M.; writing—review and editing, A.C.B., A.D.M., G.J. and M.S.; visualization, A.C.B. and A.D.M.; supervision, G.J. and M.S.; project administration, G.J. and M.S.; funding acquisition, G.J. and M.S. All authors have read and agreed to the published version of the manuscript.

Funding: The research work presented in this paper was performed within the ASTIB research project, which has received funding from the Clean Sky 2 Joint Undertaking under the European Union's Horizon 2020 research and innovation program under grant agreement CSJU—GAM REG 2014–2015.

Data Availability Statement: Data sharing is not applicable to this article.

Conflicts of Interest: The authors declare no conflict of interest. The funders had no role in the design of the study; in the collection, analyses, or interpretation of data; in the writing of the manuscript, or in the decision to publish the results.

References

1. Vachtsevanos, G.; Lewis, F.; Roemer, M.; Hess, A.; Wu, B. *Intelligent Fault Diagnosis and Prognosis for Engineering Systems*; John Wiley & Sons, Inc.: Hoboken, NJ, USA, 2006; ISBN 9780470117842.
2. Brown, D.W.; Georgoulas, G.; Bole, B.M. Prognostics Enhanced Reconfigurable Control of Electro-Mechanical Actuators. In Proceedings of the Annual Conference of the Prognostics and Health Management Society, PHM 2009, San Diego, CA, USA, 27 September–1 October 2009.
3. Balchanos, M.; Mavris, D.; Brown, D.W.; Georgoulas, G.; Vachtsevanos, G. Incipient failure detection: A particle filtering approach with application to actuator systems. In Proceedings of the 2017 13th IEEE International Conference on Control & Automation (ICCA), Ohrid, Macedonia, 3–6 July 2017; pp. 64–69. [CrossRef]
4. Ismail, M.A.; Windelberg, J.; Bierig, A.; Spangenberg, H. A potential study of prognostic-based maintenance for primary flight control electro-mechanical actuators. In Proceedings of the Recent Advances in Aerospace Actuation Systems and Components, Toulouse, France, 16–18 March 2016; pp. 193–200.

5. Dalla Vedova, M.D.L.; Germanà, A.; Berri, P.C.; Maggiore, P. Model-Based Fault Detection and Identification for Prognostics of Electromechanical Actuators Using Genetic Algorithms. *Aerospace* **2019**, *6*, 94. [CrossRef]
6. Autin, S.; De Martin, A.; Jacazio, G.; Socheleau, J.; Vachtsevanos, G. Results of a Feasibility Study of a Prognostic System for Electro-Hydraulic Flight Control Actuators. *Int. J. Progn. Health Manag.* **2020**, *12*, 1–18. [CrossRef]
7. Balaban, E.; Saxena, A.; Narasimhan, S.; Roychoudhury, I.; Goebel, K.F.; Koopmans, M.T. Airborne electro-mechanical actuator test stand for development of prognostic health management systems. *Annu. Conf. Progn. Health Manag. Soc. PHM* **2010**, 2. Available online: https://papers.phmsociety.org/index.php/phmconf/article/view/1804 (accessed on 25 February 2023).
8. Bertolino, A.C.; De Martin, A.; Jacazio, G.; Sorli, M. A technological demonstrator for the application of PHM techniques to electro-mechanical flight control actuators. In Proceedings of the 2022 IEEE International Conference on Prognostics and Health Management (ICPHM), Detroit, MI, USA, 6–8 June 2022; pp. 70–76. [CrossRef]
9. Li, D.; Lin, M.; Tian, L. Design of iron bird for a regional jet aircraft. *Proc. Inst. Mech. Eng. Part G J. Aerosp. Eng.* **2020**, *234*, 681–688. [CrossRef]
10. Spangenberg, H.; Friehmelt, H. Hardware-in-the-loop simulation with flight control actuators. In Proceedings of the Collection of Technical Papers—AIAA Modeling and Simulation Technologies Conference 2005, Grapevine, TX, USA, 9–13 January 2017; Volume 1, pp. 614–622.
11. Blasi, L.; Borrelli, M.; D'amato, E.; Di Grazia, L.E.; Mattei, M.; Notaro, I. Modeling and control of a modular iron bird. *Aerospace* **2021**, *8*, 39. [CrossRef]
12. Jensen, S.C.; Jenney, G.D.; Dawson, D. Flight test experience with an electromechanical actuator on the F-18 Systems Research Aircraft. In Proceedings of the 19th DASC. 19th Digital Avionics Systems Conference, Philadelphia, PA, USA, 7–13 October 2000. [CrossRef]
13. Malisani, S.; Capello, E.; Guglieri, G. Development of a flight mechanics simulation computer based on a flexible aircraft model for a regional aircraft. *IOP Conf. Ser. Mater. Sci. Eng.* **2021**, *1024*, 012066. [CrossRef]
14. Chiavaroli, P.; De Martin, A.; Evangelista, G.; Jacazio, G.; Sorli, M. Real Time Loading Test Rig for Flight Control Actuators Under PHM Experimentation. In Proceedings of the ASME International Mechanical Engineering Congress and Exposition, Proceedings (IMECE), Pittsburgh, PA, USA, 9–15 November 2018; Volume 1, p. V001T03A032. [CrossRef]
15. *SAE-AIR8012*; Prognostics and Health Management Guidelines for Electro-Mechanical Actuators. HM-1 Integrated Vehicle Health Management Committee, 2020. Available online: https://saemobilus.sae.org/content/air8012 (accessed on 25 February 2023). [CrossRef]
16. De Martin, A.; Jacazio, G.; Vachtsevanos, G. Windings Fault Detection and Prognosis in Electro-Mechanical Flight Control Actuators Operating in Active-Active Configuration. *Int. J. Progn. Health Manag.* **2017**, *8*, 1–13. [CrossRef]
17. Moosavi, S.S.; Djerdir, A.; Amirat, Y.A.; Khaburi, D.A. Demagnetization fault investigation in permanent magnet synchronous motor. In Proceedings of the 5th Annual International Power Electronics, Drive Systems and Technologies Conference (PEDSTC 2014), Tehran, Iran, 5–6 February 2014; pp. 617–623. [CrossRef]
18. Wang, C.; Delgado Prieto, M.; Romeral, L.; Chen, Z.; Blaabjerg, F.; Liu, X. Detection of Partial Demagnetization Fault in PMSMs Operating under Nonstationary Conditions. *IEEE Trans. Magn.* **2016**, *52*, 1–4. [CrossRef]
19. Kastha, D.; Bose, B.K. Investigation of fault modes of voltage-fed inverter system for induction motor drive. In Proceedings of the Conference Record of the 1992 IEEE Industry Applications Society Annual Meeting, Houston, TX, USA, 4–9 October 1992; pp. 858–866. [CrossRef]
20. Khanniche, M.S.; Mamat-Ibrahim, M.R. Fault Detection and Diagnosis of 3-Phase Inverter System. *Rev. Energ. Ren. Power Eng.* **2001**, *1*, 69–75.
21. Bertolino, A.C.; De Martin, A.; Fasiello, F.; Mauro, S.; Sorli, M. A simulation study on the effect of lubricant ageing on ball screws behaviour. In Proceedings of the 2022 International Conference on Electrical, Computer, Communications and Mechatronics Engineering (ICECCME), Maldives, Maldives, 16–18 November 2022; pp. 16–18.
22. García-Martínez, M.; García de Blas Villanueva, F.J.; Valles González, M.P.; Pastor Muro, A. Failure analysis of the rod-end bearing of an actuating cylinder. *Eng. Fail. Anal.* **2019**, *104*, 292–299. [CrossRef]
23. Xiao, X.; Chen, C.M.; Zhang, M. Magnet Demagnetization Observation for Permanent Magnet Synchronous Motor. In Proceedings of the 2008 International Conference on Electrical Machines and Systems, Wuhan, China, 17–20 October 2008; pp. 3216–3219.
24. Belmonte, D.; Vedova, M.; Maggiore, P. Electromechanical servomechanisms affected by motor static eccentricity: Proposal of fault evaluation algorithm based on spectral analysis techniques. In *Safety and Reliability of Complex Engineered Systems, Proceedings of the 25th European Safety and Reliability Conference, ESREL 2015*; CRC Press: Boca Raton, FL, USA, 2015; pp. 2365–2372.
25. Di Rito, G.; Schettini, F.; Galatolo, R. Model-based prognostic health-management algorithms for the freeplay identification in electromechanical flight control actuators. In Proceedings of the 2018 5th IEEE International Workshop on Metrology for AeroSpace (MetroAeroSpace), Rome, Italy, 20–22 June 2018; pp. 340–345. [CrossRef]
26. Candon, M.; Levinski, O.; Ogawa, H.; Carrese, R.; Marzocca, P. A nonlinear signal processing framework for rapid identification and diagnosis of structural freeplay. *Mech. Syst. Signal Process.* **2022**, *163*, 107999. [CrossRef]
27. Mohan, N.; Undeland, T.M.; Robbins, W.P. *Power Electronics*, 3rd ed.; John Wiley and Sons, Inc.: Hoboken, NJ, USA, 2005.
28. Gökdere, L.U.; Bogdanov, A.; Chiu, S.L.; Keller, K.J.; Vian, J. Adaptive control of actuator lifetime. In Proceedings of the 2006 IEEE Aerospace Conference, Big Sky, MT, USA, 4–11 March 2006. [CrossRef]

29. Antoine, J.-F.; Visa, C.; Sauvey, C.; Abba, G. Approximate Analytical Model for Hertzian Elliptical Contact Problems. *J. Tribol.* **2006**, *128*, 660. [CrossRef]
30. Lugt, P.M. *Grease Lubrication in Rolling Bearings*; John Wiley & Sons, Inc.: Hoboken, NJ, USA, 2013; ISBN 9781118353912.
31. Stachowiak, G.W.; Batchelor, A.W. *Engineering Tribology*, 4th ed.; Elsevier: Amsterdam, The Netherlands, 2014; ISBN 978-0-12-397047-3.
32. Nijenbanning, G.; Venner, C.H.; Moes, H. Film thickness in elastohydrodynamically contacts. *Wear* **1994**, *176*, 217–229. [CrossRef]
33. Bertolino, A.C.; Mauro, S.; Jacazio, G.; Sorli, M. Multibody dynamic model of a double nut preloaded ball screw mechanism with lubrication. In Proceedings of the ASME International Mechanical Engineering Congress and Exposition, Proceedings (IMECE), Virtual, 1–5 November 2021; Volume 7B-2020.
34. Balan, M.R.D.; Stamate, V.C.; Houpert, L.; Olaru, D.N. The influence of the lubricant viscosity on the rolling friction torque. *Tribol. Int.* **2014**, *72*, 1–12. [CrossRef]
35. Zhou, H.X.; Zhou, C.G.; Feng, H.T.; Ou, Y. Theoretical and experimental analysis of the preload degradation of double-nut ball screws. *Precis. Eng.* **2020**, *65*, 72–90. [CrossRef]
36. Bertolino, A.C.; Jacazio, G.; Mauro, S.; Sorli, M. Investigation on the ball screws no-load drag torque in presence of lubrication through MBD simulations. *Mech. Mach. Theory* **2021**, *161*, 104328. [CrossRef]
37. Damiens, B.; Venner, C.H.; Cann, P.M.E.; Lubrecht, A.A. Starved Lubrication of Elliptical EHD Contacts. *J. Tribol.* **2004**, *126*, 105. [CrossRef]
38. Van Zoelen, M.T.; Venner, C.H.; Lugt, P.M. Prediction of film thickness decay in starved elasto-hydrodynamically lubricated contacts using a thin layer flow model. *Proc. Inst. Mech. Eng. Part J J. Eng. Tribol.* **2009**, *223*, 541–552. [CrossRef]
39. Zhou, Y.; Bosman, R.; Lugt, P.M. A Model for Shear Degradation of Lithium Soap Grease at Ambient Temperature. *Tribol. Trans.* **2018**, *61*, 61–70. [CrossRef]
40. Bacci, A.; Bertolino, A.C.; De Martin, A.; Sorli, M. Multiphysics modelling of a faulty rod-end and its interaction with a flight control actuator to support PHM activities. In Proceedings of the IMECE 2021 Volume 7: Dynamics, Vibration, and Control, Virtual, 1–5 November 2021.
41. Orchard, M.E.; Vachtsevanos, G.J. A particle-filtering approach for on-line fault diagnosis and failure prognosis. *Trans. Inst. Meas. Control* **2009**, *31*, 221–246. [CrossRef]
42. Roemer, M.J.; Byington, C.S.; Kacprzynski, G.J.; Vachtsevanos, G.; Goebel, K. Prognostics. In *System Health Management: With Aerospace Applications*; Wiley: Hoboken, NJ, USA, 2011; ISBN 9780470741337.
43. Arulampalam, M.S.; Maskell, S.; Gordon, N.; Clapp, T. A Tutorial on Particle Filters for Online Nonlinear/NonGaussian Bayesian Tracking. *IEEE Trans. Signal Process.* **2002**, *50*, 174–188. [CrossRef]
44. Acuña, D.E.; Orchard, M.E. Particle-filtering-based failure prognosis via sigma-points: Application to Lithium-Ion battery State-of-Charge monitoring. *Mech. Syst. Signal Process.* **2017**, *85*, 827–848. [CrossRef]
45. Acuña, D.E.; Orchard, M.E. A theoretically rigorous approach to failure prognosis. In Proceedings of the 10th Annual Conference of the Prognostics and Health Management Society 2018 (PHM18), Philadelphia, PA, USA, 24–27 September 2018.
46. Bishop, C.M. *Pattern Recoginiton and Machine Learning*; Springer: Berlin/Heidelberg, Germany, 2006; ISBN 978-0-387-31073-2.
47. De Martin, A.; Jacazio, G.; Sorli, M. Enhanced Particle Filter framework for improved prognosis of electro-mechanical flight controls actuators. In Proceedings of the PHM Society European Conference, PHME 2018, Utrecht, The Netherlands, 3–6 July 2018; Volume 4.
48. Saxena, A.; Celaya, J.; Balaban, E.; Goebel, K.; Saha, B.; Saha, S.; Schwabacher, M. Metrics for evaluating performance of prognostic techniques. In Proceedings of the 2008 International Conference on Prognostics and Health Management, Denver, CO, USA, 6–9 October 2008; pp. 1–17. [CrossRef]

Article

Influence of Electric Wing Tip Propulsion on the Sizing of the Vertical Stabilizer and Rudder in Preliminary Aircraft Design

Alexander Albrecht [1,*], Andreas Bender [1], Philipp Strathoff [2], Clemens Zumegen [2], Eike Stumpf [2] and Andreas Strohmayer [1]

[1] Institute of Aircraft Design, University of Stuttgart, 70569 Stuttgart, Germany
[2] Institute of Aerospace Systems, RWTH Aachen University, 52062 Aachen, Germany
* Correspondence: albrecht@ifb.uni-stuttgart.de; Tel.: +49-711-685-62409

Abstract: During preliminary aircraft design, the vertical tail sizing is conventionally conducted by the use of volume coefficients. These represent a statistical approach using existing configurations' correlating parameters, such as wing span and lever arm, to size the empennage. For a more detailed analysis with regard to control performance, the vertical tail size strongly depends on the critical loss of thrust assessment. This consideration increases in complexity for the design of the aircraft using wing tip propulsion systems. Within this study, a volume coefficient-based vertical tail plane sizing is compared to handbook methods and the possibility to reduce the necessary vertical stabilizer size is assessed with regard to the position of the engine integration and their interconnection. Two configurations, with different engine positions, of a hybrid-electric 19-seater aircraft, derived from the specifications of a Beechcraft 1900D, are compared. For both configurations two wiring options are assessed with regard to their impact on aircraft level for a partial loss of thrust. The preliminary aircraft design tool MICADO is used to size the four aircraft and propulsion system configurations using fin volume coefficients. These results are subsequently amended by handbook methods to resize the vertical stabilizer and update the configurations. The results in terms of, e.g., operating empty mass and mission fuel consumption, are compared to the original configurations without the optimized vertical stabilizer. The findings support the initial idea that the connection of the electric engines on the wing tips to their respective power source has a significant effect on the resulting torque around the yaw axis and the behaviour of the aircraft in case of a power train failure, as well as on the empty mass and trip fuel. For only one out of the four different aircraft designs and wiring configurations investigated it was possible to decrease the fin size, resulting in a 53.7% smaller vertical tail and a reduction in trip fuel of 4.9%, compared to the MICADO design results for the original fin volume coefficient.

Keywords: aircraft design; distributed propulsion; hybrid-electric flight; critical loss of thrust; vertical stabilizer

Citation: Albrecht, A.; Bender, A.; Strathoff, P.; Zumegen, C.; Stumpf, E.; Strohmayer, A. Influence of Electric Wing Tip Propulsion on the Sizing of the Vertical Stabilizer and Rudder in Preliminary Aircraft Design. *Aerospace* **2023**, *10*, 395. https://doi.org/10.3390/aerospace10050395

Academic Editor: Dimitri Mavris

Received: 13 February 2023
Revised: 27 March 2023
Accepted: 18 April 2023
Published: 25 April 2023

1. Introduction

Aircraft design is a complex and interrelated field of research. Every modification of a specific component in the design process has an influence on other components or the aircraft/platform itself. This leads to an iterative design process, ideally resulting in one converged optimum design for the given top level aircraft requirements (TLARs). Small, evolutionary technology changes mostly result in designs close to already known configurations, emphasizing today's well-chosen concepts. This changes for revolutionary technology steps, such as the implementation of distributed (electric) propulsion (DEP). Several designs, differing radically from today's configurations, have already been investigated in order to use the full technological potential enabled through the use of DEP on aircraft. Globally, there have been many research attempts trying to make use of DEP's aerodynamic potential. The NASA X-57 design uses the velocity increase downstream of a propeller to reduce the wing surface [1]. Wing tip propellers are also used by other configurations in order to reduce the wake-induced drag, such as the NASA Pegasus [1], the Airbus/TBM/Safran EcoPulse [2] and the European research project FUTPRINT50 [3]. All of these projects feature designs with unconventional propeller positions for their propulsion system integration. This raises the question of the suitability of a volume coefficient-based vertical tailplane (VTP) sizing which is based on statistics of existing conventional aircraft without DEP. This is also valid for the configurations investigated in this paper.

The study presented is part of the GNOSIS project, which aims for a holistic evaluation of the potential of propulsion system electrification for commercial passenger aircraft with seat capacities in the range of 9 to 50 seats. In the first project phase, the focus is on the conceptual design of a 19-seater aircraft. The following study therefore compares a volume coefficient-based sizing of the VTP during preliminary aircraft design with handbook methods. This is performed for a partial turboelectric 19-seater incorporating wing tip propulsion with regard to the impact of the inboard engine position and the interconnection of the propulsors and thereafter the impact of a reduction in the required VTP size on the aircraft.

According to [4], the VTP is usually sized by two major flight conditions. One being the operation with one engine inoperative and the other with maximum cross-wind capability. The focus of the study presented here is the investigation of the usability of the given VTP in case of engine failure-induced thrust asymmetry and the possible reduction in VTP size and its impact on the preliminary aircraft design. As investigated by Hoogreef and Soikkeli [5], as well as Vechtel and Buch [6] and proven by Schneider et al. [7] in a full-scale flight test, the directional stability may also be provided by the use of differential thrust. Therefore, the influence of the directional stability on the vertical tail size is neglected in this study.

For conventional configurations, the dimensioning contributor for a critical loss of thrust (CLT), or "one engine inoperative" (OEI) scenario has been the control authority of the VTP around the yaw axis. Particularly for DEP configurations, the limiting CLT condition might also be given by the maximum aileron authority in certain configurations. For the study presented in this paper, only the VTP's yaw control authority shall be assessed, as the considered positions of the propulsion system integration lead to an aircraft, for which the resulting yaw moments are significantly larger than the roll moments [6].

The contents of the paper are structured as follows. First the preliminary aircraft design tool used and the methodology for calculation of the aircraft and vertical tail parameters are explained. Then, a short overview of the considered aircraft configurations is given. Thereafter, the results of the optimization of the vertical stabilizer size for the different aircraft are given. It is shown how the fin size changes depending on the propeller positions and electrical interconnections. Moreover, certain architectures lead to uncontrollable aircraft designs in case of a critical loss of thrust, refraining from the additional implementation of components and hence additional weight. The calculated VTP size is then used to determine the impact on the aircraft using a mission data analysis of the individual configurations, hence accounting for the interrelated effects of a possible reduction in VTP size on the preliminary aircraft design.

A Short Discussion of "Critical Loss of Thrust"

With the change in the EASA CS23 from Amendment 4 [8] to Amendment 5 [9], the definition of the engine failure has changed. The former and well defined "failure of the critical engine" has been replaced by "critical loss of thrust" (CLT). On purpose, this term has not been specified any further. Therefore, the need to find a solution in order to assess distributed propulsion configurations and to define a "critical loss of thrust" scenario in accordance with EASA CS23.2115 [9] arises. An evaluation conducted by NASA with regard to the respective U.S. American specifications (which are almost identical to the European specifications), NASA experts have identified several certification gaps concerning both the propulsion system and the whole aircraft certification in their analysis [10]. In order to overcome the propulsion system-related gaps, they recommend to conduct a Markov analysis to identify the most likely powertrain failure. Especially for hybrid configurations, Markov analysis can lead to the result that all powertrain components contribute to a system failure probability of less than 10×10^{-9} [11]. Therefore, the results of the Markov analysis suggest to neglect a deeper investigation of further CLT scenarios. A different approach to define the most critical scenario is found in Jézégou et al. [12]. First, top level aircraft functions (TLAFs) are identified. Next, the impact of a failure is correlated with these TLAFs in order to assess the criticality of the failure. For a design incorporating different energy generation paths, additionally the investigation should include multiple and possible cascading failures in order to identify the most critical scenario and the influence on the TLAFs. A TLAF interaction investigated in this paper is the suitability of the vertical tailplane (VTP) to counteract the resulting CLT torque, having been sized using conventional handbook methods within a preliminary aircraft design process. Based on the wiring possibilities depicted in Figure 1, a failure in the combustion engine also results in a failure of the connected electrical propulsor. Therefore, the impact of the CLT scenario on the required VTP size has to be taken into account within the aircraft design process.

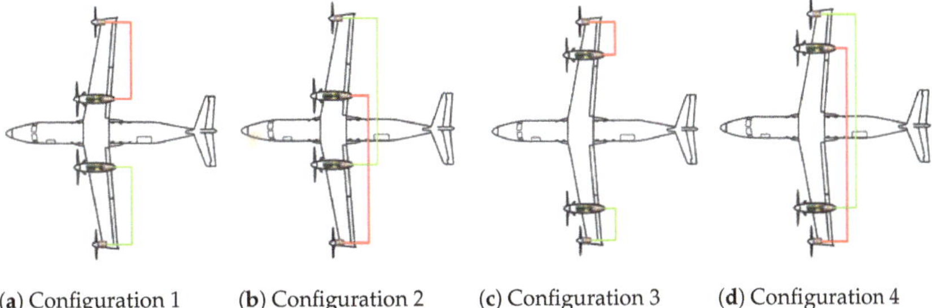

(**a**) Configuration 1 (**b**) Configuration 2 (**c**) Configuration 3 (**d**) Configuration 4

Figure 1. PT2025 aircraft with on-wing (Configuration 1) and cross wiring option (Configuration 2), as well as the PT2025opt aircraft with on-wing (Configuration 3) and cross wiring option (Configuration 4), with the red and green lines symbolizing the independent wiring harnesses.

This is especially interesting for the provided GNOSIS configurations, as, due to their powertrain layout, the probability of a 50% thrust loss is equal to the probability for the loss of the combustion engine only, assumed to be around $\times 10^{-5}$. Therefore, in contrast to [6], the loss of two propulsors has to be investigated with regard to the torque they impose on the VTP. The most critical case for a CLT is considered to be the "go" case, where the engine failure occurs at a speed above V1 and the takeoff run has to be continued. Therefore, according to [6] the maximum thrust at $V_{mc} = 1.2 \cdot V_s$ is used within the scope of this paper.

2. Methodology

In this section, a short overview of the analysed aircraft and wiring options, as well as the used software tools and the theoretical backgrounds used in the later evaluation, is given. The conceptional aircraft design tasks are performed using the aircraft design and evaluation environment MICADO, developed by the Institute of Aerospace Systems of the RWTH Aachen University, to iterate the complete aircraft design, while the rudder sizing is performed via MATLAB®-based tools.

2.1. Overview of the Conceptual Aircraft Design Tool MICADO

For the analysis of the aircraft, the multidisciplinary-integrated conceptual aircraft design and optimization (MICADO) environment is used [13,14]. MICADO is an extended version of the university conceptual aircraft design and optimization (UNICADO) environment [15] and was developed at the RWTH Aachen University's Institute of Aerospace Systems (ILR) in 2008.

As shown in Figure 2, the MICADO aircraft design process features an iterative part including aircraft component sizing, detailed system design and a design analysis step. The iteration is repeated until selected parameters of the aircraft, e.g., maximum takeoff mass (MTOM), operating empty mass (OME), mission fuel mass and lateral position of the aircraft's centre of gravity, do not change more than user-defined margins between two consecutive iteration steps.

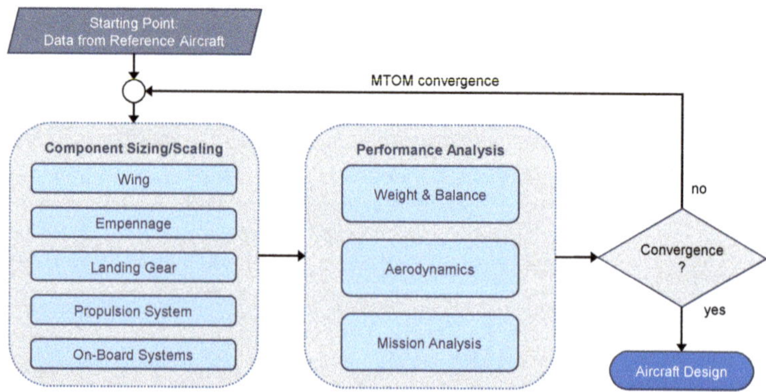

Figure 2. MICADO processing flow overview.

As comprehensive descriptions of MICADO and its tools can be found in the previously mentioned references, the following paragraphs focus on the MICADO tools and equations most important for this study.

The volume coefficient of the vertical tail determines its area. The value of the volume coefficient is either set by the user or calculated using an existing tail geometry.

Whereas mass estimation of most aircraft components, such as fuselage, landing gear, tail plane, is performed using semi-empirical handbook methods, wing mass estimation stands out as a tool based on analytical and semi-empirical methods. This tool takes into account the effect of point masses representing propulsion system components on the resulting wing structure and mass [16]. The secondary wing structure mass and mass penalties were estimated with semi-empirical methods; however, mass penalties due to aeroelastic effects were not included. Since the tail plane and electrical conductors are the focus of this study, applied methodologies for the estimation of their masses are presented in the following. The approach for estimation of the mass of the tail surfaces is taken from [17]. The fin mass is calculated in pounds according to Equation (1) and later converted to kilograms,

$$W_{fin} = 2.62 \cdot S_V + 1.5 \cdot 10^{-5} \cdot \frac{N_{ult} \cdot b_V^3 \cdot (8.0 + 0.44 \cdot \frac{MTOM}{S_{ref}})}{(t/c)_{avg} \cdot cos^2 \Lambda_{ea}} \tag{1}$$

where S_V is the fin area including the rudder, N_{ult} the ultimate load factor, b_V the span of the fin, S_{ref} the reference wing area, $(t/c)_{avg}$ the average airfoil thickness of the fin and Λ_{ea} the average sweep of the quarter chord line.

For calculation of the power cable masses according to Stückl [18], the length of the conductor using the positions of the components connected by each conductor, and the power to be transferred (in terms of current I and voltage U) by each conductor is taken into account. Moreover, for calculation of the overall conductor mass, the copper wire itself as well as insulation and sheath materials are considered. Assumptions regarding material constants and equations can be found in Table 1.

Table 1. Assumptions and equations for conductor design.

Component	Density (kg/m^3)	Thickness/ Diameter (mm)
Wire	8920	$D_{wire} = \sqrt{\frac{4 \cdot 0.0144 \cdot I[A]^{1.4642}}{\pi}}$
Insulation	930	$t_{insulation} = 0.2325 \cdot U[kV] + 1.73682$
Sheath	930	$t_{sheath} = 0.035 \cdot D_{wire}[mm] + 1$

Within the aerodynamic performance estimation module, lift and induced drag of the clean wing configuration are calculated using the German Aerospace Centre's LIFTING_LINE [19]. This program is able to consider the propeller-induced velocities on wing aerodynamics. Propeller-induced velocities are calculated for cruise conditions using a blade element momentum theory. The results are added to the LIFTING_LINE inputs. A more detailed description of this process can be found in [16]. Remaining drag components of the lifting surfaces and the remaining aircraft components are estimated using semi-empirical handbook methods. Most important for the study at hand is the estimation of the fin's viscous drag, calculated following an approach by Raymer [20]:

$$C_{D,Fin} = \frac{C_f \cdot FF \cdot Q \cdot S_{wet,Fin}}{S_{ref}} \tag{2}$$

where C_F is the flat-pate skin-friction drag coefficient of the fin, FF is the form factor of the fin, Q is a interference factor, S_{wet} is the fin's wetted area and S_{ref} is the wing's reference area.

The entire design and convergence process with the modules employed for this study is shown in Figure 2 above.

2.2. Reference and Concept Aircraft

Based on the outcome of a market analysis, the Beechcraft 1900D was chosen as the reference aircraft. According to the results of an initial technology identification and selection process (with respect to technologies associated with aircraft propulsion system electrification), a partial turboelectric propulsion system featuring two gas turbines supplemented with two electrically driven propellers on the wing tips for a reduction in induced drag was selected as the most promising concept for the evaluation in a year-2025-scenario [21]. The basic assumption in the technology selection process was that an aircraft with electrified propulsion systems needs to fulfil identical mission requirements as current conventional aircraft in terms of its range, speed and passenger capacity in order to be recognized as a viable alternative.

Based on data of the Beechcraft 1900D and the PT6A-67D turboprop engines available to the public, such as the Pilot Operating Handbook [22] and the EASA-issued engine type certificate [23], a redesign of this aircraft was conducted using the MICADO environment [15], see Figure 3 left. In order to obtain a comparable conventional reference aircraft, some adjustments were made prior to the electrification of the Beechcraft 1900D. First, the wing was moved from a low to a high position to ensure sufficient ground clearance for the wing tip propellers. Corresponding to this, the position of the landing gear was moved from the wing to the fuselage. Lastly, the engine performance and the scaling factors of the mass estimation methodologies were adjusted according to the description in Section 2.1. The resulting aircraft from a further execution of the aircraft design loop after these changes served as the conventional reference for subsequent comparisons, depicted in the centre-left of Figure 3.

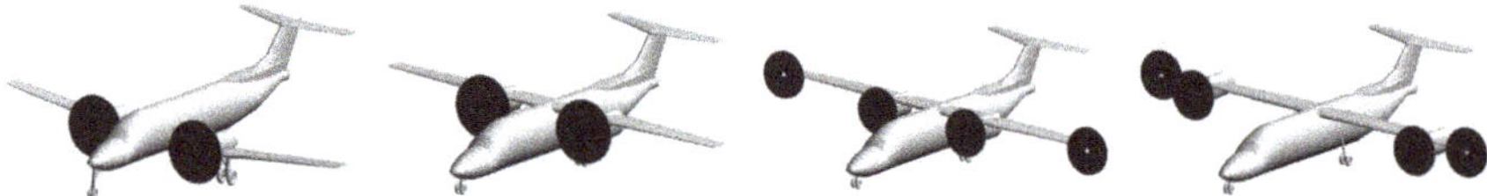

Figure 3. Redesign of a Beechcraft 1900D (**left**), modified redesign (**centre-left**), as well as non-optimized PT2025 (**centre-right**) and optimized PT2025opt (**right**) versions of the partial turboelectric concept aircraft.

In a second major modification step, the conventional propulsion system was exchanged with the partial turboelectric propulsion system, leading to the PT2025 aircraft configuration, see Figure 3 centre-right. The positions of the gas turbines of this aircraft do not change compared to the conventional reference aircraft. They are supplemented by two electric wing tip propellers that are powered directly from two generators mechanically connected to the gas turbines. Subsequently, an aerodynamic optimization of the propeller positions revealed a further outboard location of the conventionally driven propellers, leading to the derivative aircraft design PT2025opt [16]. Since MICADO considers only static loads, the estimation of the wing mass was conducted based on the sizing of the wing box structure considering the three quasi-static load cases, pull-up, gust and landing. Aeroelastic investigations were conducted by numerical flutter analysis and showed that no critical flutter instabilities occur up to 1.2 times the dive speed. Figure 3 shows the optimized aircraft configuration on the right-hand side.

For all aircraft configurations the same design mission, 510 NM trip with 100 NM diversion distance reserve and a 45 min holding, was used. Cruise altitude was set to 23,000 ft at a cruise speed of Mach 0.4.

2.3. *Considered Wiring Configurations*

The partial turboelectric powertrain architecture in this study uses four propellers, two driven by electric motors and two driven by gas turbines. Each electric motor is directly coupled via cables to its related generator, which converts part of the shaft power of the gas turbine, sparing the use of large electric energy storage devices. This configuration enables two wiring options, as suggested by [21]. An "on-wing wiring" approach, where the gas turbine on one wing is coupled to the electric motor on the same wing and a "cross wiring (x-wiring)" approach where the gas turbine is coupled to the electric motor on the opposite wing. Both wiring options can be seen in Figure 1 for the PT2025 and PT2025opt configuration, respectively. The red and green lines symbolize two separate and independent wiring harnesses, each connecting one gas turbine driven generator to one electric motor.

Some parameters of the different configurations are listed in Table 2. All four aircraft are similar to each other and have roughly the same wingspan, differing only 0.1 m. The PT2025opt aircraft are lighter than the PT2025 aircraft with the same wiring option, as the further outboard positioning of the gas turbine engines reduces the structural mass of the wing. The different positions of the gas turbine can also be seen in the conductor mass, where the difference between the "on-wing" and "cross-wing" option is larger for the PT2025opt aircraft due to the increased difference in the necessary cable lengths that can be seen in Figure 1. The increased aerodynamic efficiency of the PT2025opt aircraft, as mentioned in Section 2.2, can be seen in the trip fuel mass. The PT2025opt aircraft with cross wiring (Configuration 4) has a higher OME and MTOM than the PT2025 aircraft with on-wing wiring (Configuration 1), but uses 1.0% less fuel over the same mission.

Table 2. Parameters of the investigated concept aircraft.

	PT2025		PT2025opt	
	On-Wing (Configuration 1)	Cross-Wing (Configuration 2)	On-Wing (Configuration 3)	Cross-Wing (Configuration 4)
Wingspan	17.3 m	17.3 m	17.2 m	17.3 m
OME	4892 kg	4931 kg	4825 kg	4927 kg
MTOM	7629 kg	7671 kg	7537 kg	7653 kg
Conductor mass	31.4 kg	56.6 kg	14.2 kg	71.8 kg
Trip fuel	606 kg	609 kg	591 kg	600 kg

2.4. Handbook Methods for Vertical Tail Plane Sizing According to Roskam

For the resizing of the vertical stabilizer, methods from Part II [24] and Part VI [25] of Roskam's book series on airplane design are used. This series covers the whole design process of an aircraft, from the preliminary design phase to the detailed construction of the different components and serves as standard literature in aircraft design. Most of the formulas from Roskam used in this paper are empirical correlations mainly derived from the DATCOM study of the United States Air Force [19], where an extensive number of experiments to gather data and to derive empirical equations was performed. As the methods used in this study are intended to be used within a preliminary aircraft design and iterated intensively during the mission data interpolation, the computation performance requirements should be kept at a minimum. According to Ciliberti et al. [4] this can be achieved using handbook methods. The error contained within the DATCOM method and hence their derivatives, Roskam [24,25] being one of them, lays around −3.0% for vertical tail aspect ratios of 1.0. As the investigated aspect ratios are around 0.7, the error can be neglected for the configuration in this paper.

According to Roskam [24], a first idea of the required VTP size for conventional configurations can be estimated using the so called volume coefficient, defined as

$$\bar{V}_v = x_v \cdot S_v / S_{ref} \cdot b \quad (3)$$

where $\bar{V}_v$ is the vertical tail volume coefficient, S_v the VTP area, S_{ref} the reference wing area, b the wing span and x_v the longitudinal distance from the aircraft's centre of gravity (CG) to the VTP's aerodynamic centre. As seen in Equation (3) the formula lacks any significance concerning the controllability of the investigated design and is solely derived from the evaluation of mostly conventional configurations, already in service.

In Part II "Preliminary configuration design and integration of the propulsion system" [24], the rudder deflection required to keep the aeroplane stable in case of OEI is given as

$$\delta_r = (N_D + N_{t_{crit}}) / (q_{mc} \cdot S_{ref} \cdot b \cdot C_{n_{\delta_r}}) \quad (4)$$

with the reference wing area S_{ref}, the wing span b and the yawing moment of the remaining engine(s) $N_{t_{crit}}$. According to [24], the yawing moment resulting from the parasitic drag increase in the inoperative engine N_D for a propeller-driven aircraft with variable pitch propellers is 0.1 times $N_{t_{crit}}$. The dynamic pressure q_{mc} is calculated at the minimum control speed $v_{mc} = 1.2 \cdot v_s$, with v_s being the lowest stall speed. $C_{n_{\delta_r}}$ is the so-called control power derivative. The value of δ_r should not exceed 25° [24].

The control power derivative can be obtained via the following formula from Roskam's Part VI [25]:

$$C_{n_{\delta_r}} = -C_{y_{\delta_r}} \cdot (l_v \cdot cos\alpha + z_v \cdot sin\alpha) / b \quad (5)$$

where l_v and z_v are the horizontal and vertical distances, respectively, between the CG of the aircraft and the aerodynamic centre (AC_v) of the vertical tail and α is the angle of attack of the aircraft. The "side-force-due-to-rudder" derivative $C_{y_{\delta_r}}$ is calculated via:

$$C_{y_{\delta_r}} = C_{L_{\alpha v}} \cdot k' \cdot K_b \cdot ((\alpha_\delta)_{C_L} / (\alpha_\delta)_{c_l}) \cdot (\alpha_\delta)_{c_l} \cdot S_v / S_{ref}. \quad (6)$$

where S_v is the surface area of the vertical tail and the coefficients k', K_b, $((\alpha_\delta)_{C_L}/(\alpha_\delta)_{c_l})$, $(\alpha_\delta)_{c_l}$ and the lift curve slope of the vertical tail $C_{L_{\alpha_v}}$ can be obtained from Roskam [pp. 228–261] [25].

$$C_{L_{\alpha_v}} = 2\pi \cdot A_{v_{eff}} / (2 + \sqrt{A^2_{v_{eff}} \cdot \beta^2 / k^2 \cdot (1 + tan^2(\Lambda_{c/2}) / \beta^2) + 4}) \tag{7}$$

with the semi-chord sweep angle of the vertical tail $\Lambda_{c/2}$ and

$$\beta = \sqrt{1 - M^2_{mc}} \tag{8}$$

$$k = c_{l_{\alpha_M}} / 2\pi \tag{9}$$

where M_{mc} is the Mach number corresponding to the minimum control speed. $c_{l_{\alpha_M}}$ is the lift curve slope of the VTP at the same Mach number which can be calculated from

$$c_{l_{\alpha_M}} = c_{l_\alpha} / \sqrt{1 - M^2_{mc}}. \tag{10}$$

The effective aspect ratio of the vertical tail $A_{v_{eff}}$ in Equation (7) is obtained via

$$A_{v_{eff}} = (A_{v(f)} / A_v) \cdot A_v \cdot (1 + K_{vh} \cdot (A_{v(hf)} / A_{v(f)} - 1)) \tag{11}$$

with the vertical tail aspect ratio

$$A_v = b^2_v / S_v \tag{12}$$

where b_v and S_v are the span and the area of the vertical tail, respectively, whereas the coefficients $(A_{v(f)}/A_v)$, $A_{v(hf)}/A_{v(f)}$ and K_{vh} are obtained from Roskam [p.388–p.390] [25].

2.5. *Calculation of the Vertical Tail Plane Geometry*

The calculation of the vertical tail size is performed via a MATLAB® script that loops the whole process described in Section 2.4. The needed aircraft parameters are obtained from the aircraft designs calculated in MICADO. For atmospheric parameters, the ICAO standard atmosphere [26] at sea level is used. Depending on the wiring configuration selected (on-wing or cross wiring), the yawing moments resulting from the thrust of the remaining engines and the drag from the inoperative propellers are determined. Then, starting from a simplified original geometry of the VTP, the initial values for the coefficients and derivatives are calculated. In order to be able to utilize the aforementioned diagrams from Roskam Part VI [25], they were evaluated at discrete points and implemented as tables, splines or polynomial equations and interpolated between the given values. Using Roskam's statement that the maximum rudder deflection must not be more than 25°, Equation (4) gives the maximum value for $C_{n_{\delta_r}}$ and via Equation (5) the maximum $C_{y_{\delta_r}}$. The resulting new surface of the vertical tail plane can be derived from Equation (6).

With the assumption that the shape of the vertical tail as well as the outer profile depth and longitudinal position of the horizontal stabilizer do not change, the new vertical tail geometry and the new position of the aerodynamic centre can be calculated. These serve as an updated starting point for the calculation of the corrected coefficients and the iteration begins again. This is performed until the difference between the newly calculated surface area and the previous one is beneath the convergence limit, set to 0.1% for this study.

To carry out this investigation, some simplifications are made. First, the geometric shape of the vertical tail plane is simplified to correlate with Roskam's assumptions. The original Beechcraft 1900D, as well as the partial turboelectric aircraft described in Section 2.2, feature a vertical tail with a rather complex geometry. With the equations from Roskam it is very difficult to accurately model this shape, as they only give the surface area, the span and the sweep angle. Therefore, the vertical tail is changed to a simple trapezoid whilst maintaining the surface area and the average sweep angle of the original tail to minimize the effects of this simplification. The position of the horizontal tail and the profile depth of the top of the vertical tail are kept constant to minimize the influence on the longitudinal stability of the aircraft. The simplified and original geometry can be seen in Figure 4.

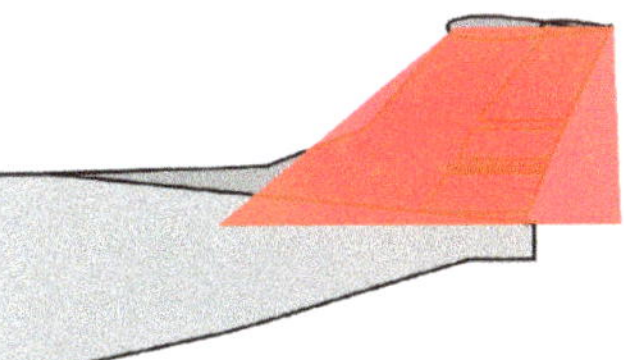

Figure 4. Original and simplified (red) geometry of the vertical tail.

The fuselage, on the other hand, is kept the same and held constant for all calculations. Only the size and position of the wing and the vertical and horizontal tail are allowed to change. The power-split between the different engines is also fixed, each propulsor contributes equally to the total thrust of the aircraft. Possible reductions in the power setting of the electric motors in case of engine loss are not taken into account in this study.

2.6. Iteration Over the Whole Aircraft

A change in the size of the vertical tail has accumulating effects on the whole aircraft—the structural mass, the CG change, and the drag of the aircraft. Therefore, a redesign of the aircraft is necessary to achieve a feasible configuration. This redesign, in turn, has an influence on the required size of the vertical tail. In order to capture these interrelated effects and to obtain a more accurate estimation of the necessary vertical tail size, the methods described in Section 2.4 are extended to take into account the changes in the overall aircraft design. This is performed in two steps. First, the MICADO environment is used to carry out a parametric aircraft design study on a range of fin volume coefficients for all four concept aircraft. Starting from a value of 0.0225, the volume coefficient is increased in steps of 0.02 up to 0.1625 and the resulting aircraft parameters, including among others the fin size, are put out into a data table. This parametric study also includes the fin volume coefficient of 0.0825 of the original Beechcraft 1900D. The data table mentioned before is used as a data source for the subsequent sizing loop of the vertical tail. Second, the iterative design loop for the sizing of the vertical tail which is executed as follows.

The initial iteration step uses the aircraft data from the database created before which belongs to the entry of the original fin volume coefficient of 0.0825. The dataset includes information on e.g., propeller positions and maximum takeoff thrust from the aircraft. Based on this, a new size for the vertical tail is calculated using the algorithm described in Section 2.5, taking into account the propeller positions and wiring options of the partial turboelectric propulsion system. The second and all subsequent iteration steps start with an interpolation of the aircraft data with respect to the previously determined size of the vertical tailplane and the aircraft data, obtained from the parameter study on different fin volume coefficients. Among others, this yields new propeller positions due to changes in wingspan and values for maximum takeoff thrust. As described for the first iteration step, this interpolated data is used to update the size of the vertical tailplane. This process is repeated until the area of the vertical tailplane does not change more than 0.1% between two consecutive iteration steps. The whole iteration process, including the steps from Section 2.5, can be seen in Figure 5.

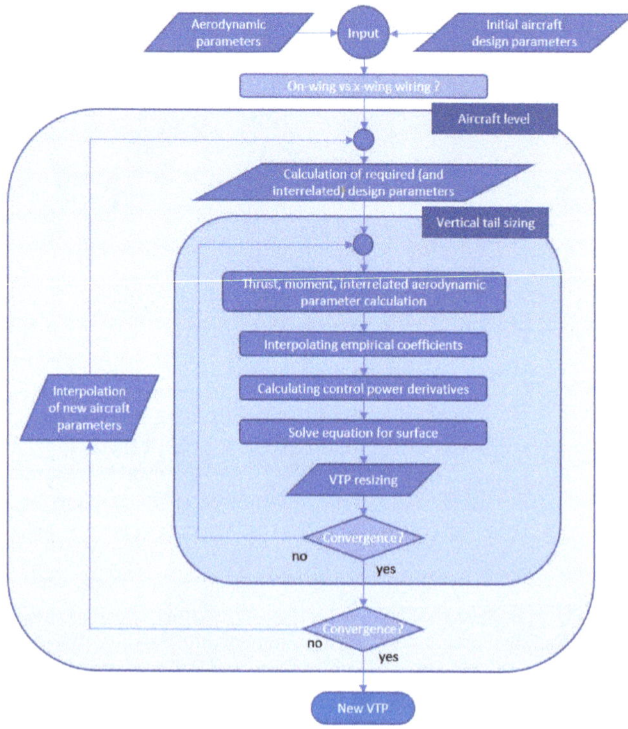

Figure 5. Whole aircraft iteration overview.

2.7. Assumptions and Restrictions

As stated in Section 2.4, the assumptions are based on the usability of handbook methods, as validated by Ciliberti et al. [4]. Therefore, the suitability of the correction method described in this paper is only valid for configurations with similar properties. These are a high-wing configuration, a comparable tail-cone geometry and a VTP aspect ratio of around 1.0. Additionally, a VTP shape close to the simplified versions of [25] should be considered. As mentioned in the introduction, according to [6] the additional rolling motion due to the local absence of power augmented lift downstream of the propellers is neglected. As the wing tip propellers reduce the induced drag, their failure will increase the induced drag slightly. As the total reduction in induced drag through the use of a wing tip propulsion is only around 5% for the investigated aircraft [16], this effect is neglected. To reduce the complexity of this study, the propeller-induced side wash interaction on the vertical tail and the possible lift increase on the wing due to the induced velocity, are neglected.

A possible solution to reduce the VTP size for the PT2025 configurations could be the implementation of an automatic thrust control system, which is able to reduce the thrust on the electric engine quickly in case of a failure of the conventional engine, according to CS23.904 [8]. Another possibility would be the interconnection of all electric machines or the implementation of a buffer battery. This way, the electric propellers could still operate if one combustion engine fails. The downside to this would be the increased weight due to the added components and the increased complexity, which is why this case was not considered in this study.

3. Results

In the following section, the results of all studies are shown. First, the calculation of the required VTP sizes for the four propeller and wiring configurations is presented, while the rest of the aircraft remains unchanged, negating the so-called snow ball effects resulting from changes to one or more aircraft components. As can be seen in Section 3.2, some configurations would require a huge VTP in order to be controllable. Therefore, only designs with realistic fin sizes are further evaluated in Section 3.3.

3.1. Sensitivity Study on the Vertical Tail's Volume Coefficient

In order to evaluate the influence of changes to the vertical tail's volume coefficient, the MICADO environment is used to carry out a systematic study on this design parameter. Figure 6 illustrates the resulting aircraft designs of the partial turboelectric aircraft with optimized propeller positions and a volume coefficient of 0.04253 and 0.14253. This also explains why it was not possible to calculate a converged aircraft design for a volume coefficient of 0.16253 or greater, as this leads to a large vertical tail and not physically feasible aircraft. The results of this study, in terms of the operating empty mass and the required trip fuel of the aircraft, can be seen in Figure 7.

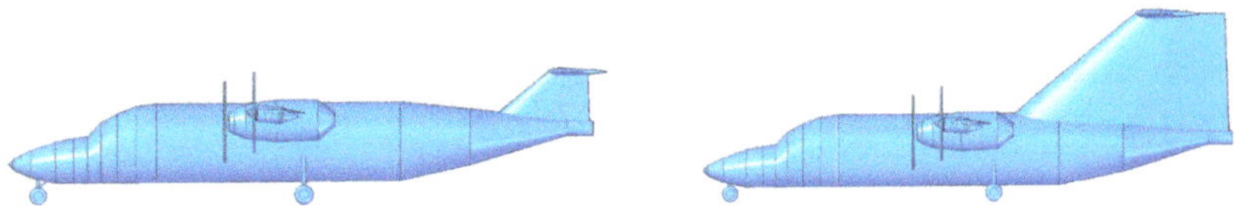

Figure 6. Partial turboelectric aircraft with optimized propeller positions and a volume coefficient of the vertical tail of 0.04253 (**left**) and 0.14253 (**right**).

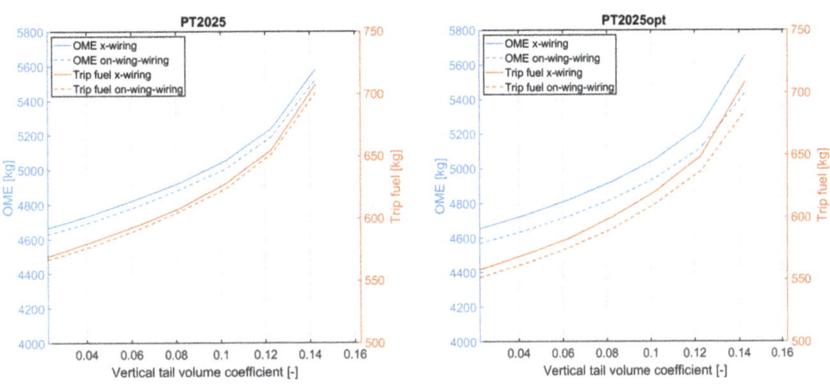

Figure 7. Influence of the vertical tail's volume coefficient on the OME (blue) and the trip fuel (red) for the PT2025 (left) and PT2025opt (right) configurations.

The OME and trip fuel were chosen to show the cascading effects caused by a change in the volume coefficient. A bigger volume coefficient results in a larger vertical tail, which increases the OME of the aircraft. This leads to a larger wing needed to generate the required lift, further increasing the empty mass. For a constant volume coefficient, the larger wing also means a larger vertical tail. This loop continues until an equilibrium is reached and explains the non-linearity of the OME seen in Figure 7. This effect is even more prominent for the mission fuel, as the fuel burn not only depends on the mass of the aircraft, but also on the drag. A larger vertical tail means a bigger wetted surface and an increased viscous drag.

The slight differences between the on-wing and cross-wiring options result from the different cable lengths and masses. This effect is more visible for the PT2025opt aircraft, as the gas turbines are located further outward on the wing. This increases the distance to the electric engine on the opposite wing, while minimizing the distance to the electric engine on the same wing.

3.2. Size of the Adjusted Vertical Tail

The results for the resizing of the vertical tail are shown for all four configurations, in form of the surface areas and corresponding volume coefficients, are given in Table 3. The difference in the surface areas of the original vertical tails for the different wiring options results from the distinct cable lengths, as described in Section 2.3. The cross-wiring option needs longer cables than the on-wing wiring and therefore has a higher OME and MTOM. The resulting aircraft needs a larger wing to account for the increased lift demand. Since MICADO is set to calculate the size of the vertical tail based on a fixed volume coefficient this results in a larger vertical tail, as shown in Equation (3).

As can be seen in Table 3, the size of the vertical tail strongly depends on the chosen wiring option and the configuration. In case of the PT2025 aircraft, the required vertical tail area for the on-wing wiring (Configuration 1) is twice the surface of the cross-wiring option (Configuration 2), since the remaining thrust in the case of a gas turbine failure is accumulated on one side of the aircraft. The large lever arm of 8.6 m of the electric engine at the wing tip, in addition to the lever arm of 2.5 m of the turboprop engine, increases the yawing moment compared to the conventional reference for which the vertical tail was originally designed. For the cross-wiring option the residual yawing moment is still larger than for the reference aircraft, as the span-wise distance between the integration positions of the electric and the turboprop engine is 2.4 times the distance between the turboprop engine and the fuselage centreline in the reference configuration. This equalizes the fact that the electric engine only produces half the thrust of the turboprop engine in the conventional reference aircraft. As expected, the more outward position of the gas turbine in the PT2025opt aircraft results in a larger difference in the VTP size between the two wiring options. Here, the vertical tail in the on-wing wiring (Configuration 3) case is almost nine times as large as the one for the cross-wiring (Configuration 4) and even supersedes the wing, which has a surface area of 32.15 m^2. The cross-wiring option, on the other hand, results in a vertical tail smaller than the original one, because the distance between the electric engines at 8.3 m and the turboprop engine at 6.0 m is smaller than the lever arm in the reference aircraft, while the electric engine also produces only half the thrust.

Table 3. Surface area and volume coefficients of the original and resized vertical tail for the four considered aircraft and wiring configurations.

		Original VTP	Resized VTP	Change
Configuration 1	S_v	6.75 m^2	27.72 m^2	+310.7%
	$\tilde{V}_v$	0.083	0.247	+197.6%
Configuration 2	S_v	6.77 m^2	13.47 m^2	+99.0 %
	$\tilde{V}_v$	0.082	0.147	+79.3%
Configuration 3	S_v	6.68 m^2	36.38 m^2	+444.6%
	$\tilde{V}_v$	0.083	0.288	+347.0%
Configuration 4	S_v	6.87 m^2	4.14 m^2	−39.7%
	$\tilde{V}_v$	0.083	0.053	−36.1%

3.3. Results of the Total Aircraft Iteration

The results of the vertical tail, as seen in Table 3, show that the on-wing wiring option is not suitable for both configurations, as it would lead to large VTPs, which significantly increase the overall weight and drag of the aircraft. The aircraft with on-wing wiring are therefore neglected in the following calculations that account for the aircraft changes and only the cross-wiring options are studied further. Additionally, MICADO could not reach convergence for aircraft with volume coefficients above 0.1425 since this leads to an aircraft that is not physically feasible, as can be seen in Figure 6.

For the PT2025 configuration, the aircraft VTP optimization was unable to reach a converged result, as the increase in mass, required thrust and change in CG, due to the larger VTP, led to a further increase in the necessary tail area. After one iteration, the updated VTP area was 20.26 m^2, already outside MICADO's calculated design space. Therefore, no results for this configuration can be shown.

For the PT2025opt configuration, the aircraft level calculation could be conducted, with the results displayed in Figure 8. In contrast to the method described in Section 3.2, the position of the horizontal tail was allowed to change, since the whole aircraft design and horizontal stability were taken into account. The resulting surface area and volume coefficient, as well as the change in OME and mission fuel are given in Table 4.

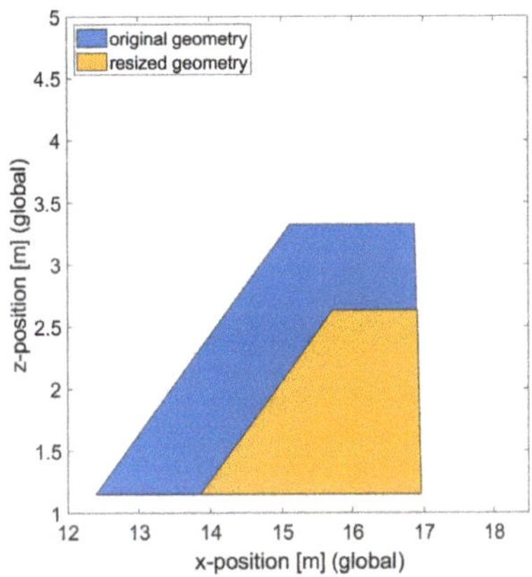

Figure 8. Original (blue) and resized (amber) vertical tail for the PT2025opt aircraft with cross wiring (Configuration 4).

Compared to the considerations of the isolated vertical tail, the resulting fin surface area is decreased. The larger distance between the aerodynamic centres of the wing and the vertical tail, as well as the changes in the whole aircraft explain the difference between the relative change of the fins surface area and the volume coefficient. The mass savings for the vertical tail of 79.6 kg offset the extra mass of the cables of 67.7 kg, resulting from the cross-wiring configuration. The cascading effects described in Section 3.1 also explain why the decrease in the OME is 182.0 kg, 102.4 kg more than the mass savings for the vertical tail. The additional drag reduction due to the smaller vertical tail leads to a combined fuel saving of 4.7% for the design mission of 510 NM.

Table 4. Comparison of VTP, OME and mission fuel of the original and optimized PT2025opt aircraft with cross wiring (Configuration 4), taking into account the iteration of the whole aircraft.

	Original Configuration 4 Aircraft	Configuration 4 Aircraft with Adjusted Fin	Change	
VTP surface	6.87 m^2	3.18 m^2	−3.69 m^2	−53.7%
Volume coeff	0.083	0.045	−0.038	−45.8%
VTP mass	146.6 kg	66.9 kg	−79.6 kg	−54.3%
OME	4926.6 kg	4744.6 kg	−182.0 kg	−3.7%
Trip fuel	600.3 kg	571.8 kg	−28.5 kg	−4.9%

4. Conclusions

The sizing of the vertical tailplane based on volume coefficients is clearly inadequate when it comes to novel powertrain concepts. The presented method allows to amend an existing preliminary design tool with a more detailed assessment of the vertical tailplane design without the need of changing the design process itself. This is performed by resizing the vertical tail with handbook methods that account for specific concepts, such as outboard positioned propellers. Within the given restrictions the research showed that the introduction of wing tip propellers has a significant influence on the sizing of the vertical tail. Out of the four configurations investigated in this study, only one configuration leads to a smaller vertical tail, reducing the VTP area by 53.7%, compared to the initial volume coefficient-based sizing of the PT2025opt configuration performed in MICADO, whilst fulfilling the necessity to counteract the residual torque resulting from the loss of one gas turbine. This leads to a reduction in OME of and 3.7% and a 4.9% lower fuel consumption over the given design mission. For the PT2025 configuration the electrification of the powertrain leads to unrealistic large vertical tailplanes that supersede the reference wing area.

The results of this work present a simple method to be integrated in the preliminary aircraft design and the necessity to do so. In our specific high-wing, T-tail configuration with wing tip propulsors, the possibility to reduce the VTP size and therefore the possibility to reduce fuel consumption and emissions of the investigated configuration was shown. As this method has the potential to improve the preliminary design of unconventional aircraft, further research should be conducted to test its suitability for different configurations, while possibly substituting the used handbook methods by advanced methods, depending on the required computation time.

Author Contributions: Conceptualization, A.A., A.B., E.S. and A.S.; methodology, A.A.; software, A.B.; data curation, P.S. and C.Z.; writing—original draft preparation, A.A., A.B, P.S. and C.Z.; writing—review and editing, A.A., A.B., P.S., C.Z., E.S. and A.S.; supervision, E.S. and A.S.; All authors have read and agreed to the published version of the manuscript.

Funding: This research was supported by Federal Ministry for Economic Affairs and Energy on the basis of a decision by the German Bundestag under Grant Agreement No. 20E1916A and 20E1916F.

Institutional Review Board Statement: Not applicable.

Informed Consent Statement: Not applicable.

Data Availability Statement: This research was conducted with data compiled by the "GNOSIS" consortium. It is yet available in various publications, this paper references. Please check the References, or contact the researchers for access to data.

Conflicts of Interest: The authors declare no conflict of interest. The funders had no role in the design of the study; in the collection, analyses, or interpretation of data; in the writing of the manuscript; or in the decision to publish the results.

Abbreviations

The following abbreviations are used in this manuscript:

CG	Centre of gravity
CS	Certification specification
CLT	Critical loss of thrust
DEP	Distributed electric propulsion
EASA	European Union Aviation Safety Agency
ICAO	International Civil Aviation Organization
MICADO	Multidisciplinary Integrated Conceptual Aircraft Design and Optimization
MTOM	Maximum take-off mass
NASA	National Aeronautics and Space Administration
OEI	One engine inoperative
OME	Operating mass empty
TLAF	Top level aircraft functions
TLAR	Top level aircraft requirements
UNICADO	University Conceptual Aircraft Design and Optimization
VTP	Vertical tailplane

References

1. Jansen, R.; Bowman, D.C.L.; Clarke, S.; Avanesian, D.; Dempsey, D.P.; Dyson, D.R.W. NASA Electrified Aircraft Propulsion Efforts. *Aircr. Eng. Aerosp. Technol.* **2019**, *92*, 667–673. [CrossRef]
2. Airbus. EcoPulse. A New Approach to Distributed Propulsion for Aircraft. 2022. Available online: https://www.airbus.com/en/innovation/zero-emission/electric-flight/ecopulse (accessed 25 January 2023).
3. Moebs, N.; Eisenhut, D.; Windels, E.; van der Pols, J.; Strohmayer, A. Adaptive Initial Sizing Method and Safety Assessment for Hybrid-Electric Regional Aircraft. *Aerospace* **2022**, *9*, 150. [CrossRef]
4. Ciliberti, D.; Della Vecchia, P.; Nicolosi, F.; de Marco, A. Aircraft directional stability and vertical tail design: A review of semi-empirical methods. *Prog. Aerosp. Sci.* **2017**, *95*, 140–172. [CrossRef]
5. Hoogreef, M.F.M.; Soikkeli, J.S.E. Flight dynamics and control assessment for differential thrust aircraft in engine inoperative conditions including aero-propulsive effects. *CEAS Aeronaut. J.* **2022**, *13*, 739–762. [CrossRef]
6. Vechtel, D.; Buch, J.P. Aspects of yaw control design of an aircraft with distributed electric propulsion. *CEAS Aeronaut. J.* **2022**, *13*, 847–860. [CrossRef] [PubMed]
7. Schneider, J.; Frangenberg, M.; Notter, S.; Scholz, W.; Fichter, W.; Strohmayer, A. Integration of propelled yaw control on wing tips: a practical approach to the Icaré solar-powered glider. *CEAS Aeronaut. J.* **2022**, *13*, 861–876. [CrossRef]
8. European Union Aviation Safety Agency. *Certification Specifications for Normal-Category Aeroplanes: CS-23, Amendment 4*; European Union Aviation Safety Agency: Cologne, Germany 2015.
9. European Union Aviation Safety Agency. *Certification Specifications for Normal-Category Aeroplanes: CS-23, Amendment 5*; European Union Aviation Safety Agency: Cologne, Germany, 2020.
10. Schlickenmaier, H.; Voss, M.G.; Wilkinson, R.E. Certification Gap Analysis. Technical report, National Aeronautics and Space Administration, 2019. Available online: https://ntrs.nasa.gov/citations/20190033236 (accessed 1 December 2022).
11. Schlittenhardt, J. Certification Gap Analysis for Hybrid-Electric 19-Seat Aircraft. Master's Thesis, University of Stuttgart, Stuttgart, Germany, 2022.
12. Jézégou, J.; Sufyan, U. Safety and Certifiability Evaluation of Distributed Electric Propulsion Airplane in EASA CS-23 Category. *IOP Conf. Ser. Mater. Sci. Eng.* **2021**, *1024*, 012076. [CrossRef]
13. Risse, K.; Anton, E.; Lammering, T.; Franz, K.; Hoernschemeyer, R. An Integrated Environment for Preliminary Aircraft Design and Optimization. In Proceedings of the 53rd AIAA/ASME/ASCE/AHS/ASC Structures, Structural Dynamics and Materials Conference: SciTech 2012, Honolulu, Hawaii, 14 June 2012; American Institute of Aeronautics and Astronautics: Reston, VA, USA, 2012; Volume AIAA 2012-1675. [CrossRef]
14. Schültke, F.; Aigner, B.; Effing, T.; Strathoff, P.; Stumpf, E. *MICADO: Overview of Recent Developments Within the Conceptual Aircraft Design and Optimization Environment*; Deutsche Gesellschaft für Luft-und Raumfahrt-Lilienthal-Oberth eV: Cologne, Germany, 2020.
15. Schültke, F.; Stumpf, E. *UNICADO—Aufbau und Etablierung einer Universitären Flugzeugvorentwurfsumgebung: Präsentation*; Deutsche Gesellschaft für Luft-und Raumfahrt-Lilienthal-Oberth eV: Cologne, Germany, 2020.
16. Zumegen, C.; Strathoff, P.; Stumpf, E.; Schollenberger, M.; Lutz, T.; Krämer, E.; Kirsch, B.; Friedrichs, J.; Schubert, M.; Dafnis, A.; et al. Aerodynamic and structural analysis of a partial turboelectric commuter aircraft with wingtip propellers at aircraft level. In Proceedings of the AIAA AVIATION 2022 Forum, Chicago, IL, USA, Virtual 27 June–1 July 2022; American Institute of Aeronautics and Astronautics 2022. [CrossRef]
17. Kroo, I.M.; Shevell, R. *Aircraft Design: Synthesis and Analysis: Textbook Version 0.99*; Desktop Aeronautics Inc.: Standford, CA, USA, 2001.
18. Stückl, S. Methods for the Design and Evaluation of Future Aircraft Concepts Utilizing Electric Propulsion Systems. Ph.D. Thesis, Technische Universität München, München, Germany, 2016.
19. Horstmann, K.H. Ein Mehrfach-Traglinienverfahren und Seine Verwendung für Entwurf und Nachrechnung nichtplanarer Flügelanordnungen. Ph.D. Thesis, DFVLR-FB 87-51, Deutsche Forschungs- und Versuchsanstalt für Luft- und Raumfahrt, Cologne, Germany, 1988.
20. Raymer, D.P. *Aircraft Design: A Conceptual Approach*, 6th ed.; American Institute of Aeronautics and Astronautics, Inc.: Reston, VA, USA, 2018.
21. Zumegen, C.; Strathoff, P.; Stumpf, E.; van Wensveen, J.; Rischmüller, C.; Hornung, M.; Geiß, I.; Strohmayer, A. Technology selection for holistic analysis of hybrid-electric commuter aircraft. *CEAS Aeronaut. J.* **2022**, *13*, 597–610. [CrossRef]
22. Raytheon Aircraft Company. *Pilot Operating Handbook Beechcraft 1900D*; Springer: Textron Aviation Inc.: Wichita, KS, USA, 2006.
23. European Aviation Safety Agency. *Type-Certificate Data Sheet for PT6A-67 Series Engines*; European Union Aviation Safety Agency: Cologne, Germany, 2019.
24. Roskam, J. *Preliminary Configuration Design and Integration of the Propulsion System*; Airplane design/Jan Roskam; Lawrence, K., Ed.; DAR Corporation: St Lawrence, KS, USA,1997; Volume II.

25. Roskam, J. *Preliminary Calculation of Aerodynamic, Thrust and Power Characteristics;* Airplane design/Jan Roskam; Lawrence, K., Ed; DARcorporation: St Lawrence, KS, USA, 1990; Volume VI.
26. International Civil Aviation Organization. *Manual of the ICAO Standard Atmosphere—Extended to 80 kilometres/262,500 feet: Doc 7488;* International Civil Aviation Organization: Montreal, QC, Canada, 1993.

Article

Prediction of Capacity Regulations in Airspace Based on Timing and Air Traffic Situation

Francisco Pérez Moreno *, Víctor Fernando Gómez Comendador , Raquel Delgado-Aguilera Jurado , María Zamarreño Suárez and Rosa María Arnaldo Valdés

Department of Aerospace Systems, Air Transport and Airports, Universidad Politécnica de Madrid (UPM), 8040 Madrid, Spain

* Correspondence: francisco.perez.moreno@upm.es; Tel.: +34-640-37-02-40

Abstract: The Air Traffic Control (ATC) system suffers from an ever-increasing demand for aircraft, leading to capacity issues. For this reason, airspace is regulated by limiting the entry of aircraft into the airspace. Knowledge of these regulations before they occur would allow the ATC system to be aware of conflicting areas of the airspace, and to manage both its human and technological resources to lessen the effect of the expected regulations. Therefore, this paper develops a methodology in which the final result is a machine learning model that allows predicting capacity regulations. Predictions shall be based mainly on historical data, but also on the traffic situation at the time of the prediction. The results of tests of the model in a sector of Spanish airspace are satisfactory. In addition to testing the model results, special emphasis is placed on the explainability of the model. This explainability will help to understand the basis of the predictions and validate them from an operational point of view. The main conclusion after testing the model is that this model works well. Therefore, it is possible to predict when an ATC sector will be regulated or not based mainly on historical data.

Keywords: ATFCM Regulations; machine learning; air traffic; historical data; explainability

Citation: Pérez Moreno, F.; Gómez Comendador, V.F.; Delgado-Aguilera Jurado, R.; Zamarreño Suárez, M.; Arnaldo Valdés, R.M. Prediction of Capacity Regulations in Airspace Based on Timing and Air Traffic Situation. *Aerospace* **2023**, *10*, 291. https://doi.org/10.3390/aerospace10030291

Academic Editors: Spiros Pantelakis, Andreas Strohmayer and Jordi Pons-Prats

Received: 12 January 2023
Revised: 28 February 2023
Accepted: 14 March 2023
Published: 15 March 2023

1. Introduction

Increased aircraft demand in air transportation is causing congestion at certain airports and airspace capacities [1]. Moreover, this demand continues to grow year after year. EUROCONTROL has estimated that air traffic will grow by 1.9% per year until 2040 [2]. This increase in aircraft demand creates capacity problems and associated delays, and these effects result in a drop in the efficiency of the air transport system [3]. This situation is already of concern, but is expected to worsen in the coming years. Therefore, improving Air Traffic Management (ATM) efficiency is critical to cope with the increase in aircraft demand [4].

To solve the problems due to the lack of capacity of the Air Traffic Control (ATC) system, the Air Traffic Flow Capacity Management (ATFCM) system arises. Its function is to try to balance the air traffic demand and the capacity of the ATC system at the strategic, pre-tactical, and tactical levels [5]. The ATFCM system proposes short-term and long-term measures to address the effects of the lack of capacity and meteorological uncertainties, and the effectiveness of these measures will depend on the amount, accuracy, and timeliness of the information exchanged [6]. Currently, the most typical ATFCM measure is to limit the entry of certain aircraft into airspace where there are capacity problems. These are the so-called ATFCM regulations. Regulations aim to reduce the workload of Air Traffic Controllers (ATCOs) [7], but they carry associated ground delays concerning the initially planned time of certain aircraft [8]. In recent years, around 50–60% of the total delay in Europe was caused by en-route airspace problems such as en-route capacity or weather [9], reaching even more than 70% in 2018 and 2019 [9]. This makes regulations analysis very relevant and a topic of interest for the industry.

Specifically, there are up to 14 types of regulations identified by EUROCONTROL [10]. The most common regulations are those due to weather and lack of ATC capacity. The major cause of en-route ATFCM delay is the lack of ATC capacity or ATC staffing [9], meaning that in a part of the airspace, the ATC system cannot cope with the air traffic demand [11]. The weather is also a cause of regulation that causes significant delay. Due to the impact on airport and airspace capacity, and its strong influence on operations [12], adverse weather conditions can lead to large demand-capacity mismatches [13].

This great importance of regulations and their causes means that the study of ATFCM and regulations is arousing interest in the industry [14,15]. Related to this topic, the prediction of regulations due to lack of capacity has been set as the objective of this paper. Thanks to emerging digital technologies, such as [16], the proposed methodology is expected to help improve the efficiency of the ATC system [17]. In this regard, this paper is expected to help manage the ATC system's capacities through prior knowledge of the regulations due to lack of capacity. This way, it is expected that the ATC system will be able to better organise its human and technological resources.

In this paper, our focus is to study only regulations due to capacity. Some studies raise the prediction of regulation during adverse weather conditions [18] or propose solutions such as rescheduling during adverse weather conditions [19]. However, the nature of weather is random and variable [20], being of a different nature to the lack of capacity regulations. For this reason, this is not the subject of study of this paper.

As regulations often arise from imbalances between capacity and demand [21], in this paper, it has been decided to try to predict these regulations. The specific objective is to develop a machine learning model based on historical ATFCM regulations and some information on how air traffic is structured in the sector. This machine learning is thus set to predict regulations due to a lack of capacity.

To achieve this objective, in Section 2 a literature review is presented to see related research on the topic. Then, the methodology developed for the approach of this model is described in Section 3. In addition, an example of the performance and explainability of the developed model under a real operating scenario is described in Section 4. Finally, Section 5 discusses the conclusions obtained after the development of the model and the future steps to be taken in this line of research.

2. Literature Review

In this section, a review of the literature related to the topic of ATFCM regulations and their prediction will be carried out. ATFCM regulations aim to adapt air traffic demand to the capacity of the ATC system. The regulations have their consequences on an operational tier, but also on an economic tier. In [11], the adverse economic aspects of the regulations are analysed. It is estimated that the total cost can be very large due to the cost of extra fuel, crew cost, or compensation for the regulations themselves. Leading to this, the number of ATFCM regulations should be minimised.

As ATFCM regulations have to be minimised, the study of the causes of re-regulations becomes a priority. According to EUROCONTROL [10], the different causes of these imbalances can be as many as 14. However, in practice, most regulations are caused by the same reasons. Specifically, based on historical data in European airspace in 2019, the distribution of causes of regulation was as stated in Table 1:

Table 1. Main causes of ATFCM regulations.

	Percentage of Regulation Cause (%)
C (ATC Capacity)	50.90
S (ATC Staff)	15.00
W (Weather)	12.84
G (Aerdrome Capacity)	9.71
I (Industrial Action)	3.33
M (Military Activity)	3.18
T (Equipment)	2.03
P (Special Event)	1.09

Table 1 shows only those causes with a percentage higher than 1% of the total ATFCM regulations. These causes are mainly due to the lack of operational capacity or the weather. Therefore, although the EUROCONTROL studies estimate that there are many causes, there are two main causes that are responsible for ATFCM regulations: Lack of capacity and climate.

Climate is the most unpredictable factor. Some studies make analyses related to climate, such as [20], where a model is developed to predict trajectories when there are weather uncertainties, or [18], where a prediction of aircraft in the sector and regulations in adverse weather conditions is made. However, this variability makes weather regulations very difficult to analyse and estimate. In [22], a study is carried out on the possibility of estimating weather regulations. In this reference, a machine learning model based on components such as wind, humidity, and temperature is used. Although the results seem to be positive, there is still a lot of work to be completed.

On the other hand, the study of capacity regulation is more widespread. There is a belief that it will be possible to estimate, and therefore anticipate, capacity regulations and their effects. For this reason, more studies are being carried out on this subject.

Some articles focus on studies of the influence of capacity regulation, such as [23]. This paper develops a study of the applicability of machine learning models to predict the delay caused by capacity regulations. This research concludes that it is beneficial to use data-driven machine learning models to predict these delays, rather than using causal relationships.

Another article that focuses on the study of capacity regulations is [14]. Here, a study is made of how capacity regulations can help reveal restorative mechanisms for tactical planning. A methodology for defining network states has been developed based on these capacity regulations.

In addition to the analysis of the capacity regulations themselves, there is also interest in their prediction by allowing the ATFCM service to anticipate their effects. This line of research is represented by [24], where a machine learning model is developed that is capable of estimating capacity regulations using variables such as the capacity itself, the number of aircraft, or the expected workload of the controllers.

This research is currently gaining importance due to the development of machine learning models and the increasing imbalance between capacity and demand. Therefore, this paper is in this line. The aim of this paper is the same as the one of [24]. Both papers attempt to predict capacity regulations. Therefore, the objective is different from the rest of the publications analysed:

- The aim of this paper is different from that of [14,23], because in these two publications, regulations are analysed as a component, although the final target of analysis is the delays in [23] and the definition of network states in [14].
- The aim of [22] is to predict regulations by weather, so the model is different from the one in this paper, and the theoretical background will be different.

Therefore, from the publications analysed, the only one that shares this objective is [24]. The main difference is the scope of the model, and thus the composition. The objective of [24] is the prediction of regulations by capability but on a tactical or pre-tactical time horizon. In this publication, the scope is to predict regulations in a strategic horizon. This makes the theoretical background, and therefore the machine learning models developed, different, and even complementary.

After a review of the literature, one can also conclude the contributions of this paper to the prediction of regulations.

This paper aims at predicting regulations, as the other analysed papers. However, it tries to predict capacity regulations by means of a different and novel approach. Previous models are based solely or mainly on what the traffic is like at the time of prediction or what it is expected to be like. This is important, but the regulations will depend on many external aspects such as the situation of the ATC system. From an operational point of view, the time component is very important, as here, behaviour patterns can be found that do not depend solely on traffic.

In addition, the traffic flow distribution also allows for studying the influence of traffic without taking into account each individual aircraft. This will make it possible to find behavioural patterns in the traffic structure in general.

Finally, this model for predicting regulations will allow progress to be made in the study and prediction of their effects, which is what is really important. From an operational point of view, in a control room, 10-min time ranges can be used to evaluate the possible effect of the regulation, since the final real delay is always different from the ATFCM delay that an aircraft has, and this is seen in 10-min periods in order to have a margin to estimate this difference. Therefore, from an operational point of view, the 10-min window is used as the analysis parameter. Therefore, the approach followed has been based on real operational knowledge and will also bring these investigations closer to real operation.

3. Methodology

Once the motivation and objectives of this paper have been stated, the methodology in which the prediction of regulations in airspace is presented. This paper builds on some previous works, such as [23], where the possibility of predicting delay in the presence of airspace capacity regulations in the airspace is discussed. However, in this paper, the aim is to go further, looking directly at the cause rather than the consequence.

In [22], the authors also propose a prediction of ATFCM regulations, but this prediction is of weather-based regulations. Even so, part of the methodology is applicable. These same authors adapt this prediction of regulations to capacity regulations in [24]. This model is based on predictions of capacity regulations, which are also based on the traffic situation and evolution in the sector, as well as on the controller's workload. To complete the study of this work, it has been decided to use complementary variables to this study. Specifically, it has been decided to eliminate the variables related to the workload of the controllers, as this variable is subjective and subject to a certain model [25]. In addition, it has been decided to structure the air traffic variables in the main traffic flows. Therefore, the model will have a limited number of variables, but the idea is that these variables will give a complete picture of how the traffic is structured.

For this reason, in this paper, the aim is to make a prediction of capacity regulations based on objective data. The data which will be used in the machine learning model is:

- The expected date. In other models, the time component is small or null. However, in air traffic, patterns of behaviour can be found. It is expected that patterns can also be found in the occurrence of capacity regulation in time. Given this hypothesis, the temporal analysis is considered fundamental in this paper.
- Static traffic predictions. CRIDA organises traffic into main traffic flows based on flight plans. Therefore, the machine learning model will be adapted to be able to structure the air traffic into these flows and obtain a picture of the traffic at a given time.

Therefore, the proposed methodology will be based on three main areas, and Figure 1 shows the scheme that will be followed for the development of the machine learning model.

As schematised in Figure 1, the model is based on three information inputs:

- Air Traffic database: This database shall be composed of all aircraft flight plans. These flight plans are used to organise the traffic by its main air traffic flows, identifying each trajectory with a traffic flow.
- Regulations database: This database will only be present for training, as it will be the result of the model prediction.
- Date selection: Although it is not a database, the fundamental input to the model will be the date on which the prediction wants to be made.

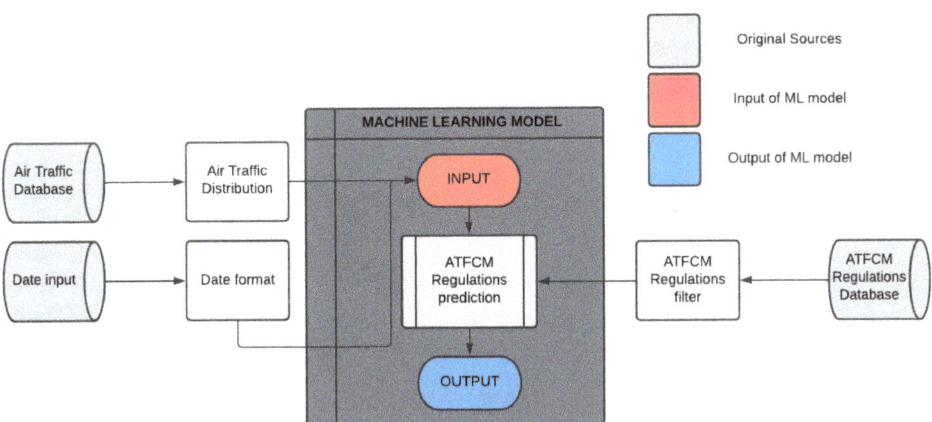

Figure 1. Scheme of the methodology of the machine learning model establishment.

In the following, it is detailed how each of the branches of the model is worked with and how they will be combined to form the complete machine learning model.

3.1. ATFCM Regulations Database

Firstly, the format of the output is explained, as the entire regulatory prediction model will depend on it. As the aim of this paper is to predict ATFCM regulations based on a strong temporal component, it is fundamental how these regulations are determined.

The first step is to filter the regulations to be predicted. In this paper, regulations based on the lack of ATC capacity will be predicted. Based on the causes proposed by [10], the following causes shall be considered:

- ATC Capacity
- ATC Staff
- Airport Capacity

These causes are related to the lack of capacity of the ATC system or of certain airports.

In addition, it has been decided to organise the ATFCM regulations in 10-min periods and in days of the year. In this way, all the time that the airspace was regulated will be arranged in a matrix as in Figure 2. The rows of the matrix will be the day of the year under analysis, and the columns will be the 10-min period, with the start of the period marked as the column name.

In this matrix, it will be noted in which periods and days when the sector will be regulated (1) and when it will not be regulated (0). With this format, a simple picture of when the sector will or will not be regulated emerges. Once the format of the output has been determined, the input variables, both the time and traffic databases, need to be adapted.

Furthermore, by arranging the data in this way, it can be established that the best-fitting machine learning model is a binary classification model. Classification problems are already widespread in the industry [26,27], and binary classification is the most widespread type of model. In a binary classification model, the training and test data are distributed into two labels, 0 (in this case, the sector is not regulated) and 1 (in this case, the sector is regulated). Therefore, by training the model with only two labels, the model will predict only the two labels. This type of model has the advantage of facilitating training by having fewer classes, and of being a model with simpler explainability than a model with a larger number of classes to classify.

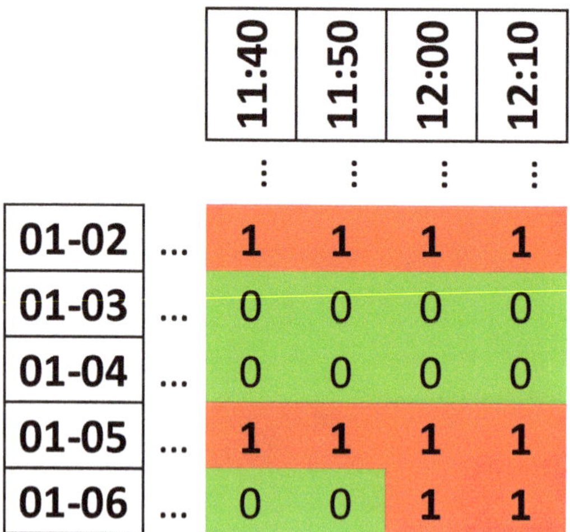

Figure 2. ATFCM Regulations array.

This will facilitate the development and implementation of this model. In addition, a Random Forest model has been chosen to carry out this classification as it is an algorithm that works well in problems of a different nature and that allows a correct analysis of the explainability of the model [28].

3.2. Date Input

The format of the model output has been determined, and consequently what type of machine learning model will be used. The next step is to determine the input variables and adapt them to the model output.

The most important part of the model is the time-based variables. The machine learning model will try to analyse patterns in the historical data to try to predict when a sector will be regulated, so it is essential to have a date-based input variable format. The input variables will be based on the information that can be extracted from the matrix in Figure 2. In particular, the following information will be extracted.

- Month: The date format indicated in the rows of the matrix will be mm-dd. Therefore, knowing the month to which any row belongs is straightforward. The month is a very useful variable, as the operation in certain sectors can be very seasonal, an example being the Balearic Islands [29].
- Day of the week: Like the month, the day of the week can be a very interesting variable for the search for patterns. This information can also be obtained relatively easily by formatting the rows of the matrix, knowing the year of operation, and with the help of a calendar.
- Period of the day: The latest information will be obtained from the columns of the matrix in Figure 2. This information is the period of the day. In the matrix in Figure 2, there will be 144 rows, starting with the period 00:00–00:10 and ending with the period 23:50–00:00. Therefore, by numbering from 1 to 144, it is possible to identify in which period of the day the sector will be regulated or not. This variable is also considered interesting from an operational point of view as it will allow us to know if there are periods of the day in which the sector is more likely to be regulated.

Using this information, it is considered that a complete temporal analysis can be established and that the simplicity of the machine learning model is maintained.

3.3. Air Traffic Database

The last step in determining the input variables is to add to the temporal variables that allow the air traffic to be evaluated to the time variables. To do this, the first step is to structure the traffic according to its main traffic flows using the CRIDA methodology.

Once the traffic has been structured into flows, matrices will be filled in with information on this traffic, estimated before the operation. These matrices are presented in Figure 3.

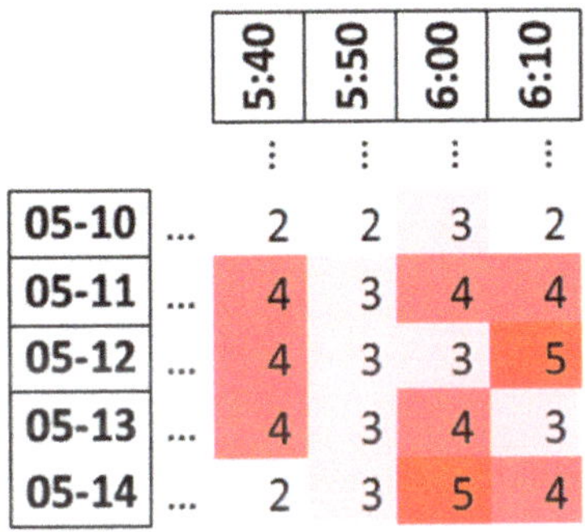

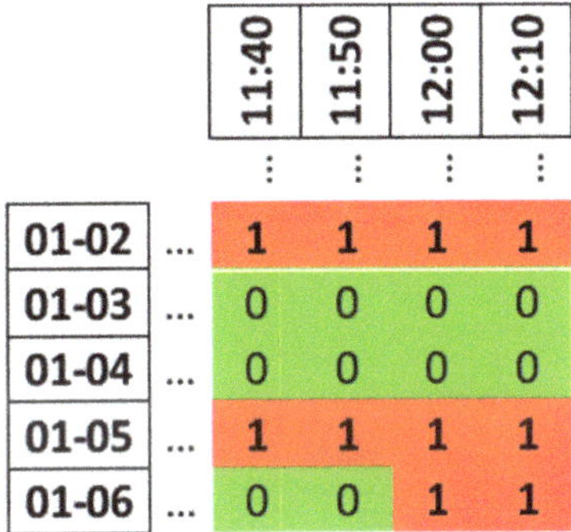

Figure 3. Air Traffic arrays.

In matrices such as the one on the left, the following information shall be included:

- Number of aircraft within the sector: Simply by using the traffic information, it is possible to estimate the number of aircraft that will be in the sector at a given time.
- Number of flows containing aircraft: Through the allocation of flows made for each aircraft, it will also be possible to identify the number of flows that will be in the sector at the time of analysis.
- Matrices such as the one on the right will include:
- Which flows contain aircraft: In addition to knowing how many flows will contain aircraft, it is also possible to know what these flows will be. Specifically, a matrix such as the one on the left of Figure 3 will be made for each flow, indicating when there will be aircraft in this flow and when there will not. In this way, it is intended to produce a map of the sector at the time of the prediction with previous information.

With all the information analysed, the input information of the model is completed. Therefore, the complete format of the model is presented in Table 2 with the addition of an example use case. In this case, an example has been made in which the variables are aleatory. It does not represent a real case, but this example is used to observe the format of the data for the generation of the machine learning model. The values of the variables will vary depending on the actual operating scenario:

Table 2. Format of machine learning model.

Input								Output
Period of Day	Day of Week	Month	Number of Aircraft	Number of Flows	Flow1	...	FlowN	Regulated/Not Regulated
26 04:10–04:20	3 Wednesday	2 February	7	3	1	...	0	1

With Flow1, . . . , FlowN being the identifiers of the flows within the sector, and N is the total number of flows. It shall be identified with 0 when there are no aircraft within the sector and with 1 when there are aircraft. Similarly, 0 shall identify when the sector is not regulated and 1 when it is regulated.

3.4. Machine Learning Model Evaluation

With the machine learning model in place, part of the methodology is to determine which methods will best evaluate the performance of the model. Evaluating a machine learning model is a fundamental step. In machine learning applications in ATM, it is as important to correctly evaluate the model as it is to emphasise its explainability [30]. The methods for evaluating and validating the model, based on a binary classification, are presented below.

The first method of evaluating the model will be Accuracy. This was the most important criterion for determining the performance of a classification algorithm, which showed the percentage of the proper classification of the total set of the experimental record [31]. The Accuracy formula is:

$$Accuracy = \frac{TP + TN}{TP + TN + FP + FN} \tag{1}$$

In Table 3, the meaning of *TP*, *TN*, *FP*, and *FN* is explained:

Table 3. Accuracy description.

TP (True Positive)	Cases where the sector is regulated and the model predicts that it will be regulated
TN (True Negative)	Cases in which the sector will be not regulated and the model predicts that it is not regulated
FP (False Positive)	Cases where the sector is not regulated but the model predicts that it is regulated
FN (False Negative)	Cases where the sector is regulated but the model predicts that it is not regulated

In this case, T and F are "True" and "False". This classification is used in the literature to represent which elements of the training set are correctly (True) or not correctly (False) evaluated by the machine learning model or not. The terms "Negative" and "Positive" refer to whether the sector is not (Negative) or is (Positive) regulated.

This indicator will give a picture of the overall performance of the application. Based on applications of a similar nature, an Accuracy threshold of 0.85 out of 1 is set [32]. As Accuracy is defined as correctly classified cases among total cases, the maximum Accuracy is 1 (corresponding to all the elements having been correctly classified). In the literature studied, a value that is considered to say that a machine learning model is good at classifying is that it classifies 85% of the cases correctly [32]. However, since it is measured over one, this 85% corresponds to 0.85.

This limit is independent of the training of the model. The model is trained and tested independently. Subsequently, the Accuracy of the model is obtained and compared with the defined threshold. If the Accuracy of the model is greater than 0.85, the model is considered valid. If the Accuracy of the model is less than 0.85, the model is not valid and cannot be used in real applications.

To evaluate the two labels independently, the Recall, Precision, and F1-score parameters for each of the classes are added to the analysis for each of the classes [33] (0 when the sector is not regulated and 1 when the sector is regulated). Recall indicates the ability of the algorithm to accurately detect when the sector will be regulated or not. The Precision indicates the ability of the algorithm to detect the categories. The F1-score is a harmonic mean of the Recall and the Precision [31]. The formulae are [34]:

$$Recall = \frac{TP}{TP + FN} \tag{2}$$

$$Precision = \frac{TP}{TP + TN} \tag{3}$$

$$F1 - score = \frac{2 * Recall * Precision}{Precision + Recall} \quad (4)$$

In addition, to represent this information in a visual and summarised form, the confusion matrix is presented [35].

Moreover, emphasis will be placed on the explainability of the model. This analysis will provide insight into the learning and prediction process of the developed model. In this case, the analysis of explainability was carried out with graphs made in the Shapley Additive exPlanations (SHAP) library. This library is used for the explainability of machine learning models and its use is widespread in the industry [36,37].

4. Results

Once the methodology used to develop the machine learning model has been explained, the model is tested in a real application case. For its testing, it has been decided to choose the data of the LECMPAU sector in the year 2019. Specifically, the machine learning models have been trained with the matrix developed for 2019. In this matrix, there are 144 columns and 365 rows. With this, 52,560 data have been obtained to train and test the model. Specifically, 80%, 42,048 data, have been used to train the model. The rest has been used for model testing.

The operational data have been obtained based on ENAIRE radar traces and have been provided to the authors after processing and validation by the company CRIDA. The company CRIDA has also provided the data to the authors with the necessary regulatory data, after proper processing and validation.

4.1. Analysis of LECMPAU Air Traffic and ATFCM Regulations

Before starting to test the model and analyse its explainability, it is important to understand the operation in the Pamplona Upper (LECMPAU) sector, as well as the behaviour of the regulations in the sector. To study traffic behaviour, the methodology developed by CRIDA has been used to organise traffic into flows. The flows obtained from the methodology are presented in this paper as part of the input variables of the machine learning model will be whether there are aircraft in each flow or not. The flows are presented in Figure 4.

The air traffic flows identified have been divided into four main groups. The flows that cross the sector from north to south are of great importance in the operation of the sector, as they contain flights that normally depart from or go to the Madrid-Barajas airport. Other flows of great interest are those crossing the sector from east to west, as these will normally be associated with flights departing from or going to Barcelona-El Prat. With this, it can also be said that LECMPAU is a sector whose operation is very complex, as it includes operations around the two largest airports in Spain, making it a sector of great interest.

Additionally, there is another group of flows that cross the sector diagonally from the west of the sector to the north or vice versa. Flights belonging to these flows are more variable and difficult to classify into a single flight type.

With these air traffic flows, an attempt to characterise the traffic in the sector simply is made, taking into account the different trajectories that can be flown. In addition, it is important to characterise the regulations in the LECMPAU sector, as it will be possible to see certain patterns in the appearance of these regulations that can be used to validate the results of the machine learning model.

Overall, there were 150 regulations in LECMPAU in 2019. These regulations had a mean regulation time of 121.9 min, and a standard deviation of 71.16. To be more specific, the boxplot of the regulation time is presented in Figure 5.

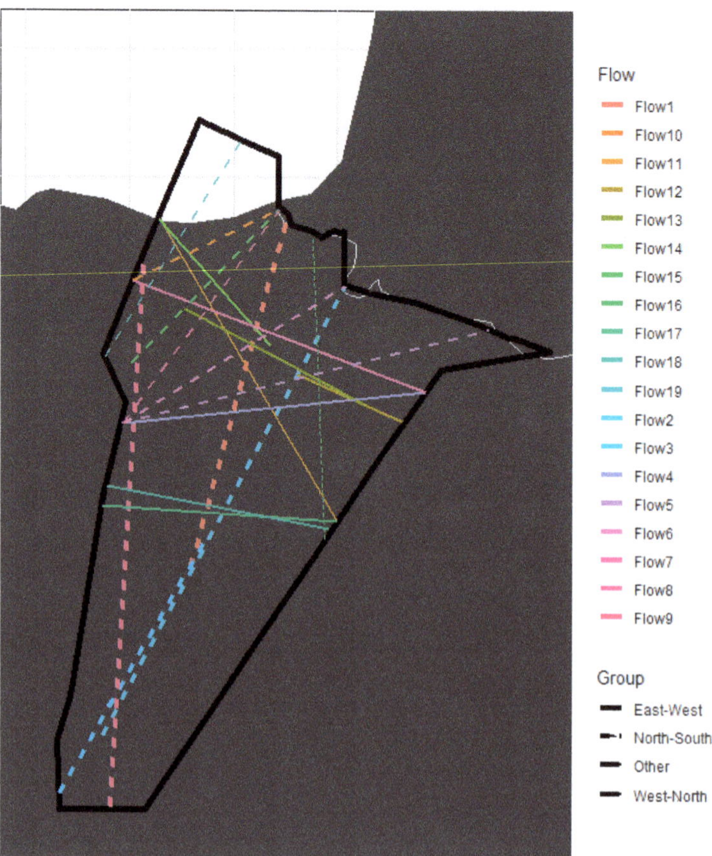

Figure 4. LECMPAU Air Traffic Flows.

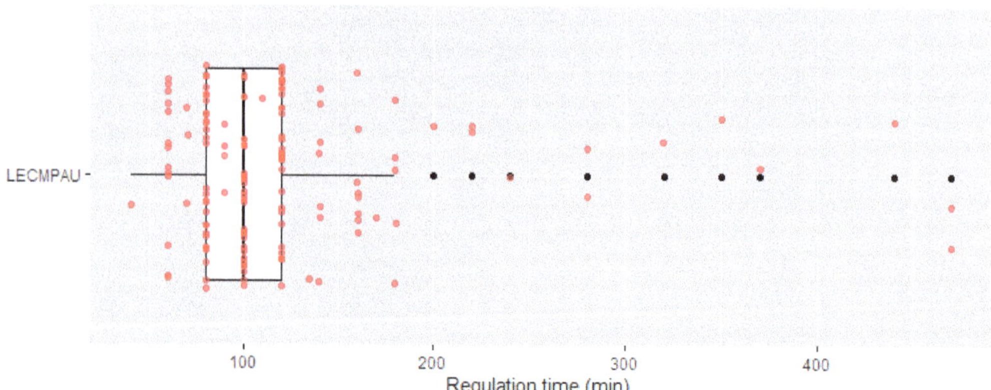

Figure 5. Boxplot of Regulation time in LECMPAU.

In the boxplot, it can be seen that most of the regulations have a regulation time between 50 and 200 min. There are cases of outliers due to both deficit and excess. With this data, only 18,285 min in the whole year will be regulated. This is 3.5% of the total time analysed. As the model will tell whether the sector is regulated or not, there is likely an imbalance in the sample to be analysed, as LECMPAU is much longer unregulated than regulated.

Rather than the duration of these regulations, it is of interest to know how they are distributed over time. This information is presented in Figure 6, which shows histograms of the regulations according to their month (a), day of the week (b), and start time (c).

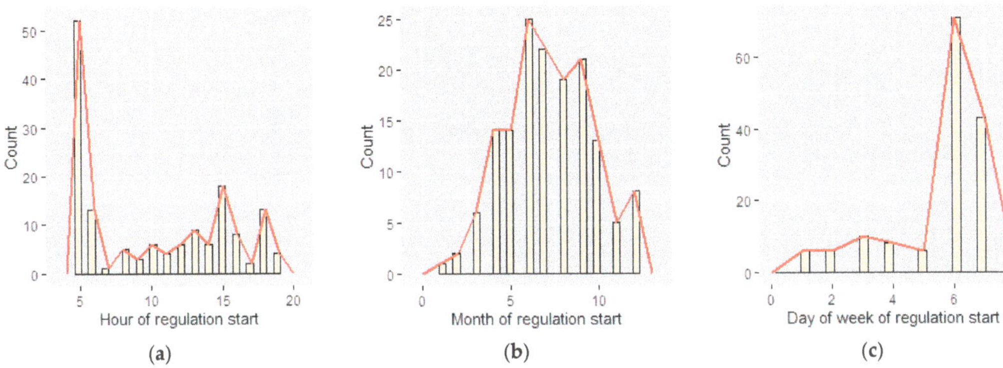

Figure 6. Temporal analysis of regulations in LECMPAU, of hour of regulation (**a**), month of regulation (**b**) and Day of the week of the regulation (**c**).

The temporal analysis indicates that most of the regulations will occur in summer. This is normal, as Barcelona and Mallorca are common holiday destinations, and flights from the US to these airports will pass through LECMPAU via flows crossing the sector from east to west. This increase in traffic will lead to more capacity regulations. As for the day of the week, most of the regulations will occur on weekends, mainly on Saturdays. This is also natural, as weekend traffic will be higher than Monday to Friday. Finally, the hourly analysis indicates that regulation will be centred at 05:00, with three secondary peaks of regulation at 06:00, 15:00, and 18:00. These are the most common times for business and holiday flights.

With this information, it is possible to find patterns in the behaviour of regulations in 2019, so it is possible that the machine learning algorithm, which is mainly based on temporal variables, will find these patterns and manage to act satisfactorily. The results of the model are presented below.

4.2. Machine Learning Model Evaluation

Once how both traffic and regulations behave in LECMPAU in 2019 has been studied, the machine learning model that will attempt to predict when the sector will be regulated is evaluated. For this model, the total dataset has been divided into 80% for training and 20% for testing. This ratio is often used by machine learning models developed for different academic fields [38]. Firstly, the Accuracy of the model is shown to test the model in general.

$$Accuracy = 0.868 \quad (5)$$

The Accuracy is above 0.85, so the model seems to work correctly according to the established standards. However, to evaluate the model more specifically, the Confusion Matrix is presented in Figure 7 and the indicators related to the Confusion Matrix are in Table 4. The confusion matrix is a visual indicator that simply indicates the number of cases where the model predicts whether the sector is regulated (1) or not (0) and compares it with the actual labels. This indicator is complementary to those defined in Section 3.4 and presents the same information visually.

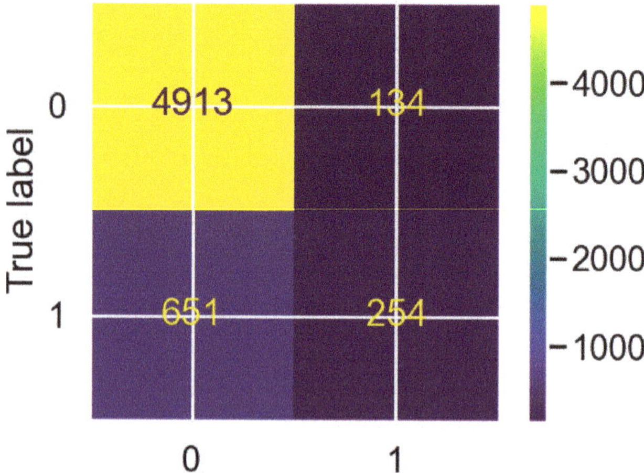

Figure 7. Confusion matrix of unbalanced machine learning model.

Table 4. Classification report of unbalanced machine learning model.

	Precision	Recall	F1-Score
0 (Not regulated)	0.88	0.97	0.93
1 (Regulated)	0.65	0.28	0.39

These indicators show that the model only predicts well when the sector is not regulated. Since the sample is so unbalanced, and most of the time the sector is not regulated, the model normally predicts that the sector will not be regulated. In doing so, the model is mostly correct, giving an Accuracy above the minimum. However, the model is influenced by the imbalance of the sample and the indicators of when the sector is regulated are well below what is considered correct.

As the sample is highly unbalanced, it has been decided to balance it with Synthetic Minority Oversampling TEchnique (SMOTE) to generate minority class samples [39]. This model creates samples of the minoritarian class based on the behaviour of contiguous elements (neighbours). In particular, the creators of the method state: "synthetic samples are generated in the following way: Take the difference between the sample under consideration and its nearest neighbor. Multiply this difference by a random number between 0 and 1 and add it to the feature vector under consideration. [40]".

With this sample balancing, the Accuracy of the model is:

$$Accuracy = 0.917 \quad (6)$$

Accuracy has increased by 6%. This, in advance, makes the model better beforehand. However, it is necessary to check that the model acts correctly when predicting for each of the classes. For this purpose, Figure 8 shows the Confusion Matrix, and Table 5 the Classification report.

These indicators are all above 0.9, exceeding the 0.85 set, so the performance of the model is very good both in predicting that the sector will be regulated and in predicting that it will not be regulated.

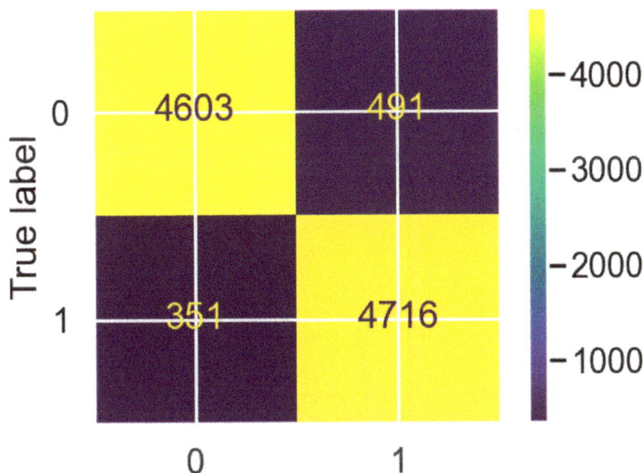

Figure 8. Confusion matrix of balanced machine learning model.

Table 5. Classification report of balanced machine learning model.

	Precision	Recall	F1-Score
0 (Not regulated)	0.93	0.90	0.92
1 (Regulated)	0.91	0.93	0.92

The balanced machine learning model seems to have found behavioural patterns in seasonality or air traffic and correctly predicts when the sector will be regulated. Therefore, with these results obtained, this model can be validated for the case of the LECMPAU sector.

4.3. Explainability of Machine Learning Model

Once the model has been validated and the indicators are found to be above the preset minimum, a study of the model's explainability is carried out.

The explainability of the model makes it possible to learn what the learning process of the machine learning model is like. It also shows the behavioural patterns found in the data and how it arrives at the predictions it makes. The Python SHAP library is used to perform this explainability analysis, which allows us to obtain graphs of explainability graphs. This method of describing the explainability of a machine learning model was introduced in [41] and is explained as "SHAP values attribute to each feature the change in the expected model prediction when conditioning on that feature. They explain how to get from the base value E[f(z)] that would be predicted if we did not know any features to the current output f(x) [41]" by the creators of this algorithm.

The first graph obtained is the influence of the temporal parameters on the prediction of the model. The SHAP algorithm starts from an initial expected value and adjusts the actual value of the prediction according to the labels of the input variables. Therefore, it is possible to analyse what effect the input variables will have, whether they will make the sector regulated (adding to the estimated initial value) or will make it unregulated (subtracting from the estimated initial value). Figure 9 presents the effect of the period of the day (a), the day of the week (b), and the month of the year (c) on the final prediction.

It can be seen how Figure 9 is very similar to the histograms shown in Figure 6. Figure 9 represents the influence of each of the training sample data, so the distribution of the training sample will be adjusted as seen above. Regarding the period of the day, approximately up to about period 40 of the day (06:30), this variable allows predicting the sector as regulated. On the other hand, from period 90 to 120 (15:00–20:00), there are two peaks where this variable helps the sector to be regulated. This presents an analogy with Figure 6c where at 15:00 and 18:00 there are two peaks in the time of occurrence of regulation.

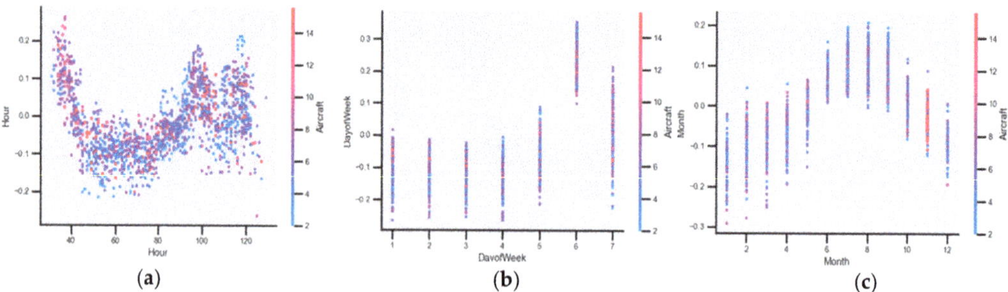

Figure 9. Effect of date-based variables on the output of the model. Analysis of hour (**a**), day of the week (**b**) and month (**c**).

Furthermore, the day of the week will only serve for the sector to be regulated sometimes on Fridays and Sundays, and always on Saturdays. This also represents a clear analogy with Figure 6b, where it can be seen that most of the regulations in LECMPAU appear on Saturday and Sunday.

As for the months of the year, the trend is also similar to the histogram results in Figure 6a. The months from June to September will be those in which the model tends to regulate the sector, while during the rest of the year, the model will tend not to regulate the sector.

Based on these results, it can be said that the model has been able to analyse time trends and make a personalised prediction based on the time data. The result allows us to verify that the approach of the model, mainly focused on an analysis of temporal patterns, seems to be correct. Furthermore, Figure 10 shows the same type of graphs, but in this case for the number of aircraft (a) and the number of flows (b) in the sector. This analysis is executed to check whether the model has also been able to find patterns in the traffic data of the sector.

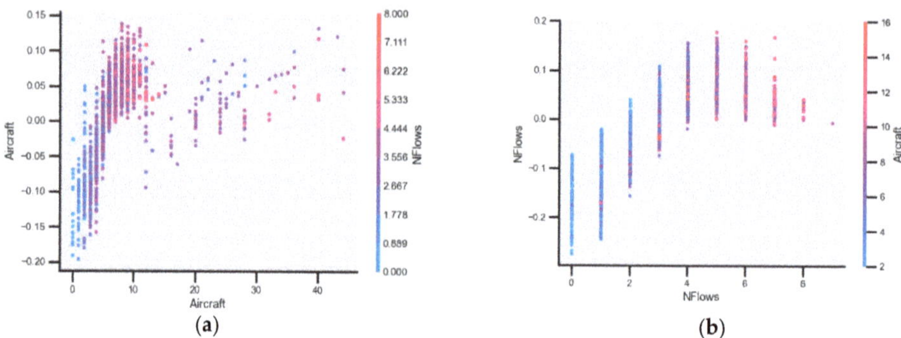

Figure 10. Effect of a number of aircraft and flows on the output of the model. Analysis of number of aircraft (**a**) and number of flows (**b**).

These variables representing air traffic behaviour also seem to behave quite intuitively. As for the number of aircraft, it is observed that when there are hardly any aircraft in the sector (from 0 to 5), the model helps to predict that the sector will not be regulated, whereas from five aircraft upwards, the model tends to predict that the sector will be regulated. This makes a lot of sense from an operational point of view, as the sector is most likely to be regulated when there is a considerable amount of traffic. The number of flows has a similar trend, with the frontier at four flows. From here, this variable will help the prediction that the sector is regulated.

In addition, the relative importance of the model is presented in this explanatory study. The SHAP algorithm can also identify which variables will be most important in the prediction of both classes. The graph is presented in Figure 11.

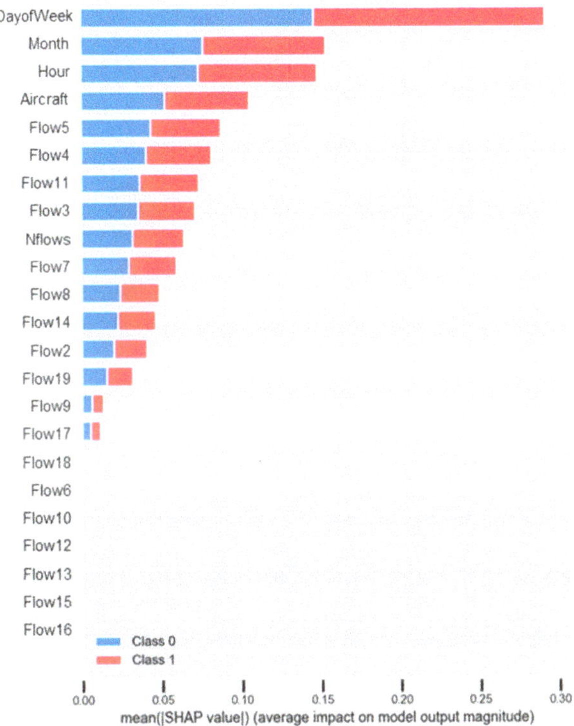

Figure 11. The relative importance of Machine Learning model.

The main variables of the algorithm are time variables. The greatest relative importance is that of the timing variables. This leads to the conclusion that the model is based on time-based components, and that it is possible to predict when the sector will be regulated or unregulated based on the date of analysis. The following variable is the number of aircraft in the sector and the presence of various traffic flows. In particular:

- Flow5: Flow across the sector from the west to the northeast.
- Flow4: Flow across the sector from west to east
- Flow1: Flow across the sector from north to south.
- Flow3: Flow across the sector from north to south.

Among the flows, there are the main flows of each of the previously classified groups. There are two representatives of the flows that cross the sector from north to south, as traffic from Madrid-Barajas is the most influential in terms of traffic in LECMPAU.

To conclude the explanatory analysis, two examples of predictions and how they are influenced by different variables are presented. Figure 12 shows an example where the sector will not be regulated, and Figure 13 shows an example where it will be regulated.

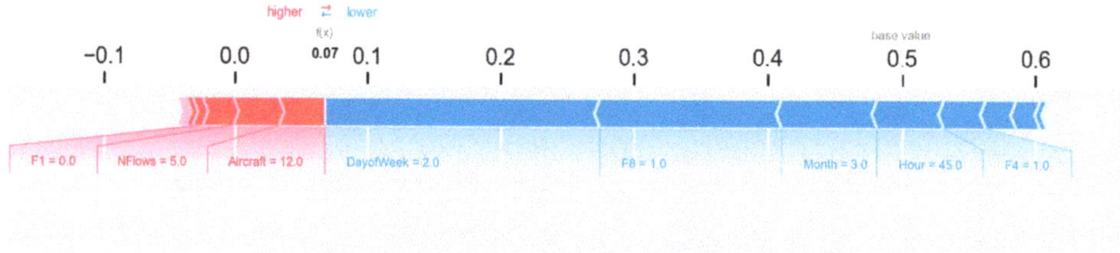

Figure 12. SHAP Explanation of not regulated example.

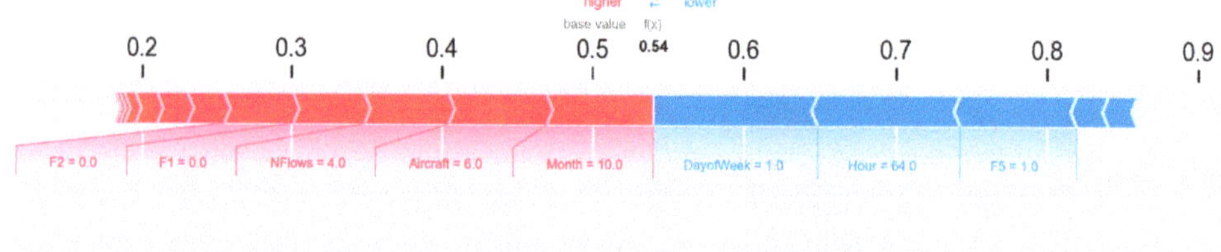

Figure 13. SHAP Explanation of regulated example.

The variables that cause the sector to be regulated are shown in red, and those that cause the sector to be not regulated are shown in blue. In the first example, on a Tuesday in March, even though there are 12 aircraft in 5 flows in the sector, the time component has too much influence on the prediction. On the other hand, in the second example, it can be seen how a flight on a Monday in October is regulated. In this case, the time component is more complex. The period of the day and the day of the week tend not to regulate the sector, but the month of the year compensates for this tendency. The latter variable, with the help of the traffic component of the sector, means that the sector is expected to be regulated.

This explainability analysis gives an idea of how the algorithm behaves and on which variables it bases its predictions on. Thanks to this analysis, it is possible to validate the model and the methodology developed from an operational point of view. This, together with the results obtained, which are above the established standards, means that the methodology is considered a success in LECMPAU.

4.4. Comparison with Other Machine Learning Models

This section compares the results of this model with the results of other models with the same objective. Following a literature review, it is concluded that [24] is the closest reference to the model being used in this paper. Firstly, the indicators of the model are presented to compare the overall performance. In Table 6, the maximum and minimum values are presented. This may allow a comparison of evaluation results for models with the same objective.

Table 6. Minimum and maximum indicators of [24] model.

	Accuracy	Precision	Recall	F1-Score
Minimum	0.7788	0.7051	0.8031	0.7757
Máximum	0.9725	0.9375	1	0.9677

This indicates that the two models have a very similar tier of performance. In Table 6, the minimum and maximum are presented, because in [24] several algorithms are tested.

In terms of explainability, this model is mainly influenced by the time components. In the model of [24], the main variable is the timestamp. In this respect, the two models coincide. The next most important variables are the Entry Count for the next 60 min, the capacity, and the Entry Count for the next 20 min. From this point on, workload and traffic distribution variables start to appear. The Entry Count is analogous to the number of aircraft, and the traffic distribution variables would be similar to the flow distribution in this model.

With these two comparisons, although with different approaches, the two models arrive at similar results. This allows us to conclude that the model developed in this paper is correct, having been tested against a robust model with prestige within academia.

5. Conclusions and Future Work

Once the machine learning model has been validated, it can be concluded that the results are satisfactory and that it is indeed possible to predict when a sector will or will not be regulated based on mainly temporal components. This model will not decrease the number of regulations directly, as it simply evaluates and predicts based on the situation within the sector and the time of year. The realisation of this model can present advantages in the management of the technological and human resources of the ATC system. By applying this model before the actual operation, the ATC system will be able to see which sectors will be regulated at which time and be able to dedicate more or fewer resources to the more or less regulated areas, allowing the ATC system to be more efficient in its labour.

Several conclusions can be drawn from the design of the model and its application.

- The fact that the model is based on a modular methodology is a great advantage. By having three modules independent of each other, variations are possible. Modules could be added, or the existing ones could be changed. This gives the model great flexibility and makes the developed methodology robust.
- The model has been able to predict when the regulations will appear in the sector, mainly based on the temporal analysis. This fulfils the objective of the development of this work.
- The model explainability analysis of the model is very useful. In an aviation application, it is not enough to have a model whose test results are correct, it is also necessary that the learning process is correct, and this learning process has been validated thanks to the explainability analysis.
- The number of total flows is a fixed parameter. The NFlows parameter indicates the number of flows in which there are aircraft within the total number of flows. In situations where there are a large number of flows, the model will depend on a number of other parameters that will have more influence. This is the ninth most important (Figure 11). For this reason, the analysis of the model when there are a large number of flows, which could be interesting from an operational point of view, may lead to unrepresentative and misleading results, as the model relies mainly on other variables.
- 10-min windows have been chosen because this is enough time for regulations to appear or not to appear in the airspace. Smaller time windows would introduce noise into the sample by not allowing time for the model labels to change. Furthermore, the aim of the regulation study is to see, in future studies, the effects that these regulations may have, and from an operational point of view, in a control room, this range is used to evaluate the possible effect of the regulation. With larger time windows, insufficient granularity would be achieved.

In addition to the general conclusions of the model, the computational time allows conclusions to be drawn about the feasibility of the model, and the possibility of its implementation in a real case study. This model has been trained and tested in a time of 4 min, having been developed on a general computer. This time is practically immediate for such a large volume of data. When running a new application of the model, the prediction of a full day (144 elements) has been completed in just 2 s. These application times mean that the model can be implemented in a real tool. Moreover, although the results obtained have been good, there is interest in continuing this line of research. Future lines of research that would be interesting for the full development of the model are set out below.

- Application to more ATC sectors. This model has been successfully validated in a single ATC sector, with very specific traffic conditions and patterns at the onset of regulations. An application to other sectors could allow the validation of this methodology and model in different circumstances.
- Study of additional variables. Although the traffic study is a complement to the model, based mainly on the time component, it is necessary to take it into account. Therefore, adding or considering variables other than those from the study may be of interest, as it would allow a broader view.
- Application of the model to specific air traffic flows. At present, regulations affecting an entire sector are being studied. However, regulations may only affect air traffic flows within the sector. It would be interesting to develop a model based on the current one that is able to predict whether an air traffic flow will be regulated or not.

In general, in order to have an operational validation of the methodology, more comparative studies will be needed. Such studies will be pursued in the future when more data is available to apply the methodology in different sectors, or when another methodology is available to compare the results with.

Author Contributions: Conceptualization, V.F.G.C.; methodology, F.P.M.; software, R.D.-A.J.; validation, R.D.-A.J.; formal analysis, M.Z.S.; investigation, R.M.A.V.; writing—original draft preparation, F.P.M.; writing—review and editing, M.Z.S.; supervision, R.M.A.V.; project administration, V.F.G.C. All authors have read and agreed to the published version of the manuscript.

Funding: This research was funded by ENAIRE.

Acknowledgments: Acknowledgement to ENAIRE and CRIDA for the collaboration and funding of the project in which this research paper has been developed. I would also like to express my gratitude to CRIDA for providing the data necessary to carry out the work and obtain the results.

Conflicts of Interest: The authors declare no conflict of interest.

Abbreviations

ATM	Air Traffic Management
ATC	Air Traffic Control
ATCO	Air Traffic Controller
ATFCM	Air Traffic Flow Capacity Management
TP	True Positive Elements
TN	True Negative Elements
FP	False Positive Elements
FN	False Negative Elements
SHAP	SHapley Additive exPlanations
SMOTE	Synthetic Minority Oversampling Technique
LECMPAU	Pamplona Upper ATC sector

References

1. Cecen, R.K. Multi-objective optimization model for airport gate assignment problem. *Aircr. Eng. Aerosp. Technol.* **2021**, *93*, 311–318. [CrossRef]
2. EUROCONTROL. European Aviation in 2040. 2018. Available online: http://www.eurocontrol.int/publication/long-term-forecast-annual-numbers-ifr-flights-2040 (accessed on 30 October 2022).
3. Aydoğan, E.; Demirel, S. The omnidirectional runway with infinite heading as a futuristic runway concept for future free route airspace operations. *Aircr. Eng. Aerosp. Technol.* **2022**, *94*, 1180–1187. [CrossRef]
4. Zhang, Y.; Su, R.; Li, Q.; Cassandras, C.; Xie, L. Distributed Flight Routing and Scheduling for Air Traffic Flow Management". *IEEE Trans. Intell. Transp. Syst.* **2017**, *18*, 2681–2692. [CrossRef]
5. Gatsinzi, D.; Nieto, F.J.S.; Madani, I. ECAC Use Case of Optimised Pre-tactical Time of Arrival Adjustments to Reduce Probability of Separation Infringements. *IFAC-Pap.* **2018**, *51*, 186–192. [CrossRef]

6. Kistan, T.; Gardi, A.; Sabatini, R.; Ramasamy, S.; Batuwangala, E. An evolutionary outlook of air traffic flow management techniques. *Prog. Aerosp. Sci.* **2017**, *88*, 15–42. [CrossRef]
7. Dmochowski, P.A.; Skorupski, J. Air Traffic Smoothness. A New Look at the Air Traffic Flow Management. *Transp. Res. Procedia* **2017**, *28*, 127–132. [CrossRef]
8. Dalmau, R.; Gawinoski, G.; Arnoraud, C. Comparison of various temporal air traffic flow management models in critical scenarios. *J. Air Transp. Manag.* **2022**, *105*, 102284. [CrossRef]
9. EUROCONTROL. Network Operations Report 2019. 2020. Available online: https://www.eurocontrol.int/publication/annual-network-operations-report-2019 (accessed on 1 November 2022).
10. EUROCONTROL. ATFCM Users Manual 2022. Available online: https://www.eurocontrol.int/publication/atfcm-users-manual (accessed on 1 November 2022).
11. Delgado, L.; Gurtner, G.; Bolić, T.; Castelli, L. Estimating economic severity of Air Traffic Flow Management regulations. *Transp. Res. Part C: Emerg. Technol.* **2021**, *125*, 103054. [CrossRef]
12. Rodríguez-Sanz, Á.; Cano, J.; Fernández, B.R. Impact of weather conditions on airport arrival delay and throughput. *Aircr. Eng. Aerosp. Technol.* **2021**, *94*, 60–78. [CrossRef]
13. De Oliveira, M.; Eufrásio, A.R.; Guterres, M.; Murça, M.R.; Gomes, R. Analysis of airport weather impact on on-time performance of arrival flights for the Brazilian domestic air transportation system. *J. Air Transp. Manag.* **2021**, *91*, 101974. [CrossRef]
14. Sanaei, R.; Lau, A.; Gollnick, V. A study of capacity regulations to define European air traffic management network states. *Transp. Plan. Technol.* **2021**, *44*, 337–355. [CrossRef]
15. Schultz, M.; Lubig, D.; Asadi, E.; Rosenow, J.; Itoh, E.; Athota, S.; Duong, V.N. Implementation of a Long-Range Air Traffic Flow Management for the Asia-Pacific Region. *IEEE Access* **2021**, *9*, 124640–124659. [CrossRef]
16. Xue, D.; Hsu, L.-T.; Wu, C.-L.; Lee, C.-H.; Ng, K.K. Cooperative surveillance systems and digital-technology enabler for a real-time standard terminal arrival schedule displacement. *Adv. Eng. Inform.* **2021**, *50*, 101402. [CrossRef]
17. Zeng, L.; Wang, B.; Tian, J.; Wang, Z. Threat impact analysis to air traffic control systems through flight delay modeling. *Comput. Ind. Eng.* **2021**, *162*, 107731. [CrossRef]
18. Jardines, A.; Soler, M.; García-Heras, J. Estimating entry counts and ATFM regulations during adverse weather conditions using machine learning. *J. Air Transp. Manag.* **2021**, *95*, 102109. [CrossRef]
19. Kammoum, M.A.; Rezg, N. An efficient hybrid approach for resolving the aircraft routing and rescheduling problem. *J. Air Transp. Manag.* **2018**, *71*, 73–87. [CrossRef]
20. Pang, Y.; Zhao, X.; Yan, H.; Liu, Y. Data-driven trajectory prediction with weather uncertainties: A Bayesian deep learning approach. *Transp. Res. Part C: Emerg. Technol.* **2021**, *130*, 103326. [CrossRef]
21. Xu, Y.; Dalmau, R.; Melgosa, M.; Montlaur, A.; Prats, X. A framework for collaborative air traffic flow management minimizing costs for airspace users: Enabling trajectory options and flexible pre-tactical delay management. *Transp. Res. Part B Methodol.* **2020**, *134*, 229–255. [CrossRef]
22. Mas-Pujol, S.; Salami, E.; Pastor, E. Predict ATFCM weather regulations using a time-distributed Recurrent Neural Network. In Proceedings of the 2021 IEEE/AIAA 40th Digital Avionics Systems Conference (DASC), San Antonio, TX, USA, 3–7 October 2021; pp. 1–8.
23. Sanaei, R.; Lau, A.; Linke, F.; Gollnick, V. Machine Learning Application in Network Resiliency based on Capacity Regulations. In Proceedings of the 2019 IEEE/AIAA 38th Digital Avionics Systems Conference (DASC), San Diego, CA, USA, 8–12 September 2019; pp. 1–6. [CrossRef]
24. Mas-Pujol, S.; Salamí, E.; Pastor, E. RNN-CNN Hybrid Model to Predict C-ATC CAPACITY Regulations for En-Route Traffic. *Aerospace* **2022**, *9*, 93. [CrossRef]
25. Tobaruela, G.; Schuster, W.; Majumdar, A.; Ochieng, W.Y.; Martinez, L.; Hendrickx, P. A method to estimate air traffic controller mental workload based on traffic clearances. *J. Air Transp. Manag.* **2014**, *39*, 59–71. [CrossRef]
26. Malakis, S.; Psaros, P.; Kontogiannis, T.; Malaki, C. Classification of air traffic control scenarios using decision trees: Insights from a field study in terminal approach radar environment. *Cogn. Technol. Work.* **2019**, *22*, 159–179. [CrossRef]
27. Reitmann, S.; Shultz, M. An Adaptive Framework for Optimization and Prediction of Air Traffic Management (Sub-)Systems with Machine Learning. *Aerospace* **2022**, *9*, 77. [CrossRef]
28. Speiser, J.L.; Miller, M.E.; Tooze, J.; Ip, E. A comparison of random forest variable selection methods for classification prediction modeling. *Expert Syst. Appl.* **2019**, *134*, 93–101. [CrossRef] [PubMed]
29. Moreno, F.P.; Comendador, V.F.G.; Jurado, R.D.-A.; Suárez, M.Z.; Janisch, D.; Valdes, R.M.A. Dynamic model to characterise sectors using machine learning techniques. *Aircr. Eng. Aerosp. Technol.* **2022**, *94*, 1537–1545. [CrossRef]
30. Rudd, K.; Eshow, M.; Gibbs, M. Method for Generating Explainable Deep Learning Models in the Context of Air Traffic Management. In *Machine Learning, Optimization, and Data Science: 7th International Conference, LOD 2021, Grasmere, UK, 4–8 October 2021*; Springer International Publishing: Cham, Switzerland, 2022; pp. 214–234. [CrossRef]
31. Aghdam, M.Y.; Tabbakh, S.R.K.; Chabok, S.J.M.; Kheyrabadi, M. Optimization of air traffic management efficiency based on deep learning enriched by the long short-term memory (LSTM) and extreme learning machine (ELM). *J. Big Data* **2021**, *8*, 1–26. [CrossRef]
32. Pham, D.; Alam, S.; Duong, V. An Air Traffic Controller Action Extraction-Prediction Model Using Machine Learning Approach. *Complexity* **2020**, *2020*, 1659103. [CrossRef]

33. Khan, W.A.; Ma, H.-L.; Chung, S.-H.; Wen, X. Hierarchical integrated machine learning model for predicting flight departure delays and duration in series. *Transp. Res. Part C Emerg. Technol.* **2021**, *129*, 103225. [CrossRef]
34. Khan, S.; Thorn, J.; Wahlgren, A.; Gurtov, A. Intrusion Detection in Automatic Dependent Surveillance-Broadcast (ADS-B) with Machine Learning. In Proceedings of the IEEE 2021 IEEE/Aiaa 40th Digital Avionics Systems Conference (DASC), San Antonio, TX, USA, 3–7 October 2021.
35. Li, J.; Sun, H.; Li, J. Beyond confusion matrix: Learning from multiple annotators with awareness of instance features. *Mach. Learn.* **2023**, *112*, 1053–1075. [CrossRef]
36. Xie, Y.; Pongsakornsathien, N.; Gardi, A.; Sabatini, R. Explanation of Machine-Learning Solutions in Air-Traffic Management. *Aerospace* **2021**, *8*, 224. [CrossRef]
37. Hickey, J.M.; Di Stefano, P.G.; Vasileiou, V. Fairness by Explicability and Adversarial SHAP Learning. In *Machine Learning and Knowledge Discovery in Databases: European Conference, ECML PKDD 2020, Ghent, Belgium, 14–18 September 2020*; Springer International Publishing: Cham, Switzerland, 2021; pp. 174–190. [CrossRef]
38. Gholamy, A.; Kreinovich, V.; Kosheleva, O. Why 70/30 or 80/20 Relation between Training and Testing Sets: A Pedagogical Explanation. 2018. Available online: https://scholarworks.utep.edu/cgi/viewcontent.cgi?article=2202&context=cs_techrep (accessed on 30 November 2022).
39. Pan, T.; Zhao, J.; Wu, W.; Yang, J. Learning imbalanced datasets based on SMOTE and Gaussian distribution. *Inf. Sci.* **2020**, *512*, 1214–1233. [CrossRef]
40. Chawla, N.V.; Bowyer, K.W.; Hall, L.O.; Kegelmeyer, W.P. SMOTE: Synthetic Minority Over-sampling Technique. *J. Artif. Intell. Res.* **2002**, *16*, 321–357. [CrossRef]
41. Lundberg, S.; Lee, S. A Unified Approach to Interpreting Model Predictions. In Proceedings of the 31st Conference on Neural Information Processing Systems, Long Beach, CA, USA, 4–9 December 2017.

Article

A Model-Based Prognostic Framework for Electromechanical Actuators Based on Metaheuristic Algorithms

Leonardo Baldo *, Ivana Querques, Matteo Davide Lorenzo Dalla Vedova and Paolo Maggiore

Department of Mechanical and Aerospace Engineering, Politecnico di Torino, 10129 Turin, Italy; ivana.querques@studenti.polito.it (I.Q.); matteo.dallavedova@polito.it (M.D.L.D.V.); paolo.maggiore@polito.it (P.M.)
* Correspondence: leonardo.baldo@polito.it

Abstract: The deployment of electro-mechanical actuators plays an important role towards the adoption of the more electric aircraft (MEA) philosophy. On the other hand, a seamless substitution of EMAs, in place of more traditional hydraulic solutions, is still set back, due to the shortage of real-life and reliability data regarding their failure modes. One way to work around this problem is providing a capillary EMA prognostics and health management (PHM) system capable of recognizing failures before they actually undermine the ability of the safety-critical system to perform its functions. The aim of this work is the development of a model-based prognostic framework for PMSM-based EMAs leveraging a metaheuristic algorithm: the evolutionary (differential evolution (DE)) and swarm intelligence (particle swarm (PSO), grey wolf (GWO)) methods are considered. Several failures (dry friction, backlash, short circuit, eccentricity, and proportional gain) are simulated by a reference model, and then detected and identified by the envisioned prognostic method, which employs a low fidelity monitoring model. The paper findings are analysed, showing good results and proving that this strategy could be executed and integrated in more complex routines, supporting EMAs adoption, with positive impacts on system safety and reliability in the aerospace and industrial field.

Keywords: EMA; prognostics; PHM; model-based; metaheuristic; MEA; FDI

Citation: Baldo, L.; Querques, I.; Dalla Vedova, M.D.L.; Maggiore, P. A Model-Based Prognostic Framework for Electromechanical Actuators Based on Metaheuristic Algorithms. *Aerospace* **2023**, *10*, 293. https://doi.org/10.3390/aerospace10030293

Academic Editor: Spiros Pantelakis

Received: 26 January 2023
Revised: 8 March 2023
Accepted: 10 March 2023
Published: 16 March 2023

1. Introduction

Aviation subsystem architectures are evolving substantially as a result of the more electric aircraft (MEA) concept, which, among other improvements, calls for the gradual replacement of hydraulic and electro-hydraulic actuators (EHA) with electro-mechanical actuators (EMA). In fact, it is believed that this paradigm shift would result in significant weight reductions, substantial life cycle cost (LCC) savings [1,2], lower repercussion on the environment, and, last but not least, increased reliability of the entire aircraft system [3].

Flight control surfaces in commercial aircrafts are currently actuated using FBW (fly-by-wire) technology: the pilot commands are translated into low-power electrical signals that are then managed by a computer and passed to hydraulic servovalves, which finally drive the appropriate aerodynamic surface and close the position control loop. The result is an electrical control, which, however, still leverages hydraulic power [4,5]. The core concept is aiming at an all-in-one electrical solution that can meet the necessary safety criteria [6,7], encompassing the most power-demanding aircraft subsystems. The authors in [8] provided a brief overview of the usage of EMAs and electro-hydraulic actuators (EHA) on the most widespread aircraft platforms. In what is considered the forefather of the electric aircraft par excellence, the Boeing 787, EMAs and EHAs are already taking the place of hydraulic actuators. The latest versions of the Airbus A350 and A380 follow the same principle, but EMAs are still only used for secondary flight controls (e.g., flaps, slats, spoilers), while EHAs are installed for both primary and secondary flight controls. It is clear that some challenges still limit a seamless replacement of EMAs in place of hydraulically driven actuation devices [7,9] for primary flight controls. This article follows the ideas presented in [10] and aims to propose a possible step forward by exploiting prognostics. De facto, it is critical to evaluate the implications of the substitution of an hydraulic subsystem with its electrical alternative, in terms of the usage, implementation, monitoring, and the equipment's reliability and safety. In the case of hydraulic systems, a potential failure (for example, a pressure drop caused by a leak) can be detected far before a load is demanded by appropriate pressure sensors.

Electrical system failures give rise to completely different problems from the power electronics point of view [11], as well as from the actuation one: new safety concerns are raised because no preventive mitigation plan can be implemented to reduce the impact of the fault itself if no additional auxiliary system is envisioned. As a consequence, the system must be exceedingly fault-tolerant. On top of that, EMAs show some issues that are less influential for hydraulic actuators, such as EMC, the mechanical jamming of the overall subsystem, and overheating problems, due to the high currents. A possible solution could be represented by hardware redundancy; however, this would result in weight increases, as well as incompatibilities with actuation requirements [12], therefore reducing the benefits of MEA principles.

Prognostics main selling point stands in the capability of detecting and identifying component early failures and track down their progression during the equipment use. This result brings a lot of positive outcomes with it; one of the most important is definitely the possibility of exploring innovative types of maintenance strategies (CBM, opportunistic and predictive maintenance [13–15]). On top of that, prognostics and health management (PHM) strategies could really be useful to back EMAs up, in order for them to reach the required safety standards for safety-critical applications, such as primary flight control actuation. Prognostics can be then seen as an effective mean to assist EMAs, thanks to its ability to identify hidden faults and prevent the related potential hazardous or catastrophic failure conditions. Prognostics sphere of influence stands in the monitoring and tracking of component or system parameters during their operation [16]: in this way, by checking the operational values and physical outputs, incipient failures can be detected resulting in the improvement of mission readiness, upgrade of RAMS capabilities, and a reduction of LCCs [17,18]. There is an ongoing substantial effort in the research community that focuses on the development of PHM systems for EMAs: a very detailed literature review on the matter can be found in [19]. Prognostics can be categorized based on many different criteria. The most general one is linked to the way the data used for the comparison is generated: data-driven [20], model-based [21,22], or hybrid. The method followed by the authors in this work is strictly model-based, due to the lack of enough real-life data to build the prognostic framework upon. The developed prognostic strategy is envisioned within an operational scenario and, as such, a general concept of operation (ConOps) is proposed, along with the high-level failure detection and identification (FDI) methodology. After that, a detailed explanation of the employed metaheuristic search algorithms (MSAs) and a brief overview of the models is reported. Two different models are used in this work: an high fidelity one (RM—reference model) and a low fidelity counterpart (MM—monitoring model), the latter being in the core of the prognostic framework. This work continues the ongoing effort started in [23], where a similar strategy was applied on brushless BLDC trapezoidal motors: in this case, the employed motor is a sinusoidal PMSM motor. Finally, the results and comparisons between the algorithms are reported.

2. Related Work

Metaheuristic algorithms for prognostics are only partially approached due to the limitations linked to the computational cost. However, there are many studies that already focused on MSAs to solve prognostic challenges. For instance, the authors already applied a similar strategies to a BLDC motor [23]. The literature on MSA is vast and covers almost every field of academic and industrial research. Furthermore, it must be said that the field of mathematical optimization using MSA is an extremely fast paced research area where new algorithms applied on very different application are being published continuously.

We focused our literature review on MSA application in the prognostic field. In [24], a PHM framework for lithium-ion batteries has been developed using an improved PSO. PSO has been used also in [25] for power transformers PHM and in [26] for wind turbine gearbox RUL estimation. The same application on lithium-ion batteries has been the focus of [27,28], which proposed the use of teaching–learning-based optimization (TLBO) for multiple degradation factors condition and hunger game search (HGS), respectively. The authors in [29] proposed a fuel cell health estimation strategy using, among other technique, a very popular MSA: the cuckoo search algorithm (CSA), inspired by cuckoo species parasitism. In [30], the authors used PSO to optimize the objective function related to rotating component prognostics. MSAs can also be used to optimize neural network parameters, as performed in [31], with landslide mapping, prediction, and prognostics or in [31], where CSA was employed.

As far as aerospace applications are concerned, a few works have been found and examined. A very interesting approach has been proposed in [32], where a combination of genetic algorithms (GAs) and a bio inspired artificial immune system has been used to schedule predictive maintenance tasks in a PHM framework. The same authors approached GAs and variable neighbourhood search (VNS) to solve the same problem in [33]. Aircraft motion planning issues have been the focus of [34], where a review of different population-based MSAs has been carried out, showing that PSO is the most used approach. A very relevant study [35] proposed ant colony optimization (ACO), in order to integrate PHM in aircraft maintenance planning, from a CBM perspective.

After a detailed literature review, which focused the fundamental queries "Prognostics", "Metaheuristic Algorithm", "Bio-inspired", and a wide range of secondary keyworkds, no work that focused the prognostic area applied to the EMA/aerospace domain using MSA has been found, to the best of our knowledge.

3. Materials and Methods

The proposed PHM strategy is built around the MM, whose simulations can be run almost in real time [36,37]. The other inputs for the prognostic strategy are, of course, the signals coming from the sensors mounted on the physical system (as shown in Figure 1).

However, real-life actuator data is difficult to obtain, and the failures are very rare indeed. Therefore, in this study, data have been generated through the aforementioned RM, which has been used as a NTB. The authors set up the RM and generated the failure data; the prognostic algorithm was then employed to detect and identify the failures and, after its run, the results were compared and the errors were calculated. Figure 1 shows the overview of the overall methodology. On the right, the reader can notice data coming from sensors or from the database created by the NTB.

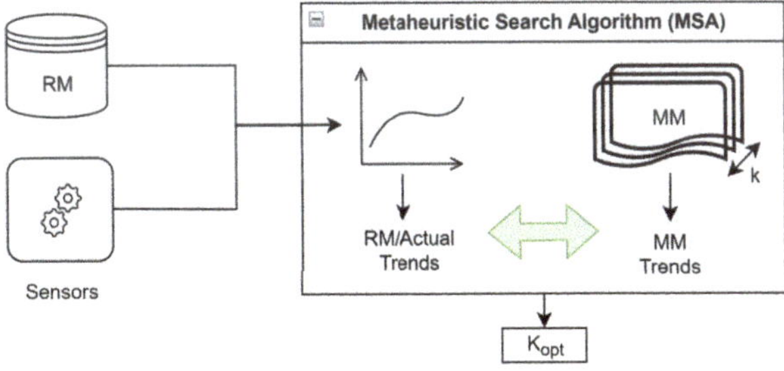

Figure 1. FDI methodology overview [10].

The RM and MM are two physical-based models, developed in Simulink starting from equations implementing the real physical behaviours: actual equations are implemented in Simulink blocks (e.g., EM motor equations, dynamic equations, etc.). A more in depth description will be provided in Section 3.2. The models are defined by a set of top level parameters (TLPs), grouped inside a $\overrightarrow{k}$ vector, as better explained later on. Each component of the vector is linked and can be traced back to a failure in the EMA. By altering these parameters, it is possible to inject failures inside the models. For all intents and purposes, the TLPs change the simulation boundary conditions; hence, it is possible to simulate the actuator behaviours in different conditions. Both models have been verified and experimentally validated under nominal and non nominal settings, proving the models' capacity to estimate actual trends with high precision.

During a prognostic routine, the MM is then iteratively run with different sets of TLPs. In fact, in each simulation run, the TLPs are updated following the results obtained by the optimization using the MSA. The algorithm tries to find the monitoring TLPs that alter the MM outputs in order to reduce the error between the predicted and actual trends (real or simulated). A real-time solution is not feasible because of the total running time of the high number of simulation runs, as stated better later on. The error is defined using a suitable fitness function (Equation (4)). As a result, after the optimization process is complete, the output of the MM is extremely similar to the actual one. After finding a satisfactory match, the system's health is evaluated by comparing the related TLPs with the related failures. Therefore, it is possible to execute a basic but effective FDI process by examining TLPs to determine whether a failure is occurring and establish what kind of failure it is.

As stated before, and as [38] explained, the TLPs are grouped inside a normalised vector $\overrightarrow{k}$ (Equation (1)): each component k_i is strictly linked to a physical system failure and it can assume values between 0 and 1 to characterize different failure magnitudes.

$$\overrightarrow{k} = [k_1\ k_2\ k_3\ k_4\ k_5\ k_6\ k_7\ k_8], \tag{1}$$

- k_1: Dry friction. When $k_1 = 1$, the resulting friction is the nominal value multiplied by three.
- k_2: Backlash. When $k_2 = 1$, the backlash magnitude is the nominal value multiplied by one hundred.
- k_3, k_4, k_5: Short circuit (SC). Being a three-phase motor, each coefficient is linked to a short circuit in one phase.
- k_6, k_7: Static eccentricity. These coefficients are linked to the modulus and phase of the eccentricity in the rotor. Under nominal conditions, the phase corresponds to 0 rad, so $k_7 = 0.5$.
- k_8: Proportional gain drift (PGD). $k_8 = 1$ is linked to an increase of 50 per cent in the proportional gain, while $k_8 = 0$ determines a 50 per cent decrease. The nominal value is $k_8 = 0.5$.

Table 1 shows in a schematic way each TLP along with the effects on the system at the relative maximum and minimum values. Note that maximum eccentricity refers to a situation where the eccentricity modulus is equal to the rotor air gap width.

As far as the failure implementation is concerned, each k coefficient is used in the MM and in the RM as a parameter that affects the model behaviour in different ways. For instance, k_1 is multiplied with the static or dynamic friction coefficients within the coulombian friction implementation. k_2 coefficient multiplies the global backlash value inside the model, appropriately modelled by a Simulink block. On the other hand, k_3, k_4, and k_5 represent the percentages of each phase affected by short circuit. These coefficients modify the wirings resistance (in a linear way) and inductance (in a quadratic way), hence affecting the electrical modelling of the system, as reported in [39]. The outcome of rotor eccentricity is modelled according to the following formulas [40], shown in Equation (2):

$$\begin{aligned} c_1 &= c_{1,0} \cdot (1 + k_6 \cdot cos(\theta_m + k_7)) \\ c_2 &= c_{2,0} \cdot (1 + k_6 \cdot cos(\theta_m + k_7 + \tfrac{2\pi}{3})) \\ c_3 &= c_{3,0} \cdot (1 + k_6 \cdot cos(\theta_m + k_7 - \tfrac{2\pi}{3})), \end{aligned} \tag{2}$$

where c_1, c_2, and c_3 are the resulting back-EMF coefficient for the three phases, starting from the nominal coefficients $c_{1,0}$, $c_{2,0}$, and $c_{3,0}$. On the other hand, θ_m is the mechanical rotor angle. Finally, k_8 simply multiplies the proportional gain in the controller block.

The nominal vector (i.e., the vector whose components would lead to a nominal behaviour of the EMA) is, hence, the one reported (Equation (3)):

$$\overrightarrow{k} = [0\ 0\ 0\ 0\ 0\ 0\ 0.5\ 0.5], \tag{3}$$

The fitness function is defined as:

$$e_{tls} = \sum_i \frac{(I_{MM,i} - I_{RM,i})^2}{\left(\frac{\Delta I_{RM,i}}{\Delta T}\right)^2 + 1} \cdot \Delta T, \tag{4}$$

where $I_{MM,i}$ and $I_{RM,i}$ are the MM and RM current outputs respectively at the instant time i [41]. Furthermore, the quadratic error between these two currents is evaluated to consider both the error in time and in the signal. Finally, the first order derivative is calculated numerically using Matlab's gradient command and divided by the integration step ΔT. Finally, to avoid the time dependence of the error, it is multiplied by the simulation integration step itself ΔT. The current values referred to the two models are considered at each instant of time i.

Table 1. Top level parameters meaning and effects.

TLP	Physical Failure	Effect at ($k_i = 0$)	Effect at ($k_i = 1$)
k_1	Dry friction	No effect (Nominal friction)	300% of nominal friction
k_2	Backlash	No effect (Nominal backlash)	100 times nominal backlash
k_3	Short circuit (Phase A)	No effect (No SC on Phase A)	Complete SC on phase A
k_4	Short circuit (Phase B)	No effect (No SC on phase B)	Complete SC on phase B
k_5	Short circuit (Phase C)	No effect (No SC on phase C)	Complete SC on phase C
k_6	Eccentricity modulus	No effect (No eccentricity)	Maximum Eccentricity
k_7	Eccentricity phase	−180°	180°
k_8	PGD	50% of nominal proportional gain	150% of nominal proportional gain

3.1. Employed Algorithms

The authors used a specific kind of optimization method that falls into the category of metaheuristic bio-inspired algorithms. Optimization algorithms' primary goal is to reduce or maximize an objective function, also known as a fitness function, by modifying the so-called decision variables: in this case the $\overrightarrow{k}$ vector's components, which are bound by established constraints [42]. Although heuristic techniques cannot always guarantee the optimal result, they aim to produce satisfactory solutions, or solutions that are at least close to the optimal outcome, at a reasonable processing cost. MSAs use a variety of strategies to identify effective solutions to optimize problems starting from a large population of acceptable candidates, while making few assumptions about the problem being optimized [43]. Furthermore, the algorithms are bio-inspired, since they have underlying mechanisms based on biological processes. In fact, natural adaptation might be viewed as a type of optimization. In this work, the authors have approached on the evolutionary (EA) and swarm intelligence (SI) algorithms. Following a careful literature review as reported in Section 2, we selected three different MSA:

- PSO, since it resulted as one of the most used ones;
- DE, to represent the evolutionary algorithm category;
- GWO, which we selected among new algorithms.

3.1.1. Evolutionary Algorithms

Evolutionary algorithms draw their inspiration from the natural evolutionary behavior. They are described by a few key characteristics and parameters:

- Population: The solution "pool", which is initialized at the start of the process;

- Variety: The population must be varied enough to explore the solution space effectively;
- Heredity: This values is linked to the capability of passing a characteristic to the offspring;
- Selection: For artificial algorithms, selection must only occur in the desired direction, which is a key parameter to ensure that only the best solutions will be reproduced.

Individuals serve as a representation of the different solutions, and a score, known as the fitness value, is assigned to them by means of Equation (4), calculated by analyzing the phenotype, or set of traits, of the subjects in question.

Differential Evolution. One of the most famous EA is the differential evolution strategy [43,44], which is also one of the tested algorithms in this work. The main concept of this algorithm follows the genetic principles.

Figure 2 shows the logical process behind this optimization algorithm. The process starts with a step called mutation: three individuals (three vectors in this case) are selected randomly from the population and a fourth vector is created by calculating the difference (the evolution is differential) between the first two vectors, multiplying it by a mutation factor, and then adding the third one. The recombination or crossover phase begins at this stage, when the altered parameters of the starting vector are combined with those of the so-called target vector to produce the trial vector. The trial and target vectors are compared during the selection phase. If the trial's score exceeds the target, it will be used as the next target; otherwise, it will be discarded. In the next generation, the vector with the highest fitness value will be maintained. The pseudo-code for this algorithm is reported in Figure 3.

Genetic algorithms are another popular EA optimization method (GA). In [23,45], a detailed investigation of the implementation of such algorithms on comparable challenges for PHM techniques was performed. However, in previous tests, the global GA performance (considering the same metrics employed in this work) results were inferior, with respect to other metaheuristic optimization methods [38].

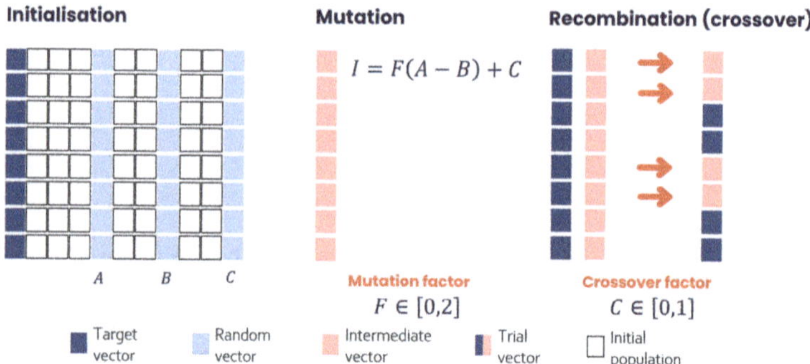

Figure 2. A schematic representation of the Differential Evolution Algorithm.

```
START
\\ Parameters definition
    Set the population dimension N
    Set the vector k dimension D
    Set the mutation factor F in [0,2]
    Set the crossover factor C in [0,1]
\\ Initialise
    t=0; %counter
    Create NxD random inidividuals
    WHILE (stopping criterion)
        FOR (i=1:N)
        \\ Mutation
            Random choice of three vectors x1, x2, x3
            v=x1+F(x2-x3)
        \\ Crossover
            Random extraction of an index y between 0 and D
            %the trial vector takes at least one value of the v vector
            FOR z=1:D
                Random extraction of an index w between 0 e 1
                IF w<C or z=y
                    u(i)=v(i)
                ELSE
                    u(i)=x(i)
                END IF
            END FOR
        \\ Selection
            IF f(u(i))<f(x(i))
                x(i, t+1)=u(i)
            ELSE
                x(i, t+1)=x(i,t)
            END IF
        END FOR
        t=t+1
    END WHILE
END
```

Figure 3. Pseudo-code for DE algorithm, as taken from [41].

3.1.2. Swarm Intelligence Methods

These methodologies are part of the bio-inspired category because they draw their inspiration from the behavior that may be observed when more than one entity (e.g., animal) interacts with other individuals. As stated before, the biological behaviours linked to the interaction between individual can result in a sort of optimization, in the sense that the resources are shared and, thanks to the mutual collaboration, a solution is reached quicker. The result is a sort of information structure (which would not be present if the problem is approached by a single entity) that can be translated into mathematical formulations and, hence, used in optimization problems.

The difference with EA is that, in this case, the process is interaction-driven by the entities' behaviours (i.e., collective intelligence) and not controlled by genetic processes. The ability to exchange information and send or receive feedbacks from the other individuals inside the group is what makes the collective intelligence so powerful. In other words, SI methods are the result of a very attentive observation of real-life actions between animals. As said before, two SI methods have been approached in this study: particle swarm optimization (PSO) and gray wolf optimization (GWO). In this case, each individual is seen as a $\overrightarrow{k}$ vector and, hence, as a possible solution.

Particle Swarm Optimization. Being a SI algorithm, PSO [46] draws its inspiration from the movement of bird flocks or fish schools. In fact, starting from a population of potential solutions (i.e., particles) and moving them throughout the search space, this methodology can solve optimization problems by following rigid mathematical formulas. As said before, the optimization is guaranteed by the fact that there is a capillary and diffused intelligence: the movement of each particle, and hence, its path, is affected by each particle's local best known position and the best known locations in the search space (these are known because the knowledge is shared by particles). That is precisely why the swarm is able to iteratively identify the optimum solutions by information sharing. Some initial parameters shall be defined, such as the population size, particle initial placements and speed, and particle inertia. After the initialization set up, each particle is given a random neighborhood, and by travelling, the best overall position is discovered. The position associated with the optimal global location are updated, so that each particle knows it. A detailed examination of the solution space is possible, thanks to the velocities' inherent stochastic component [47].The pseudo-code for this algorithm is reported in Figure 4. More information on the algorithm implementation can be found in [41]. The employed code routine iteratively runs until 200 iterations are reached or until the error between two successive runs is less than 10^{-9}.

```
START
\\ Parameters definition
    Set the population dimension N
    Set the vector k dimension D

\\ Initialise
    Initialise the position of each particle (vector k)
    Initialise the speed of each particle (vector k)
    Initialise the best position of each particle (vector k)

    WHILE (stopping criterion)
        Update the best position of each particle (vector k)
        Update the best global position
        Update position and velocity for each particle (vector k)
    END WHILE
END
```

Figure 4. Pseudo-code for PSO algorithm, as taken from [41].

Grey Wolf Optimization. If PSO was generically inspired by birds' and fishes' movement, GWO [48] has a more precise inspiring animal: wolves. This optimization technique follows the idea of the rigid hierarchical scales among grey wolf population's members. After choosing the size of the wolf pack and the initial positions of each "animal", an initial population hierarchy is defined by looking at the fitness function values. The decision of the number of wolves is crucial, as both the accuracy of the algorithm and the execution time are affected by this parameter. The higher the fitness value, the higher the hierarchical position in the scale. In this way, the individuals with lower scores will be less influential in the optimization process, while the better-positioned animals will lead the process. In other words, each wolf represents a distinct solution to the problem.

The main difference with PSO lies in the fact that, in this case, the wolves cannot communicate their position to the other members of the pack. In order to find the best solution, an adequate population initialization and search method must be adopted [49]. The pseudo-code for this algorithm is reported in Figure 5. The interested reader can find more information on the algorithm implementation in [41]. The size of the population has been set to 50 individuals. The optimization is stopped when the number of iterations reached is 150 or the tolerance of 10^{-9} is reached within two successive runs.

```
START
\\Parameters definition
    Set the population dimension N
    Set the vector k dimension D

\\Initialise
    Initialise the position of each individual
    Score each solution
    Categorize all solutions

    WHILE (stopping criterion)
        Update the position of each individual
        Update the hierarchy
    END WHILE
END
```

Figure 5. Pseudo-code for GWO algorithm, as taken from [41].

3.2. Models

The general architecture of an aeronautical EMA is extremely complex, characterized by multibody interactions and multiple non linearities and can be summed up as follows: a controller, an electric motor (usually PMSMs or BLDC motors), a gearbox, and an intricate network of sensors for measuring the currents, vibrations, voltages, temperatures, etc. As a result, the system is made up of multiple interconnected hardware and software components. It is also required to investigate their cross-interactions, so that the control system can appropriately fulfill its functions of monitoring, fault detection, and the assessment of a possible divergence route.

As stated before, two models have been assembled in Simulink and then validated, thanks to experimental test rigs. The models are physical-based and each component in real life is treated and modelled thanks to the appropriate equations. In other words, each block encloses the mathematical formulation by means of real formulas (e.g., electromagnetic formulas for the motor, dynamic equations for the physical systems, etc.).

If the RM has been built with a very high degree of precision, the MM has been conceived with approximations and lumped parameters, so that an almost real-time use is more feasible; the computational cost is, hence, reduced. The thorough description of the model is way beyond the scope of this work. The interested reader should see [36,37,50,51], as far as the RM and MM models are concerned, respectively. Furthermore, as explained in [36,37,50,51], the models have been built to represent an existing and experimental test bench located in Politecnico di Torino laboratories. In order to match the models and experimental results, the test bench structure has been reported in Simulink through simulation blocks. The experimental test bench has been pivotal in the model development: in fact, validation is essential to test models simulation performances [52].

The main parameters used in the models are, hence, taken from components datasheets and technical documents. The hardware configuration is based on the "S120 AC/AC Trainer Package" by Siemens. The motor parameters are reported in Table 2. The PID controller parameters have been calculated according to experimental tests to match the real and simulated behaviours.

For reasons of clarity, a simple logical scheme of the RM model is shown in Figure 6, while the four main blocks are quickly analysed below:

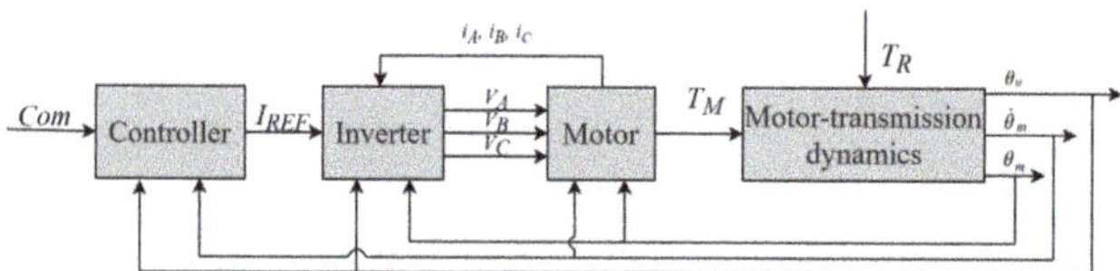

Figure 6. RM model structure as taken from [41].

- Controller: This block is essentially composed of a PID controller. In fact, even if there are much more advanced and sophisticated control logics (e.g., [53]), PID controllers are still the way to go and they are still chosen even in complex systems, as they are easy to implement and tune. PID controllers are composed of three separated branches, where the proportional, differential, and integral action are calculated. The controller aim is comparing the command signals with the actual signal obtained from the motor transmission dynamics block, hence closing the control loop. In this particular case, both position and speed can be monitored. This block outputs the reference current I_{REF}, obtained from the motor torque thanks to the torque constant, which is finally passed to the inverter.
- Inverter: This block contains Clarke-Park equations, and it provides the motor block with the three voltages (one for each phase) for the PMSM motor by performing the corresponding pulse width modulation (PWM). A very complicated physics-based process is handled by Simscape, a specific Simulink library, capable of providing electrical simulation packages. The main actions inside this block are the calculation of the electrical angle starting from the motor position, the splitting of I_{REF} into the three phase currents (with Clarke-Park equations), the PWM process, and the calculation of the three phase voltages, using the fed-back currents.
- Sinusoidal BLDC motor: This block is able to simulate the electrical and magnetic interactions inside a PMSM. It contains Simscape elements, and it manages three main processes:
    1. The calculation of the counter-electromotive force coefficient c_j for each phase. This is achieved with the multiplication of the back EMF coefficients (obtained with experimental test campaigns) with three sine waves 120° out of phase from each other.
    2. The implementation of the motor resitive-inductive circuit. A set of mathematical equations (Equation (5)) that model the three star connected LR branches is solved and phase currents (i_j) are, hence, calculated. The resistance R_j and inductance L_j of the motor are taken from equipment data sheets.

$$\begin{gathered} \sum_{j=1}^{3} i_j = 0 \\ V_j - c_j\omega = R_j i_j + L_j \frac{\mathrm{d}i_j}{\mathrm{d}t}, \end{gathered} \tag{5}$$

    where i_j and V_j are the currents and voltages across a single j phase.

    3. The calculation of the motor available torque. Three different electromotive coefficients are used to calculate the motor torque along with the relative phase currents:

$$T_m = \sum_{j=1,2,3} i_j c_j, \tag{6}$$

- Motor transmission dynamics: this final block compares the available torque with the external requested torque and solves a second-order dynamical system (Equation (7) comprehensive of multiple non linearities, such as dry friction and backlash [52]). The outputs of this block are the motor position and speed, which are looped back to the controller.

$$T_m - T_l = J_m \frac{\mathrm{d}^2\theta_m}{\mathrm{d}t^2} + C_m \frac{\mathrm{d}\theta_m}{\mathrm{d}t}, \tag{7}$$

where T_m represents the motor torque, and T_l represents the external torque. On the other hand, J_m represents the assembly inertia, C_m is the viscous friction coefficient, and θ_m is the motor position.

Table 2. PMSM motor parameters. The motor part number is S 1FK7060- 2AC71-1CA0 provided by Siemens. The numbers 60 K and 100 K refer to overtemperature values of 60 K and 100 K.

Characteristic	Value
Rated speed (100 K)	2000 rpm
Number of poles	8
Rated torque (100 K)	5.3 Nm
Rated current	3.0 A
Static torque (60 K)	5.00 Nm
Static torque (100 K)	6.0 Nm
Stall current (60 K)	2.55 A
Stall current (100 K)	3.15 A
Efficiency	90.00

Following the results of a detailed EMA failure mode effect and criticality analysis (FMECA) found in literature [9], it was decided to take into account five distinct failures. These failures show a medium high or high probability and/or criticality for the overall EMA: dry friction, backlash, short circuit, eccentricity, and proportional gain drift. On top of that, they show a quite slow propagation rate, which is desirable when designing failure detection and identification methodologies. For each failure mode, one or more k_i components have been assigned and two different magnitude levels have been simulated: high magnitude ($k_i = 0.75$) and low one ($k_i = 0.25$). In order to complete the analysis, also a condition with multiple failure affecting the EMA has been taken into account: the $\overrightarrow{k}$ vector, used in this case is the one shown in Equation (8).

$$\overrightarrow{k} = [0.0133\ 0.05\ 0.003\ 0\ 0\ 0.012\ 0.5\ 0.35], \tag{8}$$

A more in-depth explanation on the failure implementation and the modelling of each failure mode can be found in [40,41].

4. Results

In this section, the mean percentage error and the mean computational cost for each single failure and for the multiple failure situation are presented (Figures 7 and 8). The error has been calculated following Equation (9).

$$Err[\%] = \frac{100}{\sqrt{6.5}} \cdot \sqrt{\sum_{i=1}^{6}(k_i - k_{i,RM})^2 + k_{6,RM}(k_7 - k_{7,RM})^2 + (k_8 - k_{8,RM})^2}, \tag{9}$$

As stated before, for each failure, a low magnitude and a high magnitude condition have been considered (high magnitude ($k_i = 0.75$) and low one ($k_i = 0.25$)).

These data have been obtained through ten runs of each algorithm for a total of 300 runs to have a minimal statistical relevance. The employed computer is based on a Intel Core i7-8750H CPU @ 2.2 GHz with a RAM of 16 GB and a dedicated GPU NVIDIA GeForce GTX 1050. DE shows a slightly lower error than the other algorithms, as far as the low intensity failures are concerned. On the other hand, GWO seems to be the precise solution for the high level failures (e.g., short circuit and proportional gain). Finally, the PSO algorithm turns out to be the most accurate for backlash faults in general, high static eccentricity, and low proportional gain. Apparently there is not an algorithm able to outperform the others in every situation; this was expected as metaheuristic algorithms, despite being very versatile, are not the panacea for every problem. However, if we look at the computational cost, since a first glance, it is clear that PSO is the most efficient algorithm.

In fact, apart in case of the static low eccentricity failure, this optimization technique shows computational burdens almost halved, with respect to the other algorithms, with an average 25 min to detect and identify the failures. If we now consider the other two remaining algorithms, GWO is slightly faster than DE, but the last one seems to provide more stable and repetitive results.

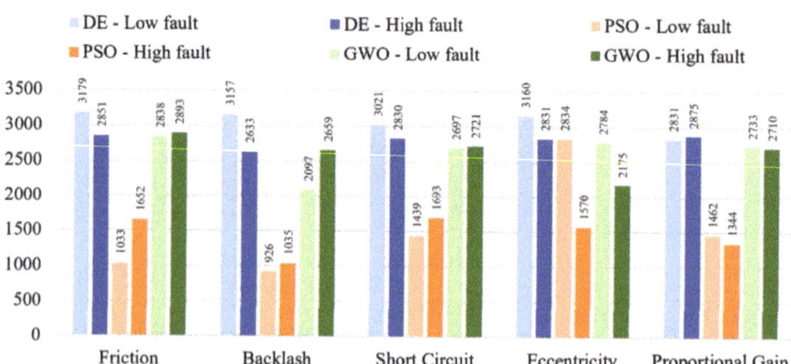

Figure 7. Comparison between different algorithms: mean percentage error [%]. For each failure mode and for each algorithm, the failure magnitudes are selected as follows: high magnitude ($k_i = 0.75$) and low one ($k_i = 0.25$).

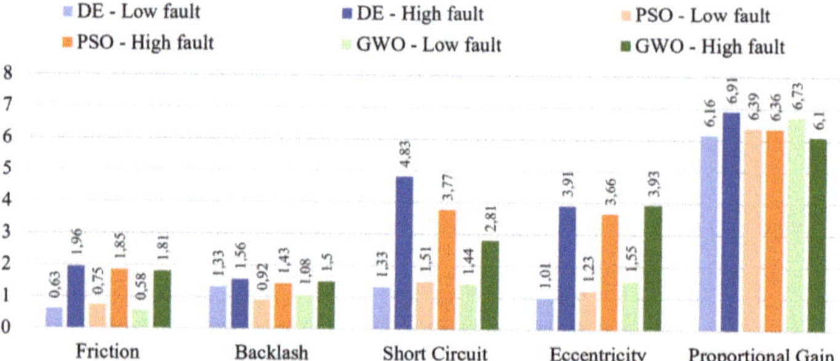

Figure 8. Comparison between different algorithms: computational cost [s]. For each failure mode and for each algorithm, the failure magnitudes are selected as follows: high magnitude ($k_i = 0.75$) and low one ($k_i = 0.25$).

The authors then used a predefined performance coefficient (PC) (first defined in [38] and reported in Equation (10)) to take into the account both the accuracy (mean percentage error) and the computational cost in a single parameter, thus providing an objective criterion useful to choose the best solution overall.

$$PC_i = 100 \cdot \left(1 - \frac{t_i \cdot err_i}{\sum_{i=1}^{3} t_i \cdot err_i}\right), \tag{10}$$

As regards Equation (10), the formula expresses the PC for the i-th algorithm, t_i is the average computational time of the i-th algorithm for the considered fault, and err_i is the average percentage error of the i-th algorithm. The rationale behind the formula is as follows: the denominator is added to make the result non-dimensional, while the multiplication and the subtraction are added to transform it in a percentage value. The results of this further investigation are reported in Table 3, along with other data to sum up the analysis. For reasons of brevity, the reported values of average percentage error and computational costs are obtained by an average between the two magnitude failures. As expected by linking the best results, in terms of accuracy and efficiency, PSO shows the highest figures for every failure. This happens not so much because of the error, as to the significantly lower computational cost related to the PSO-based solution.

On the opposite side of the ranking, there is clearly the DE algorithm. This is due to the very high computational cost.

Table 3. Different optimization algorithms outcomes with single failures [10]. The values related to best performing algorithm (PSO) are highlighted in bold, along with the relative PCs.

Failures	DE			PSO			GWO		
	Time (s)	Err. (%)	PC (%)	Time (s)	Err. (%)	PC (%)	Time (s)	Err. (%)	PC (%)
Friction	3015	1.30	56.97	1342.5	1.30	**80.76**	2865.5	1.20	62.26
Backlash	2895	1.45	50.21	980.5	1.18	**86.28**	2378	1.28	63.49
Short Circuit	2925.5	3.08	52.32	1566	2.64	**78.12**	2709	2.13	69.54
Eccentricity	2995.5	2.46	62.30	2202	2.45	**72.45**	2479.5	2.75	65.24
Prop. Gain	2853	6.54	58.61	1403	6.38	**80.14**	2721.5	6.42	61.24
Total	2936.8	2.96	56.75	1498.8	2.79	**79.24**	2530.7	2.75	64.00

The multiple failure condition results are reported in Table 4. Once again, it is clear that PSO is the leading algorithm, both in terms of efficiency and accuracy. Interestingly, it has to be highlighted that the multiple failure condition requires less than the single failure technique to reach a result. This was probably due to the stochastic operating principles of the algorithms. In fact, as the $\vec{k}$ vector contains random values (Equation (8)), rather than just one component dissimilar from its nominal value, it is believed that the algorithms are able to approach the error tolerance more quickly.

Table 4. Different optimization algorithms outcomes with multiple failures [10]. The values related to best performing algorithm (PSO) are highlighted in bold, along with the relative PCs.

	Time (s)	Err. (%)	PC (%)
DE	1777.0	4.21	60.95
PSO	1131.4	3.37	**80.09**
GWO	1816.6	4.33	58.94

5. Discussion

By looking at the results, PSO is the leading algorithm, providing the best results for the single and multiple failure condition. It is deemed that this algorithm is able to outperform the others, due to the main constructive principle behind it: unlike GWO, there is a strong collaboration and sharing of information between particles. Each particle contributes to the creation of the aforementioned collective intelligence by sharing the information on the best position it has found in the solution space and, hence, the likelihood to find always better solution rises drastically. This does not happen with GWO, which has a strong hierarchical law. DE showed incomparably worse results, and its implementation is quite challenging, too. It has to be noted that, even though the algorithms are metaheuristic and they pose few assumption on the application, they are very sensible to the problem statement; thus, an algorithm could behave very badly on a specific situation, but could provide excellent results when approaching an even slightly different problem.

The most challenging failure to detect and identify is the proportional gain drift. This particular phenomenon can be led back to the fact that a controller issue affects every single aspect of the actuation system; hence, the prognostic system struggles to identify the specific problems. Percentage errors for this failure mode, albeit higher than the ones relative to the other failures, are restrained around 8%, which is a satisfactory result, indeed.

After checking the results of the different algorithms, the authors envisioned a wider and more comprehensive concept of operations that could provide a practical application of the prognostic checks. The proposed monitoring framework could be easily implemented in a check routine that could be run when the aircraft is on the ground and during the flight to check the EMA subsystem health status. For instance a monitoring procedure can be run while the aircraft is at the gate waiting for the passengers, during maintenance checks, even 24-h or walk-around checks. Additionally, throughout the flying phase, a test procedure can be carried out at predetermined intervals. In this way, EMA health status is assessed and the mission safety is guaranteed. As seen before the proposed checks can not be run in real-time due to the needed number of simulation runs; on the other hand, a real-time monitoring capability is deemed superfluous and unnecessary as failures usually have a progression curve that can be monitored. The proposed prognostic strategy is applicable to a wide range of actuators and do not require the installation of any additional sensors, which is a very researched requirement. A prognostic and health management computer (PHMC) in the avionic bay is needed to perform the calculations.

6. Conclusions

In this paper, a set of three different metaheuristic search algorithms have been considered for failure detection and identification purposes. This work was aimed at testing new algorithms to identify some suitable ones. After having introduced the scenario in which this kind of PHM system could provide useful insights, a brief description of the models have been reported. Particular attention has been given to failure description and implementation, as well as the general methodology overview. A detailed description of the three algorithm has also been reported, in order to enable the reader to understand the most important aspect of each MSA, without the need of additional resources. Finally, the results are shown in a schematic way, highlighting the methodology performance with the use of purpose built coefficients. Further research should focus on the ideal interval between prognostic checks, in order not to jeopardize mission safety and not overloading the PHMC. On top of that, the running time of these routines can definitely be improved with other models or with more performing algorithms. This is why future work should consider other MSAs and other failures inside the EMA, as well as investigate on the unexpected result, regarding the multiple failure condition running time. Finally, if we compare the results of this study (specifically for PMSM-based EMAs) with the outcomes obtained in [23] for BLDC-based EMAs, it is clear that PSO is confirmed as the best performing algorithm, even though the core of the system has changed.

Author Contributions: Conceptualization, M.D.L.D.V.; methodology, M.D.L.D.V.; software, M.D.L.D.V.; validation, I.Q.; formal analysis, I.Q.; investigation, I.Q.; resources, I.Q.; data curation, I.Q.; writing—original draft preparation, L.B.; writing—review and editing, L.B.; visualization, L.B.; supervision, M.D.L.D.V.; project administration, M.D.L.D.V.; funding acquisition, P.M. All authors have read and agreed to the published version of the manuscript.

Funding: This research received no external funding.

Data Availability Statement: Not applicable.

Conflicts of Interest: The authors declare that they have no known conflict of interest nor competing financial interests or personal relationships that could have appeared to influence the work reported in this paper.

Abbreviations

The following abbreviations are used in this manuscript:

MEA	More Electric Aircraft
EMA	Electro-Mechanical Actuator
PHM	Prognostic and Health Management
PMSM	Permanent Magnet Syncrhonous Motor
DE	Differential Evolution
PSO	Particle Swarm Optimization
GWO	Grey Wolf Optimization

NTB	Numerical Test Bench
LCC	Life Cycle Costs
EHA	Electro-Hydraulic Actuators
CBM	Condition-Based Maintenance
RAMS	Reliability, Availability, Maintenability, and Safety
TLBO	Teaching–Learning-Based Optimization
HGS	Hunger Game Search
VNS	Variable Neighbourhood Search
ACO	Ant Colony Optimization
CSO	Cuckoo Search Optimization
MM	Monitoring Model
RM	Reference Model
SC	Short Circuit
FDI	Failure Detection and Identification
ConOps	Concept of Operation
TLP	Top Level Parameter
MSA	Metaheuristic Search Algorithm
EA	Evolutionary Algorithm
SI	Swarm Intelligence
GA	Genetic Algorithm
BLDC	BrushLess Direct Current
PID	Proportional Integral Derivative
PWM	Pulse Width Modulation
FMECA	Failure Mode Effect and Criticality Analysis
PC	Performance Coefficient
PHMC	Prognostic and Health Management Computer
EMF	Electro-Motive Force

References

1. Vercella, V.; Fioriti, M.; Viola, N. Towards a methodology for new technologies assessment in aircraft operating cost. *Inst. Mech. Eng. Part J. Aerosp. Eng.* **2021**, *235*, 879–892. [CrossRef]
2. Hölzel, N.B.; Gollnick, V. Cost-benefit analysis of prognostics and condition-based maintenance concepts for commercial aircraft considering prognostic errors. In Proceedings of the Annual Conference of the Prognostics and Health Management Society, PHM, Coronado, CA, USA, 18–24 October 2015.
3. Berri, P.C.; Dalla Vedova, M.D.; Mainini, L. Computational framework for real-time diagnostics and prognostics of aircraft actuation systems. *Comput. Ind.* **2021**, *132*, 103523. [CrossRef]
4. Gp Capt Atul, G.; Rezawana Islam, L.; Tonoy, C. Evolution of Aircraft Flight Control System and Fly-By-Light Flight Control System. *Int. J. Emerg. Technol. Adv. Eng.* **2013**, *3*, 12.
5. Sutherland, J.P. Fly-by-Wire Flight Control Systems. *Air Force Flight Dyn. Lab Wright-Patterson Air Force Base Ohio* **1968**, *3*, 2250–2459.
6. Telford, R.D.; Galloway, S.J.; Burt, G.M. Evaluating the reliability & availability of more-electric aircraft power systems. In Proceedings of the 2012 47th International Universities Power Engineering Conference (UPEC), Uxbridge, UK, 4–7 September 2012; pp. 1–6. [CrossRef]
7. Qiao, G.; Liu, G.; Shi, Z.; Wang, Y.; Ma, S.; Lim, T.C. A review of electromechanical actuators for More/All Electric aircraft systems. *Inst. Mech. Eng. Part J. Mech. Eng. Sci.* **2018**, *232*, 4128–4151. [CrossRef]
8. Wang, C.; Fan, I.S.; King, S. Failures Mapping for Aircraft Electrical Actuation System Health Management. In Proceedings of the PHM Society European Conference, London, UK, 27–29 May 2022; Volume 7, pp. 509–520. [CrossRef]
9. Balaban, E.; Saxena, A.; Bansal, P.; Goebel, K.F.; Curran, S. A Diagnostic Approach for Electro-Mechanical Actuators in Aerospace Systems. In Proceedings of the 2009 IEEE Aerospace Conference, Big Sky, MT, USA, 7–14 March 2009; pp. 1–13.
10. Baldo, L.; Querques, I.; Dalla Vedova, M.D.L.; Maggiore, P. Prognostics of aerospace electromechanical actuators: Comparison between model-based metaheuristic methods. In Proceedings of the 12th EASN International Conference on Innovation in Aviation & Space for Opening New Horizons, Barcelona, Spain, 18–21 October 2022.
11. Wileman, A.J.; Aslam, S.; Perinpanayagam, S. A road map for reliable power electronics for more electric aircraft. *Prog. Aerosp. Sci.* **2021**, *127*, 100739. [CrossRef]
12. Kuznetsov, V.E.; Khanh, N.D.; Lukichev, A.N. System for Synchronizing Forces of Dissimilar Flight Control Actuators with a Common Controller. In Proceedings of the 2020 23rd International Conference on Soft Computing and Measurements, SCM 2020, St. Petersburg, Russia, 27–29 May 2020. [CrossRef]

13. Benedettini, O.; Baines, T.S.; Lightfoot, H.W.; Greenough, R.M. State-of-the-art in integrated vehicle health management. *Proc. Inst. Mech. Eng. Part J. Aerosp. Eng.* **2009**, *223*, 157–170. [CrossRef]
14. Lee, J.; de Pater, I.; Boekweit, S.; Mitici, M. Remaining-Useful-Life prognostics for opportunistic grouping of maintenance of landing gear brakes for a fleet of aircraft. In Proceedings of the PHM Society European Conference, London, UK, 27–29 May 2022; Volume 7, pp. 278–285.
15. Scott, M.J.; Verhagen, W.J.C.; Bieber, M.T.; Marzocca, P. A Systematic Literature Review of Predictive Maintenance for Defence Fixed-Wing Aircraft Sustainment and Operations. *Sensors* **2022**, *22*, 7070. [CrossRef]
16. Swerdon, G.; Watson, M.J.; Bharadwaj, S.; Byington, C.S.; Smith, M.; Goebel, K.; Balaban, E. A Systems Engineering Approach to Electro-Mechanical Actuator Diagnostic and Prognostic Development. In Proceedings of the Machinery Failure Prevention Technology (MFPT) Conference, Dublin, Ireland, 23–25 June 2009.
17. Rosero, J.A.; Ortega, J.A.; Aldabas, E.; Romeral, L. Moving towards a more electric aircraft. *IEEE Aerosp. Electron. Syst. Mag.* **2007**, *22*, 9380100. [CrossRef]
18. Garcia Garriga, A.; Ponnusamy, S.S.; Mainini, L. A multi-fidelity framework to support the design of More-Electric Actuation. In Proceedings of the 2018 Multidisciplinary Analysis and Optimization Conference, Atlanta, GA, USA, 25–29 June 2018; American Institute of Aeronautics and Astronautics: Reston, VA, USA, 2018. [CrossRef]
19. Yin, Z.; Hu, N.; Chen, J.; Yang, Y.; Shen, G. A review of fault diagnosis, prognosis and health management for aircraft electromechanical actuators. *IET Electr. Power Appl.* **2022**, *16*, 1249–1272. [CrossRef]
20. Kaplan, H.; Tehrani, K.; Jamshidi, M. A Fault Diagnosis Design Based on Deep Learning Approach for Electric Vehicle Applications. *Energies* **2021**, *14*, 6599. [CrossRef]
21. Sutharssan, T.; Stoyanov, S.; Bailey, C.; Yin, C. Prognostic and health management for engineering systems: A review of the data-driven approach and algorithms. *J. Eng.* **2015**, *2015*, 215–222. [CrossRef]
22. Berri, P.C.; Dalla Vedova, M.D.L.; Mainini, L. Learning for predictions: Real-time reliability assessment of aerospace systems. *Aiaa J.* **2022**, *60*, 566–577. [CrossRef]
23. Dalla Vedova, M.D.; Berri, P.C.; Re, S. Metaheuristic Bio-Inspired Algorithms for Prognostics: Application to on-Board Electromechanical Actuators. In Proceedings of the 2018 3rd International Conference on System Reliability and Safety, ICSRS 2018, Barcelona, Spain, 24–26 November 2018. [CrossRef]
24. Ma, Y.; Yao, M.; Liu, H.; Tang, Z. State of Health estimation and Remaining Useful Life prediction for lithium-ion batteries by Improved Particle Swarm Optimization-Back Propagation Neural Network. *J. Energy Storage* **2022**, *52*, 104750. [CrossRef]
25. Li, A.; Yang, X.; Dong, H.; Xie, Z.; Yang, C. Machine Learning-Based Sensor Data Modeling Methods for Power Transformer PHM. *Sensors* **2018**, *18*, 4430. [CrossRef] [PubMed]
26. Gougam, F.; Chemseddine, R.; Benazzouz, D.; Benaggoune, K.; Zerhouni, N. Fault prognostics of rolling element bearing based on feature extraction and supervised machine learning: Application to shaft wind turbine gearbox using vibration signal. *Proc. Inst. Mech. Eng. Part J. Mech. Eng. Sci.* **2021**, *235*, 5186–5197. [CrossRef]
27. Rodrigues, L.R.; Coelho, D.B.P.; Gomes, J.P.P. A Hybrid TLBO-Particle Filter Algorithm Applied to Remaining Useful Life Prediction in the Presence of Multiple Degradation Factors. In Proceedings of the 2020 IEEE Congress on Evolutionary Computation (CEC), Glasgow, UK, 19–24 July 2020; pp. 1–8. [CrossRef]
28. Jin, Z.; Li, X.; Yu, D.; Zhang, J.; Zhang, W. Lithium-ion battery state of health estimation using meta-heuristic optimization and Gaussian process regression. *J. Energy Storage* **2023**, *58*, 106319. [CrossRef]
29. Chen, K.; Laghrouche, S.; Djerdir, A. Health state prognostic of fuel cell based on wavelet neural network and cuckoo search algorithm. *ISA Trans.* **2021**, *113*, 175–184. [CrossRef]
30. Zhong, J.; Long, J.; Zhang, S.; Li, C. Flexible Kurtogram for Extracting Repetitive Transients for Prognostics and Health Management of Rotating Components. *IEEE Access* **2019**, *7*, 55631–55639. [CrossRef]
31. Hong, H.; Tsangaratos, P.; Ilia, I.; Loupasakis, C.; Wang, Y. Introducing a novel multi-layer perceptron network based on stochastic gradient descent optimized by a meta-heuristic algorithm for landslide susceptibility mapping. *Sci. Total Environ.* **2020**, *742*, 140549. [CrossRef]
32. Ladj, A.; Benbouzid-Si Tayeb, F.; Varnier, C. An integrated prognostic based hybrid genetic-immune algorithm for scheduling jobs and predictive maintenance. In Proceedings of the 2016 IEEE Congress on Evolutionary Computation (CEC), Vancouver, BC, Canada, 24–29 July 2016; pp. 2083–2089. [CrossRef]
33. Ladj, A.; Tayeb, F.B.S.; Varnier, C. Hybrid of metaheuristic approaches and fuzzy logic for the integrated flowshop scheduling with predictive maintenance problem under uncertainties. *Eur. J. Ind. Eng.* **2021**, *15*, 675–710. [CrossRef]
34. Wu, Y. A survey on population-based meta-heuristic algorithms for motion planning of aircraft. *Swarm Evol. Comput.* **2021**, *62*, 100844. [CrossRef]
35. Rodrigues, L.R.; Gomes, J.P.P.; Ferri, F.A.S.; Medeiros, I.P.; Galvão, R.K.H.; Nascimento Júnior, C.L. Use of PHM Information and System Architecture for Optimized Aircraft Maintenance Planning. *IEEE Syst. J.* **2015**, *9*, 1197–1207. [CrossRef]
36. Berri, P.C.; Dalla Vedova, M.D.L.; Maggiore, P. A Simplified Monitor Model for EMA Prognostics. In Proceedings of the MATEC Web of Conferences, Osaka, Japan, 21–22 September 2018; EDP Sciences: Les Ulis, France, 2018.
37. Berri, P.C.; Dalla Vedova, M.D.L.; Maggiore, P.; Viglione, F. A simplified monitoring model for PMSM servoactuator prognostics. In Proceedings of the MATEC Web of Conferences, Osaka, Japan, 21–22 September 2018; EDP Sciences: Les Ulis, France, 2019; p. 04013.

38. Dalla Vedova, M.D.; Berri, P.C.; Re, S. A comparison of bio-inspired meta-heuristic algorithms for aircraft actuator prognostics. In Proceedings of the 29th European Safety and Reliability Conference, ESREL, Hannover, DE, USA, 22–26 September 2019. [CrossRef]
39. Kim, B.W.; Kim, K.T.; Hur, J. Simplified Impedance Modeling and Analysis for Inter-Turn Fault of IPM-type BLDC motor. *J. Power Electron.* **2012**, *12*, 10–18. [CrossRef]
40. Berri, P.C. Design and Development of Algorithms and Technologies Applied to Prognostics of Aerospace Systems. Ph.D. Thesis, Politecnico di Torino, Torino, Italy, 2021.
41. Querques, I. *Prognostics of On-Board Electromechanical Actuators: Bio-Inspired Metaheuristic Algorithms*; Technical Report; Politecnico di Torino: Torino, Italy, 2021.
42. Wahde, M. *Biologically Inspired Optimization Methods: An Introduction*; WIT Press: Ashurst, UK, 2008.
43. Ahmad, M.F.; Isa, N.A.M.; Lim, W.H.; Ang, K.M. Differential evolution: A recent review based on state-of-the-art works. *Alex. Eng. J.* **2022**, *61*, 3831–3872. [CrossRef]
44. Storn, R.; Price, K. Differential evolution–a simple and efficient heuristic for global optimization over continuous spaces. *J. Glob. Optim.* **1997**, *11*, 341–359. [CrossRef]
45. Aimasso, A.; Berri, P.C.; Dalla Vedova, M.D. A genetic-based prognostic method for aerospace electromechanical actuators. *Int. J. Mech. Control* **2021**, *22*, 195–206.
46. Kennedy, J.; Eberhart, R. Particle swarm optimization. In Proceedings of the ICNN'95—International Conference on Neural Networks, Perth, WA, Australia, 27 November–1 December 1995; Volume 4, pp. 1942–1948.
47. Darwish, A. Bio-inspired computing: Algorithms review, deep analysis, and the scope of applications. *Future Comput. Inform. J.* **2018**, *3*, 231–246. [CrossRef]
48. Mirjalili, S.; Mirjalili, S.M.; Lewis, A. Grey wolf optimizer. *Adv. Eng. Softw.* **2014**, *69*, 46–61. [CrossRef]
49. Kumar, V.; Kumar, D. An astrophysics-inspired Grey wolf algorithm for numerical optimization and its application to engineering design problems. *Adv. Eng. Softw.* **2017**, *112*, 231–254. [CrossRef]
50. Berri, P.C.; Dalla Vedova, M.D.L.; Maggiore, P.; Scanavino, M. Permanent Magnet Synchronous Motor (PMSM) for Aerospace Servomechanisms: Proposal of a Lumped Model for Prognostics. In Proceedings of the 2018 2nd European Conference on Electrical Engineering and Computer Science (EECS), Bern, Switzerland, 20–22 December 2018; pp. 471–477.
51. Berri, P.C.; Dalla Vedova, M.D.; Maggiore, P. A lumped parameter high fidelity EMA model for model-based prognostics. In Proceedings of the 29th European Safety and Reliability Conference, ESREL 2019, Hannover, Germany, 22–26 September 2019; Research Publishing Services: Singapore, 2020; pp. 1086–1093. [CrossRef]
52. Baldo, L.; Berri, P.C.; Dalla Vedova, M.D.L.; Maggiore, P. Experimental Validation of Multi-fidelity Models for Prognostics of Electromechanical Actuators. In Proceedings of the PHM Society European Conference, Turin, Italy, 6–8 July 2022; Volume 7, pp. 32–42.
53. Bendjedia, M.; Tehrani, K.A.; Azzouz, Y. Design of RST and Fractional Order PID Controllers for an Induction Motor Drive for Electric Vehicle Application. In Proceedings of the 7th IET International Conference on Power Electronics, Machines and Drives, Manchester, UK, 8–10 April 2014. [CrossRef]

Article

Comprehensive Comparison of Different Integrated Thermal Protection Systems with Ablative Materials for Load-Bearing Components of Reusable Launch Vehicles

Stefano Piacquadio [1,*], Dominik Pridöhl [1], Nils Henkel [2], Rasmus Bergström [3], Alessandro Zamprotta [3], Athanasios Dafnis [1] and Kai-Uwe Schröder [1]

1 Institute for Structural Mechanics and Lighweight Design, RWTH Aachen University, Wüllnerstraße 7, 52062 Aachen, Germany
2 Faculty of Mechanical Engineering, RWTH Aachen University, Templergraben 55, 52062 Aachen, Germany
3 Pangea Aerospace S.L., Avinguda Número 1, 20, 08040 Barcelona, Spain
* Correspondence: stefano.piacquadio@sla.rwth-aachen.de

Abstract: Economic viability of small launch vehicles, i.e., microlaunchers, is impaired by several factors, one of which is a higher dry to wet mass ratio as compared to conventional size launchers. Although reusability may reduce launch cost, it can drive dry and/or wet mass to unfeasibly high levels. In particular, for load-bearing components that are exposed to convective heating during the aerothermodynamic phase of the re-entry, the mass increase due to the presence of a thermal protection system (TPS) must be considered. Examples of such components are aerodynamic drag devices (ADDs), which are extended during the re-entry. These should withstand high mechanical loading, be thermally protected to avoid failure, and be reusable. Ablative materials can offer lightweight thermal protection, but they represent an add-on mass for the structure and they are rarely reusable. Similarly, TPS based on ceramic matrix composite (CMC) tiles represent an additional mass. To tackle this issue, so-called integrated thermal protection systems (ITPS) composed of CMC sandwich structures were introduced in the literature. The aim is to obtain a load-bearing structure that is at the same time the thermally protective layer. However, a comprehensive description of the real lightweight potential of such solutions compared to ablative materials with the corresponding sub-structures is, to the authors' knowledge, not yet presented. Thus, based on the design of an ADD, this work aims to holistically describe such load bearing components and to compare different TPS solutions. Both thermal and preliminary mechanical designs are discussed. Additionally, a novel concept is proposed, which is based on the use of phase change materials (PCMs) embedded within a metallic sandwich structure with an additively manufactured lattice core. Such a solution can be beneficial due to the combination of both the high specific stiffness of lattice structures and the high mass-specific thermal energy storage potential of PCMs. The study is conducted with reference to the first stage of the microlauncher analysed within the European Horizon-2020 project named Recovery and Return To Base (RRTB).

Keywords: reusable launch vehicle; thermal protection system; integrated thermal protection system; ablative material; ceramic matrix composite; phase change material; lattice structure; additive manufacturing

Citation: Piacquadio, S.; Pridöhl, D.; Henkel, N.; Bergström, R.; Zamprotta, A.; Dafnis, A.; Schröder, K.-U. Comprehensive Comparison of Different Integrated Thermal Protection Systems with Ablative Materials for Load-Bearing Components of Reusable Launch Vehicles. *Aerospace* **2023**, *10*, 319. https://doi.org/10.3390/aerospace10030319

Academic Editors: Spiros Pantelakis, Andreas Strohmayer and Jordi Pons-Prats

Received: 31 January 2023
Revised: 18 March 2023
Accepted: 20 March 2023
Published: 22 March 2023

1. Introduction

1.1. Context

Microlaunchers are considered a strategic asset to achieve high frequency, tailored access to space for small satellites [1]. The addition of first stage reusability lowers costs and improves business sustainability. However, the structural mass does not scale linearly with the launchers' size, leading to microlaunchers having comparatively higher structural coefficients with respect to conventional launchers, such as Ariane 5, Soyuz, and Falcon

9 [2]. This affects the launchers' performance and thus the economic viability. To address this challenge, novel lightweight solutions are required for primary structural components. Additionally, to reduce wet and dry mass as well as complexity, passive re-entry and landing concepts can be considered. This, among others, is the aim of the feasibility studies led within the framework of the research project "Recovery and Return To Base" (RRTB) funded through the European Horizon 2020 programme [3]. In particular, the project aims to perform a re-entry flight phase without the use of retro-propulsion to achieve deceleration. Four mechanically actuated aerodynamic drag devices (ADDs), as shown in Figure 1, are extended from the rocket body to achieve the desired ballistic coefficient and obtain a sufficient deceleration during the re-entry and descent flight phase.

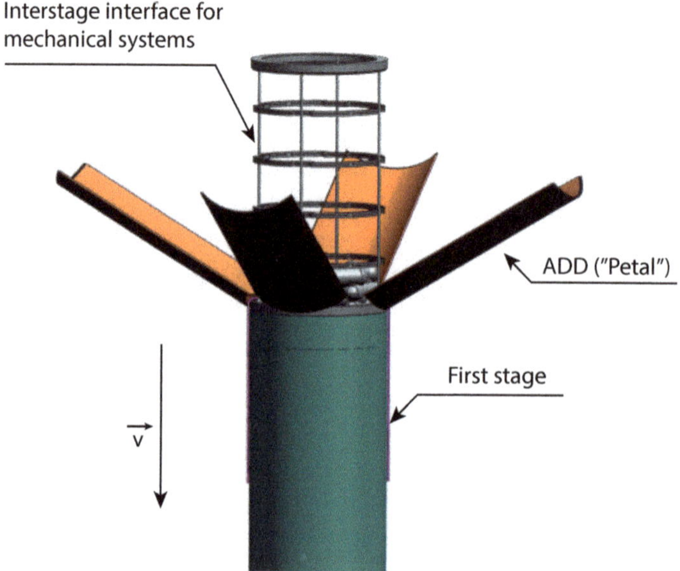

Figure 1. Architectural CAD sketch of the aerodynamic drag devices in open configuration.

The ADDs are extended before the beginning of the re-entry flight phase and are then exposed to high convective heat fluxes and high dynamic pressures. Similar drag devices were investigated in the European project RETALT for a conventional size reusable launcher. As described by Marwege et al. [4], the design was judged unfeasible because of excessive reaction forces and moments, which would have led to high structural mass of the components. Although similar issues are faced for the design of ADDs for the mission considered in the RRTB project, different sizes and different design methodologies allow for not excluding the concept a priori. The design of such components requires a holistic approach and is mainly influenced by the thermal protection system (TPS) design, the structural design, and the design of the extension mechanisms. Each element intrinsically influences the design of the others. This work concentrates on the design of the TPS and offers a consideration on lightweight design potential of different thermal protection solutions and their effect on the overall structural mass. In particular, ablative materials as well as integrated TPS (ITPS) solutions are considered. ITPSs are load bearing structures made of materials with a high operative temperature and a high thermal insulation capability. The geometry is designed to increase the thermal resistance and have a high specific stiffness, thus aiming to obtain holistic mass reductions.

1.2. Ablative and Integrated Thermal Protection Systems

Ablative materials represent a high TRL solution for medium to high heat fluxes, typical of a ballistic re-entry [5]. Due to the endothermic reactions that take place during ablation, such materials offer high mass-specific thermal protection properties when compared to TPS based on sensible heat storage, i.e., ceramic tiles. However, the obvious disadvantage is that ablative materials are intrinsically expendable. Thus, re-application or replacement of the entire structure is often necessary to achieve reusability of the component. Furthermore the ablative TPS represents a non load-bearing add-on to the structural mass.

Passive TPS based on tiles, often made of high temperature metallic alloys or ceramic matrix composites (CMCs), are commonly employed for reusable launch vehicles (RLVs) [6]. Their application is limited by the maximum operative temperature of the material, thermally induced stresses, and outward thermal deflection. As indicated by Dorsey et al. in [7] as well as by Le and Goo in [8,9], excessive deflection can cause transition from a laminar flow on the outer wall of the TPS to a turbulent one. This is correlated with an increase of the heat transfer coefficient between the wall and hot gas, which thus causes an increase in the convective heat flux that the TPS outer wall experiences. Such an increase is coupled to even higher deformations and stresses, which can lead to failure. In particular, as evidenced by Heidenreich et al. [10], the higher the maximum operative temperature is, the lower the specific tensile strength of the material. CMCs received increasing attention due to the relatively constant mechanical properties at different temperatures and for retaining the highest specific tensile strength at temperatures above 1000 °C ([10–12]). In parallel to such material development, several efforts were made by authors in the literature to obtain passive thermal protection systems with high thermal and mechanical load bearing capability. With this goal, Blosser et al. [13] and Fischer et al. [14] developed structures consisting of a metallic honeycomb sandwich (based on nickel-superalloys and gamma titanium-aluminide, respectively) on the outer surface and a fibrous insulation encapsulated between the sandwich and the vehicle interior.

Bapanapalli et al. [15] and Gogu et al. [16] first proposed the concept of an ITPS. It is based on a corrugated core sandwich panel, hosting a fibrous insulation within the webs. In recent years, several authors investigated the use of CMCs as a structural material for ITPS, with different core topologies [17]. Le et al. [18] indicated that the choice of a particular core topology and material combination is not trivial and depends on the thermal and mechanical load profiles, the specific mechanical properties of the material as a function of expected temperature, and the obtainable effective thermal conductivity.

The specific stiffness and strength of most materials decrease with increasing temperature. Although CMCs are suitable for operation at high temperature (above 1000 °C), the specific mechanical properties are, in absolute terms, lower than the ones of high temperature alloys, i.e., Ni-based superalloys or titanium aluminides. Additionally, operation at high temperature, i.e., approaching the radiative equilibrium temperature at the outer face sheet, while reducing the sensibly stored thermal energy, is cause for high thermal gradients in the out-of-plane direction of the sandwich structure. These gradients, in turn, increase the thermally induced stresses with respect to operation at lower temperatures. Therefore, when considering load-bearing components that require a lightweight, reusable TPS, the choice of material and configuration is not trivial. For a load-bearing component, a reduction of the wall temperature can be beneficial by allowing the use of materials with high specific mechanical properties.

Use of Phase Change Materials for Integrated Thermal Protection Systems

In this perspective, latent heat storage, i.e., melting of a so-called phase change material (PCM), is more efficient than sensible storage in mass-specific terms. Indeed, the thermal mass required to store heat via phase change is lower than the one needed to sensibly store the same amount of energy. However, only few authors ([19,20]) have investigated the use of PCMs for TPS. As explained by Nazir et al. [21], encapsulation and thermal conductivity improvement of PCMs proved to be the main challenge hindering the widespread application of such materials.

Recent literature ([22,23]) introduced the use of additively manufactured lattice structures to address both issues. These cellular solids were found to deliver high effective thermal conductivity, improving the thermal energy storage capability of the material. Additionally, sandwich structures built with such lattice cores exhibit attractive specific mechanical properties, simplify the junction of the core with the face sheets, and allow for the reduction of issues related to delamination [24]. Due to these favorable thermal and mechanical properties, lattice core sandwich structures with embedded PCMs are attractive for use in load-bearing lightweight TPS.

To address the aforementioned challenges in ITPS design, considering both the thermal and the mechanical behaviour is of fundamental importance. Thus, in this work, we perform a comparison between the thermal response of the three introduced TPS concepts for use on the ADD of the RLV considered in the RRTB project:

1. A phenolic impregnated carbon ablator (PICA) ablative TPSis analysed by means of a solver based on the one-dimensional finite volume method. The thermal mass is optimised via a root finding algorithm.
2. The CMC-based ITPS is composed of a corrugated core sandwich structure made of C/SiC and Saffil® insulation. The aforementioned solver (with ablation terms deactivated) is used to analyse it. A constrained optimisation algorithm based on sequential least squares programming (SLSQP) implemented in Python® is used to optimise the core and face sheets geometry for minimal thermal mass.
3. The solution based on lattice core-PCM sandwich structures is analysed via implementing a homogenisation technique based on the semi-analytical model proposed by Hubert et al. [22] and on the application of mixture rules, as reported in [23]. The PCM behaviour is modelled with use of the apparent heat capacity method, implemented in COMSOL® Multiphysics.

After treatment of each concept's thermal design, a preliminary structural design for each considered solution is described. In the end, a comprehensive evaluation of both the thermal response results and the associated mechanical design is given. The results are compared in terms of total mass.

2. Governing Equations

2.1. Ablation

The internal energy balance of the solid and pyrolysis gas takes the form of a classic conduction equation combined with a source term that arises from the pyrolysis gas flow:

$$\frac{\partial}{\partial t}(\rho C_p T) + \nabla \cdot \left(-\bar{\bar{\lambda}} \cdot \nabla T\right) + \nabla \cdot \left(\dot{\boldsymbol{m}}_g'' h_g\right) = 0 \tag{1}$$

In this equation, t denotes time, ρ and C_p the solid density and specific heat capacity, respectively, T the temperature, $\bar{\bar{\lambda}}$ the thermal conductivity tensor, $\dot{\boldsymbol{m}}_g''$ the area specific pyrolysis gas mass flow rate vector, and h_g the enthalpy of pyrolysis gas.

If one assumes that a control volume V moves at speed $\boldsymbol{v}_{gr}$, through use of the Gauss' theorem, one obtains [25]:

$$\frac{d}{dt}\iiint_V \rho C_p T \,\mathrm{d}V + \iint_A -\bar{\bar{\lambda}} \cdot \nabla T \cdot \mathrm{d}\boldsymbol{A} + \iint_A \dot{\boldsymbol{m}}_g'' h_g \cdot \mathrm{d}\boldsymbol{A} - \iint_A \rho C_p T \boldsymbol{v}_{gr} \cdot \mathrm{d}\boldsymbol{A} = 0. \tag{2}$$

The terms in Equation (2) can be interpreted as follows from left to right: time change in internal energy of the control volume, conductive heat flux, convective heat flux due to pyrolysis gas movement, and convective heat flux due to grid movement.

With reference to Figure 2, the one-dimensional energy balance can be written as:

$$\frac{\mathrm{d}}{\mathrm{d}t}\int_V \rho C_p T dz + \left(\dot{m}_g'' h_g\right)_{right} - \left(\dot{m}_g'' h_g\right)_{left} + \left(\lambda\frac{\partial T}{\partial z}\right)_{right} - \left(\lambda\frac{\partial T}{\partial z}\right)_{left} - \left(\rho C_p T u_{gr}\right)_{right} + \left(\rho C_p T u_{gr}\right)_{left} = 0, \quad (3)$$

where z is a stationary coordinate and u_{gr} is the grid velocity.

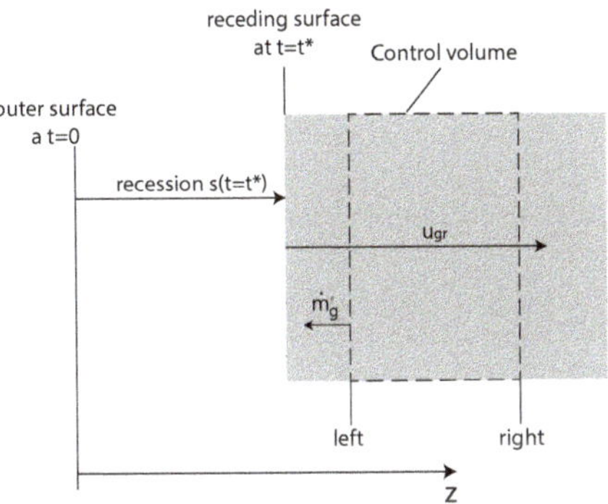

Figure 2. Schematic of a one-dimensional control volume describing the coordinates and labeling used in Equation (3).

The rate of decomposition of the material is temperature-dependent and can be described via an Arrhenius equation:

$$\frac{\mathrm{d}\rho_i}{\mathrm{d}t} = -c_i\left(\frac{\rho_i - \rho_{c,i}}{\rho_{v,i} - \rho_{c,i}}\right)^{n_{r,i}} \mathrm{e}^{-\frac{A_i}{T}}. \quad (4)$$

As the material can be made of multiple components, the index i indicates the ith component with its own decomposition law; c is called the pre-exponential factor, ρ, ρ_v, and ρ_c the current, virgin, and char density, respectively, n_r the reaction order, and A_i the scaled activation energy. The density of the material is obtained via the weighted sum of each phase:

$$\rho = \sum_{i=1}^{N_p} \Gamma_i \rho_i, \quad (5)$$

where i represents the considered phase, N_p is the number of present phases, Γ_i is the volume fraction of the ith phase, and ρ_i is the density of the ith phase.

The virgin material is generally assumed to be impermeable, thus forcing all gas to leave through the outer surface [26]. All the material that did not remain solid, i.e., transitioning from virgin to char status, is in gaseous form. Thus, for an internal control volume, as depicted in Figure 2, the pyrolysis gas mass flux is related to the decomposition from virgin to charred material, described in Equation (4), via mass conservation:

$$\frac{\partial \dot{m}_g''}{\partial z} = -\frac{\mathrm{d}\rho}{\mathrm{d}t} = -\sum_i \Gamma_i \frac{\mathrm{d}\rho_i}{\mathrm{d}t}. \quad (6)$$

The char ablation needs to be considered in the mass conservation when considering a control volume that includes the receding surface. At the surface of the ablative thermal protection system, chemical reactions such as oxidation and nitridation take place [27]. Three components take part in this reaction:

1. The hot boundary layer gases of the flow,
2. The surface (mostly charred) solid material,
3. The pyrolysis gas emerging from the depths of the decomposing layer.

If chemical equilibrium is assumed, the products of the reaction and their associated properties such as enthalpy can be determined. Programs such as Mutation++ [27] or CEA [28] can compute this equilibrium by minimisation of Gibbs energy. As an input, the ratios of the three components listed above are needed. To obtain these, the mass flux of the charred material $\dot{m}_c''$ and pyrolysis gas at the wall $\dot{m}_{g,w}''$ are non-dimensionalised:

$$B_c' = \frac{\dot{m}_c''}{\rho_e u_e C_M} \tag{7}$$

$$B_g' = \frac{\dot{m}_{g,w}''}{\rho_e u_e C_M}, \tag{8}$$

where ρ_e and u_e are the boundary layer edge gas density and velocity, respectively; C_M is the local Stanton number for mass transfer. As a result of the surface chemistry calculation, one receives:

$$B_c' = f\left(T_w, B_g', p\right) \tag{9}$$

$$h_w = f\left(T_w, B_g', p\right), \tag{10}$$

where T_w is the wall temperature, h_w the enthalpy of the gaseous surface reaction products, and p the pressure [29]. The enthalpy flux that is carried away from the material can then be calculated using $(\dot{m}_c + \dot{m}_{g,w})h_w$. The charred material mass flux $\dot{m}_c$ can be connected to the surface recession rate $\dot{s}$ via:

$$\dot{m}_c = \rho_w \dot{s}, \tag{11}$$

where ρ_w is the density of the solid material at the wall.

2.2. Energy Equation for ITPS

For a passive, non-ablative TPS, the energy equation can be written as a special case of the already described one for an ablative material, in which ablation does not take place.

$$\frac{\partial}{\partial t}\left(\rho_{eff} C_{p_{eff}} T\right) + \nabla \cdot \left(-\bar{\bar{\lambda}}_{eff} \cdot \nabla T\right) = 0, \tag{12}$$

where ρ_{eff} is the effective density of the material, $C_{p_{eff}}$ its effective specific heat capacity, and $\bar{\bar{\lambda}}_{eff}$ is the effective thermal conductivity tensor. However, a special treatment is needed to describe the behaviour of the ITPS embedding a PCM. This is done by means of the apparent heat capacity formulation [30]. The term corresponding to the latent heat of fusion is included as an additional non-linear term in the definition of the heat capacity of the material:

$$\rho_{eff} C_{p_{eff}} = \frac{\partial H}{\partial T} = C_{eff} + L\frac{\partial \alpha_l}{\partial T}, \tag{13}$$

where C_{eff} is the actual heat capacity, H is the enthalpy, L is the PCM latent heat of fusion, and α_l is the liquid fraction at the melting front.

2.3. Material Properties

In this section, the treatment for the material properties of each considered concept is described.

2.3.1. Ablative Material

Difficulties in the determination of the local material properties arise from the thermal properties of the ablative TPS that depend on the degree of char β. As the heat shield material decomposes, the material properties, such as thermal conductivity or heat capacity, change. One commonly used approach in literature [31,32] to model this change is to prescribe fully virgin and fully charred material properties and interpolate based on the weight fraction of virgin and charred material.

The extent of the decomposition reaction β can be calculated through

$$\beta = \frac{\rho_v - \rho}{\rho_v - \rho_c}, \tag{14}$$

where v refers to the virgin and c to the charred material; β is therefore 0 when the whole material is virgin and 1 for a fully charred state. Because of the assumption that a denser material contributes to the material properties to a higher degree, the weight fraction w_v of virgin material is introduced:

$$w_v = \frac{\rho_v}{\rho_v - \rho_c}\left(1 - \frac{\rho_c}{\rho}\right) = \frac{\rho_v}{\rho}(1 - \beta). \tag{15}$$

The char weight fraction w_c is then

$$w_c = 1 - w_v = \frac{\rho_c}{\rho}\beta. \tag{16}$$

The heat capacity c_p for instance is computed using:

$$c_p(T, w_v) = w_v C_{p,v}(T) + w_c C_{p,c}(T) = w_v C_{p,v}(T) + (1 - w_v) C_{p,c}(T). \tag{17}$$

2.3.2. Corrugated Core ITPS

The three-dimensional structure of the ITPS sandwich cores is homogenised to allow a reduced treatment. A unit cell of a corrugated sandwich panel and its defining dimensions are sketched in Figure 3. The corrugated sandwich consists of a top face sheet (TFS) with thickness t_T and a bottom face sheet (BFS) with thickness t_B separated by the core thickness t_C. These are connected by webs of thickness t_W at an angle of corrugation Θ. The voids in-between are filled with a high temperature insulation material. The pattern repeats with multiples of the unit cell length $2p$.

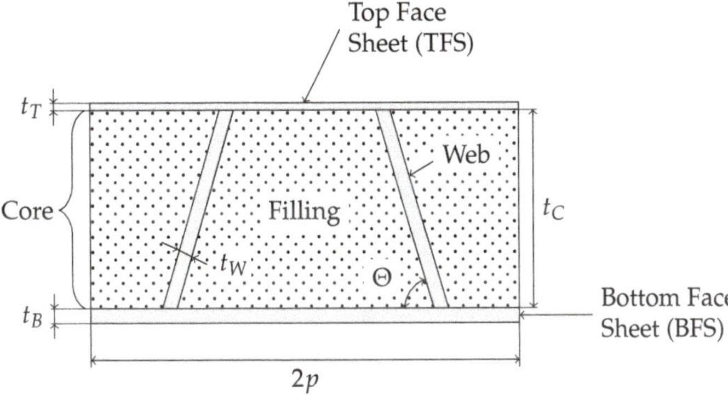

Figure 3. Dimensioned sketch of a corrugated sandwich unit cell.

Effective properties for the core are derived from rule of mixtures. As density is based on the volume fractions of the respective materials, a volume rule of mixtures is chosen for this property. In contrast, heat capacity is defined by the respective mass fractions, which leads to a mass rule of mixtures. The respective areas of web and filling are simply the volumes divided by the core thickness t_C. As this factor is common to both volumes, there is no difference between areal and volumetric homogenisation for this case. Thus, homogenisation equations for effective specific heat capacity $C_{p,eff}$, effective density ρ_{eff}, and effective thermal conductivity λ_{eff} of the corrugated core are obtained as follows [16]:

$$C_{p,eff} = \frac{C_{p,W}\rho_W t_W + C_{p,F}\rho_F(p\sin\Theta - t_W)}{p\sin\Theta} \tag{18}$$

$$\rho_{eff} = \frac{\rho_W V_W + \rho_F V_F}{V_C} = \frac{\rho_W t_W + \rho_F (p \sin\Theta - t_W)}{p \sin\Theta} \quad (19)$$

$$\lambda_{eff} = \frac{\lambda_W A_W + \lambda_F A_F}{A_C} = \frac{\lambda_W V_W + \lambda_F V_F}{V_C} = \frac{\lambda_W t_W + \lambda_F (p \sin\Theta - t_W)}{p \sin\Theta}. \quad (20)$$

Here, the indices W and F refer to properties of the web and filling materials, respectively.

2.3.3. Lattice Core ITPS with Embedded PCM

Orthotropic cell geometry leads to an orthotropic effective thermal conductivity tensor of the composite. Several types of lattice cores exist. The ones most investigated in the literature are the cubic ones inspired by Bravais crystals; see Figure 4. Excluding the *bcc* cell, which exhibits an isotropic morphology, cubic arrangements of these cells exhibit orthotropic behaviour.

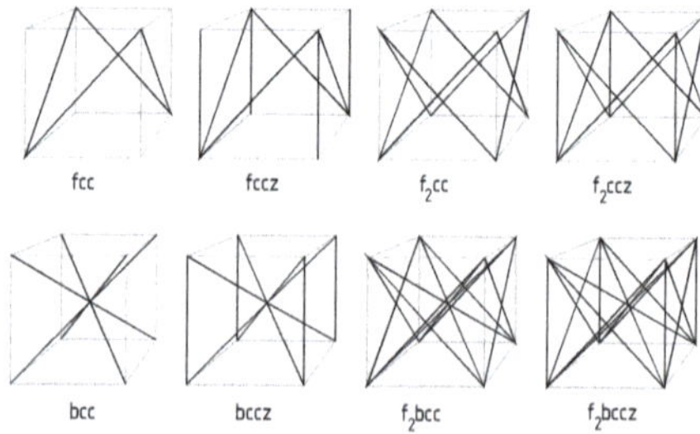

Figure 4. Possible lattice types for cubic unit cells.

The effective thermal conductivity tensor must be written in the form

$$\bar{\bar{\lambda}}_{eff} = \begin{bmatrix} \lambda_{xy} & 0 & 0 \\ 0 & \lambda_{xy} & 0 \\ 0 & 0 & \lambda_z \end{bmatrix}, \quad (21)$$

where x and y are the in-plane coordinates and z the out plane one in Figure 4, and the three-dimensional problem can be reduced to a two-dimensional one. The respective values for the thermal conductivity are obtained using a semi-analytical correlation for the definition of the relevant contributions in the equation, as according to Hubert et al. [22].

$$\lambda_i = \lambda_{PCM}\epsilon + \lambda_s G_i (1 - \epsilon), \quad (22)$$

where i indicates a generic principal direction, λ_{PCM} is the thermal conductivity of the PCM, λ_s is the thermal conductivity of the metallic lattice structure, G_i is a dimensionless term that addresses the topology of the cell and is different for different directions, and ϵ is the porosity of the lattice structure. The effective density, effective specific heat, and effective latent heat of fusion can be easily obtained via the mixture rule as described in detail by Piacquadio et al. [23].

3. Solver for Ablative TPS and Corrugated Core ITPS

Using the governing equations described above, a software tool based on the finite volume method (FVM) for the calculation of the thermal response of both ablative and corrugated core ITPS is implemented in Python®. The implementation in Python® allows a simplified connection of the realised solver with different optimisation packages, which allow one to optimise the thermal mass. This way, an easy and accurate comparison of the two options is achievable. For this reason, the tool is named **Hot-Str**ucture and **A**blative **R**eaction **Shi**eld **P**rogram (Hot-STARSHIP). Figure 5 shows a flowchart of the program.

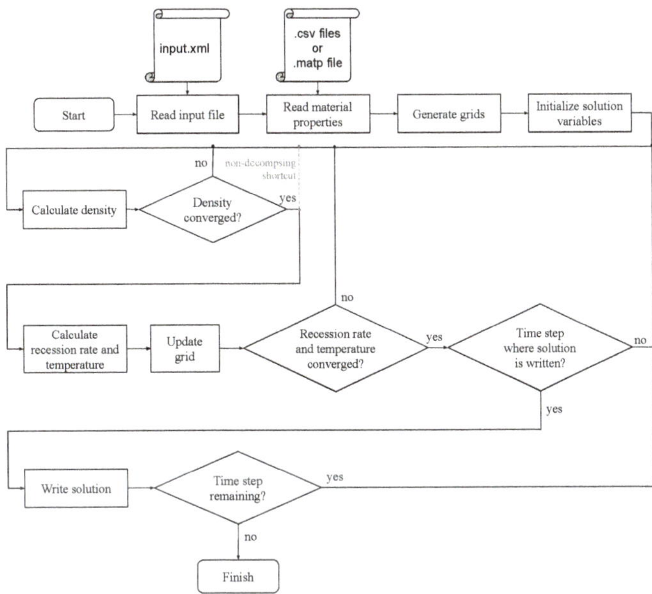

Figure 5. Flowchart of the developed FVM solver. The dashed line indicates a shortcut for the non-decomposing case.

3.1. Verification

To verify the accuracy of the solver, the solution for non-ablative and non-decomposing materials can be compared with available analytical solutions. The solver is able to accurately predict the thermal response of a passive TPS. The details of this verification procedure can be found in Appendix A. As there is no known analytical solution for the decomposing case, a comparison with other programs is performed. For this purpose, the theoretical ablative composite for open testing (TACOT) material is used, and the results presented by Lachaud et al. in [33] are used for verification. For the considered verification sample, a 7% higher recession was obtained by Hot-STARSHIP compared to the software presented in [33] (PATO or Amaryllis). The final difference between the results of the two solvers for the back-face temperature is about 12 K, and the average difference is even lower as the two curves cross each other (Appendix A, Figure A2). This is considered to be accurate enough for fast preliminary design purposes and for mass optimisation.

3.2. Optimisation

Due to its simplicity and the implementation in Python®, the Hot-STARSHIP solver can be easily connected to a constrained optimisation algorithm. The constraints are given via maximum achievable temperatures at the boundaries.

For the ablative material case, the constraint is given only by the maximum back-face temperature of the material. As this is in contact with the support structure, which is often made of carbon fibre-reinforced polymers (CFRP), the maximum temperature is set to 80 °C. For the mass optimisation, in this case, it is sufficient and faster to solve a root-finding problem for the back-face temperature. As a thin TPS, while being lighter, will have higher back-face temperatures, the minimum mass, i.e., the minimum thickness *th*, is the one that matches the maximum allowable back-face temperature. The root-finding problem can be expressed as:

$$\text{Find } th \text{ so that } \quad f(th) = T - 353.15\,\mathrm{K} \stackrel{!}{=} 0 \tag{23}$$

where *th* is the material thickness and T is the back-face temperature. As each function evaluation of $f(th)$ translates to a whole run of the transient solver, a fast convergence of the algorithm is key. As the function $f(th)$ is univariate, algorithm 748 from Python's scipy package with a convergence order of 2.7 is chosen [34,35].

For the corrugated core ITPS, two temperature constraints must be fixed. The first one, i.e., TFS temperature, depends on the maximum operative temperature of the face sheet material. The second one, i.e., BFS temperature, is dependent on the underlying components. A common choice is to fix this to a maximum of 373 K (100 °C). This work considers C/SiC based CMCs for the corrugated core ITPS. The maximum operative temperature is fixed at 1400 K. The geometric variables that influence the thermal response are fitted into a vector of variables $\boldsymbol{x} = [t_T, t_C, t_B, t_W, p, \Theta]^T$, with reference to Figure 3. The minimisation problem then takes the following form:

$$\min_{\boldsymbol{x}} f(\boldsymbol{x}) \qquad \text{with } \boldsymbol{g}(\boldsymbol{x}) < 0 \tag{24}$$

where $f(\boldsymbol{x})$ is the function to be minimised and $\boldsymbol{g}$ is a vector-valued function of constraints. The function f can be written as:

$$f(\boldsymbol{x}) = \rho_T t_T + \frac{\rho_W t_W + \rho_F(p \sin\Theta - t_W)}{p \sin\Theta} t_C + \rho_B t_B \tag{25}$$

The constraint functions split the domain of $\boldsymbol{x}$ into a feasible range that fulfills the constraints function and an infeasible one.

$$\boldsymbol{g}(\boldsymbol{x}) = \begin{bmatrix} T_B - 373\,\text{K} \\ T_{max} - 1400\,\text{K} \end{bmatrix} \overset{!}{<} 0 \tag{26}$$

For this study, the sequential least squares programming algorithm (SLSQP) from the Python® scipy package [36] is chosen. It has a high success rate up to problem dimensions of 20 compared to similar methods [37]. It is therefore well-suitable for the size of this problem.

4. Parametric Study of the Lattice Core-PCM ITPS

The use of a PCM is considered here to allow for a reduction of the wall temperature and thus the use of metallic alloys with high specific mechanical properties. The PCM should not be employed with a thermal insulation purpose as, in fact, a high thermal conductivity is needed to improve the thermal energy storage potential and thus reduce the wall temperature. The amount of PCM, i.e., the thickness of the lattice core, is directly related to the thermal response and is thus considered as a geometric parameter. To obtain thermal protection without mass increase, it is inefficient to consider a sandwich structure with a single core in which the PCM is embedded and for which the constraint is the same of a common ITPS, i.e., a maximum bottom face sheet temperature of 100 °C. Exploiting the design flexibility offered by additive manufacturing, a multi-material, hierarchical sandwich structure, as schematically shown in Figure 6, made of two stacked sandwiches can be designed. In the outer sandwich core, the PCM is embedded, whereas the inner one has a thermal insulation functionality.

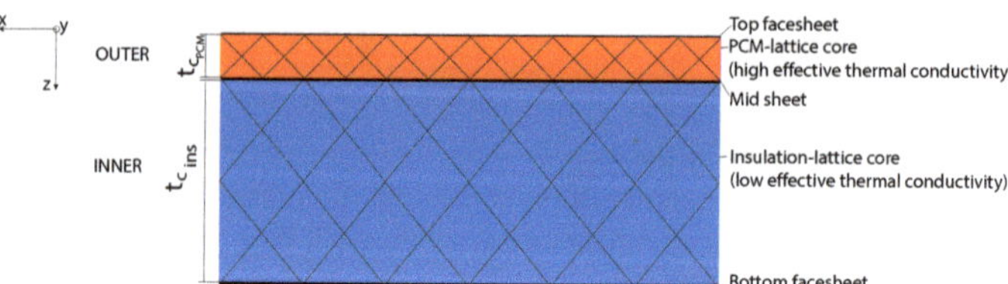

Figure 6. Schematic of a hierarchical lattice core ITPS, in which the PCM is embedded in the outermost core (orange), and a fibrous high temperature insulation is embedded in the innermost core (blue).

The considered material for the lattice structure in which the PCM is embedded is CuCr1Zr, a commonly used alloy for additively manufactured components in the aerospace field. The top face sheet is made of high temperature nickel alloy, i.e., Inconel®718. The center face sheet, lattice core of the insulation layer, and bottom face sheet are, for simplicity of treatment, considered to be made of Inconel 718 as well.

The different parts can be joined via brazing in different steps or via bi-metallic additive manufacturing. Several geometrical parameters influence the thermal performance of the structure, namely unit cell topology, unit cell size, strut radius, and porosity. For cubic unit cells, if the porosity, the unit cell, and the cell size are fixed, the strut radius is obtained as a dependent variable.

The results already present in the literature ([22,23]) allow for the reduction of the number of geometric parameters that must be varied to evaluate the thermal response of the lattice structure-PCM composite. Indeed, the f_2ccz cell is consistently the unit cell that exhibits the highest out-of plane thermal conductivity for a given porosity. This is not true for the bcc unit cell, which shows the lowest thermal conductivity. Thus, the chosen unit cell topologies are trivially f_2ccz for the PCM core and bcc for the insulation core. Similarly, the porosity of the lattice structure for the insulation core should be as high as possible to reduce both mass and effective thermal conductivity, thus leading to a trivial choice. The same is not true for the PCM core. The effective thermal conductivity should be high enough to improve the thermal energy storage of the PCM, but, as the conductivity increases with diminishing porosity, it should be kept as low as possible to keep mass at a minimum. For this reason, the porosity is varied in a range, as reported in Table 1. Similarly, the PCM core thickness defines the PCM mass available and thus the thermal response. This is therefore also a parameter to be varied in the study. The insulation core thickness is varied as well, as it influences the effective thermal resistance. The geometric parameters and their range are summarised in Table 1.

Table 1. Geometrical parameters of the lattice structures with reference to Figure 6.

Core	Unit Cell	Porosity [-]	Core Thickness t_C [mm]
PCM (outer)	f_2ccz	(0.95–0.8)	(5–20)
Insulation (inner)	bcc	0.95	(10–50)

In addition to the geometrical parameters, the thermo-physical properties of the PCM should be considered. A comparably high thermal diffusivity is beneficial for obtaining a fast expansion of the melting front. However, this is not beneficial to the overall thermal protection purpose. Additionally, a material with comparably high latent heat of fusion should be chosen. The melting point also defines the thermal response of the structure. Finally, density of the material has an obvious influence on the lightweight potential of the component. Therefore, it is clear that the material choice does not have a trivial indication. The parametric study performed in this work includes a plausible domain of geometrical variables and and different materials. In particular, the PCMs listed in Table 2 are considered to cover a wide range of melting point, latent heat of fusion, and thermal diffusivity. The listed properties are considered at room temperature. The listed materials are all compatible with Inconel or materials with higher nobility.

Table 2. Thermophysical properties of the studied PCMs with different melting points.

Material	Density [kg/m^3]	Specific Heat Capacity [J/(kg K)]	Thermal Conductivity [W/(m K)]	Thermal Diffusivity [mm^2/s]	Melting Point [°C]	Latent Heat of Fusion [kJ/kg]
Erythritol	950	1900	0.4	0.22	134	213
LiCl(37%)-LiOH	1550	2400	1.1	0.29	262	485
KCl(61%)-$MgCl_2$	2110	900	0.8	0.42	435	351
Li_2CO_3(22%)-Na_2CO_3(16%)-K_2CO_3	2340	2000	1.9	0.40	580	288

The solution of the problem associated with the composite of metallic lattice core and embedded PCM is not implemented in Hot-STARSHIP. Instead, the commercial solver COMSOL® Multiphysics is used, which is based on the finite element method (FEM). It implements the apparent heat capacity method described in Section 2.2. The homogenisation approaches described in Section 2.3.3 are used for both the PCM and insulation core.

5. Results and Discussion

5.1. Boundary Conditions

The ADD is extended via a group of mechanisms with two anchoring points on the structure. One at the root, the other at half of the ADD's longitudinal length, as schematically shown in Figure 7. Rigid body elements are used to connect the fixation points to the structure. All translational and two rotational degrees of freedom (DOFs) are restricted. Only the rotational DOF about the y-axis is not.

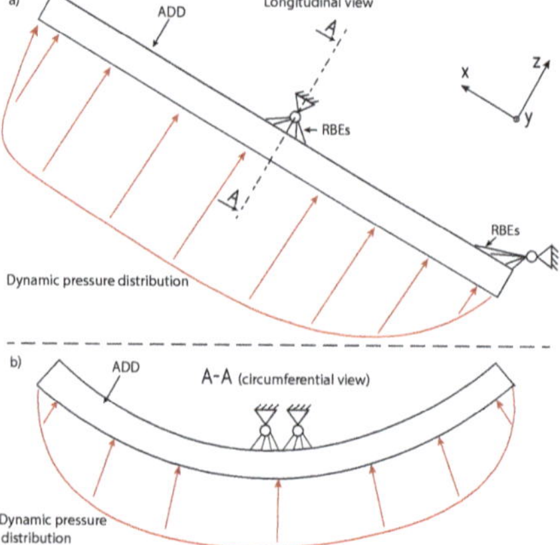

Figure 7. Mechanical boundary conditions (in black) and mechanical load (in red) acting on the ADD: (**a**) in a longitudinal view of the ADD, (**b**) in a circumferential cut (A–A) view.

The input convective heat flux is obtained from the reference mission analysed in the Recovery and Return to Base project of the Horizon 2020 programme. In this work, the focus lies on the ADDs of the first stage analysed within the project.

During ascent, the four identical quarter shells (Figure 1) form a cylindrical shell and function together as primary structure of the launcher's interstage. During atmospheric re-entry, the ADDs are extended to act as aerodynamic decelerators and therefore experience high mechanical and thermal loads.

Based on the re-entry mission analysis, computational fluid dynamic (CFD) simulations were performed to obtain the heat flux distribution on the rocket body at the point of maximum heat flux of the trajectory, which corresponds to an altitude of 35 km and

MACH = 8 speed. To obtain the heat flux variation as a function of time, the trajectory data is analysed with the Sutton–Graves formula [38]. The heat flux distribution on the ADDs as a function of time and longitudinal position along the component is obtained via interpolation and is reported in Figure 8a. The time $t = 0$ corresponds to the moment of the point of the descent trajectory at which the input convective heat flux at stagnation point first reaches $1\,\mathrm{kW/m^2}$. In a similar way, the pressure distribution is obtained and is reported in Figure 8b.

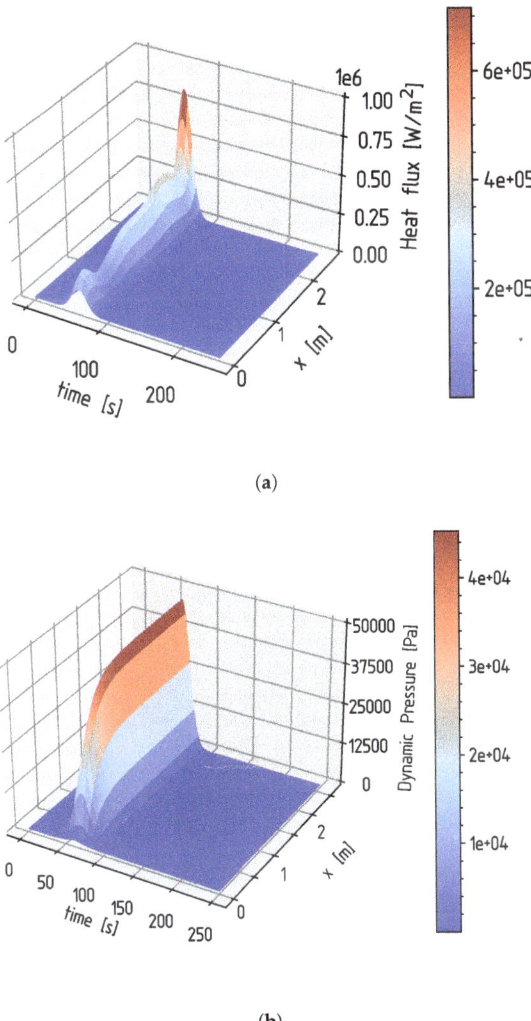

Figure 8. (**a**) Heat flux distribution along the ADD longitudinal coordinate as a function of time; (**b**) dynamic pressure distribution along the ADD longitudinal coordinate as a function of time.

The overall simulation time for the transient thermal analysis is 250 s. Although the convective heating approaches zero after approximately 100 s from the considered initial condition, additional simulation time is considered to take into account heat diffusion within the structure. Although, after the hypersonic and supersonic phases of the flight, convective cooling takes place on the body, the conservative assumption is made that only radiative cooling takes place. To simplify the representation of the thermal analysis, only a section of the ADD is considered in the following. The time curve corresponding to the local maximum of 400 kW/m^2 at the longitudinal position of 2.2 m on the ADD is considered. The pressure distribution is applied on the whole component.

The thermal boundary conditions of the problem are schematically shown in Figure 9. It should be noted that the simulation of the ablative TPS differs from the ITPS cases because of the presence of blowing of the ablation products. The input convective heat flux is corrected via a blowing-corrected heat transfer coefficient, which is calculated as in [39]. The value for the term $\rho_e u_e C_H$ still needs to be assumed and is conservatively defined to be 0.3 as in [33]. The output radiative heat flux is obtained assuming heat transfer with the environment at room temperature. The emissivity of the TFS of the ITPS is assumed to be 0.8. The emissivity of the ablative material depends on the char grade and is obtained from empirical data implemented in the material model.

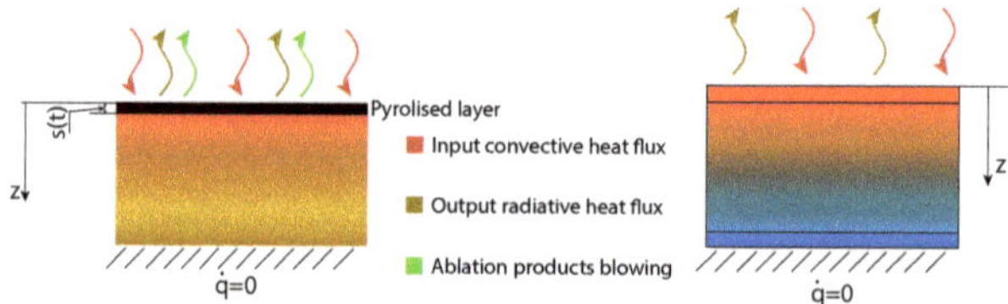

Figure 9. Applied thermal boundary conditions for (**left**) an ablative material and (**right**) for an homogenised ITPS with different cores.

5.2. *Thermal Response of the Ablative TPS*

The analysed material is PICA [40]. It exhibits a low recession rate, however, it also has a relatively high thermal conductivity. The root finding algorithm described in Section 3 is used to obtain the minimal thickness of the material. Figure 10a shows the temperature evolution at different points within the material as a function of re-entry time. Figure 10b shows the recession as a function of re-entry time. In Figure 10a, z is considered the thickness coordinate, which is fixed in space, i.e., the origin lies on the outer edge of the virgin ablative material. For this reason, several temperature curves end abruptly, indicating that the material at the corresponding coordinate ablated away at the given time point. The recession s is obtained by the subtraction of the initial thickness of the virgin material and the position of the moving ablating surface, as shown in Figure 9. The minimum thickness obtained is 47 mm, and a recession of 14.3 mm takes place. The additional material that does not ablate until the end of re-entry is necessary to respect the imposed constraint at the back-face temperature. Due to the relatively high thermal conductivity of PICA, much more material is needed for a proper insulation. The areal weight is 10.75 kg/m^2.

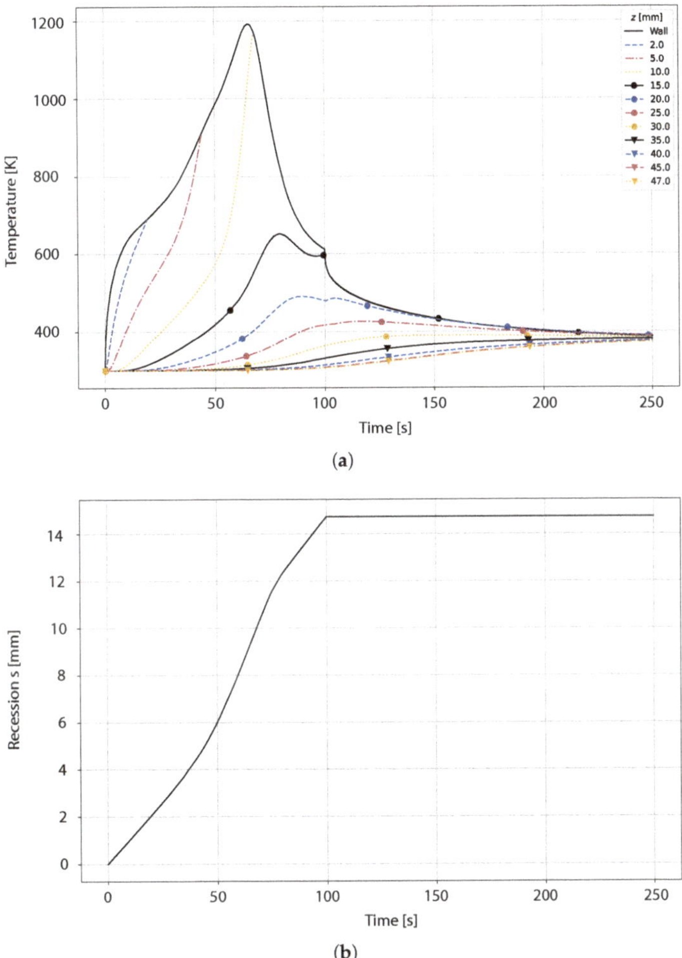

Figure 10. (**a**) Temperature evolution during re-entry. Wall indicates the receding outer surface, whereas the other temperature curves are at fixed z coordinate; (**b**) material recession of the ablative PICA TPS during re-entry.

5.3. Thermal Response of the Corrugated Core ITPS

While the focus of this work aims at comparing ablative TPS with a novel ITPS based on metallic lattice core sandwich structures with PCMs embedded, the comparison with CMC-based corrugated ITPS is useful, as this technology is established in the literature. The considered structural material is an existing C/SiC composite obtained via chemical vapour infiltration whose properties are homogenised based on [10]. The thermal conductivity parallel to the fibre orientation is considered on the webs, whereas the one orthogonal to the fibre is considered for the face sheets. The filling material considered is the Saffil® fibrous insulation felt.

The SLSQP optimisation algorithm described in Section 3 is used to obtain the geometrical parameters of the component, which are reported in Table 3. The obtained overall areal weight is 23.7 kg/m^2. Figure 11 shows the temperature evolution under the same boundary conditions previously analysed. It can be noticed that a thermal gradient of 1100 K is present between the top face sheet and the bottom face sheet. This indicates that the optimisation reached its goals, achieving a component with a very low effective thermal diffusivity. This allows the re-radiation of a wide amount of the convective heat input. This design is beneficial from the thermal protection design point of view. However, due to the combination of high stiffness of C/SiC and high thermal gradient, thermo-mechanical stresses can become a concern, given the low specific strength of CMCs.

Table 3. Optimized geometric parameters for the corrugated core ITPS.

Parameter	Value
t_T	1.7 mm
t_C	35 mm
t_W	1 mm
p	25 mm
θ	60°

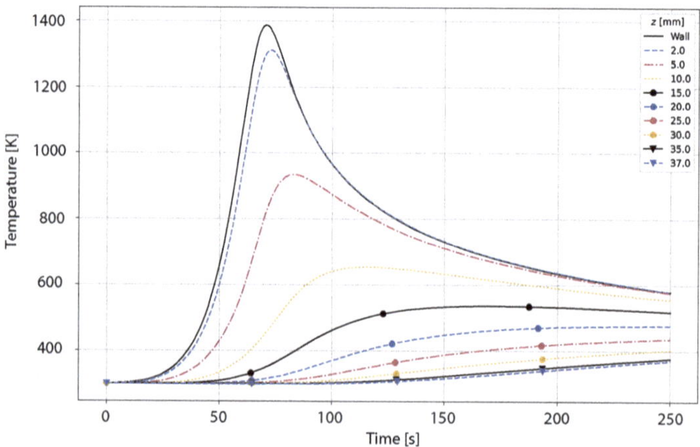

Figure 11. Temperature evolution at different points in the out-of-plane direction (z) for the optimised CMC-based corrugated core ITPS.

5.4. Thermal Response of the Lattice Core-PCM ITPS

In Figure 12, the wall temperature variations during the re-entry trajectory for different PCMs are shown. The geometric parameters are fixed to allow comparability between the results ($t_{c_{PCM}} = 10\,\text{mm}$, $\epsilon = 0.9$, $t_{c_{ins}} = 40\,\text{mm}$). The other parameters are fixed a priori, as reported in Table 1.

It can be noticed that a low temperature peak at the top face sheet corresponds to the eutectic mixture Li_2CO_3(22%)-Na_2CO_3(16%)-K_2CO_3, which, however, exhibits a much higher melting point. This indicates that the thermal behaviour is ascribed to only sensible heat storage. This indicates that the material is not suitable for lightweight latent heat thermal energy storage, as its thermal behaviour is only related to the high thermal mass.

Erythritol, which is the lightest material and also exhibits the lowest melting point, is not suitable for the application. Although a low melting point is advantageous, the low latent heat of fusion compared to other materials makes it an inappropriate choice. The KCl-$MgCl_2$ mixture exhibits a comparably high latent heat, which is shown via the flattening of the temperature curve around its melting point. However, the melting point is higher than that of the LiCl-LiOH mixture, which also shows the highest latent heat of fusion. Thus, the material choice for further consideration in the geometric parametric study falls on the LiCl-LiOH mixture.

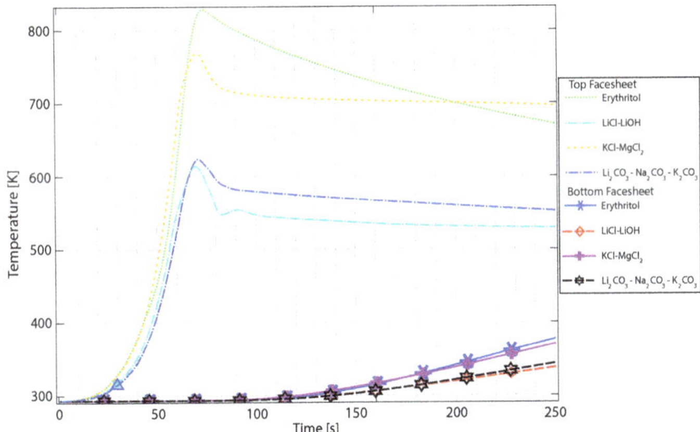

Figure 12. Temperature evolution on top and bottom face sheets for different PCMs.

In the following Figures 13–15, the parametric study for different geometrical parameters is described. Figure 13 shows the wall temperature (top face sheet) evolution for different PCM core thicknesses and porosities of the lattice structure. It can be noticed that the thickness has the highest influence on the thermal behaviour. Diminishing returns in terms of wall temperature reduction are observed with increasing thickness. On the other hand, for a small core thickness, the effect of varying porosity, and thus varying effective thermal conductivity, is marginal. However, for increasing core thicknesses, the effective thermal conductivity becomes more relevant. Indeed, the difference between wall temperature peaks at different porosities increases for the same core thickness.

One can notice that the peak of the temperature curve corresponding to a wall thickness of 20 mm and porosity $\epsilon = 0.95$ is higher than the one corresponding to a core thickness 10 mm and porosity $\epsilon = 0.9$. Even in this case, diminishing returns are observed. Higher peak temperature reductions are observed, e.g., between $\epsilon = 0.95$ and $\epsilon = 0.9$ than between $\epsilon = 0.85$ and $\epsilon = 0.8$.

If one considers mass as a limiting constraint, no trivial optimum exists. To minimise mass, porosity should be as high as possible, as the lattice core material is heavier than the PCM material. The core thickness has a cubic relationship with the bending stiffness of the structure. Therefore. it can not be a priori minimised.

One should notice that all configurations considered are effective in reducing the wall temperature with respect to the case of sensible thermal energy storage of, e.g., a corrugated core ITPS. Indeed even for a core thickness of only 5 mm, the wall temperature reaches a peak of maximum 797 K (524 °C) , which is well below the maximum operative temperature of both Inconel 718 and CuCr1Zr alloys. Therefore, a valid range of core thickness between 5 mm and 10 mm can be considered for application. All in all, a sweet spot can be identified at a core thickness of 10 mm and a porosity of $\epsilon = 0.9$. In such a configuration, the wall temperature does not drastically overshoot the melting point of the PCM.

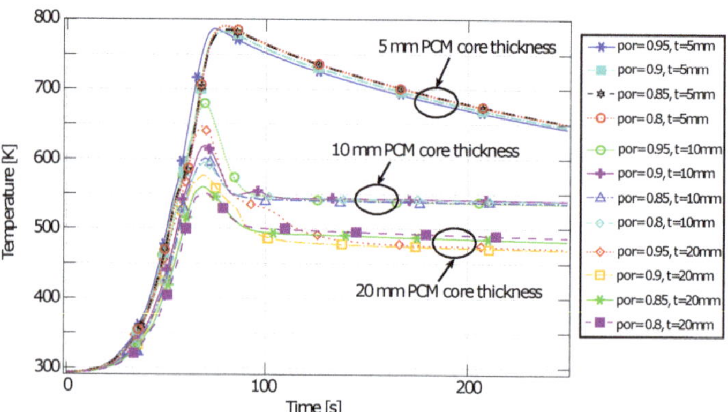

Figure 13. Temperature evolution for different PCM core thicknesses ($t_{c_{PCM}}$) and porosities (ϵ) of the lattice structure.

Figure 14 shows the temperature curves for the top face sheet and the bottom face sheet for different insulation core thicknesses. The PCM core geometrical features are fixed at the identified optimum. The porosity of the insulation core is fixed at $\epsilon = 0.95$ to obtain a low effective thermal conductivity. Increasing the thickness leads to higher thermal resistance, which can be observed with the progressively flattening temperature curves of the bottom face sheet. The temperature of the top face sheet is marginally influenced by the insulation core, as the temperature curves for such a point of the component are dominated by the latent heat thermal energy storage. Considering the bottom face sheet temperature constraints previously described, the case of an insulation core thickness of 10 mm should be discarded. On the other hand, an insulation core thickness of 50 mm does not bring appreciable differences with respect to the thinner 40 mm case. Therefore, it should also be discarded. Table 4 summarizes the final material choice and the identified valid geometrical parameters range. Finally, Figure 15 shows the temperature curves of different positions along the out-of-plane direction for the case of a hierarchical sandwich with a PCM core thickness of 10 mm, $\epsilon = 0.9$, and a thickness of the insulation layer of 40 mm.

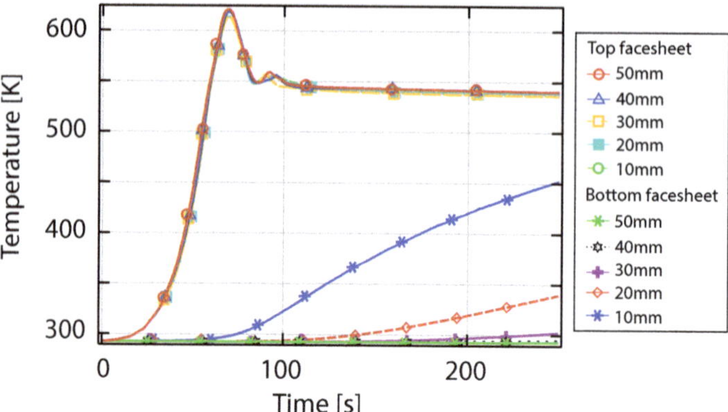

Figure 14. Top and bottom face sheet temperature evolution for different thicknesses of the insulation layer ($t_{c_{ins}}$).

Table 4. Geometrical and physical properties of the lattice core-PCM ITPS layers, with reference to the schematic in Figure 6.

Layer	Component	Material	Thickness [mm]	Volume Fraction	Density [kg/m^3]	Areal Weight [kg/m^2]
1	Top face sheet (TFS)	Inconel 718	1	1	8170	8.17
2	PCM lattice core	CuCr1Zr	(5–10)	0.1	8900	(4.45–8.9)
2	PCM	LiCl-LiOH	(5–10)	0.9	1550	(6.97–13.95)
3	Center face sheet	Inconel 718	1	1	8170	8.17
4	Insulation lattice core	Inconel 718	(20–40)	0.05	8170	(8.17–16.34)
4	Insulation	Saffil®	(20–40)	0.95	96	(1.82–3.64)
5	Bottom face sheet (BFS)	Inconel 718	1	1	8170	8.17
	Total					(45.92–67.34)

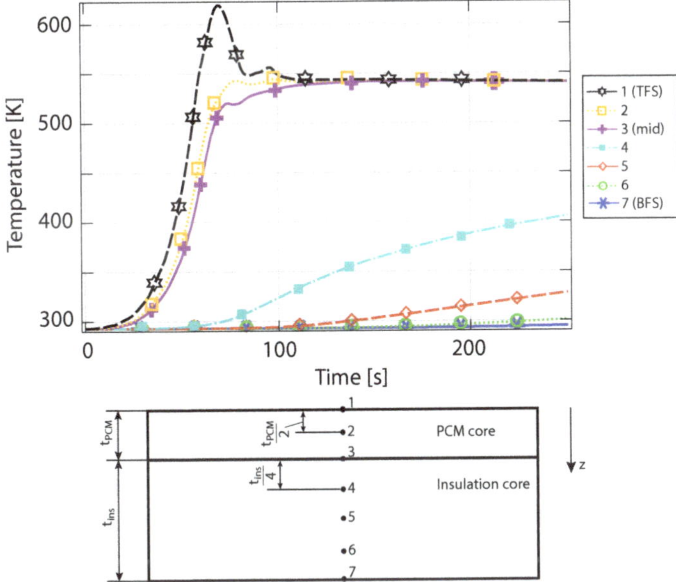

Figure 15. Temperature profile of the hierarchical sandwich structure ($t_{c_{PCM}}$ 10 mm, $\epsilon = 0.9$, $t_{c_{ins}}$ 40 mm) with schematic description of the evaluation points considered.

One can observe that the overall areal weight of the obtained composite is higher than both solutions previously considered. This is mainly due to the high density of the structural materials in face sheets and lattice cores. However, a proper treatment of the overall mass cannot ignore the contribution of the structural design to the mass budget. This is described in what follows.

5.5. Preliminary Structural Design

It was shown that an ablative material can offer the best thermal protection capability at the minimum mass from a thermal design point of view. This conclusion is not trivial when considering the structural performance. In particular, regarding a heavily loaded structural element, such as the ADD discussed in this work, the structural mass can represent the highest contribution to the overall mass budget. Considering a sandwich ITPS could therefore be advantageous for overall mass reduction. Having both the thermal and the structural mass integrated within one component, no add-on mass such as in the case of the ablative TPS is present. Additionally, from the operative point of view of an RLV, a reusable passive TPS is considered advantageous compared to ablative materials.

However, ITPSs not only face mechanical loads due to the dynamic pressure, but additional thermo-mechanical ones. Therefore, a mechanical analysis of the considered solutions is necessary to assess the overall lightweight potential of the considered solutions. The thermal loads applied are the ones corresponding to the time step at which the highest thermal gradient is present. The mechanical load is the same for all three structures, namely the maximum dynamic pressure distribution (see Figure 7).

In the following, the two thermally optimised ITPS concepts from above are analysed using FEM simulations under mechanical and thermal loads. For the ablative TPS concept, a CFRP sandwich structure is designed iteratively to function as a load-bearing structure attached to the inside of the ablative layer of the ADD. This allows for a comparison of structural performance as well as total mass of the concepts. The following configurations were chosen from the previous sections:

1. For the corrugated core ITPS solution, no modification of the design is made, and the final geometrical configuration obtained from the thermal optimization (see Section 5.3) is analysed under mechanical and thermal loads.
2. The considered configuration of the lattice core-PCM ITPS is the one on the higher end of the geometrical ranges considered in Section 5.4 (i.e., $t_{c_{PCM}} = 10\,\mathrm{mm}$, $\epsilon = 0.9$, $t_{c_{ins}} = 40\,\mathrm{mm}$).
3. The mechanical analysis of the load-bearing structure for the ablative PICA TPS analysed in Section 5.2 is used to iteratively optimise the CFRP laminate. The goal of the optimization is to obtain a layup that does not exhibit material failure under the mechanical loads.

The three configurations are tested under the same mechanical boundary conditions described in Section 5.1 (see Figure 7).

5.5.1. Load-Bearing Structure Carrying the Ablative TPS

The dynamic pressure load acts on the component mainly via bending. Thus, the most promising lightweight design concept is that of a sandwich structure. To maximize the load bearing capability to mass ratio, the sandwich is designed with an aluminium honeycomb core between CFRP face sheets. As schematised in Figure 7, the bending load due to pressure acts bi-directionally on the component, i.e., bending it around the longitudinal (x-)axis and the circumferential (y-)axis. While an increase of the aluminium core thickness increases the stiffness of the structure, stresses in the face sheets are increased as well. Therefore, X-shaped aluminium reinforcements are introduced on the inner face sheet of the ADD against bending deformations, as shown in Figure 16.

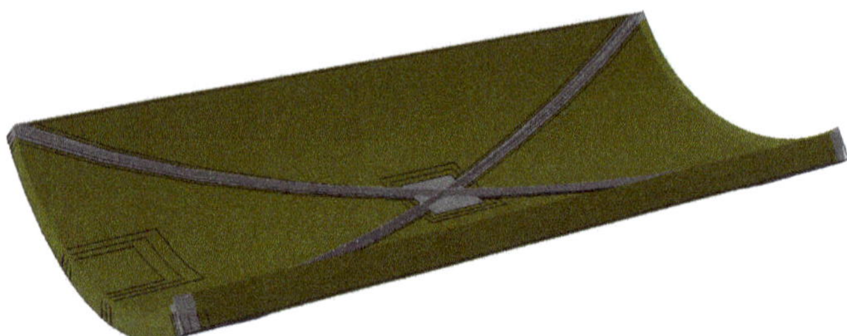

Figure 16. Structural design of the load-bearing composite sandwich structure carrying the ablative TPS with reinforcements in form of an X-shaped frame and beams along the outer edges.

The layup and dimensions of the structure were determined in an iterative design process using geometrically non-linear static analyses in Abaqus® 2020 on a mesh consisting of a linear shell and beam elements. Typical material data for unidirectional (UD) T700 prepreg material and aluminium honeycomb are used (see Appendix B). Due to the optimised ablative TPS, virtually no thermal loads act on the load-bearing structure underneath. Therefore, only dynamic pressure loads (see Figure 8b) are considered. In the laminate, the number and orientation of the UD layers was iterated, as well as the cross-section of the X-shaped reinforcements and the frame along the ADD's perimeter.

The obtained design has face sheet laminates with a thickness of 1.75 mm each and a core thickness of 50 mm, resulting in a mass of the structure of 35 kg. For the evaluation of stresses in the composite material, the Tsai–Wu failure criterion [41] is utilised with the goal of maintaining the criterion in all layers below 1.0. For sake of brevity, only the failure criterion values in the most critical composite layer with the highest failure criterion overall are reported (see Figure 17). As the maximum value of 0.92 occurs at border of the ideally stiff boundary condition, stresses in the real component are assumed to be lower than calculated here.

This preliminary structural design study was performed to obtain a benchmark design whose mass can then be compared to the lattice core and corrugated core ITPS solutions. It can be concluded that it was feasible to find a lightweight design for the load-bearing structure when subjected to dynamic pressure loads. The reported mass of 35 kg is that of the load-bearing structure only. The mass of the PICA TPS layer amounts to 32 kg per ADD, thus resulting in an overall mass of 67 kg (neglecting bonding, attachment points, inserts, etc.).

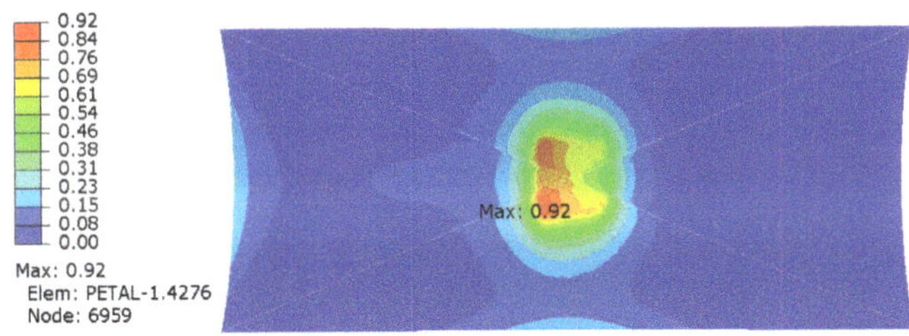

Figure 17. Tsai–Wu failure criterion in the critical composite layer of the load-bearing structure.

5.5.2. Corrugated Core ITPS

For the corrugated core ITPS, the optimised configuration obtained from the thermal study is considered in the mechanical analysis. A quasi-isotropic laminate layup is considered. The same mechanical boundary conditions described for the honeycomb structure of the ablative TPS are applied (see Figure 7), and the commercial solver ANSYS® APDL is used. To take into account the thermal deformation, a coupled thermo-mechanical analysis is performed. The thermal solver is used to obtain the temperature field on the whole structure for the time point at which the maximum outer facesheet temperature is reached, which also corresponds to the maximum thermal gradient. The analysis leads to the results shown in Figure 18. The material properties are reported in Table A5. Due to the lack of established failure criteria for CMCs, the Von Mises equivalent stress on the component is reported. Widespread failure in several parts of the component is detected. The material tensile strength of 260 MPa is exceeded in several points of the structure, even far from the constraints where a local, artificial increase in stress is observed. This is mainly due to the high thermal gradient acting on the structure. The sandwich structure offers a high bending stiffness, which, although advantageous for the mechanical loading of the component, leads to high thermally induced stresses. Although the material exhibits a low coefficient of thermal expansion (CTE), the thermal stresses still exceed the allowable values in several sections of the component. Future design involving different fibre orientations that achieve a three-dimensional tailoring of the CTE might mitigate the incurred failures while retaining high bending stiffness.

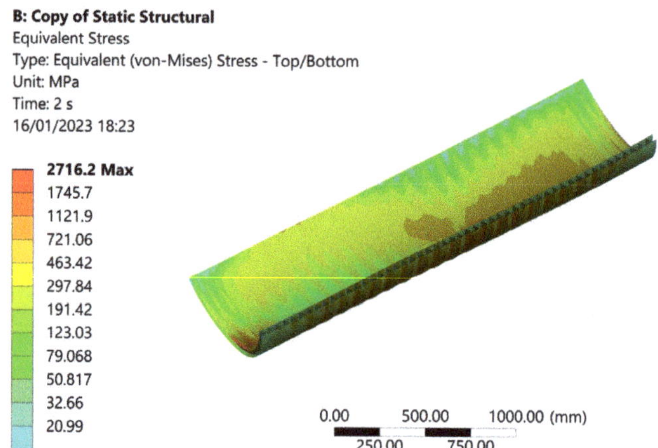

Figure 18. Von Mises stress [MPa] for the corrugated core ITPS subjected to coupled pressure and thermal loads.

5.5.3. Lattice Core-PCM ITPS

The lattice core/PCM solution faces milder thermo-mechanical loads compared to the corrugated core ITPS (cf. Section 5). The maximum top face sheet temperature reaches 618 K, whereas the inner face sheet remains at the initial temperature of 300 K (see Figure 15). Due to the direct bonding of different materials, their difference in coefficient of thermal expansion (CTE) leads to high thermo-mechanical stresses. To calculate these stresses, the temperature field through the thickness of the ADD is required. Therefore, in a first step, we calculate the temperature field by a steady-state heat transfer simulation in Abaqus® 2020.

The thermo-mechanical model consists of linear shell elements to model the three face sheets and linear beam elements for the lattice struts. The structural analysis is split in two load steps:

First, the previously calculated temperature field is applied to the model with respect to its stress-free initial state at a temperature of 298 K (room temperature). The CTEs of the materials are assumed constant over the entire temperature range. For the FE model of the lattice-PCM ITPS solution, the same constraints are used as in the simulations of the other two TPS concepts (see Figure 7).

In a second load step, the dynamic pressure during re-entry is applied to the outer face sheet of the sandwich structure in addition to the persisting thermo-mechanical loads. The resulting deformation is plotted in Figure 19, showing a maximum displacement of 42 mm.

The results of the structural analysis for both load cases are summarised in Table 5. It can be observed that the greater part of the deformations and stresses arises from the thermal gradients, not from the additional pressure load. The comparison with the allowable yield stresses leads to the conclusion that the thermally optimised design is not feasible from a mechanical design point of view.

In all layers made from Inconel 718 (i.e., face sheets and inner lattice core), stresses do not exceed the allowable to a level that could not be managed by further optimization of geometric parameters.

Figure 19. Deformation of the outer face sheet in millimetres on the lattice/PCM model for combined thermal and pressure loading. Cutouts represent the areas with local effects around the nodal constraints that were ignored in stress evaluations.

Table 5. Results of the mechanical simulation for the lattice core/PCM solution (see evaluated area in Figure 19).

Load Case	Max. Displacement [mm]	Outer Face Sheet Max. von Mises Stress [MPa]	Inner Face Sheet Max. von Mises Stress [MPa]	Inner Lattice Core Max. Principal Stress [MPa]	Outer Lattice Core Max. Principal Stress [MPa]
Thermal	35.6	1196	1197	744	698
+Pressure	42	1195	1177	752	707
Allowable		1035	1035	1035	310

The lattice in the outer core is made from an additively manufactured CuCr1Zr alloy with a yield stress of 310 MPa. The yield stress in the CuCr1Zr struts is exceeded by a factor of more than two in both load cases. It can therefore be concluded that the design resulting from the thermal optimization is not feasible from a mechanical engineering point of view. This requires the choice of a different material for the outer lattice core that has high strength and at the same time good thermal conductivity, e.g., tungsten.

5.6. Final Mass Estimation

The three presented ADD concepts were initially optimised for thermal performance (Sections 5.2–5.4) and then analysed in terms of mechanical performance (cf. Sections 5.5.2 and 5.5.3). The load-bearing structure for the ablative TPS was iterated to obtain a feasible design of the composite sandwich layup (Section 5.5.1). For a holistic comparison of the concepts in the context of their application in a reusable microlauncher, their masses are an important performance indicator and are therefore compiled in Table 6.

Table 6. Mass comparison of the three TPS concepts.

Ablative TPS			CMC Corrugated Core			Lattice Core / PCM		
Component	Mass [kg]	Areal Density [kg/m^2]	Component	Mass [kg]	Areal Density [kg/m^2]	Component	Mass [kg]	Areal Density [kg/m^2]
						Face sheets	71	23.7
CFRP sandwich	35	11.7	CMC	60	20	Lattice core	99	33
PICA TPS	32	10.7	Insulation	11	3.7	PCM	42	14
Σ	67	22.4	Σ	71	23.7	Σ	212	70.7

It can be seen that the ablative TPS with the load-bearing CFRP sandwich structure is the lightest concept, followed by the CMC corrugated core. The proposed hierarchical lattice core / PCM solution has a significantly higher mass, at a factor of three compared to the other concepts. It must be noted that the CMC corrugated core as well as lattice core/PCM solution were not optimised for structural performance but only for thermal performance. Therefore, their masses as well as the resulting stresses cannot be taken as an absolute measure for performance of the concepts. Rather, the comparably high mass/poor mechanical performance of the ITPS solutions should be seen as an indication that further optimization of these structures is required.

6. Conclusions

This work described the multidisciplinary design of an aerodynamic drag device used to allow a passive re-entry, i.e., avoiding retropropulsion, of a reusable launch vehicles' first stage. The drag device consists of four sub-components and represents, in a closed configuration, the interstage of the launcher. In the extended configuration, high thermal and mechanical loads are experienced. To achieve a lightweight design, a holistic design approach is required. Therefore, in this work, both thermal and mechanical analyses are conducted. Three different concepts are compared: One is based on an ablative thermal protection system and a CFRP-aluminium honeycomb sandwich structure. The second is a sandwich structure representing a so-called integrated thermal protection system based on a ceramic matrix composite. The third is, as well, an integrated thermal protection system, whose design is based on the use of metallic lattice structures in which a phase change material is embedded. The main results as well as the outlook for each analysed technology are summarised as follows:

- Ablative TPS solution
  - The separation of thermal and structural functions allows one to use efficient materials and construction methods for each absolved function, namely PICA for thermal protection and CFRP-aluminium honeycomb sandwich for load-bearing functionality.
  - The solution delivers the lowest overall mass.
  - It is easier to obtain a feasible solution because of the two high-TRL solutions used in this concept.
  - Reusability is a concern. Indeed, after-flight maintenance operations should include either a check of the receded amount of ablative material or a re-application. Alternatively, a fast-swap concept can be considered, directly removing and substituting both the structural element and the thermal protection system.
- ITPS-CMC corrugated core sandwich
  - The concept represents a lightweight, reusable solution for thermal protection purposes.
  - However, the thermally optimised solution does not withstand the thermo-mechanical loads.
  - Although ceramic matrix composites exhibit a low coefficient of thermal expansion, the high thermal gradients and the high stiffness lead to high thermal stresses compared to the low tensile strength of the material. Improvements in this direction are needed to allow a load bearing functionality of CMC-based TPS. Three-dimensional CTE tailoring via appropriate fibre orientation can be considered in future work.
- ITPS-Lattice core/PCM
  - The integration of a PCM drastically reduces outer wall (top face sheet) temperatures and therefore allows use of materials with high specific mechanical properties, i.e., Inconel.
  - However, thermal stresses above the yield strength of the respective materials in the different layers are identified. These can be caused by mismatch in the CTE of

the different materials and high bending stiffness. Additionally, the use of copper alloy, although beneficial to improving the thermal conductivity of the PCM, has the drawback of a low specific yield strength.

- Different material combinations can be considered in the future. In particular, given the obtained operative temperatures, titanium based alloys are good candidates for the face sheets and for the insulation core. High temperature aluminium alloys, which retain their strength up to 300°C, could be considered for the PCM core. This way, a higher lightweight potential can be obtained.
- Additive manufacturing allows for local adaptation of the structure. Local optimization of lattice unit cell parameters can allow further mass reduction with improved thermo-mechanical behaviour.

For use in a reusable microlauncher, a holistic assessment of load-bearing TPS structures is required. Specifically, the reusability requirement could make the use of ablative TPS expensive compared to heavier solutions with lower expected overhaul time and cost between launches.

Future work should aim at improved thermal analyses with better estimation of the boundary conditions, i.e., a better definition of the ambient temperature for the radiative heat exchange term of the outer surface. Such ambient temperature should be based on piecewise interpolation of ambient temperatures at different points during the flight trajectory.

Furthermore, future activities will concentrate on the multi-objective (thermal and mechanical) optimisation of the two reusable TPS solutions (CMC corrugated and lattice PCM). Only in this way can an integrated structure with good thermal and mechanical performance be obtained. More adequate material choice and combination should be considered among the parameters of the optimisation as well. Furthermore, manufacturing constraints that hinder the construction methodology need to be taken into account. Indeed, the manufacturing of CMCs is still not mature enough to monolithically realise such wide and complex components. On the other hand, the maximum size of realisable metallic structures via additive manufacturing is still small compared to the size of the component considered in this work. The joining techniques, e.g., brazing or laser welding, of different parts of the hierarchical sandwich structure may represent a bottleneck and should be thoroughly investigated. Furthermore, compatibility of the chosen PCM with the core and face sheet material combination should be evaluated case by case. The volume expansion of the PCM after melting should also be taken into account. Although technical solutions like the use of membranes or expansion chambers exist, these might affect the overall structural design. Finally, different kinds of unit cells and local tailoring of the cell parameters of lattice structures can be used to obtain a tailored coefficient of thermal expansion. This would allow one to reduce overall thermal stresses. For the high flexibility in the design process, additively manufactured lattice structures can be considered viable candidates to obtain holistically optimised structures with thermal protection functionality.

Author Contributions: Conceptualization, S.P.; Methodology, S.P.; Software, N.H. and S.P.; CFD analyses, A.Z.; Thermal analyses, S.P.; Mechanical analyses, D.P. and S.P.; Writing—original draft preparation, S.P. and D.P.; Project administration, K.-U.S., R.B., and A.D.; Funding acquisition, A.D. and R.B. All authors have read and agreed to the published version of the manuscript.

Funding: This research was funded by EU Horizon 2020 Programme, grant number 870340.

Data Availability Statement: The open source software Hot-STARSHIP is available at the following link: https://github.com/nilsh7/Hot-STARSHIP/tree/v1.0.0 (accessed on 18 of March 2023) or after contacting the corresponding author.

Acknowledgments: The authors would like to thank Giovanni Medici from DEIMOS SPACE S.L.U for the information supplied about mission analysis and trajectory. The authors would also like to thank Tobias Schalm and Maximilian Schirp-Schoenen for their redactional help in reviewing the paper.

Conflicts of Interest: The authors declare no conflict of interest.

Abbreviations

The following abbreviations are used in this manuscript:

ADD	Aerodynamic drag device
BFS	Bottom face sheet
CFD	Computational fluid dynamic
CFRP	Carbon fibre-reinforced polymer
CMC	Ceramic matrix composite
CTE	Coefficient of thermal expansion
DOF	Degree of freedom
FEM	Finite element method
FVM	Finite volume method
ITPS	Integrated thermal protection system
PCM	Phase change material
PICA	Phenolic impregnated carbon ablator
RLV	Reusable launch vehicle
SLSQP	Sequential least squares programming
TACOT	Theoretical ablative composite for open testing
TFS	Top face sheet
TPS	Thermal protection system
TRL	Technology readiness level
UD	Unidirectional

Appendix A. Verification of the Hot-STARSHIP Solver

In the test, a 5 cm thick piece of TACOT material is heated for one minute and cooled off for another minute afterwards. The parameters for this problem are depicted in Table A1.

Table A1. Parameters for ablative test.

Property	Symbol	Value
Initial length	l_0	50 mm
Initial temperature	T_{ini}	300 K
Pressure	p	101,325 Pa
Turbulent factor	λ	0.5

The aerodynamic boundary condition is used with time-varying values of the transfer coefficient $\rho_e u_e C_{H0}$ and recovery enthalpy h_r to achieve the heating and cooling phase. Note that, in contrast to the use of pre-generated B'-tables in [33], the results presented here are computed with our own B'-tables that are extracted from Mutation++ as part of the process. The time step is chosen to be 0.1 s, and the grid has a first cell thickness of 0.05 mm and a maximum growth factor of 1.03. The temperature and recession history for a calculation with Amaryllis and PATO are obtained from [33]. Figure A1 shows a comparison of surface recession s and char and gas mass flow rate ($\dot{m}_c$ and $\dot{m}_g$) as a function of time.

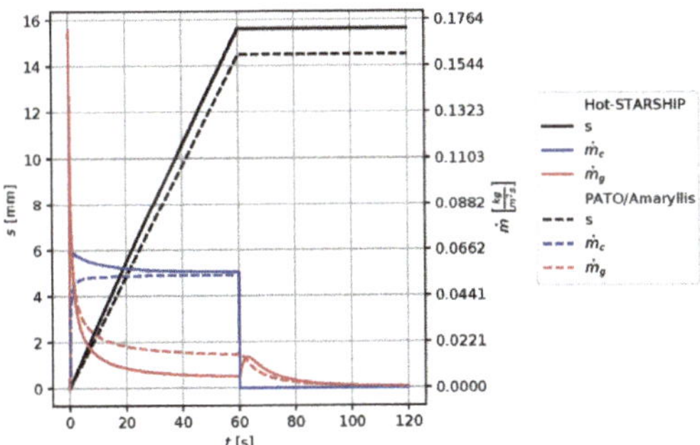

Figure A1. Comparison of Hot-STARSHIP and Amaryllis/PATO recession and mass flow rates.

Hot-STARSHIP analysis results in a 7.6% higher final recession value s (15.6 mm versus 14.5 mm). This difference can be attributed to the initial difference in char mass flow rate $\dot{m}_c$ where Hot-STARSHIP peaks, whereas PATO/Amaryllis show a smoother transient behaviour. As time progresses, the two curves approach each other. In the end, the difference in recession rate $\dot{s}$ is about 4.6%. Once the heat flux input ends, both programs conform to each other. The difference in recession amount can also be observed in the temperature plots in Figure A2. The temperatures are plotted in stationary locations. Thus, once the surface has receded to a fixed location, the location's temperature history ends and merges with the surface temperature history at that point. Both programs are in good agreement of the surface temperature. Wider differences are only observable in the first 20 s where the higher char ablation rate of Hot-STARSHIP provides more cooling. In the fixed locations, the temperature difference between both programs grows with time. As noted, part of this is because of the higher recession amount of Hot-STARSHIP. Because with higher recession amounts fixed locations are closer to the surface and the temperature gradients are large due to low conductivity, differences are observed (see also Figure A3).

Note that one of the main constraints of thermal protection system thickness, the back-face temperature at 50 mm, is in good agreement for both programs. The final difference is about 12 K, and the average difference is even lower as the two curves cross each other.

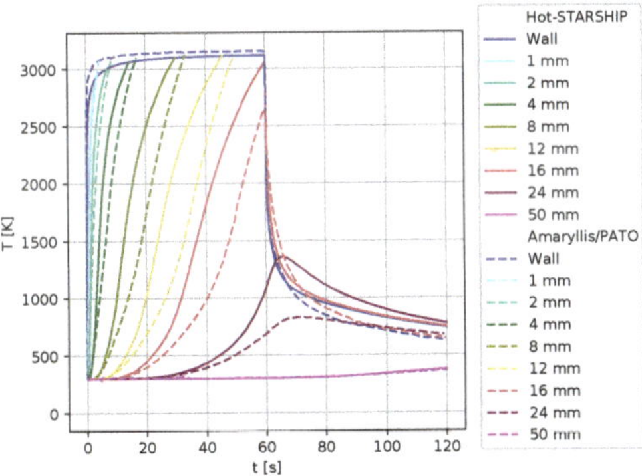

Figure A2. Comparison of Hot-STARSHIP and Amaryllis/PATO temperature curves.

Finally, Figure A3 shows internal temperature profiles shortly after the heating begins (2 s), when the heating stops (60 s), shortly after the heating stops (60.1 s) at the very end of the calculation (120 s), and some intermediate values.

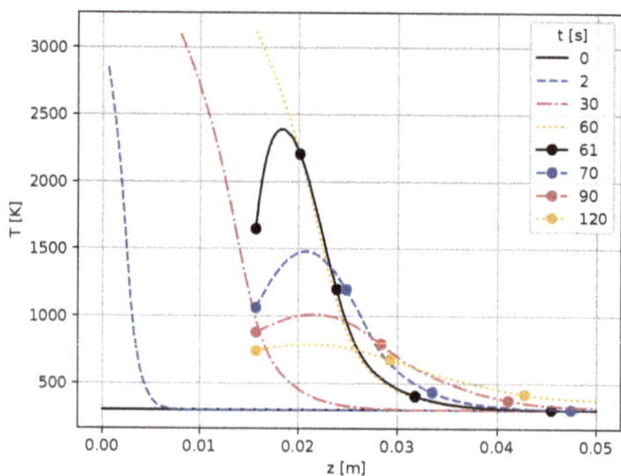

Figure A3. Temperature profiles of Hot-STARSHIP calculation at selected times.

During heating, large temperature gradients are present in the first few millimeters to centimeters, reaching values of up to 700 K/mm. This explains the seemingly large differences in Figure A2 arising from different proximity to the surface. Once the heating ends, the surface temperature drops rapidly, leading to peak temperature not at the surface, but 5 mm into the material. From this point onward, the temperature profile flattens out as dictated by the second order conductivity equation.

The differences between Hot-STARSHIP and PATO or Amaryllis might be attributed to the use of pre-generated B'-tables for PATO and Amaryllis, whereas this is not the case for Hot-STARSHIP, where the tables are computed via Mutation++. In addition, PATO and Amaryllis are "type 2" [42] solvers, whereas Hot-STARSHIP can be classified as a "type 1" solver. This means that in addition to the details resolved in Hot-STARSHIP, PATO and Amaryllis consider Darcy's law for convective transport of pyrolysis gas as well as porosity and permeability for diffusive transport [42]. Whereas Hot-STARSHIP assumes that the gas leaves instantly, solving Darcy's law as done in PATO could hold back some gas that then flows out more slowly, leading to a higher gas mass flow rate. This behaviour would also increase the cooling of the in-depth material, which explains the lower predicted temperatures of PATO and Amaryllis. On the other hand, the slightly lower surface temperature of Hot-STARSHIP can be explained by the higher char mass flow rate that provides more surface cooling.

Appendix B. Material Data

The following material data were acquired from the Ansys® database for a T700 CFRP composite material and for an aluminium honeycomb.

Table A2. UD composite stiffness properties [MPa].

E1	E2	Nu12	G12	G13	G23	α_{11}	α_{22}	α_{33}
121,000	8600	0.27	4700	3100	4700	-4.7×10^{-7}	3×10^{-5}	3×10^{-5}

Table A3. UD composite failure stresses [MPa].

Tensile X	Compression X	Tensile XY	Compression XY	Shear Strength XY
2321	−1082	29	−100	60

Table A4. Al honeycomb stiffness properties [MPa].

E1	E2	E3	Nu12	Nu13	Nu23	G12	G13	G23
1	1	255	0.49	0.01	0.01	1×10^{-6}	37	70

The material data for the CMC are reported with reference to [10]:

Table A5. CMC material properties.

Density [g/cm^3]	Tensile Str. [MPa]	Compressive Str. [MPa]	Young Modulus [GPa]
1.8	260	590	90

References

1. Niederstrasser, C.G. The small launch vehicle survey a 2021 update (The rockets are flying). *J. Space Saf. Eng.* **2022**, *9*, 341–354. [CrossRef]
2. Governale, G.; Rimani, J.; Viola, N.; Villace, V.F. A trade-off methodology for micro-launchers. *Aerosp. Syst.* **2021**, *4*, 209–226. [CrossRef]
3. Medici, G.; Bergström, R.; Martí, L.; Palumbo, N.; Hove, B.; Viladegut, A.; Paris, S.; Soepper, M.; Bhardwaj, P.; Rellakis, D.; et al. A Novel Design Approach for a Reusable VTOL Micro Launch Vehicle. In Proceedings of the *72nd International Astronautical Congress*, Dubai, United Arab Emirates, 25–29 October 2021. Available online: https://www.researchgate.net/publication/360008193_A_novel_design_approach_for_a_reusable_VTOL_Micro_Launch_Vehicle (accessed on 21 March 2023).
4. Marwege, A.; Gülhan, A.; Klevanski, J.; Hantz, C.; Karl, S.; Laureti, M.; De Zaiacomo, G.; Vos, J.; Jevons, M.; Thies, C.; et al. RETALT: Review of technologies and overview of design changes. *CEAS Space J.* **2022**, *14*, 433–445. [CrossRef] [PubMed]
5. Natali, M.; Kenny, J.M.; Torre, L. Science and technology of polymeric ablative materials for thermal protection systems and propulsion devices: A review. *Prog. Mater. Sci.* **2016**, *84*, 192–275. [CrossRef]
6. Uyanna, O.; Najafi, H. Thermal protection systems for space vehicles: A review on technology development, current challenges and future prospects. *Acta Astronaut.* **2020**, *176*, 341–356. [CrossRef]
7. Dorsey, J.T.; Poteet, C.C.; Wurster, K.E.; Chen, R.R. Metallic Thermal Protection System Requirements, Environments, and Integrated Concepts. *J. Spacecr. Rockets* **2004**, *41*, 162–172. [CrossRef]
8. Le, V.T.; Goo, N.S. Design, Fabrication, and Testing of Metallic Thermal Protection Systems for Spaceplane Vehicles. *J. Spacecr. Rockets* **2021**, *58*, 1043–1060. [CrossRef]
9. Le, V.T.; Goo, N.S. Thermomechanical Performance of Bio-Inspired Corrugated-Core Sandwich Structure for a Thermal Protection System Panel. *Appl. Sci.* **2019**, *9*, 5541. [CrossRef]
10. Heidenreich, B. C/SiC and C/C-SiC Composites. In *Ceramic Matrix Composites*; Bansal, N.P., Lamon, J., Eds.; John Wiley & Sons, Inc.: Hoboken, NJ, USA; pp. 147–216. [CrossRef]
11. Glass, D.E. Ceramic matrix composite (CMC) thermal protection systems (TPS) and hot structures for hypersonic vehicles. In Proceedings of the 15th AIAA International Space Planes and Hypersonic Systems and Technologies Conference, Dayton, OH, USA, 28 April–1 May 2008; pp. 1–36 .
12. Ferraiuolo, M.; Scigliano, R.; Riccio, A.; Bottone, E.; Rennella, M. Thermo-structural design of a Ceramic Matrix Composite wing leading edge for a re-entry vehicle. *Compos. Struct.* **2019**, *207*, 264–272.
13. Blosser, M.L.; Chen, R.R.; Schmidt, I.H.; Dorsey, J.T.; Poteet, C.C.; Bird, R.K.; Wurster, K.E. Development of advanced metallic thermal-protection-system prototype hardware. *J. Spacecr. Rockets* **2004**, *41*, 183–194. [CrossRef]
14. Fischer, W.; Bolz, J. ULTIMATE: Metallic TPS for Future RLV's. In Proceedings of the 9th AIAA/ASME Joint Thermophysics and Heat Transfer Conference, San Francisco, CA, USA, 5–8 June 2006. [CrossRef]
15. Bapanapalli, S.; Martinez, O.; Gogu, C.; Sankar, B.; Haftka, R.; Blosser, M. (Student Paper) Analysis and Design of Corrugated-Core Sandwich Panels for Thermal Protection Systems of Space Vehicles. In Proceedings of the 47th AIAA/ASME/ASCE/AHS/ASC Structures, Structural Dynamics, and Materials Conference; 14th AIAA/ASME/AHS Adaptive Structures Conference, Newport, RI, USA, 1–4 May 2006. [CrossRef]
16. Gogu, C.; Bapanapalli, S.K.; Haftka, R.T.; Sankar, B.V. Comparison of materials for an integrated thermal protection system for spacecraft reentry. *J. Spacecr. Rockets* **2009**, *46*, 501–513. [CrossRef]
17. Li, Y.; Zhang, L.; He, R.; Ma, Y.; Zhang, K.; Bai, X.; Xu, B.; Chen, Y. Integrated thermal protection system based on C/SiC composite corrugated core sandwich plane structure. *Aerosp. Sci. Technol.* **2019**, *91*, 607–616.
18. Le, V.T.; Ha, N.S.; Goo, N.S. Advanced sandwich structures for thermal protection systems in hypersonic vehicles: A review. *Compos. Part B Eng.* **2021**, *226*, 109301. [CrossRef]
19. Yendler, B.; Dang, K.; Forrest, M. A Reusable Heat Shield Using Phase Change Materials. In Proceedings of the 44th AIAA Aerospace Sciences Meeting and Exhibit, Reno, NV, USA, 9–12 January 2006. [CrossRef]

20. Cao, C.; Wang, R.; Xing, X.; Liu, W.; Song, H.; Huang, C. Performance improvement of integrated thermal protection system using shaped-stabilized composite phase change material. *Appl. Therm. Eng.* **2020**, *164*, 114529. [CrossRef]
21. Nazir, H.; Batool, M.; Osorio, F.J.B.; Isaza-Ruiz, M.; Xu, X.; Vignarooban, K.; Phelan, P.; Inamuddin.; Kannan, A.M. Recent developments in phase change materials for energy storage applications: A review. *Int. J. Heat Mass Transf.* **2019**, *129*, 491–523. . [CrossRef]
22. Hubert, R.; Bou Matar, O.; Foncin, J.; Coquet, P.; Tan, D.; Li, H.; Teo, E.H.T.; Merlet, T.; Pernod, P. An effective thermal conductivity model for architected phase change material enhancer: Theoretical and experimental investigations. *Int. J. Heat Mass Transf.* **2021**, *176*, 121364. [CrossRef]
23. Piacquadio, S.; Schirp-Schoenen, M.; Mameli, M.; Filippeschi, S.; Schröder, K.U. Experimental Analysis of the Thermal Energy Storage Potential of a Phase Change Material embedded in Additively Manufactured Lattice Structures. *Appl. Therm. Eng.* **2022**, *153*, 119091. [CrossRef]
24. Bühring, J.; Nuño, M.; Schröder, K.U. Additive manufactured sandwich structures: Mechanical characterization and usage potential in small aircraft. *Aerosp. Sci. Technol.* **2021**, *111*, 106548. [CrossRef]
25. Chen, Y.K.; Milos, F.S. Multidimensional finite volume fully implicit ablation and thermal response code. *J. Spacecr. Rockets* **2018**, *55*, 914–927. [CrossRef]
26. Chen, Y.K.; Milos, F.S. Two-Dimensional Implicit Thermal Response and Ablation Program for Charring Materials. *J. Spacecr. Rockets* **2001**, *38*, 473–481. [CrossRef]
27. Scoggins, J.B.; Leroy, V.; Bellas-Chatzigeorgis, G.; Dias, B.; Magin, T.E. Mutation++: MUlticomponent Thermodynamic And Transport properties for IONized gases in C++. *SoftwareX* **2020**, *12*, 100575. [CrossRef]
28. Gordon, S.; McBride, B.J. *Computer Program for Calculation of Complex Chemical Equilibrium Compositions and Applications. Part 1: Analysis*; NASA Technical Report, Document ID 19950013764; NASA Lewis Research Center, Cleveland, OH, USA, 1994.
29. de Mûelenaere, J.; Lachaud, J.; Mansour, N.N.; Magin, T.E. Stagnation line approximation for ablation thermochemistry. In Proceedings of the 42nd AIAA Thermophysics Conference, Honolulu, HI, USA, 27–30 June 2011; pp. 1–14. [CrossRef]
30. Bonacina, C.; Comini, G.; Fasano, A.; Primicerio, M. Numerical solution of phase-change problems. *Int. J. Heat Mass Transf.* **1973**, *16*, 1825–1832. [CrossRef]
31. Amar, A.J. Modeling of One-Dimensional Ablation with Porous Flow Using Finite Control Volume Procedure. Master's Thesis, North Carolina State University, Raleigh, NC, USA, 2006.
32. Chen, Y.K.; Milos, F.S. Ablation and thermal response program for spacecraft heatshield analysis. In Proceedings of the 36th AIAA Aerospace Sciences Meeting and Exhibit, Reno, NV, USA, 12–15 January 1998; Volume 36.
33. Lachaud, J.; Martin, A.; Eekelen, T.V.; Cozmuta, I. Ablation test-case series #2-Numerical simulation of ablative-material response: code and model comparisons. In Proceedings of the 5th Ablation Workshop, Lexington, KY, USA, 28 February–1 March 2012.
34. The SciPy Community. scipy.optimize.toms748. Available online: https://docs.scipy.org/doc/scipy/reference/generated/scipy.optimize.toms748.html (accessed on 21 March 2023).
35. Alefeld, G.E.; Potra, F.A.; Shi, Y. Algorithm 748: Enclosing Zeros of Continuous Functions. *ACM Trans. Math. Softw. TOMS* **1995**, *21*, 327–344. [CrossRef]
36. Virtanen, P.; Gommers, R.; Oliphant, T.E.; Haberland, M.; Reddy, T.; Cournapeau, D.; Burovski, E.; Peterson, P.; Weckesser, W.; Bright, J.; et al. SciPy 1.0–Fundamental Algorithms for Scientific Computing in Python. *Nat. Methods* **2019**, *17*, 261–272.
37. Varelas, K.; Dahito, M.A. Benchmarking multivariate solvers of scipy on the noiseless testbed. In Proceedings of the GECCO 2019 Companion–Proceedings of the 2019 Genetic and Evolutionary Computation Conference Companion, Prague, Czech Republic, 13–17 July 2019; pp. 1946–1954.
38. Sutton, K.; Graves, R.A.J. *A General Stagnation-Point Convective-Heating Equation for Arbitrary Gas Mixtures*; Technical Report November; NASA Langley Research Center: Hampton, VA, USA, 1971.
39. Chen, Y.K.; Milos, F.S. Ablation and Thermal Response Program for Spacecraft Heatshield Analysis. *J. Spacecr. Rockets* **1999**, *36*, 475–483. [CrossRef]
40. Tran, H.; Johnson, C.; Hsu, M.T.; Chem, H.; Dill, H.; Chen-Johnson, A.; Tran, H.; Johnson, C.; Hsu, M.T.; Chem, H.; et al. Qualification of the forebody heatshield of the Stardust's Sample Return Capsule. In Proceedings of the 32nd Thermophysics Conference, Atlanta, GA, USA, 23–25 June 1997. [CrossRef]
41. Tsai, S.W.; Wu, E.M. A general theory of strength for anisotropic materials. *J. Compos. Mater.* **1971**, *5*, 58–80. [CrossRef]
42. Lachaud, J.; Mansour, N.N. Porous-material analysis toolbox based on openfoam and applications. *J. Thermophys. Heat Transf.* **2014**, *28*, 191–202. [CrossRef]

Article

Sensitivity Analysis of a Hybrid MCDM Model for Sustainability Assessment—An Example from the Aviation Industry

Dionysios N. Markatos [1,*], Sonia Malefaki [2] and Spiros G. Pantelakis [1]

1 Laboratory of Technology & Strength of Materials, Department of Mechanical Engineering & Aeronautics, University of Patras, 26500 Patras, Greece
2 Department of Mechanical Engineering & Aeronautics, University of Patras, 26500 Patras, Greece
* Correspondence: dmark@upatras.gr

Abstract: When it comes to achieving sustainability and circular economy objectives, multi-criteria decision-making (MCDM) tools can be of aid in supporting decision-makers to reach a satisfying solution, especially when conflicting criteria are present. In a previous work of the authors, a hybrid MCDM tool was introduced to support the selection of sustainable materials in aviation. The reliability of an MCDM tool depends decisively on its robustness. Hence, in the present work, the robustness of the aforementioned tool has been assessed by conducting an extensive sensitivity analysis. To this end, the extent to which the results are affected by the normalization method involved in the proposed MCDM tool is examined. In addition, the sensitivity of the final output to the weights' variation as well as to the data values variation has been investigated towards monitoring the stability of the tool in terms of the final ranking obtained. In order to carry out the analysis, a case study from the aviation industry has been considered. In the current study, carbon fiber reinforced plastics (CFRP) components, both virgin and recycled, are assessed and compared with regard to their sustainability by accounting for metrics linked to their whole lifecycle. The latter assessment also accounts for the impact of the fuel type utilized during the use phase of the components. The results show that the proposed tool provides an effective and robust method for the evaluation of the sustainability of aircraft components. Moreover, the present work can provide answers to questions raised concerning the adequacy of the CFRP recycled parts performance and their expected contribution towards sustainability and circular economy goals in aviation.

Keywords: holistic MCDM tool; circular aviation; sustainability; CFRP recycling; aviation; sensitivity analysis; AHP; WSM; data normalization

Citation: Markatos, D.N.; Malefaki, S.; Pantelakis, S.G. Sensitivity Analysis of a Hybrid MCDM Model for Sustainability Assessment—An Example from the Aviation Industry. *Aerospace* **2023**, *10*, 385. https://doi.org/10.3390/aerospace10040385

Academic Editor: Doni Daniel

Received: 21 February 2023
Revised: 23 March 2023
Accepted: 20 April 2023
Published: 21 April 2023

1. Introduction

The aviation industry faces great sustainability challenges associated with global warming and climate change [1,2]. It has been estimated that approximately 920 million tons of CO_2 emissions were produced by the aviation industry worldwide in 2019 only [3]; a doubling or even tripling of said emissions is forecasted to occur by 2050 unless radical changes have been implemented [4]. Therefore, the development of sustainable approaches and solutions with regard to future aviation technologies and applications is of utmost importance. To this end, the utilization of low-density polymeric composites for weight reduction represents a major goal for the aviation sector, given that weight considerations are very critical compared to other transportation sectors [5,6]. In this context, carbon fiber reinforced plastics (CFRPs) have been extensively used for lightweight aircraft applications towards achieving better fuel efficiency and, consequently, lowering the associated environmental burden of the aviation sector. Despite the excellent specific properties of CFRPs, issues such as the great environmental and economic impact of their production, as well as difficulties linked to their recyclability, remain open challenges that need to be addressed [5,7]. It is worth noting that currently, approximately 98% of CFRP waste is landfilled [8]. Until today, recycled composites are not being used for mass production in aviation; only demonstrators or prototypes have been developed, targeting secondary aviation applications (e.g., seat armrests, side-wall interior panels) [9,10].

When considering the use of recycled material in aviation, the concern of circular economy (CE) principles is of great importance as it represents an integral part of sustainability. Therefore, apart from the assessment of the environmental impact as well as the economic viability of the recycled components, the dimension of circularity also needs to be examined. In addition, when focusing on high-performance applications, the technological quality features of the recycled component need to be evaluated as the components under consideration must meet specific mechanical performance limits and manufacturing requirements [11]. To this end, new tools are required to support decision-making towards CE practices and sustainability goals; in this frame, multi-criteria decision-making (MCDM) tools can be of aid in supporting decision-makers reach a satisfying solution, especially when conflicting criteria are present. MCDM belongs to a variety of techniques able to determine a preference ordering among alternative solutions whose performance is scored against a series of criteria. MCDM has been used in many fields, including the aviation sector, although the vast majority is focused on the airlines and aircraft level as it occurs from an extensive recent review paper involving MCDM-related studies in the aviation field; among the MCDM methods applied in the aviation sector, AHP, SAW, TOPSIS, ELECTRE, VIKOR, as well as hybrid methods integrating combinations of them, appear to be the most widely used ones, with AHP and TOPSIS being the first choice for decision-making [12]. However, regardless of the choice of the MCDM, it occurs that the sensitivity and robustness of the proposed tools are not systemically examined. Moreover, in cases where a robustness assessment has been conducted, it consists of a sensitivity analysis of the weights' variation, while the sensitivity of the MCDM tool to the data variation appears to be generally neglected. The latter becomes clear from the representative cited works of Table 1, incorporating MCDM methodologies within the aviation sector.

Table 1. Representative works from the aviation sector implementing MCDM methodologies.

Work	MCDM Used	Sensitivity Analysis
Hsu and Liou (2013) [13]	SWM, DEMATEL, ANP	-
Sánchez-Lozano et al. (2015) [14]	AHP and TOPSIS	-
Garg (2016) [15]	AHP, TOPSIS	Weights variation
Bae et al. (2017) [16]	AHP and TOPSIS	-
Görener et al. (2017) [17]	AHP, TOPSIS	-
Barak and Dahooei (2018) [18]	SWM, TOPSIS, VIKOR	-
Sun et al. (2018) [19]	TOPSIS, VIKOR	-
Mahtani, 2018 [20]	AHP	Weights variation

Conducting a sensitivity analysis of MCDM is particularly important in the aviation sector, given the complex and safety-critical nature of decision-making in this industry. Therefore, a data sensitivity analysis is crucial for the reliability of the tool as it helps to identify and manage uncertainty in data inputs (such as measurement error, sampling error, or missing data), leading to more accurate and reliable predictions and better-informed decisions. In this context, the implementation of a reliable and robust MCDM tool can be useful for selecting the most appropriate material, design component, and manufacturing process in the conceptual design and design phase of a product. For a given engineering application, the attention focus lies on the proper selection of criteria and metrics rather than on the selection of the most appropriate MCDM methodology [21].

In the present study, a hybrid MCDM tool, introduced by the authors in [22], to support the policy decision of selecting a sustainable material for aircraft components has been applied, and its robustness has been examined towards ensuring its reliability as a decision support tool. The research questions that will be addressed in the present work include: (1) What is the level of sustainability of virgin and recycled CFRP components, and how do they compare to each other? (2) How reliable is the assessment of sustainability through MCDM? Based on the above research questions, the work aims to support policy decisions by providing decision-makers with a reliable and robust tool that can aid in the selection of sustainable materials in the aviation industry. The studied tool combines the analytic hierarchy process (AHP) and a weighted sum model (WSM) to obtain the final output. In this context, the influence of the data normalization method, as well as the sensitivity to the weights and data variation, is evaluated. For this purpose, a case study has been considered, aiming to assess the sustainability potential of CFRP recycled composites in aviation with regard to the type of fuel utilized within aircraft operation. In this frame, kerosene, as well as liquid hydrogen from conventional and renewable sources, have been considered. The proposed MCDM tool integrates environmental, economic, and circular economy criteria, as being the most relevant aspects representing sustainability, according to the authors. The output of the model is a weighted sum that can be understood as a metric of sustainability. The results demonstrate that the proposed tool provides an effective and robust method for the evaluation of the sustainability of aircraft components.

2. Methodology

2.1. Basic Considerations

As mentioned above, a case study from the aviation industry involving recycled CFRP components has been considered to assess the robustness of the proposed tool. For the sake of the present study, the geometrical features of the considered components, with the exception of weight, are assumed to be identical. The recycled components comprising of either randomly or aligned fibers are compared against a virgin woven CFRP. To enable comparison and be in compliance with the design requirements, the stiffness of the virgin and recycled components must be identical. To this end, to compensate for the variation of stiffness among the considered components, thickness (and consequently mass) has been treated as a variable that has to be adjusted to achieve equal stiffness. Equal stiffness has been considered an appropriate criterion for the comparison of different materials/components [15]. The expected mass ratio (R_m) between the virgin and the recycled components is calculated based on the following approximate formula [23–25]:

$$R_m = \frac{m_{recycled}}{m_{virgin}} = \frac{p_{recycled}}{p_{virgin}}\left(\frac{E_{virgin}}{E_{recycled}}\right) \quad (1)$$

where m (kg) and p (kg/m^3) represent the mass and the density of the components under comparison, respectively, while E (N/m^2) is the elastic modulus of the components.

2.2. Sustainability-Related Metrics

In the present study, sustainability is understood as a matter of trade-offs among environmental, economic, and circular economy aspects. Therefore, to implement the proposed approach, both the environmental impact and costs of the whole lifecycle of the investigated components need to be assessed and integrated into the MCDM-based tool introduced in Section 2.3. Hence, lifecycle metrics linked to the environment and costs are accounted for; to this end, Life Cycle Assessment (LCA) and Life Cycle Costing (LCC) data were gathered from the relevant literature to calculate the said impact of the components. The tool also integrates a circular economy indicator which has been linked to the technological performance of the investigated components. For the sake of the current study, this is expressed through a specific property of the components, namely, specific stiffness.

Environmental impact has been linked to the emitted greenhouse gases (GHG) associated with the whole lifecycle of the components, namely raw material production, manufacturing, use phase, and recycling. GHG emissions represent the most widely reported environmental impact metric across industry and academia [11]. The economic impact of the components has been related to the costs associated with the energy requirements for the production, manufacturing, and recycling of the components or the fuel price when assessing the use phase impact of the components. The relevant environmental and economic impact results associated with production, manufacturing, and recycling are given in kgCO_2eq per specific component mass or in euros per component mass, respectively. LCA starts with the production of the primary material, i.e., carbon fibers (PAN) and epoxy resin [11,26–29]. The autoclave molding process has been chosen as the relevant manufacturing process of the virgin CFRP aviation component. For the manufacturing of recycled components, the compression molding process has been considered [30]. The environmental impact and costs of upgrade technologies of recycled carbon fibers (e.g., sizing, alignment) were not accounted for due to a lack of relevant literature data. The chosen recycling process of the CFRPs has been the fluidized bed process (FBP), as being a promising method for recovering fibers of mechanical properties comparable to these of the virgin ones [6,26]. Compared to other promising recycling methods, which are currently at a low technology readiness level (TRL) (e.g., solvolysis [31]), the FBP method is at a TRL of 6 and is found at the pilot phase. In order to calculate the process-related energy costs, the non-household price of kWh in Germany has been accounted for [32].

For the assessment of the impact of the components' mass variation on emissions and costs linked to the use phase, the type of fuel is accounted for, where fuel consumption is assumed to be proportional to the component mass [11,33]. Hence, the components have been considered a load that must be carried by aircraft during flight. In this context, the environmental and economic impact results are given in a service function unit, namely, per component mass per km, which represents a wider approach for all aircraft types and classes and types regardless of the split between passengers and cargo payloads [34]. Four types of fuels were considered, i.e., kerosene, conventionally produced liquid hydrogen, liquid hydrogen from a wind source, and liquid hydrogen from a geothermal source, where the respective environmental and cost metrics relating to these fuels have been taken from [34]. The assessment of the overall impact of the use phase was conducted considering that the average lifetime distance of Airbus A320 was approximated based on the number of flying hours for which it was designed, i.e., 60,000 flying hours over a lifespan of 25 years, and the average cruising speed, i.e., 840 km/h [35,36].

For achieving the transition towards a CE, indicators and metrics for measuring CE progress are required. Up to now, various interpretations have been proposed, e.g., [37,38]. However, said interpretations lead to a variety of metrics and indicators in both content and form [39], while many of them focus on materials preservation [40,41]. In the aviation sector, the prevailing interpretation of circularity refers to the percentage of the aircraft mass which can be recycled or reused at the End-of-Life (EoL) of the aircraft [42]. However, in the above interpretation, the performance features of the recycled products are undermined, which in our view, represent an essential parameter when using a recycled product for an aviation application. Hence, considering that the quality of the recycled material represents a decisive factor towards CE goals as quality is linked to the durability of a material, a CE metric is introduced in the present study, which is linked to a quality feature of the component under study, i.e., a mechanical property. In the context of this study, the latter is expressed through the specific stiffness of the investigated components. For the focus on an aviation application, the choice of the specific stiffness is well justified as, in most applications, the allowable design of an aircraft structure does not exceed the linear elastic region of the stress-strain curve; in the case of CFRPs, this region remains almost linear up to failure. Considering the absence of standardized circular economy indicators in the aviation sector, future studies could focus on developing more specific circular economy indicators. However, this task is beyond the scope of the present work.

2.3. Structure of the Hybrid MCDM Tool and Sensitivity Analysis

The MCDM-based tool implemented herein has been introduced by the authors in [22] as a material selection tool for the aviation sector. The said tool combines the AHP and a WSM, whose output is a weighted sum of the normalized individual indicators. The advantage of integrating the WSM into the proposed hybrid tool is that it offers a proportional linear transformation of the raw data; namely, it maintains the relative order of magnitude of the standardized scores. The latter allows for a more effective and comprehensible interpretation of the final ranking obtained, as well as for distinguishing the impact of each term on the final output. The tool integrates environmental and economic metrics related to the component under study, as well as a suitable CE indicator, as introduced in Section 2.2. Based on the definitions of Section 2.2, the WSM equation, as it has been introduced in the previous work of the authors [22], is given as:

$$S_i = K_{CEI} \cdot CEI_{Q_i} + K_C \cdot C_i + K_E \cdot E_i \quad (2)$$

where S_i is the final output value of the i component and can be considered a metric of overall sustainability and emerges as a matter of trade-off between environmental impact, costs, and circularity performance. E_i and C_i are the inversed normalized environmental and cost indicators of the i component, respectively. The inversed values have been considered due to the fact that environmental impact and costs have a negative impact on the overall sustainability index and, hence, the smaller these factors are, the higher the sustainability index becomes. CEI_{Q_i} is the normalized quality-related CEI of the i component, expressed through the specific stiffness of the considered components. K_{CEI}, K_C, and K_E stand for dimensionless weight factors and reflect the importance attributed to each term of the overall index value.

2.3.1. Factors' Weights Determination

Determination of the criteria weights is a frequent issue in many MCDM techniques. Hence, the selection of a proper weighting method is crucial in solving a multi-criteria decision problem as the weighting procedure followed may significantly influence the result; in this context, a variety of different weighting methods exist, with AHP receiving high popularity [43]. So as to define the weight factors of the above criteria, the AHP [44] was applied in [22], which is considered one of the most widely employed established decision-making methodologies [45]. AHP is based on pairwise comparisons; namely, it evaluates relationships between pairs when making group comparisons to judge which of each alternative is preferred. The main strength of AHP lies in its capability to combine it with other MCDM methodologies to obtain a flexible and tailored solution approach. The determination of the weight factors (K_{CEI}, K_C, K_E) is subjective, reflecting the priority criteria of the user for a specific application. The final ranking among the alternative components occurs through the application of the WSM. However, one of the main concerns regards the inconsistency of decision makers in pairwise comparisons owing to the large number of comparisons needed to obtain the weights [46]. In 2015, another pairwise comparison-based method, namely the best-worst method (BWM), was introduced as an appropriate alternative to AHP in MCDM problems, demonstrating some advantages over AHP, such as fewer pairwise comparisons required and hence, better consistency. The BWM determines the pairwise relative comparisons, i.e., the preference between only the best and the worst criterion over all other criteria [47,48]. For both AHP and BWM, a similar linguistic terminology is being used, i.e., the importance of the criteria is defined on the same scale, i.e., 1–9, where 1 means that two criteria are of equal importance, while 9 means that the selected criterion is extremely more important compared to another criterion, as presented in Table 2. Therefore, a direct comparison can be made under the same level of reference so as the effect of the utilized weighting method can be clearly determined.

Table 2. The AHP Scale [44].

Semantics	Grade	Reciprocal
Extremely preferred	9	1/9
Very strongly to extremely	8	1/8
Very strongly preferred	7	1/7
Strongly to very strongly	6	1/6
Strongly preferred	5	1/5
Moderately to strongly	4	1/4
Moderately preferred	3	1/3
Equally to moderately	2	1/2
Equally preferred	1	1

Although in the current work, the AHP was considered for the determination of the weight factors, BWM can be considered an effective alternative to the AHP method. However, the number of criteria (3) considered in the current study does not lead to different results as the number of pairwise comparisons as well as the system to be solved are identical for the two techniques. Yet, the sensitivity of the weighting procedure when more than three criteria (terms) are considered, and hence, a larger number of comparisons are made remains something to be investigated.

2.3.2. Assessment of the Tool Sensitivity to the Applied Normalization Technique

Normalization is a critical step in any decision-making process as it transforms heterogeneous data into data that share a common scale. In the literature, a variety of normalization techniques have been proposed, including the min-max method, the z-score, the ranking normalization, the distance to target normalization, and the proportionate normalization, which are considered the five most widely employed ones [49]. In order to obtain the normalized indicators in [22], the min-max method was implemented to rescale the range of the individual indicators between 0 and 1. The general equation of the min-max technique [49] is given as:

$$x' = \frac{x - \min(x)}{\max(x) - \min(x)} \tag{3}$$

where x' is the normalized value, x is the original value, $\min(x)$ and $\max(x)$ are the minimum and maximum values of each individual indicator, respectively. In this study, to assess the sensitivity of the results to the normalization technique utilized, two alternative normalization methods were implemented, namely z-score and proportionate normalization. Z-score normalization is a typical methodology widely used in statistics. A z-score describes the position of a raw score in terms of its distance from the mean when measured in standard deviation units. On the other hand, proportionate normalization has the advantage that each value of a dataset is divided by the total sum; in this way, the normalized values maintain proportionality, reflecting the percentage of the sum of the total indicator's values. Dividing by the sum ensures that even the smallest value, which is greater than zero, is attributed a positive normalized value, while the differences among the normalized values become narrow. Alternative normalization techniques, such as ranking normalization and distance to target normalization, were considered inappropriate for this case study. More specifically, ranking normalization is a qualitative method; therefore, a quantitative assessment of the differences among the considered alternatives is not feasible. Finally, distance to target normalization requires the definition of a desired target (deriving mainly from policy targets), which in our case, is not a straightforward one.

2.3.3. Assessment of the Tool to Criteria Weights and Data Variation

Following the assessment of the influence of the different methodologies integrated into the tool, an assessment of rank stability was conducted by accounting for weight factors and data value variations. To this end, a series of indicative weighting scenarios were considered for which smaller and larger adjustments have been made with respect to the applied weights as derived from the AHP analysis. In addition, a thorough sensitivity analysis with respect to the data value variation has been conducted. In order to test the sensitivity of the method to small changes in the values of the original data, 1000 samples were simulated by perturbing the original data by a random error. It is assumed that the errors follow a normal distribution with zero mean and standard deviation proportional to the standard deviation of the corresponding indicator of the initial data. Each of these samples was ordered according to the values of the overall sustainability index, and the mean ranking for each material, as well as the standard deviation, was calculated. All the simulations are implemented in R version 4.1.1 [50].

3. Results and Discussion

3.1. Circular Economy Indicator Calculation

In Table 3, the elastic modulus and the density of the investigated components, as taken from [30], are presented. Based on these values, the specific stiffness for each component was calculated, as well as the resulting weight, in order for the components to present equal stiffness. As mentioned in Section 2.2, the normalized specific stiffness, i.e., the stiffness ratio of the component under study to the virgin one, has been used in Equation (2) as the relevant CEI. Based on the Table 3 values, the virgin component demonstrates a higher specific stiffness compared to the recycled components, as expected, followed closely by the recycled component comprised of 50% aligned fibers. On the other hand, the recycled component comprised of randomly oriented fibers shows by far the lower quality, resulting in a considerable weight increase compared to the other two components. The poor quality of the randomly oriented recycled components highlights the need for upgrade technologies (mainly alignment) of the recycled fibers in order to be able to compete with the virgin CFRP components in terms of quality.

Table 3. Properties of The Investigated Components—Circular economy metric [data adapted from [30].

Component Type	Elastic Modulus (GPa)	Density (g/cm^3)	Specific Stiffness ($GPa/(g/cm^3)$)	Resulting Weight (kg)
Woven virgin	70	1.6	43.75	1000
Recycled aligned	60.8	1.5	40.53	1080
Recycled random	39.8	1.44	27.64	1580

3.2. Environmental and Economic Impact Indicators Calculation

Based on the obtained weight of each component, the environmental impact and costs were calculated, accounting for the whole lifecycle of the components. The results are presented in Tables 4 and 5, where data have been adapted from relevant works, as described in Section 2.2. The impact relating to the use phase of the components accounts for the different types of fuel that have been considered. The higher values, in terms of environmental impact and costs, are noted in bold.

Based on these results, it becomes clear that the virgin CFRP component presents by far the highest environmental impact and costs with regard to its production and manufacturing. This is owed to the significant energy required to produce PAN fibers as well as the considerable energy requirements of the autoclave manufacturing process. Nevertheless, the impact associated with the production and manufacturing phases contributes only to a small percentage of the overall impact, owing to the use phase impact, which clearly dominates the total lifecycle impact of the component. It is worth noting that nearly 99% of the total impact is owed to the use phase when kerosene fuel is used. A similar situation applies when liquid hydrogen from a conventional or wind source is considered; in this case, over 95% of the total impact is still owed to the use phase. However, when liquid hydrogen from a geothermal source is considered, the use phase environmental impact hardly accounts for 84% of the total impact. This remark highlights that the decarbonization of the aviation sector is expected to shift a considerable amount of the environmental burden to the production and manufacturing phases. On the other hand, the latter remark does not concern the lifecycle costs impact as the costs associated with the use of hydrogen are almost double compared to these of kerosene and over four times larger when hydrogen from renewable sources is used. This is owed to the current high cost of liquid hydrogen and especially the ones produced from renewable sources. Therefore, the use phase cost impact dominates the total lifecycle costs, regardless of the type of fuel utilized. The currently high cost of liquid hydrogen, and especially that deriving from renewable sources, may act as a prohibiting factor for the extensive use of liquid hydrogen, at least for the near future.

Table 4. Environmental Impact (LCA) metrics of The Investigated Components.

Component Type	Primary Material Production ($kgCO_2eq$-Mass)	Component Manuf. ($kgCO_2eq$-Mass)	Use Phase ($kgCO_2eq$-Mass-Lifetime Km)				Recycling ($kgCO_2eq$-Mass)
			Kerosene	Liquid Hydrogen	Liquid Hydrogen Wind	Liquid Hydrogen Geothermal	
Woven virgin	**20,440**	**103,000**	52,920,000	5,544,000	3,024,000	756,000	1540
Recycled aligned	1921	1717	57,153,600	5,987,520	3,265,920	816,480	1663
Recycled random	3549	2512	**83,613,600**	**8,759,520**	**4,777,920**	**1,194,480**	**2433**

Table 5. Economic Impact (LCC) metrics of The Investigated Components.

Component Type	Primary Material Production (€-Mass)	Component Manuf. (€-Mass)	Use Phase ($kgCO_2eq$-Mass-Lifetime Km)				Recycling (€-Mass)
			Kerosene	Liquid Hydrogen	Liquid Hydrogen Wind	Liquid Hydrogen Geothermal	
Woven virgin	**17,905**	**3340**	4,032,000	7,056,000	21,168,000	21,168,000	499
Recycled aligned	1560	1858	4,354,560	7,620,480	22,861,440	22,861,440	539
Recycled random	2882	2718	**6,370,560**	**11,148,480**	**33,445,440**	**33,445,440**	**788**

When comparing the components under consideration, the lower environmental impact belongs to the recycled component comprised of aligned fibers for which hydrogen from a geothermal source has been used. Although this component is heavier compared to the virgin one, the environmental gains derived from the production phase of the recycled material are sufficient to compensate for the increased GHG emissions of the use phase compared to the virgin one; the latter remark does not apply though to the lifecycle costs. From the above remark, it becomes clear that the environmental impact associated with the production and manufacturing of virgin CFRP components cannot be neglected, and this urges the need to turn to CFRP recycling to avoid the energy-intensive process of PAN fiber production. Moreover, the environmental gains from the implementation of liquid hydrogen from renewable sources are highlighted, although issues concerning liquid hydrogen storage, transportation and infrastructure must also be considered. Yet, for the recycled components to be competitive with the virgin ones, a comparable to virgin quality appears as a mandatory requirement. Moreover, it should be noted that other factors, such as the feasibility of upgrade technologies of the fibers, the efficiency of the recycling processes and the capabilities of remanufacturing methods to produce recycled components of high quality, as well as the availability of the recycled fibers, must be considered. The worst by far environmental and economic impact concerns the recycled component comprised of randomly oriented fibers. This makes evident that such a component cannot compete with a virgin component, especially when addressed at a high-performance application, and hence, upgrade technologies would be required.

3.3. Sensitivity Analysis Results

The individual LCA, LCC and circular economy parameters of Tables 3–5 were exploited for the calculation of the overall sustainability Index of Equation (2). Based on this calculation, a ranking occurred among the considered components, for which four different types of fuel have been accounted for.

3.3.1. Normalization Method Sensitivity Results

As described in Section 2.3.2, three different normalization methods were implemented for the values integrated into the weighted sum, which resulted in three different combinations: (a) min-max normalization, (b) z-score normalization and (c) proportionate normalization. For each of the above combinations, the sustainability index was calculated, and a ranking among the considered components was derived. In order to test the sensitivity of the final ranking to the applied normalization method, an equal weighting was considered. The rankings obtained from the three different combinations are listed in Table 6.

Table 6. Comparison of The Ranking Obtained from The Different Normalization Methods.

Component Id			Ranking Order		
No	**Component Type**	**Fuel**	**Min–Max**	**z-Score**	**Proportionate**
1	Woven virgin	Kerosene	5	5	10
2	Woven virgin	LH2 (conventional source)	1	1	1
3	Woven virgin	LH2 (wind source)	4	4	4
4	Woven virgin	LH2 (geothermal source)	3	3	3
5	Recycled aligned	Kerosene	8	8	11
6	Recycled aligned	LH2 (conventional source)	2	2	2
7	Recycled aligned	LH2 (wind source)	7	7	7
8	Recycled aligned	LH2 (geothermal source)	6	6	5
9	Recycled random	Kerosene	12	12	12
10	Recycled random	LH2 (conventional source)	9	9	6
11	Recycled random	LH2 (wind source)	11	11	9
12	Recycled random	LH2 (geothermal source)	10	10	8

Based on the obtained rankings for the three different normalization methods, min-max normalization and z-score suggested the same ranking among the components. On the other hand, proportionate normalization led to a different ranking. Nevertheless, the first four places and the last one are identical to those obtained by the first two normalization methods. All normalization methods identified the virgin component, for which liquid hydrogen from a conventional source has been considered, as the most sustainable solution. On the other hand, the recycled component comprising randomly oriented fibers showed by far the lowest index, owing to its low quality; this highlights the need for upgrade technologies to improve quality and hence promote circularity and sustainability. Moreover, it is noteworthy that the recycled aligned component, for which liquid hydrogen from a conventional source has been accounted, ranks second; the latter applies to all three normalization techniques. This can be attributed to its comparable to virgin quality, as well as to its environmental friendliness.

3.3.2. Sensitivity to Weights and Data Variation

In order to assess the sensitivity of the tool to the weights' variation, two steps have been followed. Initially, the weights have been considerably varied to assess whether the final ranking is affected by such variations and consequently assess the efficiency of the tool. To this end, the scenarios described in Section 3.3.1 have been considered. The AHP pairwise comparisons were completed by the authors based on their knowledge and expertise in the field. In each of the said scenarios, one criterion is strongly prioritized over the other two criteria. A scenario assuming an equal weighting among the criteria has also been included. The pairwise comparisons of the aforementioned scenarios and the resulting weights are demonstrated in Table 7. All scenarios were checked for consistency, indicating a consistency ratio value below the threshold value of 0.1. The consistency ratio is a metric that indicates the consistency between pairwise comparisons. The rankings obtained from the aforementioned scenarios are presented in Table 8. The min-max normalization was considered for the normalization of the initial data. The results suggested different rankings for the different scenarios considered, and thus, the proposed method was found to be sensitive to the variations of the weight derived from considerable changes in the decision maker's judgments.

Table 7. Pairwise Comparisons and Resulting Weights for Different Scenarios.

	Scenario 1—Equal Weighting			
	Environmental Impact	**Costs**	**Circularity**	**Weight Factor/Priority**
Environmental Impact	1	1	1	≈33.3%
Costs	1	1	1	≈33.3%
Circularity	1	1	1	≈33.3%
	Scenario 2—environmental impact prioritization			
Environmental Impact	1	5	3	≈66%
Costs	1/5	1	1	≈16%
Circularity	1/3	1	1	≈18%
	Scenario 3—circularity prioritization			
Environmental Impact	1	3	1/5	≈21%
Costs	1/3	1	1/5	≈10%
Circularity	5	5	1	≈69%
	Scenario 4—costs prioritization			
Environmental Impact	1	1/5	2	≈18%
Costs	5	1	5	≈70%
Circularity	1/2	1/5	1	≈12%

In the second step of the sensitivity analysis, the rank stability of the MCDM tool was evaluated by adding noise to the criteria weights. To this end, the scenario for which environmental impact was prioritized (Scenario 2) was taken as the reference scenario, and minor adjustments to the user judgments were made. Therefore, based on the AHP scale of Table 1, three alternatives to the reference scenario were considered, for which one scale above or below the reference judgments was accounted for. The pairwise comparisons of the aforementioned scenarios are presented in Table 9. The results showed that the considered minor weight adjustments did not alter the ranking order (except for an exchange between two places of the alternative scenario 3), and hence, the proposed method does not appear to be affected by such minor weight adjustments.

Table 8. Ranking Obtained from The Different Weighting Scenarios of Table 6.

Component Identifier	Ranking Order			
	Scenario 1	Scenario 2	Scenario 3	Scenario 4
virgin ker.	5	10	2	4
virgin hyd.	1	1	1	1
virgin wind	4	4	8	3
virgin geo.	3	2	7	2
aligned ker.	8	11	4	8
aligned hyd.	2	3	3	5
aligned wind	7	6	10	7
aligned geo.	6	5	9	6
random ker.	12	12	6	12
random hyd.	9	7	5	9
random wind	11	9	12	11
random geo.	10	8	11	10

Table 9. Pairwise Comparisons for The Assessment of Small Weights Variations.

	Reference Scenario			
	Environmental Impact	Costs	Circularity	Weight Factor/Priority
Environmental Impact	1	5	3	≈66%
Costs	1/5	1	1	≈16%
Circularity	1/3	1	1	≈18%
	Alternative Scenario 1			
Environmental Impact	1	4	3	≈63%
Costs	1/4	1	1	≈18%
Circularity	1/3	1	1	≈19%
	Alternative Scenario 2			
Environmental Impact	1	5	3	≈64%
Costs	1/5	1	2	≈21%
Circularity	1/3	1/2	1	≈15%
	Alternative Scenario 3			
Environmental Impact	1	6	3	≈67%
Costs	1/6	1	1/2	≈11%
Circularity	1/3	2	1	≈22%

In order to test the stability of the method to small changes in the values of the initial data, 1000 perturbated samples of the original data were simulated in each of the following cases. It is assumed that the errors that perturb the initial data follow a normal distribution with zero mean and standard deviation 0.01, 0.05, 0.1, 0.25, and 0.5 of the standard deviation of the corresponding indices of the original data. The simulated samples were normalized with the three normalization methods (min-max, z-score and proportionate), ranked with respect to the overall sustainability index, and the mean rank of each material was calculated for each normalization method and for each selected value of the standard error. In Table 10, the mean ranking of each material based on the 1000 simulated samples for the Min–Max and z-score normalization methods are presented for all the selected values of errors' standard deviation. In Table 11, the corresponding mean ranks for the proportional normalization method are presented. The proposed method is fairly stable in terms of the mean rank for each normalization method and for a relatively large value of the standard deviation of the errors.

Table 10. Mean Ranking for The Min–Max and Z-score Normalization Methods.

		Mean Rank									
		0.01		0.05		0.1		0.25		0.5	
Id No	Initial Rank	Min–Max	z-Score	Min–Max	z-Score	Min–Max	z-Score	Min–Max	z-Score	Min–Max	z-Score
1	5	4.98	5.00	4.51	4.91	4.30	4.73	4.48	4.78	4.71	4.81
2	1	1.00	1.00	1.00	1.00	1.02	1.03	1.29	1.30	2.11	2.12
3	4	4.02	4.00	4.05	3.80	4.02	3.86	4.25	4.23	4.46	4.46
4	3	3.00	3.00	3.45	3.29	3.70	3.53	4.04	3.98	4.35	4.33
5	8	7.97	8.00	7.51	7.93	7.30	7.67	6.82	7.09	6.40	6.55
6	2	2.00	2.00	2.00	2.00	2.00	1.98	2.30	2.17	2,93	2.88
7	7	7.03	7.00	7.08	6.82	7.05	6.83	6.67	6.54	6.22	6.17
8	6	6.00	6.00	6.42	6.25	6.60	6.37	6.30	6.16	5.90	5.85
9	12	12.00	12.00	11.71	11.99	11.44	11.86	11.23	11.46	11.08	11.19
10	9	9.00	9.00	9.00	9.00	9.00	9.00	8.85	8.75	8.21	8.11
11	11	11.00	11.00	11.08	10.86	11.06	10.81	11.02	10.91	10.91	10.85
12	10	10.00	10.00	10.21	10.15	10.50	10.33	10.75	10.63	10.73	10.67

Table 11. Mean Ranking for The Proportionate Normalization Method.

		0.01	0.05	0.25	0.5
		Mean Rank	Mean Rank	Mean Rank	Mean Rank
Id No	Initial Rank	Prop	Prop	Prop	Prop
1	10	10.00	9.99	9.67	8.98
2	1	1.00	1.14	1.94	3.02
3	4	4.00	4.32	4.61	4.77
4	3	3.00	3.25	4.09	4.48
5	11	11.00	11.00	10.45	9.73
6	2	2.00	1.86	2.21	3.15
7	7	7.00	6.59	5.56	5.50
8	5	5.03	5.16	4.85	5.06
9	12	12.00	12.00	12.00	11.88
10	6	5.97	5.68	5.16	5.26
11	9	9.00	8.96	9.04	8.43
12	8	8.00	8.05	8.44	7.75

In Figure 1, the mean ranks and an interval of ±one standard deviation of the rankings are presented for the different levels of noise variation and the three studied normalization methods. As it is observed in Figure 1, the larger the standard deviation of the errors, the larger the variability of each material ranking. Despite the increase in the variation of the rankings, the mean rankings seem to converge to the initial ranking.

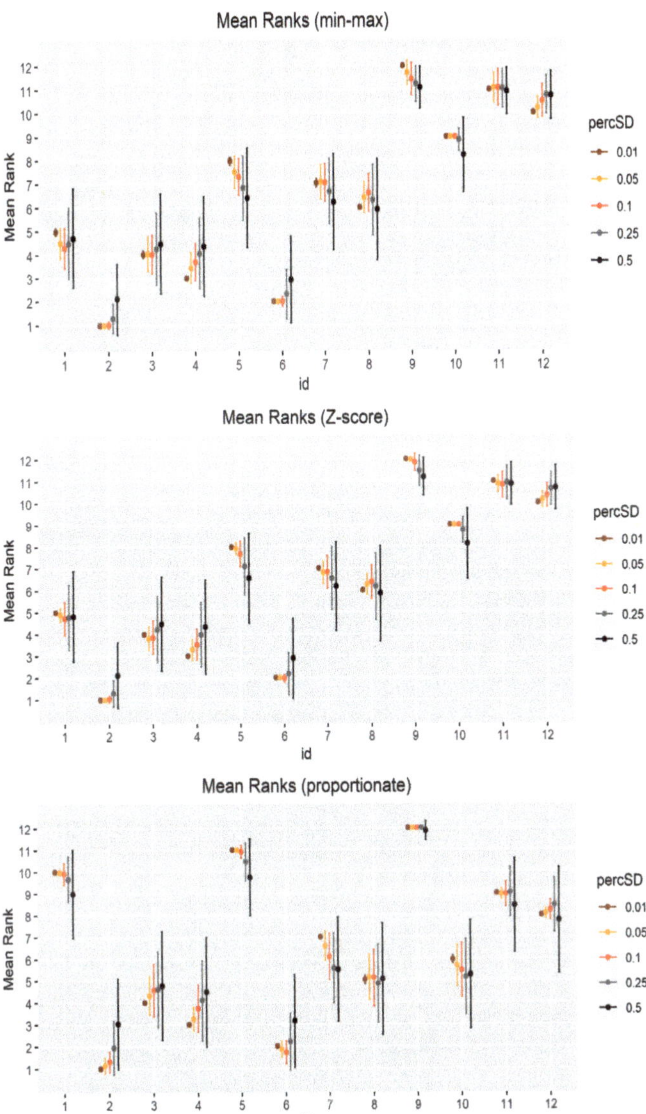

Figure 1. Mean Ranks ± One Standard Deviation of The Rankings for The Different Levels of Noise Variation (percSD).

4. Conclusions

In the present study, the robustness of a hybrid MCDM-based tool proposed by the authors for the aviation sector has been investigated. The latter is performed by accounting for a use case in which the sustainability of composite aircraft components is compared. The proposed tool combines lifecycle metrics linked to environmental, economic and circular economy aspects. Circular economy performance has been associated with a quality feature of the considered components. The tool is able to account for the type of fuel utilized during the use phase of the components. Although liquid hydrogen is not currently certified as an aviation fuel (SAF) by ASTM, its application is being actively researched and developed by various stakeholders in the aviation industry and is considered the most promising fuel option for future aircraft [1]. In this context, it is mandatory to consider and evaluate the sustainability of hydrogen as a potential fuel option for future aircraft.

The environmental impact and cost assessment of the examined components highlighted that a recycled component of near-to-virgin quality can potentially compete with a virgin component, accounting for its whole lifecycle. To this end, the utilization of liquid hydrogen from a renewable source appears necessary. Yet, to achieve a near-to-virgin quality, upgrade techniques and effective remanufacturing methods are required. Furthermore, it has been remarked that the use phase in the aviation sector dominates the overall impact; the latter signifies that the environmental emissions and costs linked to the production and manufacturing phases appear almost negligible compared to these of the use phase. However, when liquid hydrogen from a renewable source, especially from geothermy, has been accounted for, the impact of production and manufacturing comprises a considerable amount of the overall impact. The latter indicates that the decarbonization of the aviation sector may shift the environmental, at least, burden to the production and manufacturing phases. Moreover, although the environmental benefits of using liquid hydrogen are undeniable, the currently high costs of hydrogen compared to kerosene may act as a prohibiting factor for its extensive use in aviation. In addition, the impact of other aspects relating to the production, transportation and storage of liquid hydrogen must also be accounted for, although the latter assessments were outside the scope of this study.

The sensitivity of the MCDM tool to the normalization method applied, as well as to the weights and data variation, has been examined. The sensitivity analysis on the applied normalization method suggested the same ranking for the min-max and z-score methods, while the proportionate normalization method suggested a different ranking. Nonetheless, the first and the last ordered components are identical for all normalization methods. Moreover, the tool was found not to be sensitive to small variations of the weights; on the other hand, larger weights variations suggested a different ranking for the scenarios considered. Finally, the sensitivity analysis on the initial data values did not show a significant change in the final components' rankings compared to the initially obtained ones with respect to the different levels of noise variation for the three studied normalization methods. The latter remarks are quite encouraging and demonstrate the efficiency of the proposed tool as a reliable and robust decision-support tool for the aviation sector.

The goal of this work has been to enhance the reliability of the tool and bolster its credibility for making critical decisions in the aviation sector. Such decisions entail choosing suitable technologies, production and manufacturing procedures for components and materials, as well as determining the appropriate fuel for new aircraft. Future studies could focus on further validating the tool, including its sensitivity to different weights and criteria towards its practical use in the aviation industry, with input from a broader range of experts and stakeholders, ultimately contributing to more sustainable and informed decision-making.

Author Contributions: Conceptualization, D.N.M. and S.G.P.; methodology, D.N.M., S.M. and S.G.P.; software, D.N.M., S.M. and S.G.P.; validation, D.N.M., S.M. and S.G.P.; formal analysis, D.N.M., S.M. and S.G.P.; investigation, D.N.M., S.M. and S.G.P.; resources, S.G.P.; data curation, D.N.M., S.M. and S.G.P.; writing—original draft preparation, D.N.M., S.M. and S.G.P.; writing—review and editing, D.N.M., S.M. and S.G.P.; visualization, D.N.M., S.M. and S.G.P.; supervision, S.G.P.; project administration, S.G.P. All authors have read and agreed to the published version of the manuscript.

Funding: The research conducted in this paper has been funded by European Union's Horizon 2020, its research and innovation program, under grant agreement No 101058089, project EuReComp (European recycling and circularity in large composite components).

Data Availability Statement: No more data is available due to pricacy restrictions.

Acknowledgments: The research conducted in this paper has been funded by European Union's.

Conflicts of Interest: The authors declare no conflict of interest.

Nomenclature

AHP	Analytic Hierarchy Process
BWM	Best-Worst Method
CE	Circular Economy
CEI	Circular Economy Indicator
CFRP	Carbon Fiber Reinforced Plastics
ELECTRE	Elimination and Choice Translating Reality
EoL	End of Life
FBP	Fluidized Bed Process
GHG	Greenhouse Gases
LCA	Life Cycle Assessment
LCC	Life Cycle Costing
MCDM	Multi-Criteria Decision-Making
PAN fibers	Polyacrylonitrile fibers
SAW	Simple Additive Weighting
TOPSIS	Technique for Order of Preference by Similarity to Ideal Solution
TRL	Technology Readiness Level
VIKOR	VIseKriterijumska Optimizacija I Kompromisno Resenje
WSM	Weighted Sum Model

References

1. European Commission. *Flightpath 2050, Europe's Vision for Aviation*; Report of the High Level Group on Aviation Research, Directorate-General for Research and Innovation, Directorate General for Mobility and Transport; European Commission: Brussels, Belgium, 2011; p. 28.
2. Communication from the Commission to the European Parliament, the European Council, the Council, the European Economic and Social Committee and the Committee of the Regions the European Green Deal com/2019/640 final. Available online: https://eur-lex.europa.eu/legal-content/EN/TXT/?uri=COM%3A2019%3A640%3AFIN (accessed on 5 April 2022).
3. Graver, B.; Rutherford, D.; Zheng, S. *CO_2 Emissions from Commercial Aviation: 2013, 2018, and 2019*; Report of the ICCT (The International Council on Clean Transportation); ICCT: Binangonan, Philippines, 2020.
4. Gnadt, A.R.; Speth, R.L.; Sabnis, J.S.; Barrett, S.R.H. Technical and environmental assessment of all-electric 180-passenger commercial aircraft. *Prog. Aerosp. Sci.* **2019**, *105*, 1–30. [CrossRef]
5. Léonard, P.; Nylander, J. Sustainability assessment of composites in aero-engine components. In *Proceedings of the Design Society: DESIGN Conference*; Cambridge University Press: Cambridge, UK, 2020; pp. 1989–1998.
6. Meng, F.; Cui, Y.; Pickering, S.; McKechnie, J. From aviation to aviation: Environmental and financial viability of closed-loop recycling of carbon fibre composite. *Compos. Part B Eng.* **2020**, *200*, 108362. [CrossRef]
7. Markatos, D.N.; Katsiropoulos, C.; Tserpes, K.; Pantelakis, S. A holistic End-of-Life (EoL) Index for the quantitative impact assessment of CFRP waste recycling techniques. *Manuf. Rev.* **2021**, *8*, 18.
8. Zhang, J.; Chevali, V.S.; Wang, H.; Wang, C.H. Current status of carbon fibre and carbon fibre composites recycling. *Compos. B. Eng.* **2020**, *193*, 108053. [CrossRef]
9. Pimenta, S.; Pinho, S.T. Recycling carbon fibre reinforced polymers for structural applications: Technology review and market outlook. *Waste Manag.* **2011**, *31*, 378–392. [CrossRef]
10. Carbon fibers/Aerospace. Available online: https://www.compositesworld.com/articles/recycled-carbon-fibre-proves-its-potential-for-aircraft-interiors (accessed on 9 January 2023).
11. Tapper, R.J.; Longana, M.L.; Norton, A.; Potter, K.D.; Hamerton, I. An evaluation of life cycle assessment and its application to the closed-loop recycling of carbon fibre reinforced polymers. *Compos. Part B Eng.* **2020**, *184*, 107665. [CrossRef]
12. Slavica Dožić, S. Multi-criteria decision making methods: Application in the aviation industry. *J. Air Transp. Manag.* **2019**, *79*, 101683. [CrossRef]
13. Hsu, C.-C.; Liou, J.J.H. An outsourcing provider decision model for the airline industry. *J. Air Transp. Manag.* **2013**, *28*, 40–46. [CrossRef]
14. Sánchez-Lozano, J.M.; Serna, J.; Dolón-Payán, A.; Sánchez-Lozano, J.M.; Serna, J.; Dolón-Payán, A. Evaluating military training aircrafts through the combination of multi-criteria decision making processes with fuzzy logic. A case study in the Spanish Air Force Academy. *Aerosp. Sci. Technol.* **2015**, *42*, 58–65. [CrossRef]

15. Garg, C.P. A robust hybrid decision model for evaluation and selection of the strategic alliance partner in the airline industry. *J. Air Transp. Manag.* **2016**, *52*, 55–66. [CrossRef]
16. Bae, B.-Y.; Kim, S.; Lee, J.-W.; Nguyen, N.V.; Chung, B.-C. Process of establishing design requirements and selecting alternative configurations for conceptual design of a VLA. *Chin. J. Aeronaut.* **2017**, *30*, 738–751. [CrossRef]
17. Görener, A.; Ayvaz, B.; Kusakci, A.O.; Altinokc, E. A hybrid type-2 fuzzy based supplier performance evaluation methodology: The Turkish Airlines technic case. *Appl. Soft. Comput.* **2017**, *56*, 436–445. [CrossRef]
18. Barak, S.; Dahooei, J.H. A novel hybrid fuzzy DEA-Fuzzy MADM method for airlines safety evaluation. *J. Air Transp. Manag.* **2018**, *73*, 134–149. [CrossRef]
19. Sun, G.; Guan, X.; Yi, X.; Zhou, Z. An innovative TOPSIS approach based on hesitant fuzzy correlation coefficient and its applications. *Appl. Soft. Comput.* **2018**, *68*, 249–267. [CrossRef]
20. Mahtani, U.S.; Garg, C.P. An analysis of key factors of financial distress in airline companies in India using fuzzy AHP framework. *Transport. Res. Part A* **2018**, *117*, 87–102. [CrossRef]
21. Athawale, V.M.; Chakraborty, S. Material selection using multi-criteria decision-making methods: A comparative study. *Proc. Inst. Mech. Eng. Part L J. Mater. Des. Appl.* **2012**, *226*, 266–285. [CrossRef]
22. Markatos, D.N.; Pantelakis, S.G. Assessment of the Impact of Material Selection on Aviation Sustainability, from a Circular Economy Perspective. *Aerospace* **2022**, *9*, 52. [CrossRef]
23. Deng, Y. Life Cycle Assessment of Biobased Fibre-Reinforced Polymer Composites. Ph.D. Thesis, KU Leuven, Science, Engineering & Technology, Leuven, Belgium, 2014.
24. Patton, R.; Li, F. *Causes of Weight Reduction Effects of Material Substitution on Constant Stiffness Components*; SAE Technical: Warrendale, PA, USA, 2002.
25. Li, F.; Patton, R.; Moghal, K. The relationship between weight reduction and force distribution for thin wall structures. *Thin-Walled Struct.* **2005**, *43*, 591–616. [CrossRef]
26. Meng, F.; Olivetti, E.A.; Zhao, Y.; Chang, J.C.; Pickering, S.J.; McKechnie, J. Comparing life cycle energy and global warming potential of carbon fibre composite recycling technologies and waste management options. *ACS Sustain. Chem. Eng.* **2018**, *6*, 9854–9865. [CrossRef]
27. Suzuki, T.; Jun Takahashi, J. Prediction of energy intensity of carbon fibre reinforced plastics for mass-produced passenger cars. In Proceedings of the Ninth Japan International SAMPE Symposium, Tokyo, Japan, 29 November–2 December 2005.
28. Ghosh, T.; Kim, H.C.; De Kleine, R.; Wallington, T.J.; Bakshi, B.R. Life cycle energy and greenhouse gas emissions implications of using carbon fibre reinforced polymers in automotive components: Front subframe case study. *Sustain. Mater. Technol.* **2021**, *28*, e00263.
29. Dér, A.; Dilger, N.; Kaluza, A.; Creighton, C.; Kara, S.; Varley, R.; Herrmann, C.; Thiede, S. Modelling and analysis of the energy intensity in polyacrylonitrile (PAN) precursor and carbon fibre manufacturing. *J. Clean. Prod.* **2021**, *303*, 127105. [CrossRef]
30. Meng, F.; McKechnie, J.; Pickering, S.J. An assessment of financial viability of recycled carbon fibre in automotive applications. *Compos. Part A Appl. Sci.* **2018**, *109*, 207–220. [CrossRef]
31. Karuppannan Gopalraj, S.; Kärki, T. A study to investigate the mechanical properties of recycled carbon fibre/glass fibre-reinforced epoxy composites using a novel thermal recycling process. *Processes* **2020**, *8*, 954. [CrossRef]
32. EUWebsite. Electricity Price Statistics. Available online: https://ec.europa.eu (accessed on 22 October 2021).
33. Duflou, J.; Deng, Y.; Van Acker, K.; Dewulf, W. Do fibre-reinforced polymer composites provide environmentally benign alter-natives? A life-cycle-assessment-based study. *MRS Bull.* **2012**, *37*, 374–382. [CrossRef]
34. Bicer, Y.; Dincer, I. Life cycle evaluation of hydrogen and other potential fuels for aircrafts. *Int. J. Hydrogen Energy* **2017**, *42*, 10722–10738. [CrossRef]
35. Larsen, l.; Schuster, A.; Kim, J.; Kupke, M. Path planning of cooperating industrial robots using evolutionary algorithms. *Procedia Manuf.* **2018**, *17*, 286–293. [CrossRef]
36. Airliners Website. Aircraft Technical Data and Specifications. Airbus A320. Available online: https://www.airliners.net (accessed on 22 October 2021).
37. Saidani, M.; Yannou, B.; Leroy, Y.; Cluzel, F. How to Assess Product Performance in the Circular Economy? Proposed Requirements for the Design of a Circularity Measurement Framework. *Recycling* **2017**, *2*, 6. [CrossRef]
38. Rigamonti, L.; Mancini, E. Life cycle assessment and circularity indicators. *Int. J. Life Cycle Assess.* **2021**, *26*, 1937–1942. [CrossRef]
39. Corona, B.; Shen, L.; Reike, D.; Carreón, J.R.; Worrell, E. Towards sustainable development through the circular economy—A review and critical assessment on current circularity metrics. *Resour. Conserv. Recycl.* **2019**, *151*, 104498. [CrossRef]
40. Kirchherr, J.; Reike, D.; Hekkert, M. Conceptualizing the circular economy: An analysis of 114 definitions. *Resour. Conserv. Recycl.* **2017**, *127*, 221–232. [CrossRef]
41. Moraga, G.; Huysveld, S.; Mathieux, F.; Blengini, G.A.; Alaerts, L.; Van Acker, K.; de Meester, S.; Dewulf, J. Circular economy indicators: What do they measure? *Resour. Conserv. Recycl.* **2019**, *146*, 452–461. [CrossRef]
42. Zhao, D.; Guo, Z.; Xue, J. Research on scrap recycling of retired civil aircraft. In *IOP Conference Series: Earth and Environmental Science*; IOP Publishing: Bristol, UK, 2021; Volume 657, p. 012062s.
43. Zardari, N.H.; Ahmed, K.; Shirazi, S.M.; Yusop, Z.B. Introduction. In *Weighting Methods and their Effects on Multi-Criteria Decision Making Model Outcomes in Water Resources Management*; Springer Briefs in Water Science and Technology; Springer: Cham, Switzerland, 2015.

44. Saaty, T.L. *The Analytic Hierarchy Process: Planning, Priority Setting, Resource Allocation*; McGraw-Hill: New York, NY, USA, 1980; 287p.
45. Ighravwe, D.E.; Oke, S.A. A multi-criteria decision-making framework for selecting a suitable maintenance strategy for public buildings using sustainability criteria. *J. Build. Eng.* **2019**, *24*, 100753. [CrossRef]
46. Jarek, S. Removing Inconsistency in Pairwise Comparisons Matrix in the AHP. *Mult. Criter Decis. Mak.* **2016**, *11*, 63–76. [CrossRef]
47. Hasan, M.; Gulzarul, Z.; Mohammad, F. Multi-choice best-worst multi-criteria decision-making method and its applications. *Int. J. Intell. Syst.* **2021**, *37*, 1129–1156. [CrossRef]
48. Srdjevic, B.; Srdjevic, Z.; Reynolds, K.M.; Lakicevic, M.; Zdero, S. Using Analytic Hierarchy Process and Best–Worst Method in Group Evaluation of Urban Park Quality. *Forests* **2022**, *13*, 290. [CrossRef]
49. Talukder, B.; Hipel, K.W.; W. vanLoon, G. Developing Composite Indicators for Agricultural Sustainability Assessment: Effect of Normalization and Aggregation Techniques. *Resources* **2017**, *6*, 66. [CrossRef]
50. R Core Team. *R: A Language and Environment for Statistical Computing*; R Foundation for Statistical Computing: Vienna, Austria, 2023. Available online: https://www.R-project.org/ (accessed on 9 January 2023).

Article

Fast Sizing Methodology and Assessment of Energy Storage Configuration on the Flight Time of a Multirotor Aerial Vehicle †

Saad Chahba [1,*], Rabia Sehab [1,‡], Cristina Morel [1,‡], Guillaume Krebs [2,‡] and Ahmad Akrad [1,‡]

1 Pole Systèmes et Energies Embarqués pour les Transports S2ET, ESTACA Laval, 53000 Laval, France
2 GeePs Group of Electrical Engineering-Paris, UMR CNRS 8507, CentraleSupélec, Université Paris-Saclay, 91192 Gif Sur Yvette, France
* Correspondence: saad.chahba@estaca.fr; Tel.: +33-(0)176-52-08-36
† This paper is an extended version of our paper published in 12th EASN Conference on "Innovation in Aviation & Space for opening New Horizons".
‡ These authors contributed equally to this work.

Abstract: Urban air mobility (UAM), defined as safe and efficient air traffic operations in a metropolitan area for manned aircraft and unmanned aircraft systems, is being researched and developed by industry, academia, and government. This kind of mobility offers an opportunity to construct a green and sustainable sub-sector, building upon the lessons learned over decades by aviation. Thanks to their non-polluting operation and simple air traffic management, electric vertical take-off and landing (eVTOL) aircraft technologies are currently being developed and experimented with for this purpose. However, to successfully complete the certification and commercialization stage, several challenges need to be overcome, particularly in terms of performance, such as flight time and endurance, and reliability. In this paper, a fast methodology for sizing and selecting the propulsion chain components of an eVTOL multirotor aerial vehicle was developed and validated on a reduced-scale prototype of an electric multirotor vehicle with a *GTOW* of 15 kg. This methodology is associated with a comparative study of energy storage system configurations, in order to assess their effect on the flight time of the aerial vehicle. First, the optimal pair motor/propeller was selected using a global nonlinear optimization in order to maximize the specific efficiency of these components. Second, five energy storage technologies were sized in order to evaluate their influence on the aerial vehicle flight time. Finally, based on this sizing process, the optimized propulsion chain gross take-off weight (GTOW) was evaluated for each energy storage configuration using regression-based methods based on propulsion chain supplier data.

Keywords: eVTOL; multirotor aerial vehicle; sizing; optimization; hybrid energy storage; battery; hydrogen fuel cell; supercapacitor

Citation: Chahba, S.; Sehab, R.; Morel, C.; Krebs, G.; Akrad, A. Fast Sizing Methodology and Assessment of Energy Storage Configuration on the Flight Time of a Multirotor Aerial Vehicle. *Aerospace* **2023**, *10*, 425. https://doi.org/10.3390/aerospace10050425

Academic Editors: Andreas Strohmayer, Spiros Pantelakis and Jordi Pons-Prats

Received: 28 February 2023
Revised: 27 April 2023
Accepted: 28 April 2023
Published: 30 April 2023

1. Introduction

The eVTOL concept represents one of the potential solutions to remedy the traffic congestion problem in big cities across the world. In the case of French metropolitan cities, such as Paris and Marseille, an average commuter loses over 80 h every year in traffic, resulting in increased stress and anxiety (R1). Such congestion also leads to 1.85 megatons per year of CO_2 emissions into the atmosphere. In addition to having a detrimental impact on the health of commuters and the environment, it also contributes to economic loss. Therefore, it is imperative to explore new modes of transportation to facilitate faster daily commutes for passengers in urban areas and reduce traffic congestion. Several aircraft manufacturers, such as Airbus, Boeing, Lilium, and Volocopter, have actively embarked on the development of this drone taxi technology in recent years [1–5]. In addition to Uber, which estimates the launch of its air taxis (called Uber Elevate) in 2023, other companies, such as Zephyr Airworks and Airbus, are also currently taking measures to conduct tests with their electric aviation taxis. Zephyr Airworks has developed Cora, while Airbus has developed Airbus Vahana. These companies are conducting tests in various countries across the world, including the USA, Japan, Singapore, New Zealand, France, and India.

eVTOL aircraft fit into four main categories: lift plus cruise, tilt rotor, ducted vector thrust, and multicopter (Figure 1). The first three categories fall under the powered lift aircraft classification, which includes winged aircraft that are capable of both VTOL and aerodynamic lifts during forward flights. The fourth category belongs to wingless aircraft, specifically multirotor aircraft with two or more lift/thrust units that have limited or no capability for wingborne forward flights. Powered lift eVTOLs can be further categorized based on whether they use a common power plant (tilt-rotor and ducted vector thrust) or an independent power plant (lift plus cruise) for both lifting and forward flights [6–8].

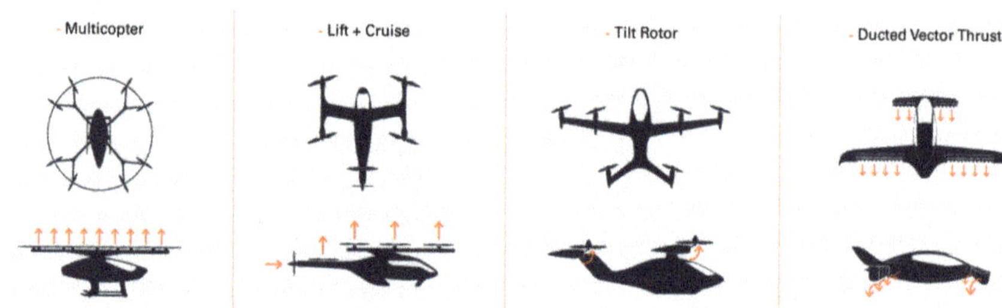

Figure 1. eVTOL propulsion configuration.

These categories are defined and characterized as follows [9,10]:

- Vectored thrust: These are powered lift eVTOL aircraft that utilize all of their lift/thrust units for both vertical lift and cruise. This is achieved by rotating (vectoring) the resultant thrust points against the direction of motion. The thrust vectoring can be accomplished in several ways: by rotating the entire wing-propulsion assembly (tilt wing), by rotating the lift/thrust unit itself (tilt fan for ducted fans and tilt prop for propellers), or by rotating the entire aircraft frame pivoted about the fuselage (tilt body or tilt frame). An example of this configuration is the Lilium jet, shown in Figure 2a, which utilizes ducted and vectored thrust. The implementation of ducted fans in the form of distributed electric propulsion (DEP) is employed [4].
- Wingless: This configuration is relatively simple and can be very efficient during vertical take-off, landing, and hovering, due to low disc-loading. However, without wings, multicopters lack cruise efficiency, which limits their application to urban air mobility (UAM) markets only. An example of this category is given in Figure 2b, named Volocopter VC2X [5]. The latter runs on nine independent batteries, powering 18 electric motor-driven variable-speed/fixed-pitch propellers. The redundancy ensures stability in the event of component failures.
- Lift plus cruise: These aircraft combine the capabilities of a multicopter for vertical take-off and landing with those of a standard aircraft for the cruise flight. This integration enables the aircraft to achieve both efficient vertical take-off and landing as well as efficient cruise performance. To optimize the range of these concepts, the propellers required for VTOL are designed with fewer blades and shorter chords to minimize drag during the cruise flight. However, the small size of the propellers used for VTOL operations presents a notable challenge in terms of noise emissions, mainly due to increased blade tip speeds. Figure 2c provides an example of this configuration, named Kitty Hawk Cora [11].
- Tilt rotor: This configuration involves either the wing and propellers or the propellers alone (tilting). This enables the propeller axis to rotate by 90 degrees as the aircraft transitions from hover to forward flight. This architecture generally allows for the

design of a more optimized propeller compared to a lift and cruise aircraft configuration. However, it comes with the trade-off of higher technical complexity and larger overall size and weight due to the inclusion of tilt and variable pitch mechanisms. Joby S4 is an example of this category; it is developed by Joby Aviation (Figure 2d) and is supposed to be commercialized by 2024 [12,13].

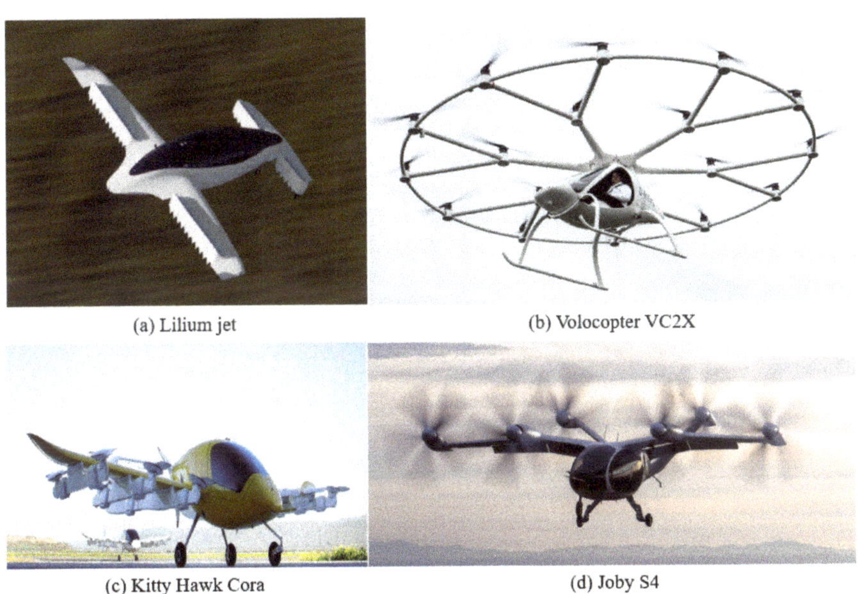

(a) Lilium jet (b) Volocopter VC2X (c) Kitty Hawk Cora (d) Joby S4

Figure 2. eVTOL categories.

As reported in [10], one of the main drawbacks of the multirotor configurations is the lack of wings, which limits their performances in long cruise flight missions. However, in the UAM mode, where the cruise phase durations are limited in comparison with extra-urban mobility, the multirotor configuration remains the best efficient solution to this transport market [10]. In this context, the Volocopter VC2X configuration, which is a non-coaxial, direct-lift one, presented in Figure 2b, will be tested in Paris, France, in 2024 [14]. Thus, the rest of the paper will focus on the multirotor wingless configuration.

One of the main steps in the eVTOL design process is to size and select the components of the propulsion system to meet the required specifications. To facilitate the assessment of proposed solutions, the development of precise and efficient sizing methodologies for the electric propulsion chain is necessary. The propulsion chain typically consists of a propeller for generating lift, a BLDC electric motor for energy conversion, an electronic speed controller (ESC) that supplies the required current to the load from the energy source, and a battery for energy storage. Multirotor design methods have been developed by Barshefsky et al. [15], Dai et al. [16], et Gur et al. [17]. In [15], the authors present a methodology that involves parameterizing the components of the propulsion chain to establish relationships between them. These relationships are then optimized to meet the specific requirements of the flight mission. In reference [16], an analytical method is proposed to estimate the optimal parameters of the propulsion system components. The approach involves modeling each component mathematically and then simplifying and decoupling the problem into smaller subproblems. By solving these subproblems, the optimal parameters for each component can be obtained. Moreover, selection algorithms are proposed based on these obtained parameters to determine the optimal combination of the propeller, motor, ESC, and battery products from their respective databases. Methodologies based on statistical data available from manufacturers for preliminary design are reported in references in [17–19]. For example, reference [17] presents a multi-disciplinary optimization (MDO) approach for designing a propulsion system based on goals such as rate of climb and loiter time. It also provides a useful modeling analysis of motors and batteries. Moreover, a sensitivity analysis is conducted on certain propeller design elements.

In this study, on the one hand, a methodology for sizing and selecting the propulsion chain components was developed. This approach combines statistical methods based on data and analytical optimization techniques, allowing to maintain an acceptable level of precision and avoid increasing the calculation algorithm complexity. The technique of optimization is used for the optimal selection of the pair motor/propeller, based on the maximization of the specific efficiency. This optimization makes it possible to select the remaining components, namely the ESC and the energy storage system. The statistical methods are considered for the multirotor aerial vehicle $GTOW$ evaluation, using the regression model for each component, based on supplier data. On the other hand, five energy storage configurations are considered, in order to evaluate their effect on the multirotor aerial vehicle performance, in particular on the flight time. These configurations are the lithium polymer battery (battery), hydrogen fuel cell (HFC), battery/hydrogen fuel cell (Bat/HFC), battery/supercapacitor (Bat/SC), and battery/supercapacitor/hydrogen fuel cell (Bat/SC/HFC). The five energy sources were sized to maximize the flight time and keep the gross take-off weight $(GTOW)$ as low as possible.

This article is organized as follows. Section 2 presents the sizing methodology flowchart, including the optimization technique used to select the optimal motor/propeller pair and the remaining components of the propulsion chain. Section 3 presents the modeling of the propulsion chain components to formulate the optimization problem and maximize the efficiency of the motor/propeller pair. Section 4 is devoted to the formulation of the motor/propeller pair optimization problem. Section 5 presents the case study for the sizing approach validation, using a reduced-scale multirotor drone with a $GTOW$ of 15 kg available in our laboratory. Section 6 presents the comparative study of the energy source configuration effect on the flight time, including the $GTOW$ evaluation based on the regression model of each component. In Section 7, the conclusion is presented.

2. Sizing Methodology

The sizing methodology is based on a combination of analytical optimization techniques and data-based techniques. It takes the following input data: the required flight time, a database of electric motor parameters, including the voltage (U_m), load current (I_m), internal resistance (R_m), and speed constant (K_v) for n examples (at this stage, the optimal motor is not known), atmospheric conditions defining the altitude, temperature, and air density, and the initial gross take-off weight ($GTOW$). Subsequently, a global non-linear optimization is performed for each motor/propeller pair using the simulated annealing algorithm (SAA). The objective of this optimization is to maximize the pair motor/propeller efficiency, also known as specific efficiency $\eta_{MP}(N/W)$. This efficiency index is widely used by industrial manufacturers, such as T-motor and Mejzlik [20,21] to measure the efficiency between the motor and the propeller. Constraints are applied to the propeller geometry, specifically the diameter (D_p) and the pitch angle φ_p. The optimized motor/propeller pair obtained from the optimization allows for checking the condition to avoid motor overheating, as stated in Equation (32). Subsequently, the maximum thrust $T_{MP_{max}}$ generated by the optimized motor/propeller pair was computed using Equation (34). By utilizing the maximum thrust T_{max} imposed by $GTOW$ as indicated in Equation (35), a filtering condition was established based on the relative error between $T_{MP_{max}}$ and T_{max}, as shown in Equation (36). This filtering condition allows for an initial selection of the motor/propeller pairs. Subsequently, the selection of the optimal motor/propeller pair was determined based on the maximum specific efficiency achieved. The sizing of the energy sources was then performed to maximize the flight time. Finally, the last step of this sizing methodology involved verifying whether the total take-off mass, obtained using statistical mass models for each component of the propulsion chain based on supplier data, was within acceptable limits. A detailed flowchart of this sizing methodology is provided in Figure 3. The specific efficiency maximization allows making a rapid and precise choice for each component in the propulsion chain. Figure 4 presents an overview of the different steps, based on which, the electric propulsion chain sizing is performed. The fact that the validation step was based on a reduced-scale multirotor drone with a *GTOW* of 15 kg explains the choice of the data scale used for the optimization of input data motors and regression models.

It is remarkable that in the developed sizing methodology, we do not consider a flight power mission in order to size the propulsion chain. However, in order to obtain a pair motor/propeller that can satisfy the take-off and cruise segment mission, the sizing methodology takes into consideration two constraints. The first one is related to the speed of rotation and the torque of the propeller reported in Equation (32). Through this constraint, the obtained motor/propeller pair is able to satisfy the cruise phase while avoiding motor overheating. The second constraint is the filter condition given in Equation (37). Through this constraint, the motor/propeller pair is able to succeed in the take-off phase.

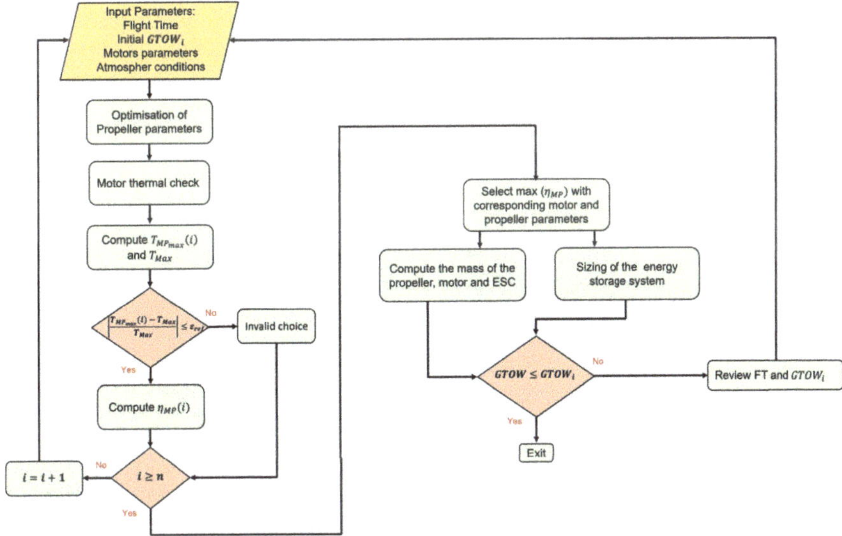

Figure 3. Sizing approach for the eVTOL multirotor flowchart.

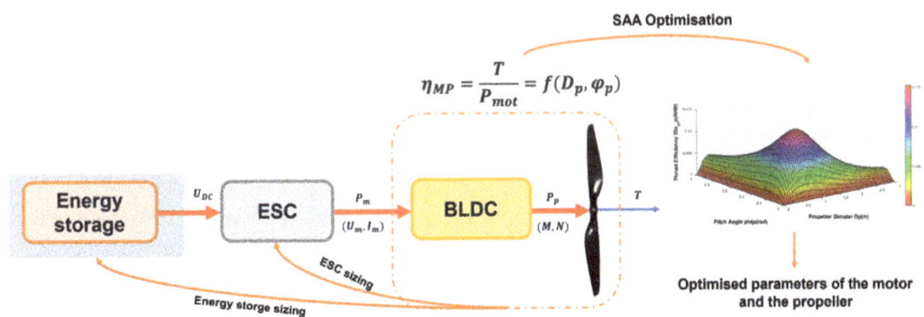

Figure 4. Overview of the sizing methodology steps.

3. Propulsion Chain Modeling

The physical model of the propulsion chain, of an eVTOL multirotor aerial vehicle (with n arms) is presented in Figure 5. Each propulsion chain is composed of a propeller, a motor with its equivalent internal resistance, an ESC represented as a simple resistor, and an energy storage system modeled as an open circuit voltage with a series resistor. In this configuration (Figure 5), the energy storage system supplies the n propulsion chains. The modeling process will aim to establish a relationship between the propulsion chain components, especially between the propeller and the motor, in order to formulate the optimization problem. In this case, each component will be modeled in a steady state.

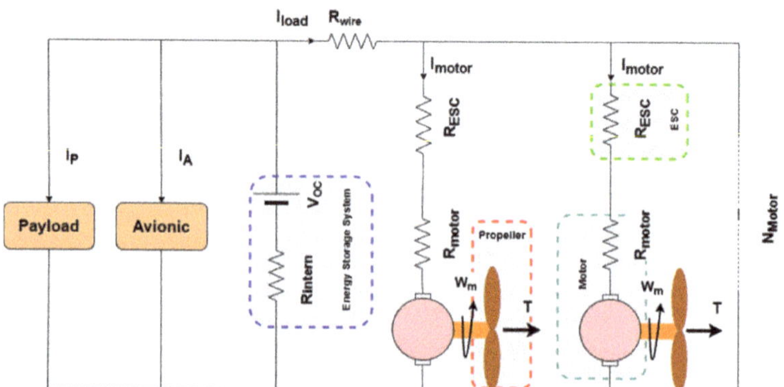

Figure 5. eVTOL multirotor physical model.

3.1. Propeller

The propeller includes many parameters to consider, such as the material of the propeller, the diameter, the shape of the blades, the pitch, and the number of blades. Currently, a variety of materials are used for propeller manufacturing, including carbon fiber (CF), nylon, plastic, and wood. The material of the propeller greatly influences its aerodynamic performance. As the rotor blades spin, the angle of attack at each spanwise region can change with reference to the original blade design [22]. Carbon fiber propellers are known for their stiffness and for being lightweight, but their downside is their high cost. Increasing the pitch and the number of blades results in higher thrust production, but it leads to a decrease in propeller efficiency, necessitating increased electrical and mechanical power. Increasing the diameter will increase the efficiency, but the ability to handle the load of the motor must also increase. When the blades are larger, with other parameters constant, they will rotate at lower velocities to produce the same lift. At this point, the induced velocity will decrease. Thereby, the efficiency of the system will increase [22]. The propeller model is described by its thrust $T(N)$ and its torque $M(Nm)$ as given by

$$\begin{cases} T = C_T \cdot \rho \cdot \left(\frac{N}{60}\right)^2 \cdot {D_p}^4 \\ M = C_M \cdot \rho \cdot \left(\frac{N}{60}\right)^2 \cdot {D_p}^5, \end{cases} \tag{1}$$

where ρ (kg/m^3), C_T, C_M, N (rpm), and D_p (m) are respectively, the air density, the thrust coefficient, the torque coefficient, the propeller velocity, and the propeller diameter. The air density, ρ, is determined by both the local temperature T_t (unit: °C) and the air pressure p, which is further determined by altitude h_hover (m). According to the international standard atmosphere model [23], we have the following expressions:

$$\begin{cases} \rho = \frac{273 \cdot p}{p_0(273+T_t)} \cdot \rho_0 \\ p = p_0 \cdot \left(1 - 0.0065 \cdot \frac{h}{273+T_T}\right)^{5.2561}, \end{cases} \tag{2}$$

where ρ_0 is the standard air density, $\rho_0 = 1.293$ (kg/m^3), at a temperature of 25 °C. The thrust and the torque coefficients are modeled using the blade element theory as presented in [22]:

$$\begin{cases} C_T = \frac{0.27\pi^3\lambda\zeta^2 K_0\epsilon}{\pi A+K_0} {B_p}^{\alpha_t}\varphi_p \\ C_M = \frac{1}{4A}\pi^2\lambda\zeta^2 B_p\left(C_{fd} + \frac{\pi A K_0^2\epsilon^2}{e(\pi A+K_0)^2}{\varphi_p}^2\right), \end{cases} \quad (3)$$

where B_p and $\varphi_p(rad)$ are, respectively, the propeller blade number and the pitch angle. They are defined using:

$$\varphi_p = \arctan\left(\frac{H_p}{\pi D_p}\right), \quad (4)$$

A,ϵ,λ,ζ,e,C_fd,et K_0 are the blade parameters, which are directly related to the propeller blade airfoil shape. Their approximate values are presented in Table A1 in the Appendix A. Since the blade airfoil shapes are similar for certain series of propellers, especially those supplied by T-motor, the blade parameters are typically fixed. Figure 7a presents the regression model of the propeller's mass M_{prop} (g), which is based on data supplied by T-motor [20] and Mejzlik [21]. The input of this model is the propeller diameter D_p optimized through the optimization methodology. Equation (5) presents the regression model of the propeller mass:

$$M_{prop} = 0.303 \cdot D_p^2 - 9.729 \cdot D_p + 105.786. \quad (5)$$

3.2. Electric Motor

Electric motors used in eVTOL applications are primarily of two types: brushed DC motors (BDC) and brushless DC motors (BLDC). BLDC motors, known for their low resistance and high efficiency, are commonly used in heavyweight multicopters. They can be further categorized into two types: outrunner (OR) and inrunner (IR), based on the rotating part of the motor. Presently, OR BLDC motors are considered a preferable choice over IR motors. OR motors have lower Kv (rpm/V) values compared to other types of BLDC motors. This means they operate at lower rotational speeds but generate higher torque, which allows for direct propeller coupling (no gearbox) [24]. K_v (rpm/V) is the speed constant, which will determine the rotation speed of the electric motor when no-load and stable voltage is supplied. This is an important element for choosing a motor that is compatible with the power supply and propellers to achieve the required speed. In Figure 6, U_m (V) is the supply voltage, I_m (A) is the current absorbed by the motor coils, R_m (Ω) is the motor equivalent resistance, T_e (Nm) is the electromotive torque produced by the motor, and N (rpm) is the shaft angular velocity. The equations describing the motor electric model are [19,24]:

$$\begin{cases} U_m = e_a + R_m \cdot I_m, \\ T_e = K_T \cdot I_m, \\ E_a = K_E \cdot N \approx \frac{N}{K_v}, \end{cases} \quad (6)$$

where K_E (Vs/rad) represents the motor back EMF constant, K_T (Nm/A) is the motor torque constant, and N is the motor rpm. K_T and K_E are related to K_v by

$$K_E = \frac{1}{K_v} = \frac{\pi}{30} \cdot K_T. \quad (7)$$

The motor output torque and the propeller torque are related by

$$M = T_e - T_0 = K_T \cdot (I_m - I_{m0}). \quad (8)$$

If the no-load current I_{m0} is neglected, the propeller torque in this case is controlled uniquely by the motor load current. From Equations (1) to (4), the supply voltage U_m and the motor current I_m are given as follows:

$$\begin{cases} I_m = \frac{\pi}{30 \cdot K_v} \cdot M + I_{m0}, \\ U_m = I_m \cdot R_m + \frac{N}{K_v}. \end{cases} \tag{9}$$

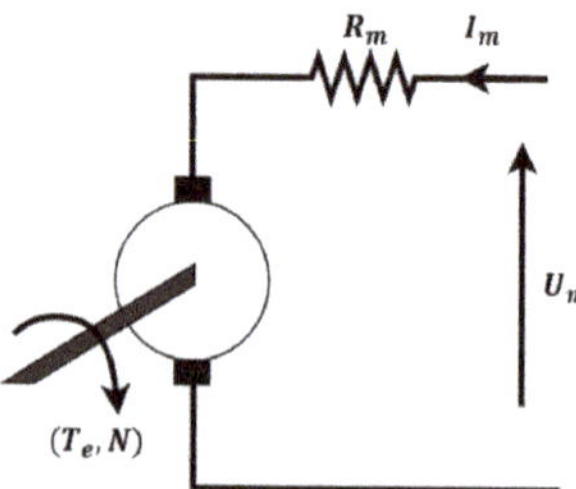

Figure 6. BLDC motor electric model.

The mass regression model of the electric motor is given in Figure 7b. The motor data used in this model were collected from T-motor and KDEDirect [25]. The model input is the motor speed constant K_v, which is defined in the motor parameters database Equation (10) presents the regression model of the motor mass:

$$M_{mot} = 0.00048K_v^2 - 1.461K_v + 840.617. \tag{10}$$

3.3. Electronic Speed Controller

The electronic speed controller (ESC) is an external device responsible for regulating the motor speed within a specific range based on the load and battery voltage. It converts the DC voltage from the battery pack into a three-phase alternating signal that is synchronized with the rotation of the rotor and applied to the armature windings. In the developed sizing methodology, the electric model of the ESC is not directly involved. However, it is essential to fix the maximum continuous current I_{emax} of the ESC, particularly during the selection and mass estimation steps [24,26]. Figure 7c presents the mass regression model of the ESC base on supplier data (collected from T-motor and KDEDirect). Equation (11) presents the regression model of the ESC mass:

$$M_{ESC} = 0.016I_{emax}^2 - 0.638I_{emax} + 42.414. \tag{11}$$

3.4. Energy Storage System

In this part, three energy storage systems are considered, namely a lithium polymer (LiPo) battery, a proton membrane exchange hydrogen fuel cell (PME), and a supercapacitor,

3.4.1. Battery

Due to the high energy density and discharge rate, eVTOL aerial vehicles use lithium polymer (LiPo) batteries. A LiPo pack consists of identical LiPo cells, each with a nominal voltage of 3.7 V and power density of $\rho_b = 140$ (Wh/kg) [24,27]. The parallel connection of battery packs raises the battery's total capacity while keeping the nominal total voltage the same. The nominal voltage of a LiPo battery is:

$$U_b = 3.7n_c, \tag{12}$$

where n_c is the number of cells connected in series in the battery pack. Each cell has a capacity C_{cs}. The total battery capacity is

$$C_b = n_p \cdot C_{cs}, \tag{13}$$

where n_p is the number of battery packs connected in parallel. As we can see from Figure 5, the motor power P_m is converted by the ESC and supplied by the battery. The battery output power P_b can be estimated by

$$P_b = N_m \cdot \frac{P_m}{\eta_e \cdot \eta_b}, \tag{14}$$

where N_m, η_e, and η_b are, respectively, the propulsion chain number, the conversion efficiency of ESC, and the battery efficiency. An oversizing of the battery is taken into account the drop in battery capacity with the discharge time, utilizing the battery efficiency. A value of $\eta_b = 0.75$ is considered suitable for battery sizing, as reported in [24].

The flight time t_{flight} (min) of the eVTOL aerial vehicle, which is equivalent to the battery discharge time, is given by

$$t_{flight} = \frac{60 \cdot \rho_b \cdot m_b}{N_m \cdot P_m} \cdot \eta_e \cdot \eta_b, \tag{15}$$

where ρ_b and m_b are, respectively, the battery power density (Wh/kg) and the battery mass (kg). Thus, for a given embedded LiPo battery mass m_b, and a load, an equivalent flight time is determined. Once the battery mass m_b (kg) is located with an objective to maximize the flight time with the $GTOW$ constraint, the battery capacity C_b (mAh) is computed using the following equation:

$$C_b = \frac{\rho_b \cdot m_b}{U_b}. \tag{16}$$

3.4.2. Hydrogen Fuel Cell

Proton exchange membrane (PEM) fuel cells offer a higher energy density than batteries, around 500 Wh/kg [28,29], in a unit that is still clean and hydrocarbon-free, mechanically simple, operates near ambient temperature, and produces no harmful emissions. In terms of fuel cell power density, there are several works estimating its improvement for a value of 800 W/kg [30]. The problems with hydrogen storage and the boil-off are also less significant in aviation compared to cars, and even lesser even for eVTOL aerial vehicles, because of the shorter duration missions, typically a few hours compared to weeks. Thus, the significant progress made in the past decade toward lighter gaseous hydrogen storage can be exploited to full advantage. A PEM pack consists of identical cells, each with a voltage E_{cell} (V) given by [30,31]:

$$E_{cell} = E_0 + \frac{R \cdot T}{2F} \cdot \ln(P_{H_2} \cdot {P_{O_2}}^{0.5}), \tag{17}$$

where E_0, R, T, F, P_{H_2}, and P_{O_2} are, respectively, the thermodynamic reversible voltage based on the higher heating value (HHV) of hydrogen (1.23 V), the universal gas constant (8.314 J/molK), the operating temperature, the Faraday constant (96,485 C/mol), the partial pressure of hydrogen (Pa), and the partial pressure of oxygen (Pa). The nominal voltage of the PEM hydrogen fuel cell stack V_{stack} (V) is given by

$$V_{stack} = N_{cell} \cdot E_{cell}. \tag{18}$$

The fuel cell area is defined as:

$$A_{cell} = \frac{P_{FC}}{(p_{cell} \cdot N_{cell})}, \tag{19}$$

where P_{FC} and p_{cell} are, respectively, the required electrical power and the power density of a single cell. For the battery case, the fuel cell output power required for the flight mission P_{FC} and the corresponding hydrogen consumption $HC(kg/h)$ can be estimated by

$$\begin{cases} P_{FC} = N_m \cdot \frac{P_m}{\eta_e \cdot \eta_{FC}}, \\ HC = \frac{P_{FC}}{LHV \cdot \eta_{FC}}, \end{cases} \tag{20}$$

where η_{FC} and LHV are, respectively, the fuel cell stack efficiency and the low heating value of hydrogen (33.3 Wh/g). At the current technology level, the efficiency of the PEMFC is approximately 40~50%, and if there is no information about the polarization curve of a single cell, this value can be used for sizing. Thus, the hydrogen fuel cell mass m_{FC} is given by [30]:

$$m_{FC} = \frac{Ncell \cdot k_A \cdot \rho_{cell} \cdot A_{cell}}{1 - \eta_{ow}} \cdot (1 + f_{BOP}), \tag{21}$$

where k_A, ρ_{cell}, f_{BOP}, and η_{ow} are, respectively, the ratio of the cross-sectional area to the electrode area of a single cell (fixed at a value of five), the area density of a single cell (fixed at 1.57 kg/m), the ratio of the BOP weight to the HFC weight (with a value that varies depending on the HFC configuration; in this paper, a value of 0.2 is considered), and the overhead fraction to account for gaskets, seals, connectors, and endplates (fixed at 0.3). The flight time in the case of a fuel cell is given by the following expression:

$$t_{flight} = \frac{60 \cdot LHV}{P_{FC}}. \tag{22}$$

For the hydrogen tank, a type 4 tank was selected among the gaseous hydrogen tanks. Liquid hydrogen is 800 times less in volume and has a higher energy density than gaseous hydrogen, but it must be kept at a low temperature, which limits its use in HFC UAVs [31,32]. A regression model that estimates the hydrogen tank mass M_{tank} (kg), based on the amount of hydrogen m_{H_2} (g) required for the flight mission, is established as shown in Figure 7d. The data are based on the tank type, e.g., such as types 3 or 4 [33], and this model can be expressed as:

$$M_{tank} = -0.000047 {m_{H_2}}^2 + 0.0367 m_{H_2} - 0.126. \tag{23}$$

3.4.3. Supercapacitor

Supercapacitors can produce much higher specific powers (multiple kW/kg) but have lower specific energy capacities (currently only a few Wh/kg) than batteries and HFC. That is, SCs cannot be used alone or in combination with an HFC. For this component, unlike the batteries, the power limitations are less constraining, since the limitations are much higher than what the load requires. The limitations are mainly related to energy. In addition, they indirectly protect the fuel cell, batteries, and DC bus. Indeed, they absorb the DC bus voltage fluctuations and can extend the battery's lifetime [34,35]. Maxwell 350F/2.7V supercapacitor technology was considered in this paper [36]. The cell characteristics are presented in Table A2. The useful energy E_{SC} available in a pack of NS_{SC} elements in series and NP_{SC} branches is calculated as follows:

$$E_{SC} = \frac{3}{8} \cdot \frac{NP_{SC}}{NS_{SC}} \cdot C_{cell} \cdot (U_{SC})^2, \tag{24}$$

where C_{cell} and U_{SC} are, respectively, the nominal capacity and the maximum voltage of a supercapacitor element. The flight time t_{flight} (min) of the eVTOL aerial vehicle (supercapacitor discharging) is given by

$$t_{flight} = \frac{60 \cdot \rho_{SC} \cdot m_{SC}}{P_{SC}}, \tag{25}$$

where ρ_{SC}, m_{SC}, and P_{SC} are, respectively, the energy density, and the mass and the power of a supercapacitor element. The mass of this component is directly estimated using data from Table A2.

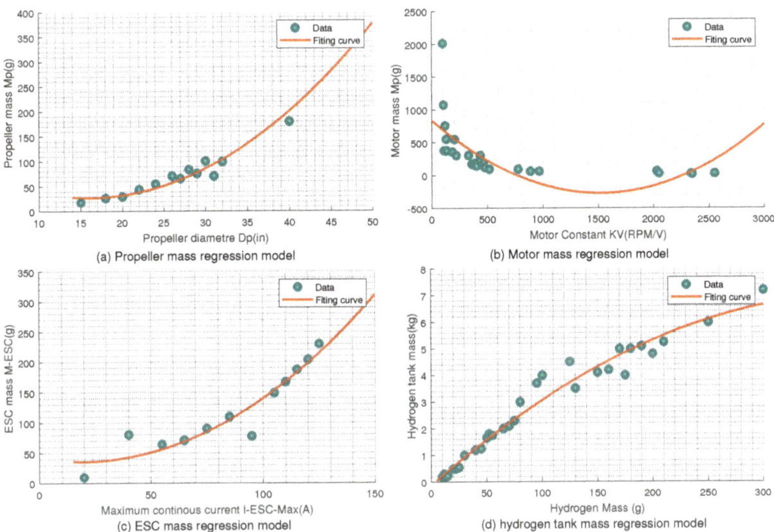

Figure 7. Regression model for each component.

4. Motor/Propeller Optimization Problem

The optimization technique is based on the simulated annealing algorithm SAA, which was introduced by inspiring the annealing procedure of the metalworking. In a general manner, the SA algorithm adopts an iterative movement according to the variable temperature parameter, which imitates the annealing transaction of the metals [37]. This algorithm is directly explored using the MATLAB global optimization toolbox. The efficiency of the pair motor/propeller η_{MP} is used as an objective function in this case. It is given by

$$\eta_{MP} = \frac{T_{flight}}{P_{mflight}}, \tag{26}$$

where T_{flight} and $P_{mflight}$ are, respectively, the propeller thrust and the motor power during the flight operation. The motor power is given by

$$P_{mflight} = U_{mflight} \cdot I_{mflight}, \tag{27}$$

from the propeller model presented in Equation (1), the propeller velocity and the propeller torque during the flight are given by

$$\begin{cases} N_{flight} = \frac{60}{D_p{}^2} \cdot \sqrt{\frac{T_{flight}}{\rho \cdot C_T}}, \\ M_{flight} = \frac{C_M \cdot D_p}{C_T} \cdot T_{flight}, \end{cases} \tag{28}$$

and from the motor model presented in Equation (9), the motor current and the motor voltage are given by

$$\begin{cases} I_{m0} \approx 0, \\ I_{mflight} = \frac{\pi \cdot C_M \cdot D_p}{30 \cdot C_T \cdot K_E} \cdot T_{flight}, \\ U_{mflight} = \frac{\pi \cdot C_M \cdot D_p}{30 \cdot C_T \cdot K_E} \cdot R_m \cdot T_{flight} + \frac{60 \cdot K_E}{D_p{}^2} \cdot \sqrt{\frac{T_{flight}}{\rho \cdot C_T}}. \end{cases} \tag{29}$$

Thus, the optimization objective function expression is given by

$$\eta_{MP} = \frac{1}{\left(\frac{\pi \cdot C_M \cdot D_p}{30 \cdot C_T \cdot K_E}\right)^2 \cdot T_{flight} \cdot R_m + \frac{2\pi \cdot C_M}{D_p} \cdot \sqrt{\frac{T_{flight}}{\rho \cdot C_T{}^3}}}. \tag{30}$$

For a fixed thrust imposed by the *GTOW*, the motor/propeller efficiency evolution in terms of propeller parameters is given in Figure 8. Through this figure, it is noticeable that the motor/propeller efficiency presents a single attraction basin, enabling the rapid identification of the point that maximizes the objective function.

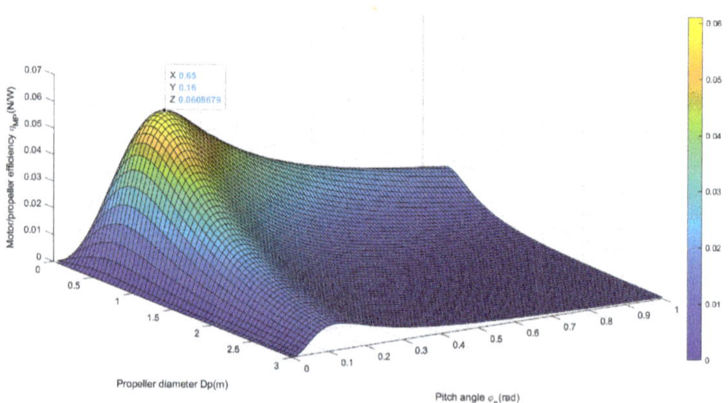

Figure 8. Motor/propeller efficiency evolution in terms of propeller parameters,

In order to avoid the motor overheating during the flight, which could influence the propulsion chain efficiency, the motor current and voltage must remain below their maximum values, U_{mMax} and I_{mMax}, imposed by the motor design:

$$U_m \leq U_{mMax} \;\&\; I_m \leq I_{mMax}, \tag{31}$$

which leads to establishing the following constraint on the propeller velocity and torque:

$$\begin{cases} N_{max} = \frac{U_{mMax} - R_{mMax} \cdot I_{mMax}}{K_E}, \\ M_{max} = \frac{30 \cdot (I_{mMax} - I_{m0}) \cdot K_E}{\pi}, \end{cases} \tag{32}$$

thus, the propeller diameter must remain below its maximum value D_{Pmax} imposed by the motor overheating avoidance condition:

$$D_p \leq D_{Pmax} = \left(M_{max}{}^4 \cdot \left(\frac{60}{N_{max}}\right)^2 \cdot \frac{1}{C_M \cdot \rho}\right)^{\frac{1}{5}}. \tag{33}$$

Thus, the optimization problem of the motor/propeller is given as follows:

$$\begin{cases} \max(\eta_{MP}) = \min(-\eta_{MP}), \\ 0 < D_p \leq D_{pMaxElec}, \\ 0 < \varphi_p \leq \pi. \end{cases} \tag{34}$$

5. Case Study

5.1. Use Case Multirotor Drone

The validation of the proposed sizing approach is conducted using data from a reduced-scale multirotor drone with a GTOW of 15 kg. An overview of the drone, composed of eight propulsion chains with U7-V2 420 KV motors and P18×6.1 propellers, is presented in Figure 9. Each set of four propulsion chains is powered by an 6S1P LiPo battery. Detailed specifications of the drone can be found in Table A3 in the Appendix A. It is worth noting that the sizing methodology does not consider the coupling effect of the coaxial configuration. This effect is disregarded for the drone used in the validation step of the sizing methodology. According to [38], at the scale level of the validation drone, the coupling effect of the coaxial configuration on propulsion efficiency does not exceed 6%. This demonstrates the effectiveness of the proposed sizing methodology. To demonstrate the efficacy of the sizing methodology, a simulation of the propulsion chain sizing approach is carried out using MATLAB code. A database comprising parameters of 45 randomly selected electric motor examples, including the motor used in the drone shown in Figure 9, is created based on data provided by T-motor. Flight mission data specific to this drone is also incorporated.

Figure 9. multirotor eVTOL drone.

5.2. Pair Motor/Propeller Optimization

The optimization method is applied to the database used for validation. For each example, the optimal propeller parameters allowing the motor/propeller efficiency maximization are located using the SA algorithm. Figure 10 gives an example of an optimized pair motor/propeller, in which the motor/propeller efficiency with the corresponding propeller parameters is presented. However, a filtering condition is required in order to select the appropriate combinations that satisfy constraints, such as the GTOW, imposed by the drone. This condition is based on the computing of the relative error ε_r between the maximum thrust $T_{MP_{max}}(N)$ generated by the optimized pair motor/propeller and the thrust imposed by the drone weight (GTOW) $T_{Max}(N)$. The maximum thrust generated by the optimized pair motor/propeller is deduced from the constraints imposed on the propeller velocity and output torque given in Equation (32):

$$T_{MP_{max}} = \left(M_{max}{}^{4} \cdot \frac{C_T{}^{5}}{C_M{}^{4}} \cdot \left(\frac{N_{max}}{60} \right)^{2} \right)^{\frac{1}{5}}. \quad (35)$$

The thrust imposed by the drone weight $GTOW$ is deduced from the acceleration a_c required during the flight by

$$T_{Max} = \frac{GTOW \cdot (g + a_c)}{N_p}, \quad (36)$$

where g and N_p are, respectively, the gravity acceleration and propulsion chain number. Thus, the relative error is given by

$$\varepsilon_r = \left| \frac{T_{MP_{max}} - T_{Max}}{T_{max}} \right|. \quad (37)$$

An error reference value $\varepsilon_{ref} = 5\%$ is fixed as the threshold value in order to make the filtering process.

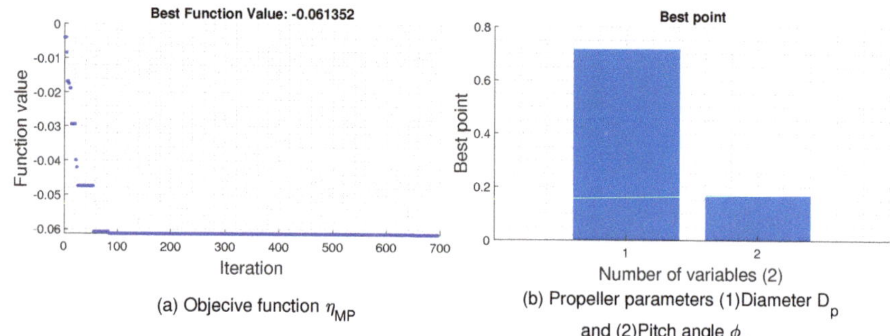

Figure 10. Motor/propeller optimization example with the objective function η_{MP} and propeller parameters D_p, φ_p.

The motor/propeller combinations, which are able to generate the thrust imposed by the specifications, are given in Table 1:

Table 1. Sizing process methodology outcome.

Combination Number	Motor Specification	Propeller Parameters	Thrust Efficiency
1	U7-V2.0 KV420	P18×6.0	0.0528
2	U7-V2.0 KV490	P20×6.7	0.0533

It is remarkable that the two combinations, in terms of the specific efficiency η_{MP}, remain equivalent, which makes the choice between them very similar. Combination 1 is used in the multirotor drone considered for validation.

6. Energy Storage Sizing and Flight Time Comparison

The sizing of the energy storage system is focused on maximizing the flight time while minimizing the GTOW. This paper considers five different energy storage configurations. The first two configurations utilize either a battery or an HFC as the primary energy source. The remaining three configurations are hybrid setups, including combinations of Bat/SC, Bat/HFC, or Bat/SC/HFC. By exploring various energy storage structures, the impact on the autonomy of the multirotor aerial vehicle can be assessed. Figure 11 provides a general configuration of the propulsion chain based on the Bat/SC/HFC hybrid setup. In the hybrid cases, it is possible to consider, during the cruise phase, the recharging of the battery, by the energy surplus of the HFC in the Bat/HFC or Bat/SC/HFC configurations, or the supercapacitor, by the energy surplus of the battery in the Bat/SC configuration; or by the HFC in the Bat/HFC/SC configuration. In both cases, oversizing of the battery or HFC is required.

After the optimal pair motor/propeller has been selected by the optimization algorithm, the sizing of the ESC part will be conditioned by the maximum current imposed by the motor. The fuselage sizing part is not considered in this paper, it is assumed to be ready. The sizing of the energy storage system makes it possible to maximize the flight time of the drone while keeping a minimum mass. For this, it is assumed that:

$$m_{copter} = m_{StorageEnergy} + m_{others}. \tag{38}$$

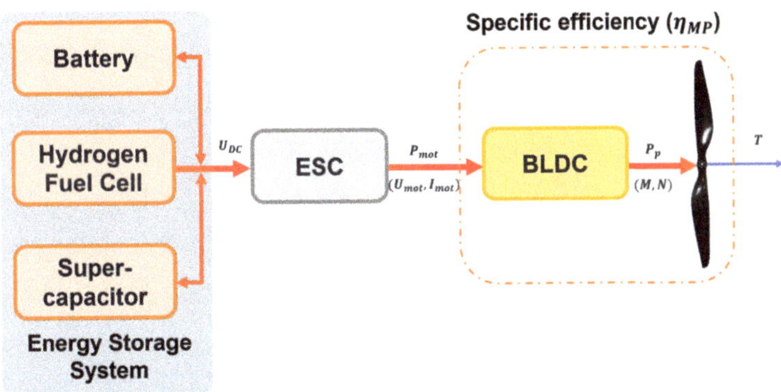

Figure 11. Multirotor propulsion chain based on a hybrid energy storage system, Bat/SC/HFC.

6.1. Sizing of the Battery and the Hydrogen Fuel Cell

When using a simple energy storage system, the sizing of the latter is performed in a way that the flight time is maximized by respecting the constraint of the $GTOW$.

6.1.1. Battery Sizing

The LiPo battery and the motor power are related by

$$P_{mot} = P_{bat} \cdot \eta_e \cdot \eta_b, \tag{39}$$

From the discharging time given in Equation (15), the multirotor aerial vehicle flight time is related to the motor/propeller-specific efficiency by

$$t_{flight} = \frac{\eta_e \cdot \eta_b \cdot \rho_b \cdot m_{bat}}{(m_{bat} + m_{others}) \cdot g} \cdot \eta_{MP}. \tag{40}$$

6.1.2. HFC Sizing

The hydrogen consumption during the flight and the motor power are related by

$$P_{mot} = \eta_e \cdot \eta_{FC} \cdot LHV \cdot HC, \tag{41}$$

From the discharging time given in Equation (21) in the case of an HFC, the multirotor aerial vehicle flight time is related to the motor/propeller-specific efficiency by

$$t_{flight} = \frac{\eta_e \cdot \eta_{FC} \cdot LHV \cdot m_{H_2}}{(m_{STACK} + m_{others}) \cdot g} \cdot \eta_{MP}. \tag{42}$$

where m_{STACK} is the stack fuel cell mass, which is given by $m_{STACK} = m_{FC} + m_{H_2} + m_{tank}$. The evolution of the flight time in terms of the battery mass or hydrogen mass is given in Figure 12a,b. For both cases, it is observable that the flight time increases at first, and then decreases as the battery mass or the hydrogen mass increases from 0 to ∞. The decrease in the flight time is caused by the decrease in the motor/propeller efficiency when the drone weight is too heavy. Usually, the flight time maximum is not reached, because the energy storage system mass is limited by the $GTOW$. Thus, the optimum weight of the battery must be sought in the permitted region given in Figure 12a,b. The optimized parameters of the battery and the hydrogen fuel cell are presented in Table A4.

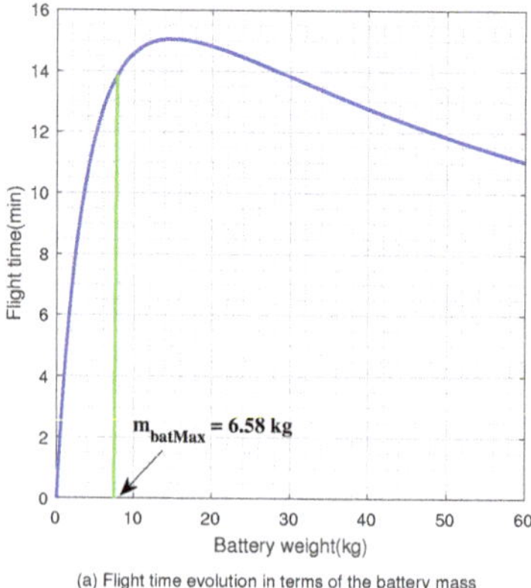

(a) Flight time evolution in terms of the battery mass

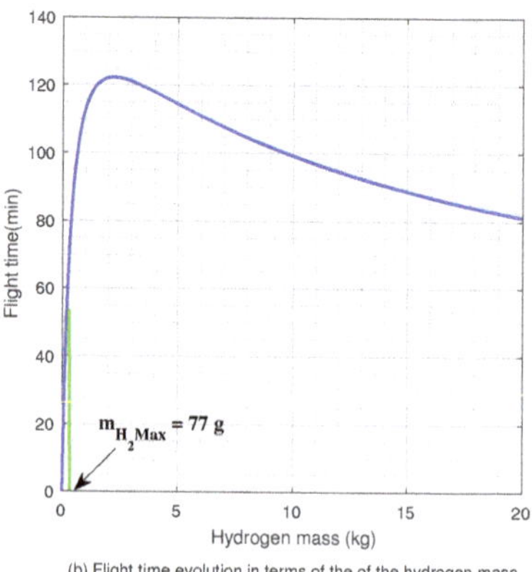

(b) Flight time evolution in terms of the of the hydrogen mass

Figure 12. Flight time evolution.

6.2. Sizing of the Energy Storage System in the Hybrid Cases

The hybridization of energy storage systems enables the enhancement of autonomy and reliability in multirotor aerial vehicles. By combining batteries, hydrogen fuel cells, and supercapacitors, the specific energy to specific power ratio of the energy storage system is significantly improved compared to the case of a single energy source. This, coupled with in-flight energy management algorithms, extends the flight time of the aerial vehicle [29,39]. Figure 13 illustrates the Ragone plot, which depicts the distribution of different energy storage systems based on their specific energy to specific power ratios [40]. From this figure, it is evident that the combinations of Bat/HFC, Bat/SC, or Bat/SC/HFC allow for an improvement in specific energy while maintaining an adequate level of specific power. In this case, the energy storage sizing process remains similar to the single-source case, where the objective is to maximize the flight time while considering the GTOW constraint. However, there are additional variables to consider, particularly the hybridization coefficient of the energy sources.

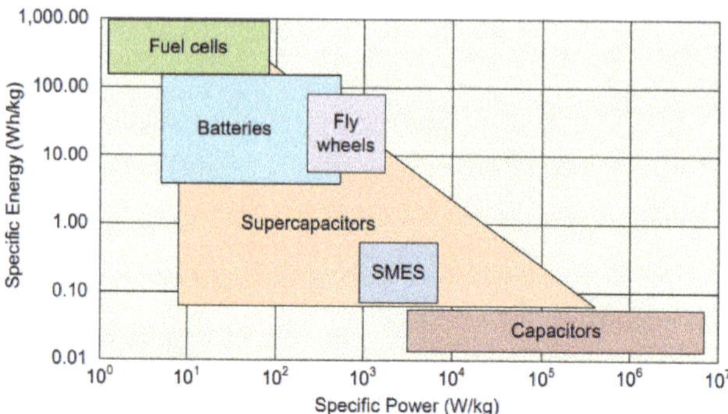

Figure 13. The Ragone chart.

6.2.1. Bat/HFC Sizing

Depending on the flight mission segment, the motor power (load) is supplied by the hydrogen fuel cell or the battery. In both cases, the battery and the hydrogen fuel cell power are related to the motor power by

$$\begin{cases} P_{bat} = x \cdot P_{mot}, \\ P_{HFC} = (1 - x) \cdot P_{mot}, \end{cases} \tag{43}$$

where x is the hybridization coefficient of the battery power to the motor power. The global flight time of the aerial vehicle t_{flight} is obtained by the contribution of the battery $t_{flightBat}$ and the hydrogen fuel cell $t_{flightHFC}$. It is given by

$$t_{flight} = t_{flightBat} + t_{flightHFC}. \tag{44}$$

From Equations (15) and (21), the flight time is given by

$$t_{flight} = \left(\frac{m_{bat} \cdot \rho_b \cdot \eta_e \cdot \eta_b}{x} + \frac{m_{H_2} \cdot LHV \cdot \eta_e \cdot \eta_{FC}}{1-x} \right) \cdot \left(\frac{\eta_{MP}}{(m_{STACK} + m_{bat} + m_{others}) \cdot g} \right). \tag{45}$$

From this equation, it is evident that the flight of the multirotor aerial vehicle is influenced by the mass of hydrogen m_{H_2}, the mass of the battery m_{bat}, and the hybridization coefficient x. The choice of this coefficient depends on the duration of the flight mission segments during which the maximum power is required. These segments typically represent less than 14% [41,42] of the total flight duration. The flight time variation with respect to the hydrogen mass, HFC stack mass, and battery mass is illustrated in Figure 14.

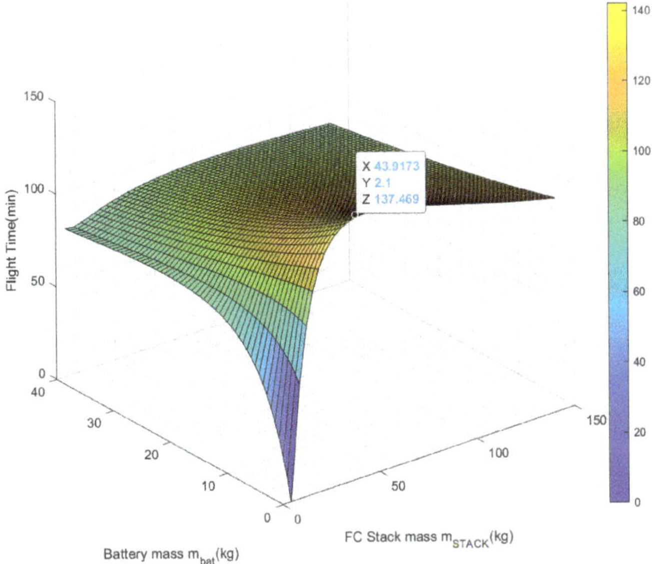

Figure 14. Flight time evolution in terms of the HFC stack mass m_{STACK} and the battery mass m_{bat}.

It is remarkable that maximizing the flight time is more favored by increasing the hydrogen mass than increasing the battery mass. This is due to the fact that as the hydrogen mass increases, the specific energy of the system also increases, resulting in an extended flight time.

It is also noticeable that the flight time evolution—as a function of hydrogen mass and the battery mass for the fixed hybridization coefficient x—presents a single basin of attraction. The maximum in this case is not attainable because the mass of the energy sources is limited by the constraint of the $GTOW$.

In order to locate the energy sources' optimal masses, which allow for maximizing the flight time with the constraint of the $GTOW$ for different values of the hybridization coefficient x, a nonlinear global optimization was carried out. The algorithm considered in this part is the same one that was used in the motor/propeller pair optimization part. The optimization problem in this case is given in

$$\begin{cases} max(t_{flight}) = min(-t_{flight}), \\ 0 < m_{bat} + m_{STACK} \leq GTOW - m_{other}. \end{cases} \quad (46)$$

Figure 15 gives an example of this optimization for a hybridization coefficient of $x = 14\%$. In this configuration, the obtained battery mass makes it possible to compute the battery capacity using the equation reported in (16). Based on the required capacity, the number of battery-parallel branches is computed. Regarding the sizing of the HFC in this case, there is the tank sizing, which is defined by hydrogen mass obtained by the optimization part, using the regression model presented in Equation (23). The stack sizing, or the fuel cell area sizing, depends on the hybridization coefficient x, by using the following equation:

$$\begin{cases} P_{FC} = (1-x) \cdot N_m \cdot \frac{P_m}{\eta_e \cdot \eta_{FC}}. \\ A_{cell} = \frac{P_{FC}}{\rho_{cell} \cdot N_{cell}} \end{cases} \quad (47)$$

Thus, the stack mass is deduced using Equation (22).

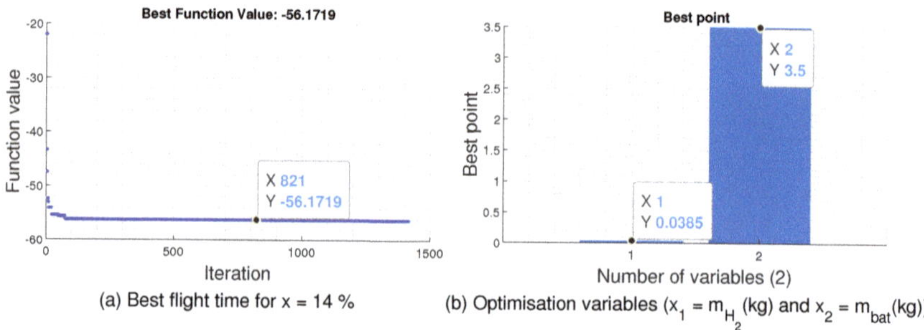

Figure 15. Example of a flight time optimization in the Bat/HFC configuration case.

Through this optimization, a flight time of $t_{flight} = 56.18$ min is obtained. The optimized parameters of the battery and the HFC are reported in Table A5.

6.2.2. Bat/SC Sizing

The sizing process of this source configuration is similar to the previous one. The supercapacitor feeds the motor in high-power segments, especially during take-off and landing. During the cruising segment, there is the possibility of charging the supercapacitor with the excess energy supplied by the battery. The power of the supercapacitor and the battery is related to the power of the motor by

$$\begin{cases} P_{SC} = x \cdot P_{mot}, \\ P_{bat} = (1-x) \cdot P_{mot}, \end{cases} \quad (48)$$

where x is the coefficient of the supercapacitor power to the motor power. The global flight time of the aerial vehicular t_{flight} is obtained by the contribution of the supercapacitor $t_{flightSC}$ and the battery $t_{flightBat}$. It is given by

$$t_{flight} = t_{flightSC} + t_{flightBat}. \quad (49)$$

From Equations (15) and (23), the flight time is given by

$$t_{flight} = \left(\frac{m_{SC} \cdot \rho_{SC} \cdot \eta_e \cdot \eta_S C}{x} + \frac{m_{bat} \cdot \rho_b \cdot \eta_e \cdot \eta_b}{1 - x} \right) \cdot \left(\frac{\eta_{MP}}{(m_{SC} + m_{bat} + m_{others}) \cdot g} \right). \quad (50)$$

The flight time evolution in terms of the battery mass and the supercapacitor mass is given by Figure 16. This evolution is obtained for a hybridization coefficient of $x = 5\%$.

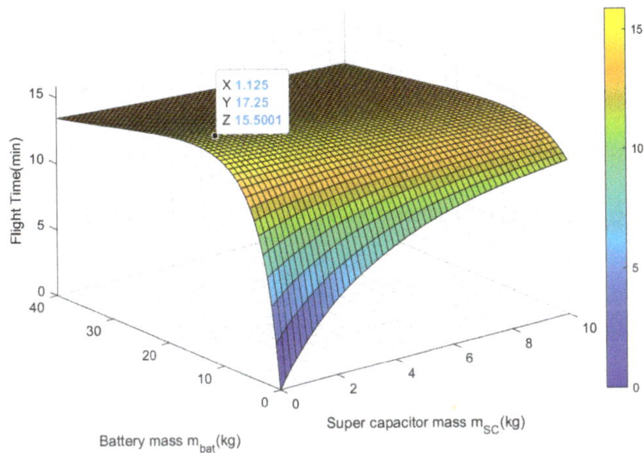

Figure 16. Flight time evolution in terms of the supercapacitor mass m_{SC} and the battery mass m_{bat}.

In the case of the Bat/SC configuration, maximizing flight time is more influenced by increasing the battery mass rather than increasing the supercapacitor mass. This is because the supercapacitor has a lower specific energy compared to the battery. As a result, when the battery mass increases, the overall autonomy of the energy storage system improves. It is also remarkable that the flight time evolution as a function of the battery mass and the supercapacitor mass, for a fixed hybridization coefficient x, presents a single basin of attraction. The maximum in this case should be reached in the permitted region imposed by the $GTOW$. The energy storage optimal masses and the hybridization coefficient, allowing the maximization of the multirotor aerial vehicle flight time, are located using a global non-linear optimization. The optimization problem in this case is given by

$$\begin{cases} max(t_{flight}) = min(-t_{flight}), \\ 0 < m_{bat} + m_{SC} \leq GTOW - m_{other}. \end{cases} \quad (51)$$

Figure 17 gives an example of this optimization for a hybridization coefficient of $x = 5\%$. Through this optimization, a flight time of $t_{flight} = 14.27$ min is obtained. The sizing of the battery remains similar to the previous case. The optimized supercapacitor mass allows obtaining the required energy based on the Maxwell cell energy density reported in Table A2 in the Appendix A. Thus, the SC capacity must satisfy the following condition:

$$C_{SC} \geq \frac{16}{3} \frac{E_{SC}}{U_{SC}}, \quad (52)$$

The optimized parameters of the battery and the SC are reported in Table A6.

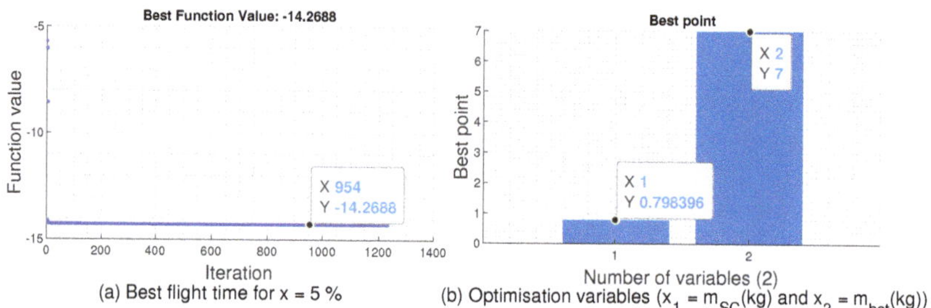

Figure 17. Example of a flight time optimization in Bat/SC configuration case.

6.2.3. Bat/HFC/SC Sizing

The Bat/SC/HFC hybrid configuration allows for the integration of three energy sources, each utilized in different flight mission segments. During the take-off and landing segments, where the power demand is highest, the supercapacitor is employed. The battery is utilized during hovering segments, while the hydrogen fuel cell is utilized during the cruise segment. The battery power, the supercapacitor power, and the hydrogen fuel cell power are related to the motor power by

$$\begin{cases} P_{bat} = x \cdot P_{mot}, \\ P_{SC} = y \cdot P_{mot}, \\ P_{HFC} = (1 - x - y) \cdot P_{mot}. \end{cases} \tag{53}$$

where x and y are, respectively, the hybridization coefficient of the battery power to the motor power and the supercapacitor power to the motor power. From Equations (15), (23) and (25), the flight time is given by

$$t_{flight} = \left(\frac{m_{SC} \cdot \rho_{SC} \cdot \eta_e \cdot \eta_{SC}}{x} + \frac{m_{bat} \cdot \rho_b \cdot \eta_e \cdot \eta_b}{y} + \frac{m_{H_2} \cdot LHV \cdot \eta_e \cdot \eta_{FC}}{1 - x - y} \right) \cdot \left(\frac{\eta_{MP}}{(m_{SC} + m_{bat} + m_{HFC} + m_{others}) \cdot g} \right). \tag{54}$$

In this case, the flight time depends on five parameters, namely supercapacitor mass m_{SC}, battery mass m_{bat}, hydrogen fuel cell mass m_{HFC}, and hybridization coefficients x and y. Figure 18a–c presents the flight time evolution for the three cases.

As the hydrogen fuel cell (HFC) mass increases (Figure 18a), the flight time shows a tendency to increase when considering the battery and supercapacitor masses. This can be attributed to the improved energy density of the energy storage system resulting from the increased hydrogen mass. Furthermore, the cruise phase typically constitutes the longest segment in a flight mission.

Regarding the effect of the SC mass on the evolution of the flight time in terms of the HFC and battery masses, as seen in Figure 18b, it is remarkable that the flight time has a tendency to decrease. This can be explained by the lower value of the SC energy density in comparison to FCs and batteries. In addition, the take-off and landing segments, where the SC is utilized, have a relatively short duration in the overall flight mission.

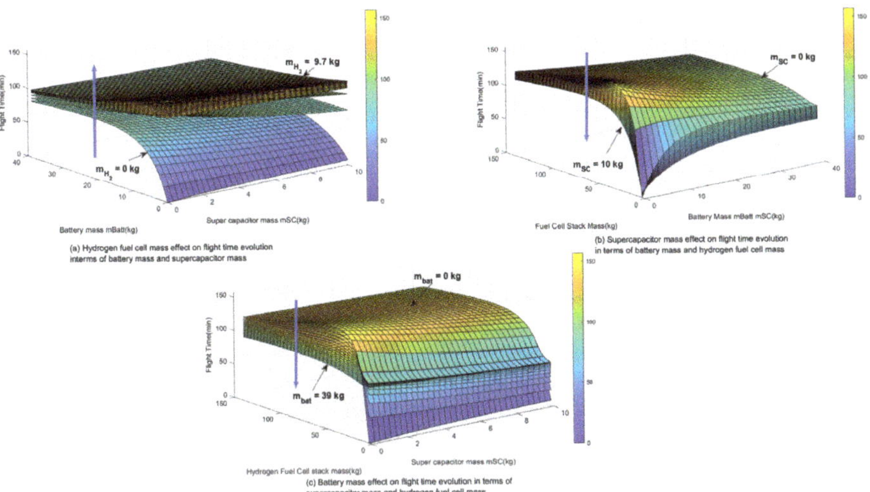

Figure 18. Flight time evolution in terms of the battery mass m_{batt}, the supercapacitor mass m_{SC}, and the hydrogen fuel cell mass m_{HFC}.

The effect of the battery mass on the evolution of the flight time, as a function of the mass of the hydrogen fuel cell and the mass of the supercapacitor, as seen in Figure 18c, remains similar to the case in Figure 18b. The flight time in this case tends to decrease as the battery mass increases. This can be attributed to the battery's low energy density and the relatively short duration of the hovering segment during which the battery is used.

In this case, the optimization process involves finding an optimal solution in terms of the three masses, with the objective of maximizing the flight time for a given level of hybridization. The optimization problem is given by

$$\begin{cases} max(t_{flight}) = min(-t_{flight}), \\ 0 < m_{bat} + m_{SC} + m_{HFC} \leq GTOW - m_{other}. \end{cases} \quad (55)$$

Figure 19 presents an optimization example for hybridization coefficients $x = 10\%$, and $y = 6\%$. In this case a, the multirotor aerial vehicle achieved a flight time of t_{flight} = 62.93 min. The sizing process for each component, in this case, follows a similar approach as in the previous cases. The optimized parameters for each component are presented in Table A7.

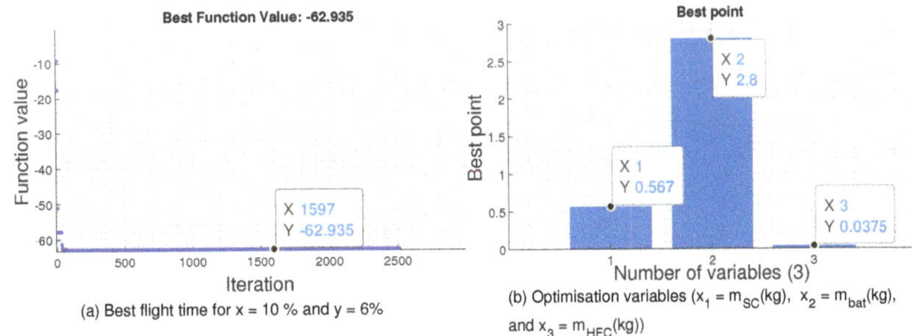

Figure 19. Example of a flight time optimization in the Bat/SC/HFC configuration case.

6.2.4. Flight Time Comparison

The flight times obtained for each energy storage system configuration in the multirotor aerial vehicle are shown in Figure 20. It is remarkable that the energy storage system configuration based on Bat/SC/HFC achieved the best flight time with a value of more than $t_{flight} = 62$ min, followed by the Bat/HFC configuration with a flight time of more than $t_{flight} = 56$ min. Both the battery-based and Bat/SC configurations achieved similar flight times on the order of $t_{flight} = 14$ min. The supercapacitor in this configuration does not have a significant influence on the flight time due to the shorter duration of the segments in which it is used. The HFC-based configuration allowed for a flight time of $t_{flight} = 30$ min. Despite the increase in the complexity of control and energy management in the Bat/SC and Bat/SC/HFC configurations, they remain the best solution for maximizing flight time.

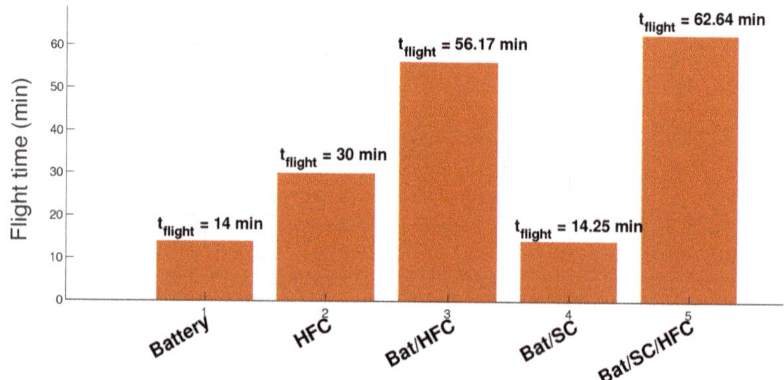

Figure 20. Flight time comparison for each energy source configuration.

6.2.5. Multirotor Aerial Vehicle *GTOW* Estimation

In this section, the multirotor aerial vehicle *GTOW* estimation is described. Regression models presented in Section 3 are utilized to estimate the masses of the propeller, motor, and ESC components. The masses of the payload and fuselage, on the other hand, are fixed. The mass of the energy storage part is computed for each configuration:

- Battery mass: The optimal battery mass is directly defined by the sizing methodology.
- HFC mass: HFC is composed of the fuel cell stack, where its mass is conditioned by the hybridization coefficient as given in Equations (21) and (47), the hydrogen tank, defined using the regression model presented in Equation (23), and the hydrogen mass, which is defined by the sizing methodology.
- SC mass: The minimum number of series cells required to achieve an output voltage of 22.2 V for the supercapacitor is 9. This corresponds to a minimum mass of 567 g. In the hybrid configurations (Bat/SC or Bat/SC/HFC), the mass of the supercapacitor is determined through optimization to maximize the flight time while adhering to the GTOW constraint. It is remarkable that the SC mass in the hybrid configurations, either in Bat/SC or in Bat/SC/HFC, is realizable as the minimum mass of the SC is well respected.

Figure 21 presents the distribution of *GTOW* for each energy storage configuration. The optimized gross take-of weight is given by GTOW = 14.9747 kg.

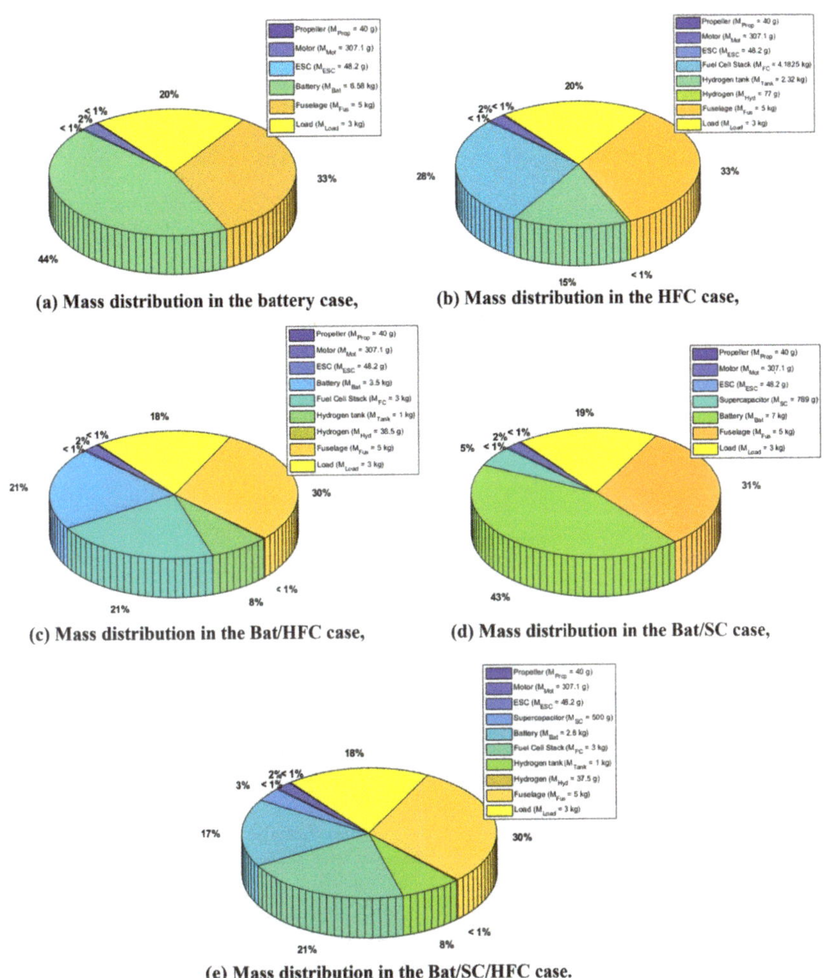

Figure 21. Multirotor aerial vehicle mass distribution.

7. Conclusions

This paper presents a rapid and robust sizing methodology, along with a comparative study on the impact of energy source configurations, on the autonomy of a multirotor aerial vehicle. The methodology focuses on selecting the optimal components for the propulsion chain of an eVTOL multirotor aerial vehicle based on a specific flight mission. The objective is to locate the optimal parameters for the propeller and motor pair, aiming to maximize the specific efficiency $\eta_{MP}(N/w)$ of the propulsion chain. To achieve this, a global nonlinear optimization using the simulated annealing algorithm (SAA) is employed, with constraints placed on the propeller's diameter and pitch angle.

The optimized parameters of the propeller/motor pair allow for the sizing of the ESC and the energy storage system, ensuring that they meet the requirements of the drone's overall mission. This methodology enables the estimation of the resulting gross take-off weight $GTOW$ of the propulsion chain, using mass regression models based on supplier data. The comparative study of different energy storage source configurations highlighted the potential of hybrid sources such as Bat/HFC or Bat/SC/HFC, in terms of autonomy and reliability, through the combination of multiple energy sources. As part of future work, some perspectives will be considered. On the one hand, a full-scale model of a multi-rotor aerial vehicle will be considered, in order to take into account the aerodynamic effect of the structure on the specific efficiency. In this case, the sizing methodology validation step will be carried out on a full scale. On the other hand, the energy management part for each energy source configuration will be considered, in order to have a more global comparison in terms of the flight time, controllability, and implementation complexity of each configuration.

Author Contributions: Conceptualization, S.C.; methodology, S.C.; software, S.C.; formal analysis, S.C.; resources, S.C., R.S., C.M., G.K. and A.A.; writing—original draft preparation, S.C.; writing—review and editing, S.C., R.S., C.M., G.K. and A.A. All authors have read and agreed to the published version of the manuscript.

Funding: This research received no external funding.

Data Availability Statement: The data are available upon request to the corresponding author, Saad Chahba: saad.chahba@estaca.fr.

Conflicts of Interest: The authors declare no conflict of interest.

Appendix A

Table A1. Carbon fiber blade parameters.

Parameter	Value
A	5
ε	0.85
λ	0.75
ζ	0.5
e	0.83
C_{fd}	0.015
α_t	0.9

Table A2. Supercapacitor supplier data.

Component	Parameters
Nominal voltage (V)	2.7
Maximum charge/discharge current (A)	840/840
Internet resistance (Ω)	0.0032
Nominal capacity (Ah)	0.2625
Specific energy (Wh/kg)	5.62
Mass (kg)	0.063
Volume (L)	13.5

Table A3. Multirotor drone specifications.

Component	Parameters
Propeller	$D_p = 18$ in; $H_p = 6$; $\varphi_p = 1.336$ rad
Motor	$KV = 420; U_m = 22.2V; I_m = 35$ A; $R_m = 0.071\ \Omega$
ESC	$I_{ESCMax} = 35$ A
Battery	$U_b = 22.2$ V; $C_b = 12$ Ah V; *C-rate*30C
Fuselage	$M_F = 5$ kg
Payload	$M_L = 3$ kg

Table A4. Battery and HFC parameters in the simple case.

Component	Parameters
Battery	$N_p = 1; N_s = 6; U_b = 22.2$ V; $C_b = 12$ Ah; *C-rate* 30C
HFC	$N_{cell} = 32; A_{cell} = 0.0486$ m^2; $V_{FC} = 22.2$ V

Table A5. Bat/HFC parameters.

Component	Parameters
Battery	$N_p = 1; N_s = 6; U_b = 22.2$ V; $C_{bth} = 6$ Ah; *C-rate* 30C
HFC	$N_{cell} = 32; A_{cell} = 0.0418$ m^2; $V_{FC} = 22.2$ V

Table A6. Bat/SC parameters.

Component	Parameters
Battery	$N_p = 1; N_s = 6; U_b = 22.2$ V; $C_{bth} = 12$ Ah; *C-rate* 30C
SC	$NP_{SC} = 1; NS_{SC} = 13; U_{SCref} = 22.2$ V; $C_{SC} = 3.42$ Ah

Table A7. Bat/SC/HFC parameters.

Component	Parameters
Battery	$N_p = 1; N_s = 6; U_b = 22.2$ V; $C_{bth} = 5$ Ah; *C-rate* 30C
SC	$NP_{SC} = 1; NS_{SC} = 9; U_{SCref} = 22.2$ V; $C_{SC} = 2.36$ Ah
HFC	$N_{cell} = 32; A_{cell} = 0.0408$ m^2; $V_{FC} = 22.2$ V

References

1. TomTom Traffic Index. Available online: https://www.tomtom.com/en-gb/traffic-index/ranking/ (accessed on 18 April 2023).
2. Airbus City Project. Available online: https://www.airbus.com/en/innovation/zero-emission/urban-air-mobility/cityairbus-nextgen (accessed on 18 April 2023).
3. Boeing Passenger Air Vehicle. Available online: https://www.boeing.com/features/frontiers/2019/autonomous-flying-vehicles/index.page (accessed on 18 April 2023).
4. Lilium Jet. Available online: https://lilium.com/jet (accessed on 18 April 2023).
5. Volocopter UAM. Available online: https://www.volocopter.com/urban-air-mobility/ (accessed on 18 April 2023).
6. Nathen, P.; Strohmayer, A.; Miller, R.; Grimshaw, S.; Taylor, J. Architectural Performance Assessment of An Electric Vertical Take-Off and Landing (eVTOL) Aircraft Based on a Ducted Vectored Thrust Concept. 2021. Available online: https://lilium.com/files/redaktion/refresh_feb2021/investors/Lilium_7-Seater_Paper.pdf (accessed on 18 April 2023).
7. Doo, J.T.; Pavel, M.D.; Didey, A.; Hange, C.; Diller, N.P.; Tsairides, M.A.; Smith, M.; Bennet, E.; Bromfield, M.; Mooberry, J. *NASA Electric Vertical Takeoff and Landing (eVTOL) Aircraft Technology for Public Services—A White Paper*; NASA Transformative Vertical Flight Working Group 4 (TVF4); National Aeronautics and Space Administration (NASA): Washington, DC, USA, 2021.
8. Osita, U.; Horri, T.; Innocente, N.; Bromfield, M.; Bromfield, M. Investigation of a Mission-based Sizing Method for Electric VTOL Aircraft Preliminary Design. In Proceedings of the AIAA SCITECH 2022 Forum 2022, San Diego, CA, USA, 3–7 January 2022. [CrossRef]
9. Pradeep, P.; Wei, P. Energy-efficient arrival with rta constraint for multirotor evtol in urban air mobility. *J. Aerosp. Inf. Syst.* 2019, 16, 263–277. [CrossRef]
10. Bacchini, A.; Cestino, E. Electric VTOL Configurations Comparison. *Aerospace* **2019**, *6*, 26. [CrossRef]
11. Kitty Cora. Available online: https://www.kittyhawk.aero/history (accessed on 18 April 2023).
12. Chauhan, S.; Martins, J.R.R.A. Tilt-wing evtol takeoff trajectory Optimisation. *J. Aircr.* **2020**, *57*, 93–112. [CrossRef]
13. Jobby Aviation S4. Available online: https://www.jobyaviation.com/ (accessed on 18 April 2023).
14. Volocopter VC2X Flight Test in Paris. Available online: https://www.volocopter.com/newsroom/volocopter-flies-at-paris-air-forum/ (accessed on 18 April 2023).
15. Bershadsky, D.; Haviland, S.; Johnson, E.N. Electric Multirotor UAV Propulsion System Sizing for Performance Prediction and Design Optimisation. In Proceedings of the 57th AIAA/ASCE/AHS/ASC Structures, Structural Dynamics, and Materials Conference, San Diego, CA, USA, 4–8 January 2016.
16. Dai, X.; Quan, Q.; Ren, J.; Cai, K.Y. An Analytical Design-Optimisation Method for Electric Propulsion Systems of Multicopter UAVs With Desired Hovering Endurance. *IEEE/ASME Trans. Mechatronics* **2019**, *24*, 228–239. [CrossRef]
17. Gur, O.; Rosen, A. Optimizing electric propulsion systems for UAVs. In Proceedings of the 12th AIAA/ISSMO Multidisciplinary Analysis and Optimization Conference, Victoria, BC, Canada, 10–12 September 2008; p. 5916.
18. Biczyski, M.; Sehab, R.; Whidborne, J.F.; Krebs, G.; Luk, P. Multirotor Sizing Methodology with Flight Time Estimation. *J. Adv. Transp.* **2020**, *2020*, 9689604. [CrossRef]
19. Ampatis, C.; Papadopoulos, E. Parametric design and Optimisation of multi-rotor aerial vehicles. In Proceedings of the 2014 IEEE International Conference on Robotics and Automation (ICRA), Hong Kong, China, 31 May–7 June 2014; pp. 6266–6271. 10.1109/ICRA.2014.6907783. [CrossRef]
20. T-Motor. Available online: https://store.tmotor.com/goods.php?id=384 (accessed on 18 April 2023).

21. Mejzlik. Available online: https://www.mejzlik.eu/technical-data/propeller-data (accessed on 18 April 2023).
22. Brandt, J.; Selig, M. Propeller Performance Data at Low Reynolds Numbers. In Proceedings of the 49th AIAA Aerospace Sciences Meeting, AIAA 2011–1255. Orlando, FL, USA, 4–7 January 2011.
23. Cavcar, M. *The International Standard Atmosphere (ISA)*; Anadolu Univ.: Eskişehir, Turkey, 2000; Volume 30.
24. Vu, N.A.; Dang, D.K.; Dinh, T.L. Electric propulsion system sizing methodology for an agriculture multicopter. *Aerosp. Sci. Technol.* **2019**, *90*, 314–326. [CrossRef]
25. KDEDirect. Available online https://www.kdedirect.com/collections/uas-multi-rotor-brushless-motors (accessed on 18 April 2023).
26. Harrington, A.M.; Kroninger, C. *Characterization of Small DC Brushed and Brushless Motors*; Aberdeen Proving Ground: Aberdeen, MD, USA, 2013.
27. Aurbach, D.; Gofer, Y.; Lu, Z.; Schechter, A.; Chusid, O.; Gizbar, H.; Levi, E. A short review on the comparison between Li battery systems and rechargeable magnesium battery technology. *J. Power Sources* **2001**, *97*, 28–32. [CrossRef]
28. Evangelisti, S.; Tagliaferri, C.; Brett, D.J.; Lettieri, P. Life cycle assessment of a polymer electrolyte membrane fuel cell system for passenger vehicles. *J. Clean. Prod.* **2017**, *142*, 4339–4355. [CrossRef]
29. Zhang, B.; Song, Z.; Zhao, F.; Liu, C. Overview of Propulsion Systems for Unmanned Aerial Vehicles. *Energies* **2022**, *15*, 455. [CrossRef]
30. Ng, W.; Datta, A. Hydrogen Fuel Cells and Batteries for Electric-Vertical Takeoff and Landing Aircraft. *J. Aircr.* **2019**, *56*, 1765–1782. [CrossRef]
31. An, J.-H.; Kwon, D.-Y.; Jeon, K.-S.; Tyan, M.; Lee, J.-W. Advanced Sizing Methodology for a Multi-Mode eVTOL UAV Powered by a Hydrogen Fuel Cell and Battery. *Aerospace* **2022**, *9*, 71. [CrossRef]
32. Thirkell, A.; Chen, R.; Harrington, I. *A Fuel Cell System Sizing Tool Based on Current Production Aircraft*; SAE Technical Paper; SAE International: Warrendale, PA, USA, 2017. [CrossRef]
33. Intelligent Energy. Available online: https://www.intelligent-energy.com/ (accessed on 18 April 2023).
34. Marzougui, T.; Neuhaus, K.; Labracherie, L.; Scalabrin, G. Optimal sizing of hybrid electric propulsion system for eVTOL. In Proceedings of the IOP Conference Series: Materials Science and Engineering, Online, 11–14 September 2022.
35. Boukoberine, M.; Zhou, Z.; Benbouzid, M. A critical review on unmanned aerial vehicles power supply and energy management: Solutions, strategies, and prospects. *Appl. Energy* **2019**, *255*, 113823. [CrossRef]
36. Maxwell. Available online: https://maxwell.com/products/ (accessed on 18 April 2023).
37. Eren, Y.; Küçükdemiral, İ.; Üstoğlu, İ. *Chapter 2—Introduction to Optimisation, Optimisation in Renewable Energy Systems*; Butterworth-Heinemann: Oxford, UK, 2017; pp. 27–74.
38. Theys, B.; Dimitriadis, G.; Hendrick, P.; De Schutter, J. Influence of propeller configuration on propulsion system efficiency of multi-rotor Unmanned Aerial Vehicles. In Proceedings of the 2016 International Conference on Unmanned Aircraft Systems (ICUAS), Arlington, VA, USA, 7–10 June 2016; pp. 195–201.
39. Bolam, R.C.; Vagapov, Y.; Anuchin, A. Anuchin. Review of Electrically Powered Propulsion for Aircraft. In Proceedings of the 2018 53rd International Universities Power Engineering Conference (UPEC), Glasgow, UK, 4–7 September 2018; pp. 1–6.
40. Rufer, A. The dream of efficient energy storage–From BESS KERS & Co to the hybrid power plant. In Proceedings of the19th European Conf. on Power Electronics and Applications (EPE'17 ECCE Europe), Warsaw, Poland, 11–14 September 2017; pp. 1–9.
41. Uber Elevate. Uber Elevate Mission and Vehicle Requirements. 2018. Volume 2018. Available online: https://s3.amazonaws.com/uber-static/elevate/Summary+Mission+and+Requirements.pdf (accessed on 18 April 2023).
42. Brown, A.; Harris, W. A Vehicle Design and Optimisation Model for On-Demand Aviation. In Proceedings of the 2018 AIAA/ASCE/AHS/ASC Structures, Structural Dynamics, and Materials Conference, Kissimmee, FL, USA, 8–12 January 2018.

aerospace

Article

Effect of Temperature on the Functionalization Process of Structural Self-Healing Epoxy Resin

Luigi Vertuccio [1], Elisa Calabrese [2], Marialuigia Raimondo [2], Michelina Catauro [1], Andrea Sorrentino [3], Carlo Naddeo [2], Raffaele Longo [2] and Liberata Guadagno [2,*]

[1] Department of Engineering, University of Campania "Luigi Vanvitelli", Via Roma 29, 81031 Aversa, Italy; luigi.vertuccio@unicampania.it (L.V.); michelina.catauro@unicampania.it (M.C.)

[2] Department of Industrial Engineering, University of Salerno, Via Giovanni Paolo II, 132, 84084 Fisciano, Italy; elicalabrese@unisa.it (E.C.); mraimondo@unisa.it (M.R.); cnaddeo@unisa.it (C.N.); rlongo@unisa.it (R.L.)

[3] Institute for Polymers, Composites, and Biomaterials (IPCB-CNR), Via Previati n. 1/E, 23900 Lecco, Italy; andrea.sorrentino@cnr.it

* Correspondence: lguadagno@unisa.it

Abstract: This work deals with developing a self-healing resin designed for aeronautical and aerospace applications. The bifunctional epoxy precursor was suitably functionalized to enhance its toughness to realize good compatibilization with a rubber phase dispersed in the hosting epoxy resin. Subsequently, the resulting mixture was loaded with healing molecules. The effect of the temperature on the epoxy precursor's functionalization process was deeply studied. Fourier trans-former infrared (FT-IR) spectroscopy and dynamic mechanical analyses (DMA) evidenced that the highest temperature (160 °C) allows for obtaining a bigger amount of rubber phase bonded to the matrix. Elastomeric domains of dimensions lower than 500–600 nanometers were found well distributed in the matrix. Self-healing efficiency evaluated with the tapered double cantilever beam (TDCB) method evidenced a healing efficiency for the system functionalized at 160 °C higher than 69% for all the explored fillers. The highest value was detected for the sample with DBA, for which 88% was found. The healing efficiency of the same sample functionalized at 120 °C was found to decrease to the value of 52%. These results evidence the relevant role of the amount and distribution of rubber domains into the resin for improving the resin's dynamic properties. The adopted strategy allows for optimizing the self-healing performance.

Keywords: self-healing efficiency; functionalization reaction; epoxy resin; mechanical properties

Citation: Vertuccio, L.; Calabrese, E.; Raimondo, M.; Catauro, M.; Sorrentino, A.; Naddeo, C.; Longo, R.; Guadagno, L. Effect of Temperature on the Functionalization Process of Structural Self-Healing Epoxy Resin. *Aerospace* **2023**, *10*, 476. https://doi.org/10.3390/aerospace10050476

Academic Editor: Khamis Essa

Received: 12 March 2023
Revised: 11 May 2023
Accepted: 16 May 2023
Published: 18 May 2023

1. Introduction

Self-healing materials can repair themselves and recover integrity using the resources inherently available or healing mechanisms activated during microfractures. The auto-repair process can be autonomic or externally assisted, but it is always triggered by damage to the material. The integration of the self-healing functionality in thermosetting resins is driven by the need to reduce the environmental impact and the speed of resource depletion. This can be achieved by consistently reducing starting materials (primary resources) and, as a consequence, energy consumption and CO_2 emissions into the atmosphere. Together with the atmosphere decarbonization contribution, the possibility to auto-repair polymeric materials significantly impacts the speed reduction of producing end-of-life plastic wastes. A strong impact on cost reduction is also expected, deriving from the longer life of materials and reduced demand for resources necessary to produce new ones. Ultimately, all this would translate into a high environmental impact reduction and energy sustainability promotion. Furthermore, together with the reduction of maintenance operations on composite components, the possibility to guarantee the structural integrity of components in spaces inaccessible for maintenance operations is strongly felt in aeronautics and aerospace.

The studies carried out until now on auto-repair polymeric materials have resulted in innumerable healing designs that recently are focused on complex systems capable of supporting multiple cycles [1–10].

Self-healing polymers can be divided as extrinsic and intrinsic [8,11]. Extrinsic polymers are based on auto-repair processes depending on external healing agents in micro/nano vessels (generally in the form of microcapsules o vascular channels). The healing agents are released to seal the damaged regions.

Capsule-based self-healing systems retain the healing agent in microcapsules. When the damage occurs and propagates in the material, the microcapsules are cracked and the self-healing mechanism is triggered, leading to a local healing event. The main systems involve the encapsulation of a liquid healing agent and the dispersion of a catalyst, active in the ring-opening metathesis polymerization (ROMP), inside the polymeric matrix [12–15]. This strategy was also used to impart self-healing features to an epoxy resin modified with CTBN rubber for application as an adhesive by Henghua Jin et al. [16]. In particular, the authors developed a toughened epoxy adhesive where self-healing is achieved via embedded microcapsules containing dicyclopentadiene monomer and Grubbs' catalyst. Vascular self-healing materials sequester the healing agent in one, two, or three-dimensional networks consisting of capillaries or hollow channels [17–20].

Intrinsic self-healing polymers are those in which the reversible bonds (dynamic covalent and noncovalent bonds) active in the material can restore the integrity of the polymer after a damage event [11,21].

Therefore, these self-healing polymers are based on the inherent reversibility of bonding of the matrix polymer. For this typology of self-healing polymers, healing events can be achieved by reversible covalent reactions [22–33], the presence of a dispersed meltable thermoplastic phase [34,35], ionomeric coupling [36–39], via molecular diffusion [40–42] or hydrogen bonding [36–39,43,44]. The intrinsic self-healing materials are less complex than capsule-based and vascular self-healing materials, avoiding the problems related to healing-agent integration, compatibility, catalyst stability, etc. [13,45]. However, most self-healing intrinsic systems have mechanical and chemical properties incompatible with those required by structural applications.

Thermosetting resins, with their combination of thermal stability, performance, and chemical resistance, are extensively used in industry. However, integrating self-healing functionality into these materials is very difficult due to their irreversible network structure and low chain mobility, which impede the chain flow necessary for the common self-healing process [46].

To address this challenge, different approaches have been experimented in literature [46]. In previous works the authors have proposed an intrinsic self-healing system consisting of an epoxy resin covalently modified by a rubber phase containing self-healing fillers [47,48]. The presence of the elastomer has allowed reducing the rigidity of the epoxy chains and promoting the activation of an auto-repair mechanism based on hydrogen bonding interactions. Furthermore, rubber-toughened thermosetting resins manifest several advantages compared with the unmodified resin. Ricciardi et al. [49] used nitrile rubber to toughen glass fiber reinforced EP composites. They found that the modified composites showed smaller delamination, although the absorbed energy was the same and the load was higher. Karger-Kocsis and Friedrich analyzed fatigue crack propagation of carboxyl-terminated acrylonitrile-butadiene rubber (CTBN) and silicon rubber (SI) modified Epoxy resin [50,51]. The incorporation of CTBN and/or SI dispersion in the EP matrix improved the resistance to fatigue crack propagation.

In order to not significantly alter the mechanical properties of the resin, intended for structural applications (such as aerospace or automotive fields), different fillers or nano-fillers have been solubilized/dispersed into the epoxy matrix [47,48].

This work focuses on the effect of the curing temperature of a rubber-toughened bifunctional epoxy resin filled with self-healing molecules to impart an auto-repair function to the resin. FT/IR analysis and dynamic mechanical analyses (DMA) were used to study the functionalization of the epoxy precursor and to estimate the amount of the elastomeric phase bonded to the epoxy precursor for different temperatures of functionalization. A more significant amount of rubber phase bonded to the matrix was found for the functionalization performed at the higher temperature of 160 °C compared to that conducted at 120 °C. Thermogravimetric analysis was employed to evaluate the beginning of the degradation temperature for all the formulated systems. Self-healing tests performed through the tapered double cantilever beam (TDCB) evidence a higher healing efficiency for the selected system functionalized at 160 °C, containing the healing molecules. A synergistic effect between the interactions determined by the active self-healing fillers and those due the presence of elastomeric domains bonded to the epoxy precursor has been hypothesized.

2. Materials and Methods

2.1. Materials

The formulated samples are composed by a hosting toughened epoxy matrix consisting of the precursor "3,4-Epoxycyclohexylmethyl-3′,4′-epoxycyclohexane carboxylate" (ECC—Empirical Formula $C_{14}H_{20}O_4$) (see the chemical structure in Figure 1a) (Gurit Holding, Wattwil, Switzerland) and the hardener agent "Methylhexahydrophthalicanhydride" (MHHPA—Empirical Formula $C_9H_{12}O_3$) (see the chemical structure in Figure 1b) (Gurit Holding, Wattwil, Switzerland), both used in a ratio 1:1. As toughening agent (5 wt% with respect to the total mixture) the liquid rubber Carboxyl-Terminated Butadiene Acrylonitrile Copolymer was employed (R, see the chemical structure in Figure 1c) supplied by Hycar-Reactive Liquid Polymers, with Mn = 3600, containing terminal carboxy groups (COOH content of 0.67×10^{-3} equiv/g of CTBN and 18 *w*/*w*% of CN). The compound triphenyl phosphine (PPh_3, Merck KGaA, Darmstadt, Germany), added in an amount of 10 wt%, was employed as a catalyst to promote the functionalization reaction of the epoxy precursor (ECC).

(a)

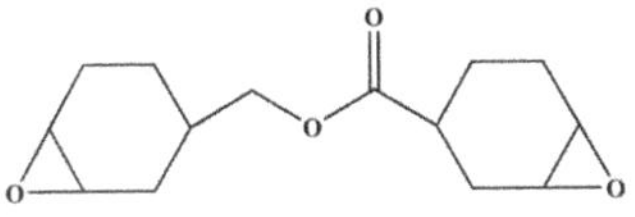

(b)

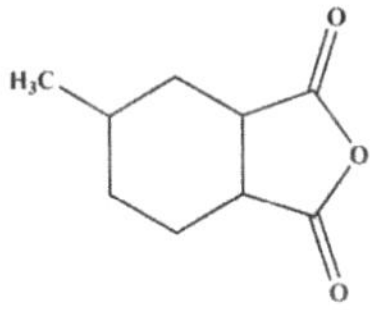

H₃C

(c)

HO—C—[(CH₂—CH=CH—CH₂)ₓ(CH₂—CH(CN))ᵧ]ₘ—C—OH

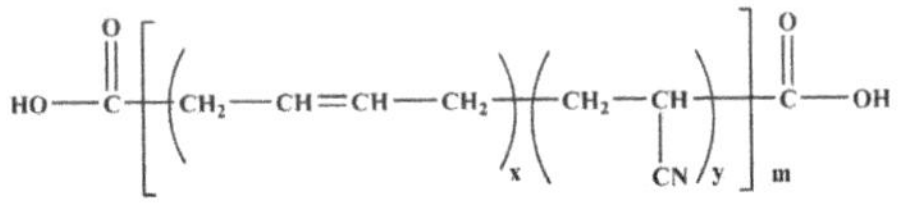

Figure 1. Chemical structures of the: (**a**) epoxy precursor (ECC); (**b**) hardener (MHHPA); (**c**) liquid rubber (R).

Table 1 shows the amount in grams of each component to prepare 23.58 g of a complete mixture Ep-R-120 or Ep-R-160.

Considering the chosen composition, the first Nucleophilic attack by triphenylphosphine (see reaction scheme in Section 3.1.1 FT-IR Analysis) opens 9.15×10^{-3} mol of oxirane rings, leaving 7.0×10^{-2} mol of oxirane rings still unreacted (not opened). After the first Nucleophilic attack of PPh_3, the unreacted oxirane rings represent 88% of the initially available rings. The moles of the terminal carboxylic groups are 0.804×10^{-3}. This amount (in mol) is slightly defective compared with the opened oxirane rings (9.15×10^{-3} mol). This choice has been made to avoid big domains of elastomeric phase in the resin (as highlighted later through SEM investigation). The question of the dimensions of the rubber domains has been a nontrivial issue that has been addressed before choosing the chemical composition of the epoxy mixture. Based on data already reported in literature [52] and authors' experience, a lower amount of triphenylphosphine (or a higher amount of rubber phase with respect to the same amount of PPh_3) determines a higher dimension of the elastomer domains, causing local inhomogeneity in the order of hundreds or units of microns with consistent local inhomogeneity.

Table 1. Amount in grams of each component to prepare 23.58 g of a complete mixture Ep-R-120 or Ep-R-160.

Component	Ep-R-120/Ep-R-160
ECC [g]	10.0
CTBN [g]	1.2
PPh_3 [g]	2.4
MHHPA [g]	10.0

The self-healing fillers 1.3-Dimethylbarbituric acid (DBA), 2-Thiohydantoin (T), and Murexide (M) (all purchased from Merck KGaA, Darmstadt, Germany) have been added in a percentage of 0.42 wt% (see Figure 2). The ability of these fillers to activate self-healing mechanisms, based on hydrogen bonding interactions, was already demonstrated in previous work with a nano-charged resin based on a tetrafunctional epoxy precursor hardened with 4, 4-diamino diphenyl sulfone [47]. The matrix was loaded with conductive nanofillers, and the healing efficiency was evaluated for the formulation with 0.5% by weight of carbon nanotubes to confer functional properties to the resin. Generally, the presence of conductive fillers confers many functional properties to the hosting polymeric matrix [53–55]. In particular, electrically conductive fillers such as CNTs, graphene-based nanoparticles, or expanded graphite have been dispersed in optimized epoxy resins to impart them self-sensing for the damage monitoring [56,57] or for activating the anti/de-icing function through the joule effect [58–60], or to make possible energy saving curing processes (electro-curing processes) of resins/composites [61], and for enhancing adhesive properties [62].

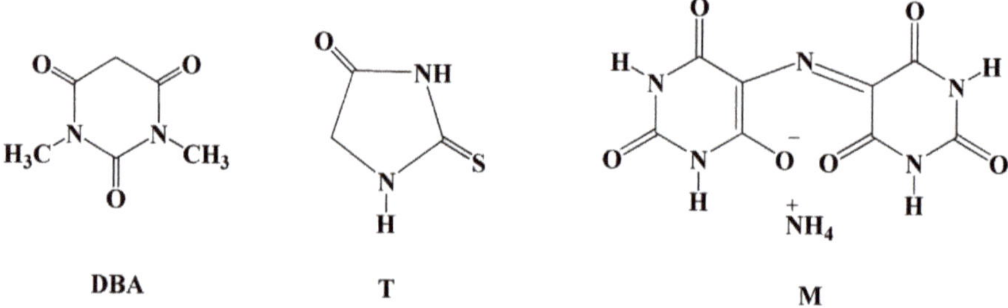

Figure 2. Chemical structure of molecules acting as self-healing filler.

2.2. *Formulation of Epoxy Samples*

2.2.1. Functionalized Epoxy Precursor

The functionalized precursor liquid blends ECC-R-120 and ECC-R-160 were composed of the epoxy precursor covalently modified with the rubber phase.

The two blends were obtained by mixing, under mechanical stirring, the epoxy precursor ECC, the elastomer R, and the catalyst PPh_3 for a time of 15 h, at temperatures of 120 and 160 °C, respectively.

2.2.2. Functionalized Epoxy Samples

The cured epoxy samples Ep-R-120 and Ep-R-160 were composed of the precursor functionalized at the temperature of 120 and 160 °C, respectively. They were obtained by mixing by magnetic stirring for 20 min at room temperature, the hardener MHHPA and the functionalized precursor (ECC-R-120 and ECC-R-160, respectively). After a degassing process for 2 h at room temperature, the mixture was polymerized in an oven by a curing cycle of 1 h at 80 °C, followed by 20 min at 120 °C and 1 h at 180 °C. To perform this curing cycle, the samples were placed in the oven for the first cure step, with the oven temperature at 80 °C. At the end of the cure treatment at 80 °C (1 h), a heating speed ramp was set to go from 80 °C to 120 °C at 10 °C/min. The sample's actual heating rate (verified with a probe) is 4 °C/min. The same thing happened for the transition from 120 °C to 180 °C.

In this paper, for comparison, the Ep sample, corresponding to the cured epoxy matrix, without the presence of the rubber phase, was prepared by mixing the ECC precursor and the hardener MHHPA and following the procedure previously described. Finally, the functionalized epoxy samples with the selected molecules, reported in Figure 2, were obtained by dispersing the self-healing filler in the functionalized precursor ECC-R-160 by ultrasonication process for 30 min at room temperature. The addition of the hardener and the curing cycle followed the same procedure described above. The obtained samples containing DBA, T, and M were labeled Ep-R-160-DBA, Ep-R-160-T, and Ep-R-160-M, respectively.

2.3. *Methods*

The preparation of the samples to carry out FT-IR spectra and the modality of their acquisition are described in Section S1 (S.M.). This last section also reported information on the thermal and dynamic mechanical analysis of the performed elaborations.

Self-healing efficiency (η) was evaluated using two different approaches. The first one consists of a fracture test performed with a tapered double cantilever beam (TDCB) geometry sample following a protocol established in literature [12], which allows to calculate the value of the self-healing efficiency by Equation (1),

$$\eta = \frac{P_{CH}}{P_{CV}} \times 100 \tag{1}$$

where P_{CH} and P_{CV} are the critical fracture load of the healed and virgin sample, respectively. The specimens were tested by INSTRON mod. 5967 Dynamometer, using a load cell of 30 KN and a 250 µm/min displacement rate. The dimensions of the tested samples are reported in Figure S1 of Section S1 of the S.M.

In the second approach, DMA tests were used to evaluate auto-repair ability. The self-healing test was carried out with a continuous dynamic flexural deformation, through which the samples, having dimensions 3 mm × 10 mm × 35 mm and a V-shaped starter notch (1 mm × 2 mm), were analyzed by applying a sinusoidal deformation with a maximum amplitude of 0.1% at a frequency of 1 Hz. An impulsive load of about 25 N induced a pre-crack in the sample. The trend of the mechanical modulus with temperature was considered as representative of the evolution of the healing process in the sample. In this case, η was calculated according to Equation (2) [48,63].

$$\eta = \frac{E_H}{E_V} \times 100 \tag{2}$$

where E_H and E_V are the storage modulus of the healed and virgin sample, respectively.

3. Results and Discussion

3.1. *Functionalization of the Epoxy Prcursor*

3.1.1. FT-IR Analysis

FT-IR investigation was performed to evaluate the best conditions to choose in the functionalization procedure. A key role is played by the catalyst triphenylphosphine (PPh_3), which is necessary to promote the reaction between epoxy groups of the matrix (ECC) and the carboxylic groups of the rubber phase (R), as depicted in the reaction scheme of Figure 3.

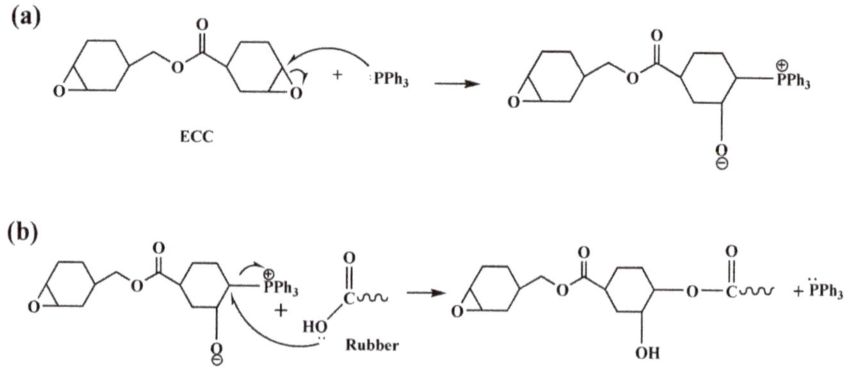

Figure 3. Reaction mechanism supposed for the functionalization of the epoxy precursor: (**a**) nucleophilic attack of PPh_3; (**b**) reaction of the intermediate with the carboxylic group of the rubber phase.

The presence of PPh_3 is a necessary step to avoid phase separation phenomena between the precursor and the elastomer.

The spectroscopic analyses were carried out on the ECC-R-120 and ECC-R-160 liquid mixtures to investigate the functionalization reaction and the effect of temperature (120 °C and 160 °C, respectively). FTIR spectra of the epoxy precursor ECC, liquid rubber R, and ECC-R blend were compared with the spectrum of the ECC-R-160 blend (see Figures 4 and 5) and the spectrum of the ECC-R-120 blend (see Figures 6 and 7). The liquid blend ECC-R corresponds to the epoxy mixture precursor/elastomer raw before the heat treatments for 15 h. Focusing the attention on Figure 4, in the range between 1650 cm^{-1} and 1850 cm^{-1} (see inset on the left), the spectrum of the rubber phase shows two absorption bands for the ester carbonyl group, as a consequence of the hydrogen bond interactions established among the molecules of the liquid rubber [64]. In particular, the band at 1738 cm^{-1} is ascribed to the free ester carbonyl group, while the band at 1710 cm^{-1} belongs to the H-bonded carbonyl group, involved in the hydrogen bond interactions with the hydroxyl groups of the same rubber molecules. In the same range of wavenumber, the spectrum of the precursor shows the ester C=O stretching band, around 1730 cm^{-1}, while the spectrum of the ECC-R blend displays broadband always at 1730 cm^{-1}, which belongs to the carbonyl groups of both the components. In the same region of wavenumber, it is possible to observe the presence of a shoulder peak at 1780 cm^{-1} that could be considered the experimental evidence of the functionalization reaction between the oxirane ring of the epoxy precursor and the carboxylic groups of the rubber phase. This effect is due to the presence of an electron-withdrawing group (hydroxyl group in β position) that can determine the shift of the carbonyl signal to higher values of wavenumber for inductive effect [65]. The peak at 1120 cm^{-1} (see inset on the right of Figure 4), assigned to the C-O stretching of the secondary alcohol generated by the opening of the epoxy group during the functionalization reaction, supports the hypothesized mechanism.

Further confirmation of the occurred functionalization is deduced by the results depicted in Figure 5, in the range of wavenumber between 3700 cm^{-1} and 3100 cm^{-1}. ECC-R-160 sample shows an absorption band at 3350 cm^{-1} ascribed to the –OH groups generated during the reaction. In addition, the absorption band at 3230 cm^{-1}, ascribed to the hydroxyls of the –COOH group of the R elastomer, disappears after the functionalization reaction.

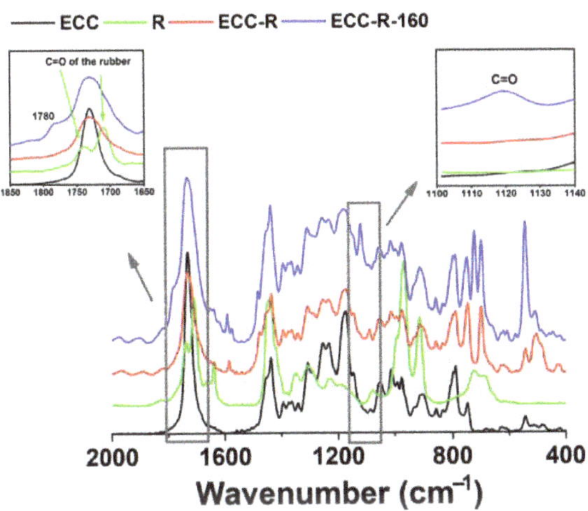

Figure 4. FTIR spectra of the precursor ECC (black curve), the liquid rubber R (green curve), the blend of ECC-R (red curve), and the blend of ECC-R-160 (blue curve), in the range 2000–400 cm^{-1}.

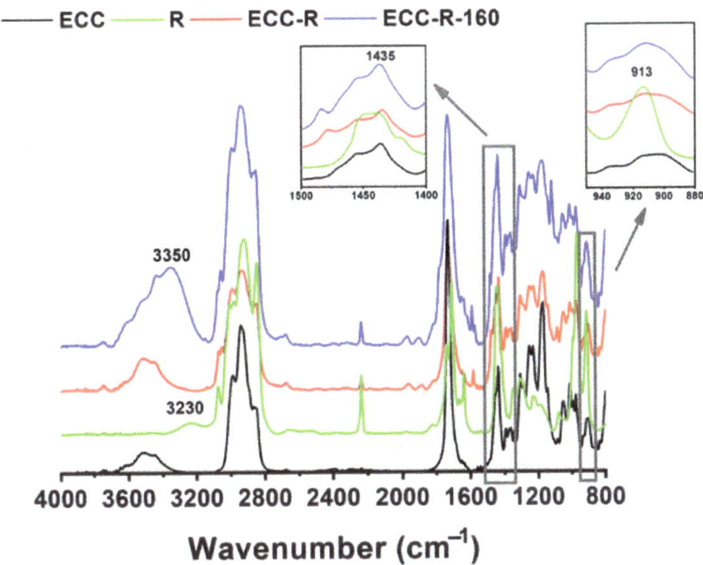

Figure 5. FTIR spectra of the precursor ECC (black curve), the liquid rubber R (green curve), the blend of ECC-R (red curve), and the blend of ECC-R-160 (blue curve), in the range 4000–800 cm^{-1}.

Similar considerations can be made for the ECC-R-120 system, as shown in Figures 6 and 7. Compared to the previous case in the spectrum of the precursor functionalized at 120 °C, the peak around 1780 cm^{-1} is not detectable (see the inset on the left of Figure 6). The reason for this is probably attributable to not very effective functionalization obtained at 120 °C. To evaluate the effectiveness of the functionalization reaction at the two different temperatures (120 °C and 160 °C), a further investigation was carried out in the range between 850 and 950 cm^{-1}, i.e., the area spectrum attributable to the oxirane ring of the precursor. As a consequence of the functionalization reactions, the peak at 913 cm^{-1}, ascribed to the oxirane group of epoxy precursor, decreases in intensity (see the right inset of Figure 5).

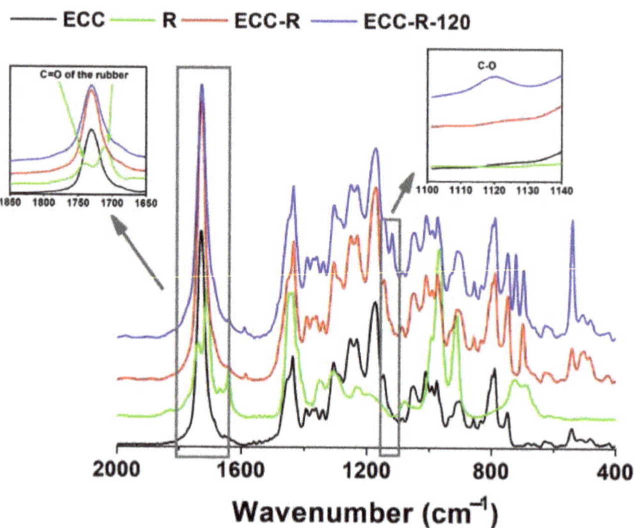

Figure 6. FTIR spectra of the precursor ECC (black curve), the liquid rubber R (green curve), the blend of ECC-R (red curve), and the blend of ECC-R-120 (blue curve), in the range 2000–400 cm^{-1}.

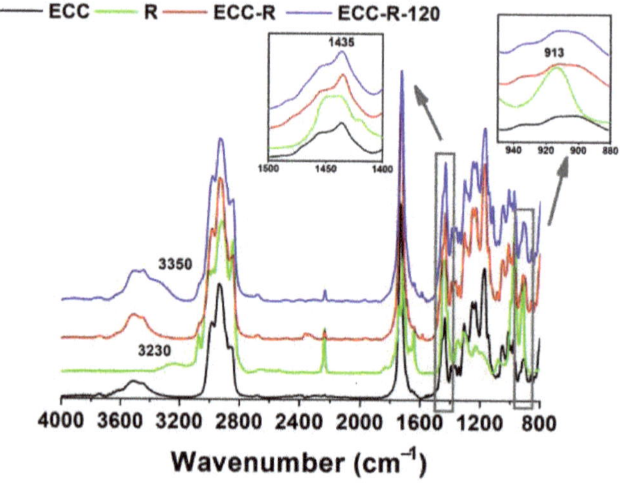

Figure 7. FTIR spectra of the precursor ECC (black curve), the liquid rubber R (green curve), the blend of ECC-R (red curve), and the blend of ECC-R-120 (blue curve), in the range 4000–800 cm^{-1}.

More in particular, the reduction of the peak relative to the oxirane group was evaluated, normalizing the peak a 913 cm^{-1} to the peak at 1435 cm^{-1} associated with the CH_2 stretching of six terms ring, which is assumed chemically unmodified during the reaction [66]. The ratio ($R = A_{peak\ 913}/A_{peak\ 1435}$) of the subtended areas was evaluated for precursor-liquid rubber system before and after the functionalization process, respectively. An algorithm based on the Levenberg–Marquardt method [67] to separate the individual peaks in the case of unresolved, multicomponent bands, was applied. To reduce the number of adjustable parameters and to ensure the uniqueness of the result, the baseline, the band shape, and the number of components were fixed. The minimum number of components was evaluated by visual in the section based on abrupt changes in the slope of the experimental line shape. The program calculated, by a non-linear curve fitting of data, the height, the full-width half height (FWHH), and the position of the individual components. The peak function was a mixed Gauss–Lorentz line shape of the form, reported in Equation (3) [68]:

$$f(x) = (1-L)H\exp\left[-4\ln(2)\left(\frac{x-x_0}{w}\right)^2\right] + LH\left[4\left(\frac{x-x_0}{w}\right)^2+1\right]^{-1} \quad (3)$$

where x_0 = the peak position; H = peak height; w = FWHH; L = fraction of Lorentz character.

The results of this deconvolution procedure for the functionalized precursor at 160 °C, in the above-mentioned ranges of wavenumbers, are shown, respectively, in Figure 8a,b. This procedure was repeated for the systems rubber-precursor before and after the functionalization process at 120 and 160 °C.

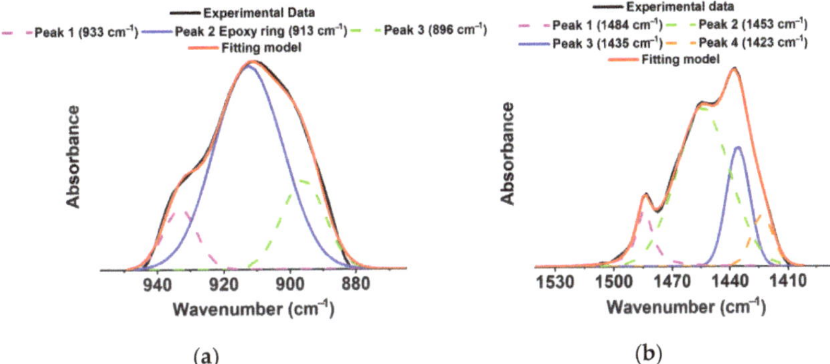

Figure 8. FT-IR spectrum of the ECC-R-160 sample; deconvolution relating to the region of the: (**a**) epoxy ring; (**b**) peak at 1435 cm^{-1} associated with the CH_2 stretching.

A reduction of 5.3% and 13.5% was found for ECC-R-120 and ECC-R-160 systems, respectively. This is further proof that the higher temperature value allows for obtaining a greater amount of bond between the carboxyl group of the rubber and the epoxy ring of the precursor, making the precursor functionalization more efficient. The most effective functionalization process affects the resin structure and consequently the thermal and mechanical properties, as described below.

3.1.2. Thermogravimetric Analyses (TGA)

Figure 9 shows the thermogravimetric curves of the ECC, ECC-R-120, and ECC-R-160 samples in air and nitrogen flow.

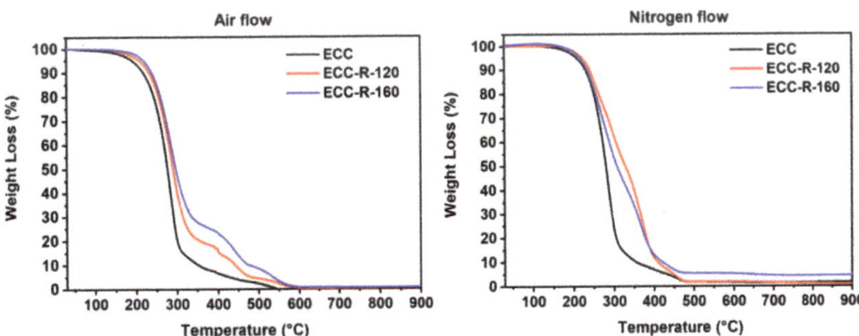

Figure 9. Thermogravimetric curves of ECC sample and the rubber functionalized mixtures ECC-R-120 and ECC-R-160 in air and nitrogen flow.

Comparing the TGA curves of the ECC-R-120 and ECC-R-160 samples with that of sample ECC, it is evident that the thermal stability of the functionalized samples is higher than that of the unfunctionalized resin ECC. Table 2 shows data from TGA analyses performed in air and nitrogen flow for the ECC, ECC-R-120, and ECC-R-160 samples. The initial degradation temperature ($T_{d5\%}$), expressed as temperature corresponding to a weight loss of 5 wt%, presents an increase of about 20 °C and 30 °C for the samples ECC-R-120 and ECC-R-160, respectively. The behaviour in nitrogen flow is the same as observed in air; in this last case, the initial degradation temperature ($T_{d5\%}$) of the samples ECC-R-120 and ECC-R-160 manifests an increase of 13 °C and 7 °C, respectively, with respect to the ECC sample.

Table 2. Data of TGA analyses performed in air and nitrogen flow for ECC, ECC-R-120, and ECC-R-160 samples.

Sample	*$T_{d5\%}$, [°C]
Air flow	
ECC	191.8
ECC-R-120	209.3
ECC-R-160	219.5
Nitrogen flow	
ECC	209.0
ECC-R-120	222.2
ECC-R-160	216.3

*$T_{d5\%}$: temperature corresponding to a weight loss of 5 wt%.

Considering the more reactive environment, the tests performed in air, the functionalized precursor presents a higher thermal stability than the raw precursor.

The functionalization reaction obtained at 160 °C allows for obtaining a modified precursor with greater thermal stability achieving a $T_{d5\%}$ value up to 220 °C.

3.2. Characterization of the Epoxy Resins

3.2.1. Dynamic Mechanical Analysis (DMA)

DMA analyses were carried out on the polymerized samples, with the aim to observe the effect of the functionalization temperature on the mechanical properties of the materials. The tests were performed on the samples Ep, Ep-R-120, and Ep-R-160. DMA investigation was carried out by evaluating the tan δ profile and the storage modulus as a function of the temperature. In Figure 10, the DMA curves of the cured functionalized epoxy matrices Ep-R-120 and Ep-R-160 were compared to the curve of the not functionalized epoxy matrix Ep.

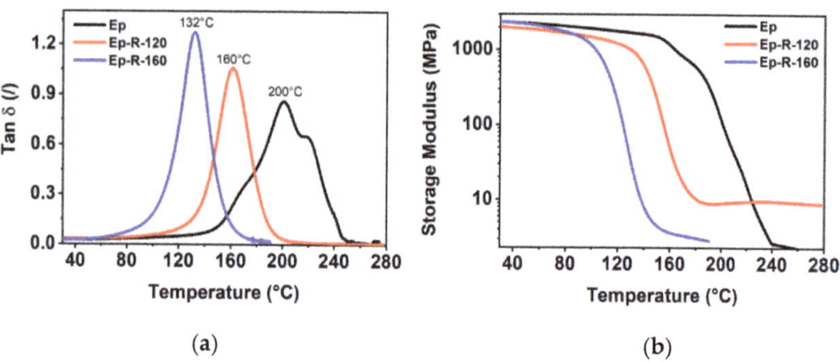

Figure 10. DMA curves: (**a**) Tan δ vs. temperature; (**b**) Storage modulus vs. temperature.

All systems present a storage modulus value (see Figure 10b) higher than 1000 MPa at room temperature and in the wide temperature range of 30 ÷ 100 °C, thus confirming their reliability if employed, for instance, as structural aeronautical parts generally working in the normal operating temperature range. The trend of the not functionalized epoxy matrix (Ep) shows a progressive decrease of modulus up to 150 °C; after that, the principal drop occurs, due to its subsequent attainment of the glass transition temperature (i.e., T_g). The mechanical behavior of this sample, shown in Figure 10a, confirms the glass transition temperature (T_g) value above the 180 °C. The highest peak in the mechanical spectrum, related to the glass transition, (i.e., α transition), is centered at 200 °C. The introduction of the rubber and the different functionalization temperature value affect both storage modulus and tan δ. The principal drop in the storage modulus is reduced to lower temperature values due to the presence of the functionalized precursor in the resin (see Figure 10b).

One of the big challenges in this work was to improve the resin's dynamic properties and reduce the matrix's rigidity, acting on its phase composition. Therefore, the design of a material containing in a rigid matrix very small domains of rubber phase at higher mobility, finely interpenetrated in the resin, has been considered. A higher mobility is expected of the chains around the elastomeric domains finely distributed in the resin, as highlighted in Section 3.3. Furthermore, DMA analysis evidences a reduction of the rigidity of the matrix (at the macroscopic level). In fact, for the sample which manifests higher healing efficiency (EP-R-160), the peak of Tan δ vs. temperature opens around 60 °C. It is worth noting that this peak is shifted to a lower temperature range. It involves a temperature range from 60 °C to 180 °C (with a max around 132 °C), where the sample without the rubber phase shows a transition from 140 °C to 250 °C. This is clear evidence of a strong toughening effect exerted by the rubber phase finely distributed in the form of small domains in the resin in a range of temperatures closest to ambient temperature.

The more effective the functionalization reaction of the epoxy precursor, the higher the toughening effect of the elastomer on the polymerized resin. This phenomenon results in a reduction of 40 °C for the Ep-R-120 system and one of 70 °C for the Ep-R-160 system leading the T_g value to 160 °C and 132 °C, respectively (see Figure 10a). These results agree with those obtained through FT-IR and TGA analyses, confirming the efficacy of the precursor functionalization performed at the temperature of 160 °C. To obtain an epoxy resin that shows good auto-repair ability, the composition Ep-R-160 was chosen as the matrix to host self-healing fillers, as the activation of auto-repair mechanism is favored in the presence of higher mobility of the polymeric chains [47,48].

3.2.2. DSC Analyses

DSC investigations were performed to evaluate the influence of the self-healing fillers on the curing process of the functionalized epoxy resin. Figure 11a shows the results of DSC analyses performed on the uncured liquid mixtures.

The presence of the auto-repair agents determines a reduction of the peak temperature (see Table 3) value and a broadening of the peak shape. This phenomenon is more evident in Figure 11b in which the fractional conversion (α) as a function of temperature is shown. The fractional conversion (α) can be expressed as Equation (4):

$$\alpha(T) = \frac{\Delta H_T}{\Delta H_{Tot}} \tag{4}$$

where ΔH_T is the partial heat of reaction at a certain temperature and ΔH_{Tot} is the total heat of reaction. The systems filled with the self-healing agents present a fractional conversion curve shifted at lower temperature and a reduced starting curing temperature value, ($T_{\alpha=0.01}$, temperature value corresponding to α = 1%), as shown in Table 3. DSC curves of the samples EP-R-160, EP-R-160-DBA, EP-R-160-2T, EP-R-160-M, before the curing process (after the functionalization reaction) and after the curing process (dashed curves) have been added in Figure S2 of Section S2 of the S.M.

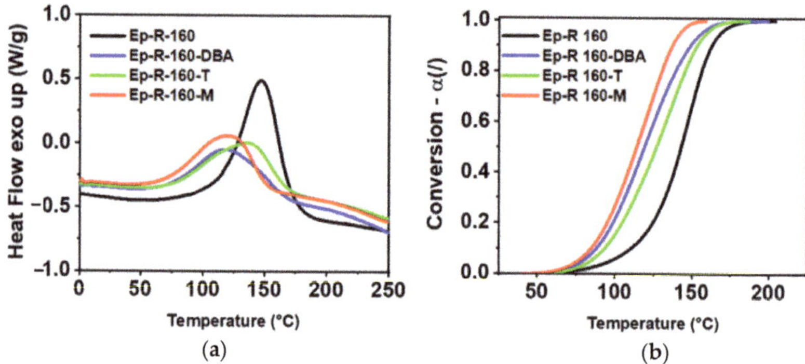

Figure 11. (**a**) DSC curves of the analyzed samples before the curing process in the oven; (**b**) Variation of the conversion (α) vs. the temperature.

Table 3. DSC results.

Sample	T_{peak} (°C)	$T_{\alpha=0.01}$ (°C)	ΔH_T (Jg^{-1})	ΔH_{Tot} (Jg^{-1})	DC (%)
Ep-R-160	147.3	73.1	0	240.75	100.0
Ep-R-160-DBA	117.7	63.3	6.6	121.5	94.6
Ep-R-160-T	135.8	68.1	7.3	138.7	94.7
Ep-R-160-M	120.3	60.6	11.1	123.1	91.0

Table 3 highlights that the presence of the fillers causes, on the one hand, a reduction of curing temperature, on the other, a decrease of the curing degree value, which remains higher than 90%, allowing to the materials to be suitable for structural applications. Usually a small amount of accelerators, such as tertiary amines or imidazoles, are used in the epoxy/anhydride systems to speed up the curing process [69]. In our case, probably, the fillers act as catalysts in the polymerization mechanism. These results could represent an advantage in reducing processing costs and energy savings.

3.2.3. Thermogravimetric Analyses (TGA)

Figure 12 shows the thermogravimetric curves of the rubber functionalized and cured epoxy mixtures Ep-R-160, Ep-R-160-DBA, Ep-R-160-T, and Ep-R-160-M in air and nitrogen flow.

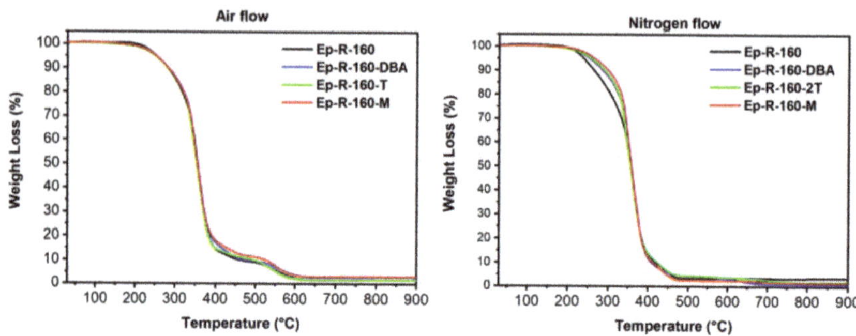

Figure 12. Thermogravimetric curves of the rubber functionalized and cured mixtures Ep-R-160, Ep-R-160-DBA, Ep-R-160-T, and Ep-R-160-M in air and nitrogen flow.

The TGA curves shown in Figure 12, compared with those of Figure 9, clearly evidence that the addition of the hardener agent MHHPA determines an increase in the initial degradation temperature ($T_{d5\%}$) of the samples, as expected for samples after the curing process. For example, for the sample Ep-R-160 (functionalized at a higher temperature), the initial degradation temperature ($T_{d5\%}$) after the curing cycle is about 253 °C in airflow and about 248 °C in nitrogen flow as shown in Table 4. The stability in the thermal degradation is also retained with the dispersion of the self-healing fillers in the formulation.

Table 4. Data of TGA analyses performed in air and nitrogen flow for the sample EP-R-160 and the same sample containing the healing fillers DBA, T, and M.

Sample	*$T_{d5\%}$, [°C]
Air flow	
Ep-R-160	252.8
Ep-R-160-DBA	251.8
Ep-R-160-T	251.1
Ep-R-160-M	252.6
Nitrogen flow	
Ep-R-160	248.4
Ep-R-160-DBA	254.3
Ep-R-160-T	258.2
Ep-R-160-M	266.3

*$T_{d5\%}$: temperature corresponding to a weight loss of 5 wt%.

3.3. Morphological Caracterization

Figure 13 shows SEM images of the etched surface of the EP-160-T, EP-160-DBA, and EP-160-M samples. The images give clear information on the size and distribution of the rubber domains in the hosting epoxy matrix. It is worth noting that to better observe the microstructure of the samples, before being analyzed by FESEM, the samples were subjected to an etching process, according to a procedure reported in literature [55].

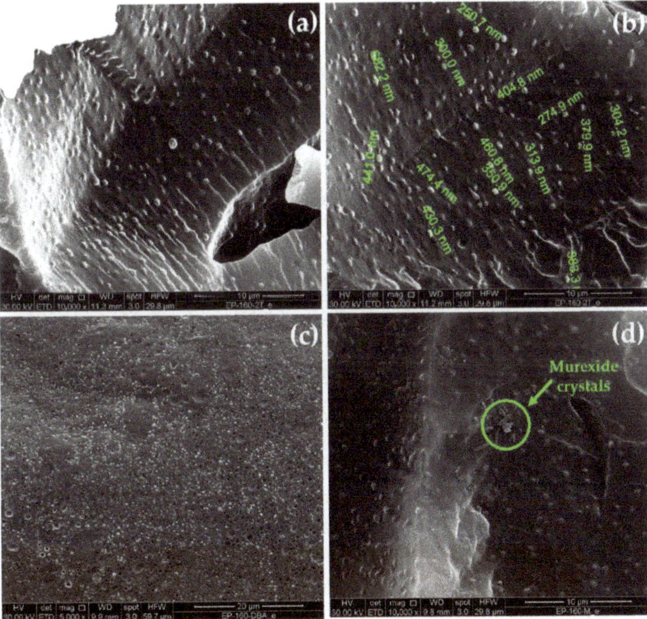

Figure 13. SEM images of the etched surface of: (**a**,**b**) the EP-160-T sample; (**c**) the EP-160-DBA sample; (**d**) the EP-160-M sample.

Chemical composition and conditions chosen for the functionalization process allow obtaining elastomeric domains of dimensions that do not exceed 500–600 nanometers, as seen in Figure 13a,b for the sample EP-160-T (where also the dimension of rubber domains are indicated on the left image), in Figure 13c for the sample EP-160-DBA, and in Figure 13d for the sample EP-160-M. In the case of M filler, very small crystallites of this component (not completely solubilized in the resin) are also observed in the solidified matrix.

Extensive studies have been carried out on the solubility of self-healing fillers in the components of the epoxy formulation before and after the curing process. Experimental tests and results are reported in Section S3 of the S.M. (see Figure S3-1–S3-4).

It is worth noting that the morphological feature obtained made it possible to maximize the interphase area between the rubber domains.

3.4. Evaluation of Self-Healing Efficiency

Ep-R-160, Ep-R-160-DBA, Ep-R-160-T, and Ep-R-160-M samples were tested to evaluate the self-healing efficiency, as described in Section 2.3 Methods.

The results evaluated with the TDCB geometry at 25 °C are shown in Figure 14. In particular, Figure 14a–c show the behaviour of load as a function of the displacement for the virgin samples Ep-R-160-DBA, Ep-R-160-T, and Ep-R-160-M (continuous line) and the same healed samples (dashed line). Figure 14d depicts the histogram of the values of healing efficiency, calculated by Equation (1), using the values of P_{CV} and P_{CH}, reported in Table 5. The introduction of the self-healing fillers causes a recovery of mechanical properties. The healing efficiency value is higher than 69% for all fillers. The highest value was detected for the sample with DBA, for which a value of 88% was found. Statistically, the tests performed on different samples with the same composition and treatments showed values with standard deviations of around 5%

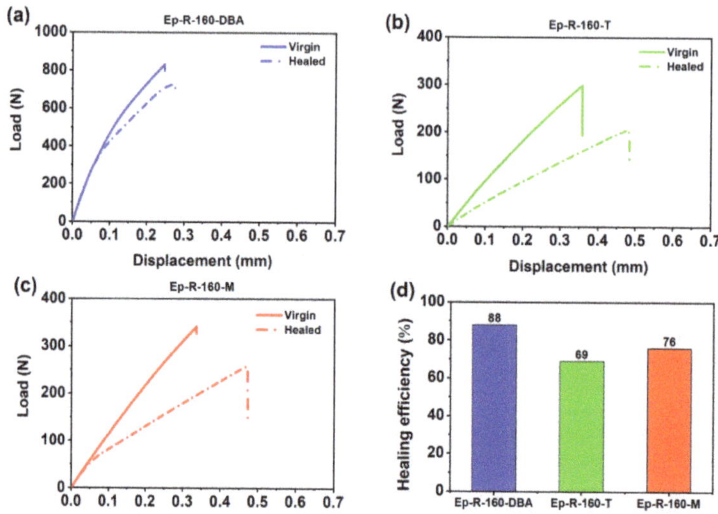

Figure 14. Load-Displacement curves for the samples (**a**) Ep-R-160-DBA; (**b**) Ep-R-160-T; (**c**) Ep-R-160-M; and (**d**) histogram illustrating the healing efficiency values.

Table 5. Critical fracture loads values of the analyzed samples.

Sample	P_{CV} (N)	P_{CH} (N)
Ep-R-160-DBA	831	730
Ep-R-160-T	299	207
Ep-R-160-M	341	259

Most likely, the high healing efficiency values are due to polar groups in the self-healing fillers, such as O-H, N-H, and C=O functional groups. These groups act as hydrogen bond donors or acceptors with the polar groups of the functionalized epoxy-precursor, establishing cumulative effects of the attractive reversible interactions. The attractive reversible interactions are probably established much more effectively in the higher mobility domains of the polymeric chains, therefore, at the interface between the rubbery domains and the matrix resin (where the degree of crosslinking is reduced).

Notably, the peculiar morphology of these samples, with rubber domains of dimensions in the order of a few hundred nanometres (see Figure 13), contributes to increasing the areas at reduced crosslinking density.

An example of possible interactions based on reversible hydrogen bonds is shown in Figure 15, which depicts the H-bond interactions built between the hydroxyl groups, acting as H-bond donor sites, of the cured epoxy resin and the carbonyl groups, acting as H-bond acceptor sites, of the DBA filler. The resulting supramolecular network activates self-healing mechanisms in the formulated materials.

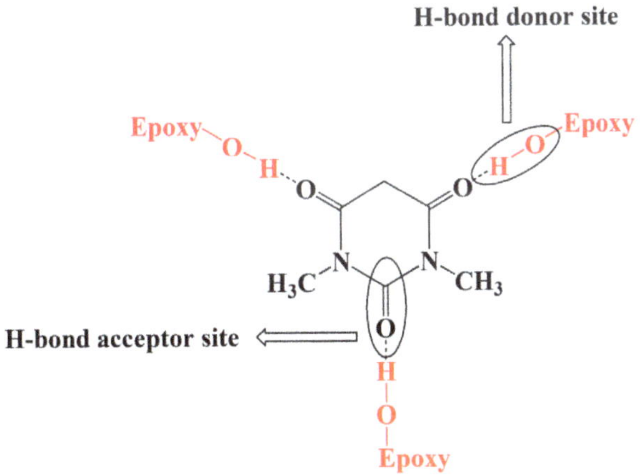

Figure 15. Picture illustrating the H-bond interactions between the hydroxyl groups of the epoxy resin (red highlighted) and the carbonyl groups of the DBA filler.

Hydrogen bonds are widely recognized to activate self-healing processes [70–72], it has mostly been applied in rubbery matrices [47,73–75]. In our samples, we have small rubber domains distributed throughout the whole sample. The employment of reversible hydrogen bonds in epoxy resins has only been reported in specially modified systems through the addition of hydrogen bonding moieties such as ureido- pyrimidinone (UPy) [76,77], barbiturate and thymine functionalized MWCNTs [48], and amide motifs [78]. Reference [48] refers to our previously published paper on self-healing resins. These papers describe self-healing resins, where barbiturate and thymine functionalized MWCNTs were embedded in a rubber-toughened epoxy formulation. The groups with hydrogen donor or acceptor sites lead to reversible MWCNTs-bridges through the matrix due to strong, attractive interactions between the rubber phase, finely dispersed in the matrix, and MWCNT walls. Healing efficiencies higher than 50% have been found for both functional groups. Dynamic mechanical analysis (DMA) evidenced an enhancement in epoxy chain movements due to the rubber phase's micro/nanodomains, enabling self-healing behavior by recovering the critical fracture load. In this paper, molecules capable of establishing strong cumulative effects due to hydrogen bonds have been dispersed in the toughened thermosetting matrix.

The authors demonstrate that the effect of the functionalization temperature should not be underestimated to evaluate the healing ability of the materials. Figure S4 of Section S4 of S.M. shows the results of self-healing tests carried out for the samples Ep-R-120 and

Ep-R-160 loaded with the DBA filler (Ep-R-120-DBA and Ep-R-160-DBA, respectively). The comparison between Ep-R-120-DBA and Ep-R-160-DBA shows that the functionalization temperature of 160 °C allows an increase in the self-healing efficiency, as expected considering the DMA results in Figure 11, where a lower value of T_g is observed for the sample functionalized at 160 °C.

DMA results of this last sample including the self-healing fillers (see Figure 16) show that the T_g values of the complete formulations are lower than the same sample without filler (132 °C, see Figure 11) of about 10 °C for Ep-R-160-DBA, Ep-R-160-T, and Ep-R-160-M.

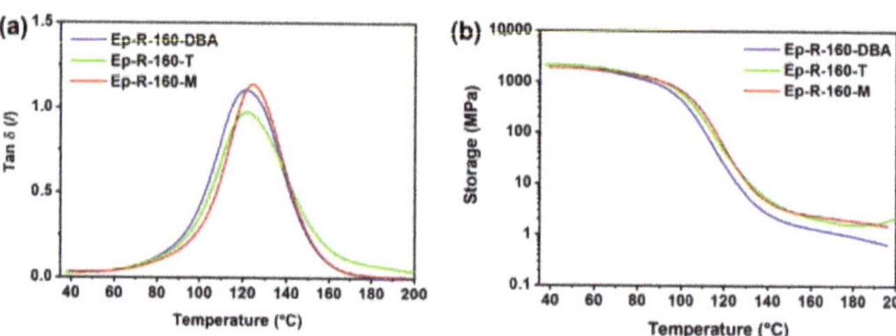

Figure 16. DMA curves: (**a**) Tan δ vs. temperature; (**b**) Storage modulus vs. temperature of the samples Ep-R-160-DBA, Ep-R-160-T, and Ep-R-160-M.

The modulus drop (see Figure 16b) is slightly anticipated for the sample containing the DBA filler for which the highest healing efficiency value is recorded.

4. Conclusions

In this study, a bifunctional epoxy precursor was functionalized with a rubber phase. Then, self-healing molecules were added to the rubber-toughened bifunctional epoxy resin to confer it auto-repair function. The functionalization process allows obtaining elastomeric domains not exceeding 500–600 nanometers. After the curing cycle of the formulated self-healing samples, the effect of the temperature of epoxy precursor functionalization on the properties of the samples was evaluated in terms of self-healing efficiency, DMA, and TGA. It was found that the higher functionalization temperature of 160 °C can better promote the reaction between the rubber phase and the epoxy precursor during the functionalization process. The more efficient interactions affect the resin structure and its thermal and mechanical properties. Self-healing efficiency analysis highlights the crucial role of functionalization temperature in increasing self-healing ability. Samples functionalized at 160 °C manifest healing efficiency higher than 69%. The highest value (88%) was detected for the sample with DBA filler. The healing efficiency of the same sample functionalized at 120 °C decreases to 52%. Results from DMA evidence that a lower value of T_g (observed for the sample functionalized at the higher temperature of 160 °C) allows for obtaining the highest healing efficiency. Thermal stability in air higher than 250 °C was observed for all formulated samples.

Supplementary Materials: The following supporting information can be downloaded at https://www.mdpi.com/article/10.3390/aerospace10050476/s1: Figure S1: Dimensions of the TDCB geometry specimen (on the left); EP-R-160 sample located in the INSTRON instrument (on the right). Numerical values of the lengths are expressed in mm; Figure S2: DSC curves of the samples EP-R-160, EP-R-160-DBA, EP-R-160-T, EP-R-160-M, before the curing process (after the functionalization reaction) and after the curing process; Figure S3-1: Photos of the self-healing fillers after incorporation into the Gurit precursor ECC at room temperature; Figure S3-2: Photos of the self-healing fillers before incorporation into the Gurit hardener MHHPA; Figure S3-3: Photos of the self-healing fillers after incorporation into the Gurit hardener MHHPA; Figure S3-4: Visualization of the complete solubilization of the self-healing fillers in the Gurit hardener MHHPA; Figure S4: (a) Load-Displacement curves for the sample Ep-R-120-DBA; and (b) histogram illustrating the healing efficiency values for the samples Ep-R-120-DBA and EP-R-160-DBA.

Author Contributions: Conceptualization, methodology, supervision, project administration, writing—original draft, writing–review and editing L.G.; Formal analysis; investigation M.R., C.N., R.L. and A.S.; Resources M.C.; Conceptualization, formal analysis, data curation, writing–original draft, investigation, writing–review and editing E.C. and L.V. All authors have read and agreed to the published version of the manuscript.

Funding: The research leading to these results was funded by the European Union's Seventh Framework Programme for Research, Technological Development and Demonstration, under Grant Agreement N 313978.

Data Availability Statement: Not applicable.

Conflicts of Interest: The authors declare no conflict of interest.

References

1. Fugolin, A.P.; Pfeifer, C.S. Engineering a new generation of thermoset self-healing polymers based on intrinsic approaches. *JADA Found. Sci.* **2022**, *1*, 100014. [CrossRef]
2. Ha, Y.-M.; Seo, H.C.; Kim, Y.-O.; Khil, M.-S.; Cho, J.W.; Lee, J.-S.; Jung, Y.C. Effects of hard segment of polyurethane with disulfide bonds on shape memory and self-healing ability. *Macromol. Res.* **2020**, *28*, 234–240. [CrossRef]
3. Lee, W.-J.; Oh, H.-G.; Cha, S.-H. A brief review of self-healing polyurethane based on dynamic chemistry. *Macromol. Res.* **2021**, *29*, 649–664. [CrossRef]
4. Li, B.; Cao, P.-F.; Saito, T.; Sokolov, A.P. Intrinsically Self-Healing Polymers: From Mechanistic Insight to Current Challenges. *Chem. Rev.* **2022**, *132*, 701–735. [CrossRef] [PubMed]
5. Montero de Espinosa, L.; Meesorn, W.; Moatsou, D.; Weder, C. Bioinspired polymer systems with stimuli-responsive mechanical properties. *Chem. Rev.* **2017**, *117*, 12851–12892. [CrossRef] [PubMed]
6. Nik Md Noordin Kahar, N.N.F.; Osman, A.F.; Alosime, E.; Arsat, N.; Mohammad Azman, N.A.; Syamsir, A.; Itam, Z.; Abdul Hamid, Z.A. The versatility of polymeric materials as self-healing agents for various types of applications: A review. *Polymers* **2021**, *13*, 1194. [CrossRef] [PubMed]
7. Thakur, V.K.; Kessler, M.R. Self-healing polymer nanocomposite materials: A review. *Polymer* **2015**, *69*, 369–383. [CrossRef]
8. Utrera-Barrios, S.; Verdejo, R.; López-Manchado, M.A.; Santana, M.H. Evolution of self-healing elastomers, from extrinsic to combined intrinsic mechanisms: A review. *Mater. Horiz.* **2020**, *7*, 2882–2902. [CrossRef]
9. Wemyss, A.M.; Bowen, C.; Plesse, C.; Vancaeyzeele, C.; Nguyen, G.T.; Vidal, F.; Wan, C. Dynamic crosslinked rubbers for a green future: A material perspective. *Mater. Sci. Eng. R Rep.* **2020**, *141*, 100561. [CrossRef]
10. Zhu, M.; Liu, J.; Gan, L.; Long, M. Research progress in bio-based self-healing materials. *Eur. Polym. J.* **2020**, *129*, 109651. [CrossRef]
11. Blaiszik, B.J.; Kramer, S.L.; Olugebefola, S.C.; Moore, J.S.; Sottos, N.R.; White, S.R. Self-healing polymers and composites. *Annu. Rev. Mater. Res.* **2010**, *40*, 179–211. [CrossRef]
12. Brown, E.N.; Sottos, N.R.; White, S.R. Fracture testing of a self-healing polymer composite. *Exp. Mech.* **2002**, *42*, 372–379. [CrossRef]
13. Guadagno, L.; Longo, P.; Raimondo, M.; Naddeo, C.; Mariconda, A.; Sorrentino, A.; Vittoria, V.; Iannuzzo, G.; Russo, S. Cure behavior and mechanical properties of structural self-healing epoxy resins. *J. Polym. Sci. Part B Polym. Phys.* **2010**, *48*, 2413–2423. [CrossRef]
14. Raimondo, M.; De Nicola, F.; Volponi, R.; Binder, W.; Michael, P.; Russo, S.; Guadagno, L. Self-repairing CFRPs targeted towards structural aerospace applications. *Int. J. Struct. Integr.* **2016**, *7*, 656–670. [CrossRef]
15. White, S.R.; Sottos, N.R.; Geubelle, P.H.; Moore, J.S.; Kessler, M.R.; Sriram, S.; Brown, E.N.; Viswanathan, S. Autonomic healing of polymer composites. *Nature* **2001**, *409*, 794–797. [CrossRef]
16. Jin, H.; Miller, G.M.; Pety, S.J.; Griffin, A.S.; Stradley, D.S.; Roach, D.; Sottos, N.R.; White, S.R. Fracture behavior of a self-healing, toughened epoxy adhesive. *Int. J. Adhes. Adhes.* **2013**, *44*, 157–165. [CrossRef]
17. Dry, C. Procedures developed for self-repair of polymer matrix composite materials. *Compos. Struct.* **1996**, *35*, 263–269. [CrossRef]
18. Dry, C.M.; Sottos, N.R. Passive smart self-repair in polymer matrix composite materials. In *Smart Structures and Materials 1993: Smart Materials*; SPIE: Bellingham, WA, USA, 1993; pp. 438–444.
19. Williams, H.; Trask, R.; Knights, A.; Williams, E.; Bond, A.I. Biomimetic reliability strategies for self-healing vascular networks in engineering materials. *J. R. Soc. Interface* **2008**, *5*, 735–747. [CrossRef]
20. Williams, H.; Trask, R.; Weaver, P.; Bond, I. Minimum mass vascular networks in multifunctional materials. *J. R. Soc. Interface* **2008**, *5*, 55–65. [CrossRef]
21. Goyal, M.; Agarwal, S.N.; Bhatnagar, N. A review on self-healing polymers for applications in spacecraft and construction of roads. *J. Appl. Polym. Sci.* **2022**, *139*, e52816. [CrossRef]
22. Chen, X.; Dam, M.A.; Ono, K.; Mal, A.; Shen, H.; Nutt, S.R.; Sheran, K.; Wudl, F. A thermally re-mendable cross-linked polymeric material. *Science* **2002**, *295*, 1698–1702. [CrossRef] [PubMed]

23. Chen, X.; Wudl, F.; Mal, A.K.; Shen, H.; Nutt, S.R. New thermally remendable highly cross-linked polymeric materials. *Macromolecules* **2003**, *36*, 1802–1807. [CrossRef]
24. Ehrhardt, D.; Mangialetto, J.; Bertouille, J.; Van Durme, K.; Van Mele, B.; Van den Brande, N. Self-healing in mobility-restricted conditions maintaining mechanical robustness: Furan–maleimide diels–alder cycloadditions in polymer networks for ambient applications. *Polymers* **2020**, *12*, 2543. [CrossRef] [PubMed]
25. Jung, S.; Oh, J.K. Well-defined methacrylate copolymer having reactive maleimide pendants for fabrication of thermally-labile crosslinked networks with robust self-healing. *Mater. Today Commun.* **2017**, *13*, 241–247. [CrossRef]
26. Murphy, E.B.; Bolanos, E.; Schaffner-Hamann, C.; Wudl, F.; Nutt, S.R.; Auad, M.L. Synthesis and characterization of a single-component thermally remendable polymer network: Staudinger and Stille revisited. *Macromolecules* **2008**, *41*, 5203–5209. [CrossRef]
27. Park, J.S.; Kim, H.S.; Hahn, H.T. Healing behavior of a matrix crack on a carbon fiber/mendomer composite. *Compos. Sci. Technol.* **2009**, *69*, 1082–1087. [CrossRef]
28. Park, J.S.; Takahashi, K.; Guo, Z.; Wang, Y.; Bolanos, E.; Hamann-Schaffner, C.; Murphy, E.; Wudl, F.; Hahn, H.T. Towards development of a self-healing composite using a mendable polymer and resistive heating. *J. Compos. Mater.* **2008**, *42*, 2869–2881. [CrossRef]
29. Peterson, A.M.; Jensen, R.E.; Palmese, G.R. Reversibly cross-linked polymer gels as healing agents for epoxy– amine thermosets. *ACS Appl. Mater. Interfaces* **2009**, *1*, 992–995. [CrossRef]
30. Peterson, A.M.; Jensen, R.E.; Palmese, G.R. Room-temperature healing of a thermosetting polymer using the Diels–Alder reaction. *ACS Appl. Mater. Interfaces* **2010**, *2*, 1141–1149. [CrossRef]
31. Plaisted, T.A.; Nemat-Nasser, S. Quantitative evaluation of fracture, healing and re-healing of a reversibly cross-linked polymer. *Acta Mater.* **2007**, *55*, 5684–5696. [CrossRef]
32. Pratama, P.A.; Sharifi, M.; Peterson, A.M.; Palmese, G.R. Room temperature self-healing thermoset based on the Diels–Alder reaction. *ACS Appl. Mater. Interfaces* **2013**, *5*, 12425–12431. [CrossRef] [PubMed]
33. Wang, D.; Chen, S.; Zhao, J.; Zhang, Z. Synthesis and characterization of self-healing cross-linked non-isocyanate polyurethanes based on Diels-Alder reaction with unsaturated polyester. *Mater. Today Commun.* **2020**, *23*, 101138. [CrossRef]
34. Hayes, S.; Jones, F.; Marshiya, K.; Zhang, W. A self-healing thermosetting composite material. *Compos. Part A Appl. Sci. Manuf.* **2007**, *38*, 1116–1120. [CrossRef]
35. Luo, X.; Ou, R.; Eberly, D.E.; Singhal, A.; Viratyaporn, W.; Mather, P.T. A thermoplastic/thermoset blend exhibiting thermal mending and reversible adhesion. *ACS Appl. Mater. Interfaces* **2009**, *1*, 612–620. [CrossRef]
36. Kalista, S.J., Jr.; Ward, T.C. Thermal characteristics of the self-healing response in poly (ethylene-co-methacrylic acid) copolymers. *J. R. Soc. Interface* **2007**, *4*, 405–411. [CrossRef]
37. Kalista, S.J., Jr.; Ward, T.C.; Oyetunji, Z. Self-healing of poly (ethylene-co-methacrylic acid) copolymers following projectile puncture. *Mech. Adv. Mater. Struct.* **2007**, *14*, 391–397. [CrossRef]
38. Varley, R.J.; van der Zwaag, S. Towards an understanding of thermally activated self-healing of an ionomer system during ballistic penetration. *Acta Mater.* **2008**, *56*, 5737–5750. [CrossRef]
39. Varley, R.J.; van der Zwaag, S. Development of a quasi-static test method to investigate the origin of self-healing in ionomers under ballistic conditions. *Polym. Test.* **2008**, *27*, 11–19. [CrossRef]
40. McGarel, O.J.; Wool, R.P. Craze growth and healing in polystyrene. *J. Polym. Sci. Part B Polym. Phys.* **1987**, *25*, 2541–2560. [CrossRef]
41. O'Connor, K.; Wool, R. Optical studies of void formation and healing in styrene-isoprene-styrene block copolymers. *J. Appl. Phys.* **1980**, *51*, 5075–5079. [CrossRef]
42. Wool, R.; O'connor, K. A theory crack healing in polymers. *J. Appl. Phys.* **1981**, *52*, 5953–5963. [CrossRef]
43. Cordier, P.; Tournilhac, F.; Soulié-Ziakovic, C.; Leibler, L. Self-healing and thermoreversible rubber from supramolecular assembly. *Nature* **2008**, *451*, 977–980. [CrossRef]
44. Montarnal, D.; Tournilhac, F.; Hidalgo, M.; Couturier, J.-L.; Leibler, L. Versatile one-pot synthesis of supramolecular plastics and self-healing rubbers. *J. Am. Chem. Soc.* **2009**, *131*, 7966–7967. [CrossRef] [PubMed]
45. Longo, P.; Mariconda, A.; Calabrese, E.; Raimondo, M.; Naddeo, C.; Vertuccio, L.; Russo, S.; Iannuzzo, G.; Guadagno, L. Development of a new stable ruthenium initiator suitably designed for self-repairing applications in high reactive environments. *J. Ind. Eng. Chem.* **2017**, *54*, 234–251. [CrossRef]
46. Zhang, F.; Zhang, L.; Yaseen, M.; Huang, K. A review on the self-healing ability of epoxy polymers. *J. Appl. Polym. Sci.* **2021**, *138*, 50260. [CrossRef]
47. Guadagno, L.; Raimondo, M.; Naddeo, C.; Vertuccio, L.; Russo, S.; Iannuzzo, G.; Calabrese, E. Rheological, Thermal and Mechanical Characterization of Toughened Self-Healing Supramolecular Resins, Based on Hydrogen Bonding. *Nanomaterials* **2022**, *12*, 4322. [CrossRef]
48. Guadagno, L.; Vertuccio, L.; Naddeo, C.; Calabrese, E.; Barra, G.; Raimondo, M.; Sorrentino, A.; Binder, W.H.; Michael, P.; Rana, S. Reversible self-healing carbon-based nanocomposites for structural applications. *Polymers* **2019**, *11*, 903. [CrossRef]
49. Ricciardi, M.; Papa, I.; Langella, A.; Langella, T.; Lopresto, V.; Antonucci, V. Mechanical properties of glass fibre composites based on nitrile rubber toughened modified epoxy resin. *Compos. Part B Eng.* **2018**, *139*, 259–267. [CrossRef]

50. Karger-Kocsis, J.; Friedrich, K. Fatigue crack propagation and related failure in modified, andhydride-cured epoxy resins. *Colloid Polym. Sci.* **1992**, *270*, 549–562. [CrossRef]
51. Karger-Kocsis, J.; Friedrich, K. Microstructure-related fracture toughness and fatigue crack growth behaviour in toughened, anhydride-cured epoxy resins. *Compos. Sci. Technol.* **1993**, *48*, 263–272. [CrossRef]
52. Ramos, V.D.; Da Costa, H.M.; Soares, V.L.; Nascimento, R.S. Modification of epoxy resin: A comparison of different types of elastomer. *Polym. Test.* **2005**, *24*, 387–394. [CrossRef]
53. Guadagno, L.; Vertuccio, L.; Foglia, F.; Raimondo, M.; Barra, G.; Sorrentino, A.; Pantani, R.; Calabrese, E. Flexible eco-friendly multilayer film heaters. *Compos. Part B Eng.* **2021**, *224*, 109208. [CrossRef]
54. Vertuccio, L.; Foglia, F.; Pantani, R.; Romero-Sánchez, M.; Calderón, B.; Guadagno, L. Carbon nanotubes and expanded graphite based bulk nanocomposites for de-icing applications. *Compos. Part B Eng.* **2021**, *207*, 108583. [CrossRef]
55. Guadagno, L.; Aliberti, F.; Longo, R.; Raimondo, M.; Pantani, R.; Sorrentino, A.; Catauro, M.; Vertuccio, L. Electrical anisotropy controlled heating of acrylonitrile butadiene styrene 3D printed parts. *Mater. Des.* **2023**, *225*, 111507. [CrossRef]
56. Böger, L.; Wichmann, M.H.; Meyer, L.O.; Schulte, K. Load and health monitoring in glass fibre reinforced composites with an electrically conductive nanocomposite epoxy matrix. *Compos. Sci. Technol.* **2008**, *68*, 1886–1894. [CrossRef]
57. Hu, N.; Itoi, T.; Akagi, T.; Kojima, T.; Xue, J.; Yan, C.; Atobe, S.; Fukunaga, H.; Yuan, W.; Ning, H. Ultrasensitive strain sensors made from metal-coated carbon nanofiller/epoxy composites. *Carbon* **2013**, *51*, 202–212. [CrossRef]
58. Guadagno, L.; Longo, R.; Aliberti, F.; Lamberti, P.; Tucci, V.; Pantani, R.; Spinelli, G.; Catauro, M.; Vertuccio, L. Role of MWCNTs Loading in Designing Self-Sensing and Self-Heating Structural Elements. *Nanomaterials* **2023**, *13*, 495. [CrossRef]
59. Prolongo, S.; Moriche, R.; Del Rosario, G.; Jiménez-Suárez, A.; Prolongo, M.; Ureña, A. Joule effect self-heating of epoxy composites reinforced with graphitic nanofillers. *J. Polym. Res.* **2016**, *23*, 189. [CrossRef]
60. Yang, P.; Ghosh, S.; Xia, T.; Wang, J.; Bissett, M.A.; Kinloch, I.A.; Barg, S. Joule Heating and mechanical properties of epoxy/graphene based aerogel composite. *Compos. Sci. Technol.* **2022**, *218*, 109199. [CrossRef]
61. Guadagno, L.; Sorrentino, A.; Delprat, P.; Vertuccio, L. Design of multifunctional composites: New strategy to save energy and improve mechanical performance. *Nanomaterials* **2020**, *10*, 2285. [CrossRef]
62. Vertuccio, L.; Guadagno, L.; Spinelli, G.; Russo, S.; Iannuzzo, G. Effect of carbon nanotube and functionalized liquid rubber on mechanical and electrical properties of epoxy adhesives for aircraft structures. *Compos. Part B Eng.* **2017**, *129*, 1–10. [CrossRef]
63. Guadagno, L.; Vertuccio, L.; Barra, G.; Naddeo, C.; Sorrentino, A.; Lavorgna, M.; Raimondo, M.; Calabrese, E. Eco-friendly polymer nanocomposites designed for self-healing applications. *Polymer* **2021**, *223*, 123718. [CrossRef]
64. Coleman, M.M.; Skrovanek, D.J.; Hu, J.; Painter, P.C. Hydrogen bonding in polymer blends. 1. FTIR studies of urethane-ether blends. *Macromolecules* **1988**, *21*, 59–65. [CrossRef]
65. Neuvonen, H.; Neuvonen, K. Correlation analysis of carbonyl carbon 13C NMR chemical shifts, IR absorption frequencies and rate coefficients of nucleophilic acyl substitutions. A novel explanation for the substituent dependence of reactivity. *J. Chem. Soc. Perkin Trans. 2* **1999**, *7*, 1497–1502. [CrossRef]
66. Li, G.; Xu, M.; Larsen, S.; Grassian, V. Photooxidation of cyclohexane and cyclohexene in BaY. *J. Mol. Catal. A Chem.* **2003**, *194*, 169–180. [CrossRef]
67. Marquardt, D.W. An algorithm for least-squares estimation of nonlinear parameters. *J. Soc. Ind. Appl. Math.* **1963**, *11*, 431–441. [CrossRef]
68. Maddams, W. The scope and limitations of curve fitting. *Appl. Spectrosc.* **1980**, *34*, 245–267. [CrossRef]
69. Mark, H.F. *Encyclopedia of Polymer Science and Technology, 15 Volume Set*; Wiley: New York, NY, USA, 2014; Volume 14.
70. Bouteiller, L. Assembly via hydrogen bonds of low molar mass compounds into supramolecular polymers. *Hydrog. Bond. Polym.* **2007**, *207*, 79–112.
71. García, S.; Fischer, H.; Van Der Zwaag, S. A critical appraisal of the potential of self healing polymeric coatings. *Prog. Org. Coat.* **2011**, *72*, 211–221. [CrossRef]
72. Peñas-Caballero, M.; Santana, M.H.; Verdejo, R.; Lopez-Manchado, M.A. Measuring self-healing in epoxy matrices: The need for standard conditions. *React. Funct. Polym.* **2021**, *161*, 104847. [CrossRef]
73. Cao, L.; Yuan, D.; Xu, C.; Chen, Y. Biobased, self-healable, high strength rubber with tunicate cellulose nanocrystals. *Nanoscale* **2017**, *9*, 15696–15706. [CrossRef] [PubMed]
74. Thangavel, G.; Tan, M.W.M.; Lee, P.S. Advances in self-healing supramolecular soft materials and nanocomposites. *Nano Converg.* **2019**, *6*, 29. [CrossRef] [PubMed]
75. Utrera-Barrios, S.; Hernández Santana, M.; Verdejo, R.; López-Manchado, M.A. Design of rubber composites with autonomous self-healing capability. *ACS Omega* **2020**, *5*, 1902–1910. [CrossRef] [PubMed]
76. Kostopoulos, V.; Kotrotsos, A.; Tsantzalis, S.; Tsokanas, P.; Loutas, T.; Bosman, A. Toughening and healing of continuous fibre reinforced composites by supramolecular polymers. *Compos. Sci. Technol.* **2016**, *128*, 84–93. [CrossRef]

77. Zhang, P.; Kan, L.; Zhang, X.; Li, R.; Qiu, C.; Ma, N.; Wei, H. Supramolecularly toughened and elastic epoxy resins by grafting 2-ureido-4 [1H]-pyrimidone moieties on the side chain. *Eur. Polym. J.* **2019**, *116*, 126–133. [CrossRef]
78. Villani, M.; Deshmukh, Y.S.; Camlibel, C.; Esteves, A.C.C. Superior relaxation of stresses and self-healing behavior of epoxy-amine coatings. *RSC Adv.* **2016**, *6*, 245–259. [CrossRef]

 *aerospace* 

Article

Additive Manufacturing of 17-4PH Alloy: Tailoring the Printing Orientation for Enhanced Aerospace Application Performance

Sandor Endre Kovacs [1], Tamas Miko [2], Enrico Troiani [3], Dionysios Markatos [4], Daniel Petho [2], Greta Gergely [2,*], Laszlo Varga [1] and Zoltan Gacsi [2]

1. Institute of Chemical Metallurgy and Foundry Engineering, University of Miskolc, 3515 Miskolc, Hungary; kovacssandorendre@gmail.com (S.E.K.); laszlo.varga1@uni-miskolc.hu (L.V.)
2. Institute of Physical Metallurgy, Metalforming and Nanotechnology, University of Miskolc, 3515 Miskolc, Hungary; tamas.miko@uni-miskolc.hu (T.M.); daniel.petho@uni-miskolc.hu (D.P.); zoltan.gacsi@uni-miskolc.hu (Z.G.)
3. Department of Industrial Engineering DIN, University of Bologna, Via Fontanelle 40, 47121 Forlì, Italy; enrico.troiani@unibo.it
4. Laboratory of Technology & Strength of Materials (LTSM), Department of Mechanical Engineering and Aeronautics, University of Patras, 26504 Patras, Greece; dmark@upatras.gr

* Correspondence: greta.gergely@uni-miskolc.hu

Abstract: Additive manufacturing (AM) is one of the fastest-growing markets of our time. During its journey in the past 30 years, its key to success has been that it can easily produce extremely complex shapes and is not limited by tooling problems when a change in geometry is desired. This flexibility leads to possible solutions for creating lightweight structural elements while keeping the mechanical properties at a stable reserve factor value. In the aerospace industry, several kinds of structural elements for fuselage and wing parts are made from different kinds of steel alloys, such as 17-4PH stainless steel, which are usually milled from a block material made using conventional processing (CP) methods. However, these approaches are limited when a relatively small element must withstand greater forces that can occur during flight. AM can bridge this problem with a new perspective, mainly using thin walls and complex shapes while maintaining the ideal sizes. The downside of the elements made using AM is that the quality of the final product is highly dependent on the build/printing orientation, an issue extensively studied and addressed by researchers in the field. During flight, some components may experience forces that predominantly act in a single direction. With this in mind, we created samples with the desired orientation to maximize material properties in a specific direction. The goal of this study was to demonstrate that an additively manufactured part, produced using laser powder bed fusion (LPBF), with a desired build orientation has exceptional properties compared to parts produced via conventional methods. To assess the impact of the build orientation on the LPBF parts' properties, one-dimensional tensile and dynamic fracture toughness tests were deployed.

Keywords: additive manufacturing; 17-4PH stainless steel; 3D printing; laser powder bed fusion; printing orientation

Citation: Kovacs, S.E.; Miko, T.; Troiani, E.; Markatos, D.; Petho, D.; Gergely, G.; Varga, L.; Gacsi, Z. Additive Manufacturing of 17-4PH Alloy: Tailoring the Printing Orientation for Enhanced Aerospace Application Performance. *Aerospace* **2023**, *10*, 619. https://doi.org/10.3390/aerospace10070619

Academic Editors: Rhys Jones and Spiros Pantelakis

Received: 28 February 2023
Revised: 23 June 2023
Accepted: 27 June 2023
Published: 7 July 2023

1. Introduction

Additive manufacturing (AM), or 3D printing, is a computer-controlled process that creates three-dimensional objects by depositing materials, usually in layers, thus enabling the creation of lighter and stronger parts and systems. The benefits of AM are numerous, allowing the creation of parts with complex geometries and low material waste, therefore providing cost reduction for high-value components while reducing lead times. In addition, parts that previously required assembly from multiple components can be fabricated as a single object with improved strength and durability. Furthermore, AM can be used to fabricate unique objects or replacement pieces for parts that are no longer produced.

Having demanding standards in terms of performance and weight reduction, the aerospace industry has been one of the first to adopt additive manufacturing since the deployment of such techniques can advance part-making methods [1,2]. Common AM applications include environmental control system ducting, custom cosmetic aircraft interior components, rocket engine components and combustor liners [3,4]. Certification requirements for these installed parts in aircrafts are discussed in *USAF Structures Bulletin EZ-19-01*, mandating a linear elastic fracture assessment for all load-bearing AM parts [5]. These requirements are based on US MIL-STD-1530D, which mandates linear elastic fracture mechanics (LEFM) and does not allow the use of S–N curves in the design/assessment of a load-bearing component on USAF aircrafts. This standard applies to the entire structure, type or procurement strategy for the entire life cycle of the aircraft [6]. AM also helps deliver complex, consolidated parts with enhanced strength, which is a prerequisite in this industry. To justify these AM parts, there are several ongoing studies on durability and damage tolerance certification [7], with a focus on the aforementioned standards, while others are taking different approaches towards the same goal [8].

What AM contributes to the already existing manufacturing methods is a whole new perspective. The layer-by-layer deposition provides product designers with more freedom to create innovative, high-performance parts. These parts can be further optimized using finite element methods (FEM) aimed specifically at AM processes [9]. Due to the fact that a minimum amount of material is used, parts are generally more cost effective and produce less waste, making the product more sustainable. Conversely, the demanding requirements of raw materials are associated with high costs, as well as an inferior surface quality and dimensional precision, necessitating the application of surface finishes to fulfill quality standards and specifications. Furthermore, the layering and multiple interfaces of additive manufacturing can cause defects in the product, whereby post-processing is needed to rectify any quality issues. Among the many 3D printing processes, such as atomic diffusion additive manufacturing (ADAM) and metal fused filament fabrication (MFFF) [10], the LPBF technique, namely, selective laser melting (SLM), is one of the most promising [11]. This method uses a continuous powder bed melted with a high-precision energy source, which is usually a solid-state laser or an electron beam [3]. During the process, this energy source scans the cross-sections layer by layer until the part is finished. This bottom-up approach allows a variety of detailed modifications to existing components, where traditional machining technologies suffer from the limitations of their operation. The issues of low surface quality and high thermal residual stresses occurring in the final parts are most of the time related to the heat source [12]. Both issues can be mitigated with the mechanical machining of the surfaces or with more complex technologies, such as hot isostatic pressing [13]. Another important parameter is the build/printing orientation that can yield different mechanical properties along different directions [14]. Since numerous components are subjected to loading along specific directions, the LPBF material does not require isotropic behavior. Furthermore, the LPBF process can accommodate various types of raw materials. Although lightweight alloys are prevalent in commercial aircraft, steels are also widely used, particularly in turbine blades, missile and rocket fittings, undercarriage components and fasteners. Of all the steel varieties, high-performance stainless steels (SSs) are particularly significant due to their remarkable ability to attain exceptional strength throughout the process of aging. The 17-4PH SS material represents one of the most common types of martensitic precipitation-hardening SSs. Owing to its high chromium, nickel and copper content, it has good corrosion resistance and high fatigue resistance, and in the aged state, the copper-based precipitates ensure high strength and high toughness, which are requirements for SS alloy applications in the aerospace industry, even in critical components, such as fan or propeller blades [15]. Typically used where designers need more reliability in their products, this alloy is great for gate valves, chemical processing equipment, pump shafts, gears, ball bearings, bushings and even fasteners. It is also one of the most common types of stainless-steel powders in the powder metallurgy and AM industry. Despite using identical raw materials, the mechanical properties of parts manufactured using different methods can vary significantly due to differences in the resulting microstructures. The properties of CP-17-4PH, as shown in Table 1, including yield strength (YS), ultimate tensile strength (UTS), hardness (HV), toughness (IE) and elongation (EL), are determined using the heat treatment process according to ASTM A564. It is worth noting that the properties can be affected by the high temperature used in the process.

Table 1. Typical mechanical properties of commercially manufactured 17-4PH SS according to ASTM A564.

Condition	YS (MPa)	UTS (MPa)	Elongation (%)	Hardness, Vickers	YS/UTS	Charpy Impact Energy (Joule)
H900	1171	1309	10	410	0.89	22
H1025	999	1068	12	349	0.94	54
H1075	861	999	13	328	0.86	61
H1150	723	930	16	292	0.78	75

Achieving a desired build orientation/direction is crucial for ensuring that a 3D-printed part can withstand the forces that will act upon it. In this work, the desired orientation was achieved using selective laser melting (SLM) and 17-4PH SS powder, which is a widely used material in aeronautical applications. The properties of the produced specimens were subsequently compared with corresponding parts made using traditional manufacturing methods and other AM processes to evaluate their potential applicability in the aviation industry. To evaluate the mechanical properties of the 3D-printed parts, several tests were conducted. Hardness measurements were performed to assess the resistance of the material to indentation and penetration, and tensile testing was performed to determine the ultimate tensile strength (UTS), yield strength (YS) and strain at break values. Impact testing was also carried out to evaluate the material toughness under high-stress conditions. Finally, density measurements were conducted to evaluate the porosity of the samples, and the fractured surfaces of the samples were analyzed to determine the fracture mechanisms.

2. Materials and Methods

The selection of process parameters in our study was a result of a comprehensive approach. Initially, we extensively reviewed the existing literature in the field to identify commonly used and well-established process parameters employed in similar studies. These parameters served as a foundation for our research. Furthermore, we conducted preliminary experiments to evaluate the performance and feasibility of different process parameters. Via these experiments, we assessed their impact on the desired outcomes and determined the most appropriate range for our specific experimental setup. Practical constraints, such as equipment limitations, were also taken into account during the finalization of parameter selection. While we recognize that there are numerous process parameters that could potentially influence the outcomes, we carefully chose a subset that we believed would yield meaningful insights within the scope of our research objective.

The base material used for this research work was water-atomized 17-4PH stainless steel powder supplied by Oerlikon. The average particle size was 40 μm, which is widely used in AM and approved by both the German Federal Office of Civil Aeronautics and the German Federal Office of Defence Technology and Procurement. The morphology of the powder is mainly spherical, which is favorable for most additive manufacturing processes due to its excellent flowability and its high apparent density. However, as shown in Figure 1, the powder contained some particles with an irregular shape, which reduced the flowability and the apparent density of the powder bed. Furthermore, this can negatively affect the compactness of the 3D-printed parts. Yet, these particles are similar in shapes and sizes to those observed in the work of P. Ponnusamy et al. [16], in which the same alloy was considered. The official certificate (specified in the datasheet of the powder alloy) and the measured (ICP or inductively coupled plasma method) chemical composition of the initial powder are shown in Table 2.

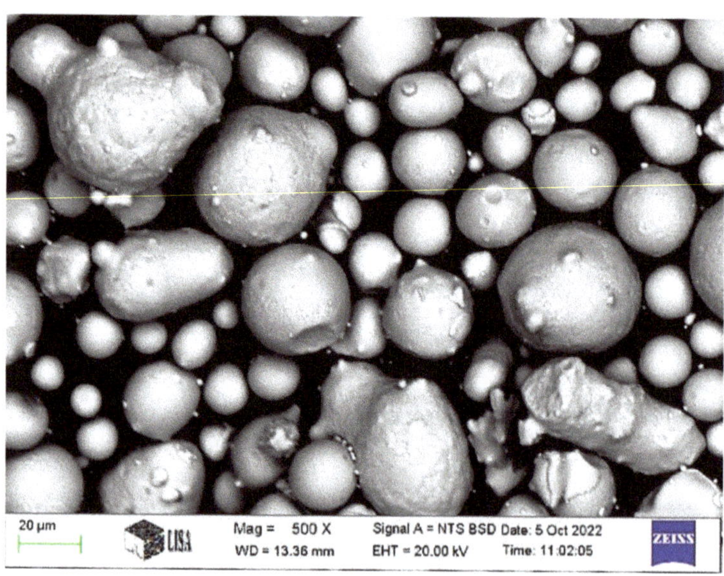

Figure 1. Morphology of 17-4 PH SS powder at 500× magnification.

Table 2. Chemical composition of 17-4 PH stainless steel powder (wt. %).

	C	Cr	Cu	Ni	Nb	Ta	Mn	Si	Fe
Measured	0.01	17.1	3.03	4.46	0.27	<0.01	-	-	Bal.
Certification	0.02	16.52	3.94	4.47	0.30	-	0.04	0.43	Bal.

The LPBF machine used in this experiment was an Orlas–Creator SLM machine, using an Yb fiber, a 1070 nm wavelength laser as an energy source, nitrogen shield gas and a ~40 µm average accuracy. The printing strategy is a scan of parallel lines with a 45° clockwise rotation for each deposited layer, thus creating a crosshatch pattern, as depicted in Figure 2. This strategy is typically used in 3D printing [16,17] to improve the strength and quality of printed parts.

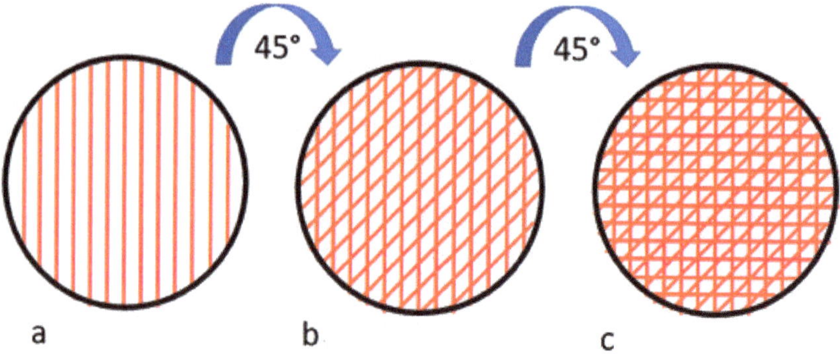

Figure 2. Crosshatch pattern achieved by 45-degree-clockwise rotation: (**a**) first deposited layer; (**b**) second deposited layer; (**c**) third deposited layer.

The selected pattern improves the adhesion between the layers and prevents delamination. Delamination is a serious production flaw that occurs when there is improper bonding between successive layers, leading to permanent deformation due to residual stresses. In order to execute this approach, it is necessary to adjust the settings of the 3D printer software to position the heat source at a 45-degree angle relative to the previous scanning direction. This configuration involves a complete 360-degree rotation every eight consecutive layers. This method enables the machine to maintain control over the desired structural integrity, resulting in improved mechanical properties and a more seamless layer construction. Additionally, the powder feed was supplied using a rubber coater, which moved in counterclockwise circular motions. Tensile specimens in the shape of rectangular dog bones and Charpy specimens were printed, adhering to the dimensions depicted in Figure 3. In the case of the Charpy specimens, the V notch was machined after the printing.

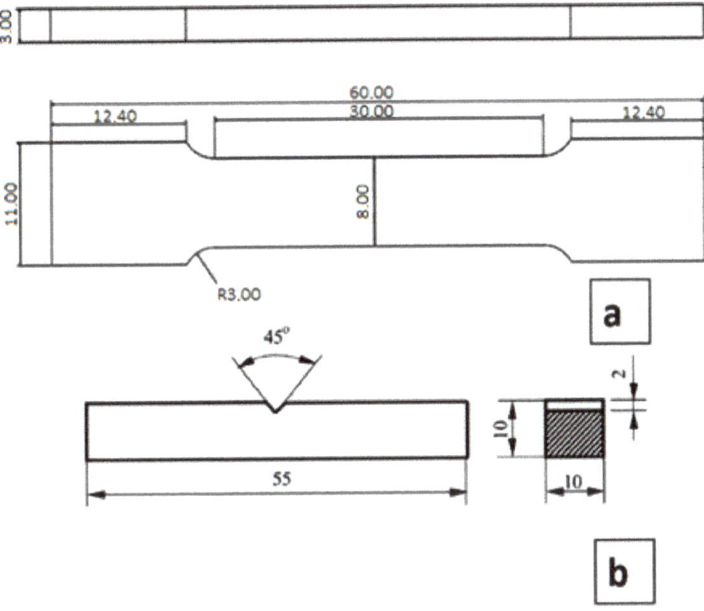

Figure 3. Dimensions and shapes of the printed (**a**) and tensile (**b**) Charpy specimen.

All printed samples were made on a horizontal (0°) orientation as it has been proven to be the favorable orientation for this type of testing [18]. Six as-printed tensile specimens and Charpy specimens were solution-heat-treated (annealed) at 1038 °C for 30 min in argon atmosphere and then quenched with water. Three of the annealed specimens were subsequently aged at 480 °C for 60 min in argon atmosphere. The heating rate was 10 °C/min in both heat treatment processes. The density of the samples was determined by the Archimedes method. The hardness of the samples was measured using the Vickers method (HV10) [19] on a Wolpert UH930. Instron 5982 equipment was utilized for the tensile tests at room temperature. Tensile tests were conducted according to ASTM E8/E8M standard [20]. Charpy tests were conducted according to ASTM E23-16b [21], using a Schenck-Trebel of 300 J capacity. Three tests were carried out on each condition. The strain rate was 3 mm/min. The microstructure and the broken surface of the tested samples were investigated using optical microscopy (OM) and scanning electron microscopy (SEM) with C. Zeiss Axio Image and a C. Zeiss EVO MA 10 equipment, respectively. The experimental flowchart of the additively manufactured 17–4PH stainless steel is shown in Figure 4.

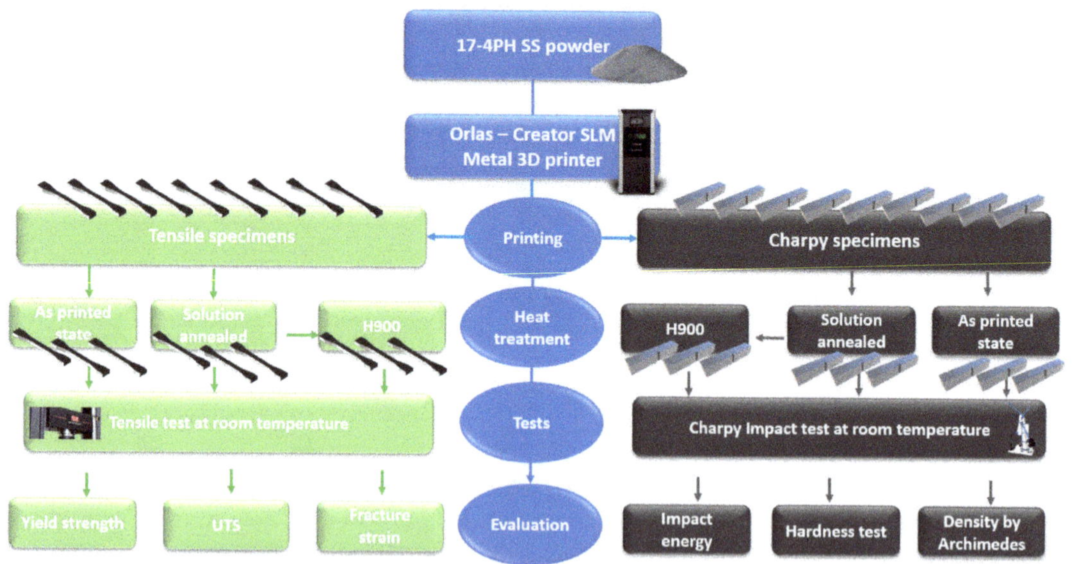

Figure 4. Experimental procedure conducted on the (AM)17-4PH stainless steel tensile test specimens and Charpy specimens. The arrows represent the life cycle of each test piece.

3. Results

3.1. Density and Microstructure

The bulk density of the porosity-free, CP-wrought 17-4PH stainless steel was measured to be 7.75 g/cm^3 [22]. The density of the AM products shows generally a lower value compared to the CP metals. In conventional processes (CP), a substantial amount of plastic deformation is applied, which helps eliminate the porosity present in the initial cast ingot. In contrast, in the field of additive manufacturing (AM), the use of long pressureless sintering or hot isostatic pressing (HIP) can reduce porosity, but it is inevitable to have some remaining porosities in the structure. This can significantly reduce the deformability as well as the strength and toughness of the material. The overall density results of the produced samples can be seen in Table 3, where the measured density and the calculated relative density values were also determined. The results show that the porosity lies between 1.5 and 2.7%. This is close to the literature's data for sintering processes [23–25]. The applied heat treatments did not have any further effects on the density.

Table 3. Density of LPBF 17-4PH SS Charpy samples.

State	Sample	ρ Measured (g/cm^3)	ρ Relative (%)
as printed	1.	7.54	97.30
	2.	7.63	98.46
	3.	7.52	96.98
annealed	4.	7.58	97.76
	5.	7.56	97.61
	6.	7.56	97.58
H900	7.	7.55	97.40
	8.	7.54	97.27
	9.	7.57	97.70

As depicted in Figure 5, the micrograph reveals the presence of two types of porosity within the microstructure. The majority of these voids exist on a microscale, with diameters less than 1–2 μm. However, there are a few voids that possess larger sizes but are less abundant. This type of porosity arises from inadequate melting during the 3D printing process, which will be further discussed in the analysis of fractured surfaces. Typically, the larger voids are observed along the outer rim of the specimen (Figure 5b). This phenomenon is well known in the realm of LPBF techniques as the process involves precise edge definition. With each layer, a closing contour is added, typically one (originally three). Although this contour addition is a machine-specific parameter and unavoidable, it does not significantly impact the mechanical properties.

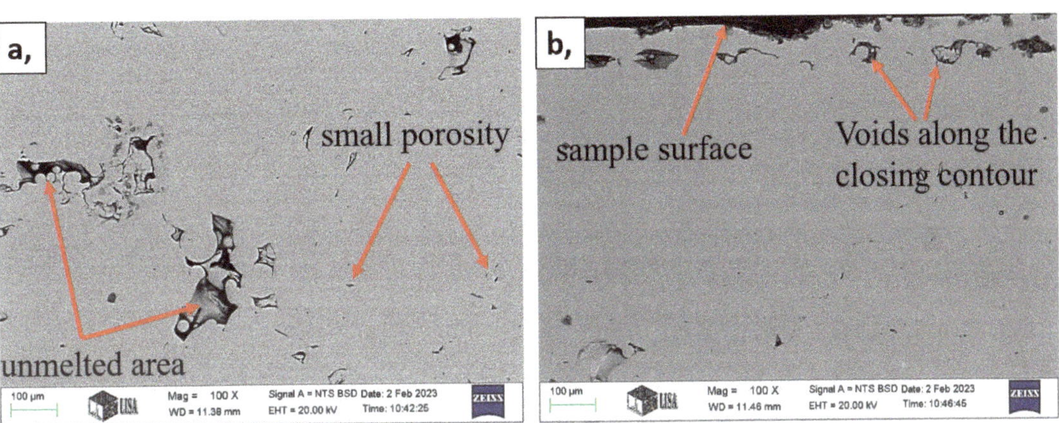

Figure 5. Typical porosity morphology of the LPBF 3D-printed samples: (**a**) inside the sample; (**b**) close to the sample surface.

3.2. Hardness and Toughness

The hardness and toughness values of the tested Charpy samples under different heat-treated conditions are presented in Figure 6. The results indicate that the as-printed state exhibits the lowest hardness and the highest toughness. The average hardness of the as-printed samples is insufficient for most engineering applications. However, the advantage of this alloy lies in its ability to be hardened by heat treatment. Prior to the aging heat treatment, annealing is performed. This heat treatment aims to maintain all the alloying elements in a solid solution state achieved by rapid cooling from the austenitic phase. Typically, this is the softest state of the material. However, as demonstrated by the results, the hardness of the annealed samples is higher than that of the as-printed state. According to the ASTM A693 standard, the maximum hardness is obtained in the annealed state, with a Rockwell hardness of 38 (HRC) or a corresponding Vickers hardness of 348 (HV). Our measured value of 314 (HV10) slightly falls below this maximum value. In practical terms, the hardest state of this alloy can be achieved by the H900 aging treatment. In this state, the hardness values usually range between 388 HV and 458 HV [26–28]. In our case, the average hardness of the aged samples is close to the maximum value of 445 (HV10). Regarding the measured impact energy (IE) values, an opposite trend can be observed.

This observed trend can be attributed to the fact that harder samples have a higher susceptibility to fracture when subjected to high-impact forces, resulting in lower impact energy (IE) values. The as-printed state exhibits the highest IE value of 22 Joules. However, after the annealing process, this value decreases to 9 joules. The lowest IE value of 6 joules was measured in the H900 state. In comparison to the results reported in the literature and the standard for the CP 17-4PH alloy, this impact energy is relatively low. The presence of unmelted areas within the printed microstructure (Figure 7) is undoubtedly the reason behind this discrepancy. These areas significantly reduce the toughness of the material [31].

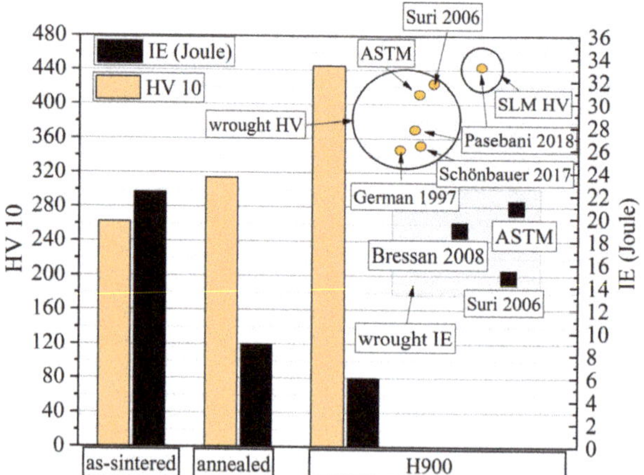

Figure 6. Hardness and impact energy at different heat-treated states [26–30]. Values from the same fields were marked with rectangle and circles.

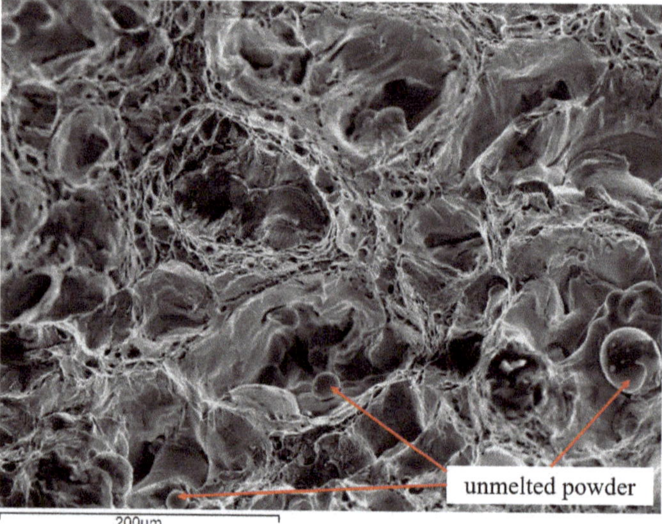

Figure 7. SEM picture on a broken surface of a Charpy specimen.

3.3. Tensile Properties

The typical tensile curves for the three different heat-treated conditions of the investigated material are depicted in Figure 8 and Table 4. It was determined that the heat treatment had a notable impact on the plastic deformation behavior of the material, as evidenced by the shape of the curves. In the case of the as-printed sample, a slight strain hardening effect can be observed between the yield strength (YS) and ultimate tensile strength (UTS). This curve exhibits the highest level of uniform plastic deformation and the lowest stress values. Following the UTS, a short period of non-uniform plastic deformation, known as necking, can be observed. Conversely, the annealed sample displays a different pattern. While the strain hardening period is short, the necking period is longer. The curve for the H900 sample exhibits a distinct curvature. After yielding, the tensile stress shows a sustained elongation phase. No further necking is observed after the UTS, and the sample fractures rapidly. The trend observed in the hardness and toughness values is also reflected in the tensile properties. The strength of the samples increases as a result of the heat treatment, while the plastic deformability experiences a significant decrease.

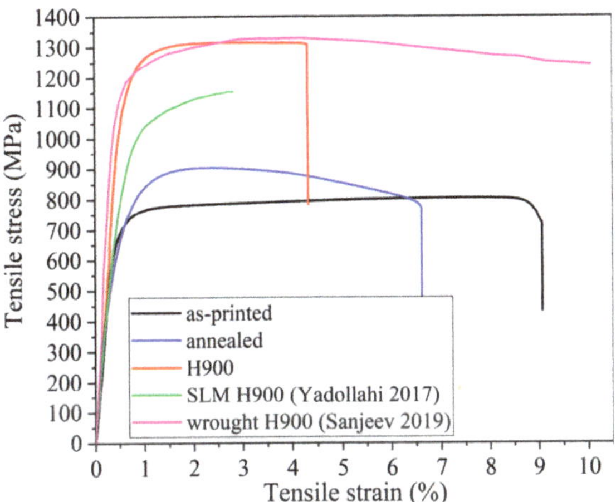

Figure 8. Tensile curves of 17-4ph LPBF samples [31,32].

Table 4. Tensile properties of LPBF specimens in various conditions.

CONDITION	YS (MPA)	UTS (MPA)	EL (%)	YS/UTS
AS PRINTED	712 ± 6	796 ± 8	9.4 ± 0.6	0.89
ANNEALED	714 ± 5	904 ± 6	6.1 ± 0.4	0.79
H900 (AGED)	1190 ± 16	1309 ± 13	3.7 ± 0.7	0.91

The values for yield strength (YS), ultimate tensile strength (UTS) and strain at break obtained from the tensile curves, along with the H900 value, are presented in Figure 9. The figure also includes data from the literature on CP- and AM-produced samples in different aged states. By comparing the results of the CP and AM alloy, as depicted in the diagram, it can be observed that the highest UTS can be achieved by the H900 treatment (1309 MPa), which is nearly equivalent to that of traditional manufacturing methods (1310 MPa). Generally, AM processes are often promoted as post-production free processes, but the as-built condition lacks the necessary mechanical properties for practical use. A comparison was made with similar results reported in the literature. In their respective studies, H.R. Lashgari et al. [33] and F.R. Andreacola et al. [14] applied heat treatment to their specimens, resulting in UTS values close to 1300 MPa, while the as-printed mechanical strength was approximately 900 MPa.

Figure 10 provides optical microscopy (OM) and scanning electron microscopy (SEM) micrographs of the fracture surfaces of the tensile specimens. Upon comparing the fracture surfaces, it becomes evident that the fracture mechanism differs. It is apparent that all the fractured surfaces exhibit a notable presence of porosity accompanied by unmelted powder. This is clearly visible in Figure 10d, although a similar amount can be observed on all other investigated surfaces. As indicated by the tensile curves, the curves corresponding to the as-printed and H900 aged samples display a short period of necking in comparison to the curve of the annealed sample. A short necking period implies that the fracture occurs without significant plastic deformation. This can be observed in Figure 10a,c,d,f, where cleavage fracture surfaces are depicted. Conversely, the tensile curve of the annealed sample exhibits a prolonged necking period. This substantial amount of plastic deformation is visible on the ductile-type fractured surfaces shown in Figure 10b,e.

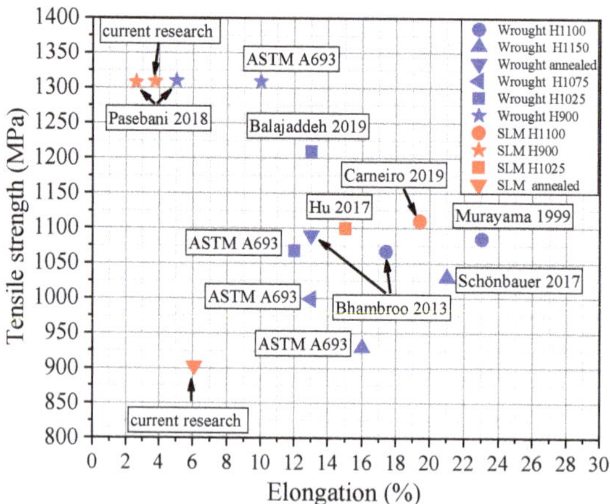

Figure 9. Tensile properties of the different values reported in the literature. CP parts are marked with blue, while AM parts are marked with red [27,29,34–38].

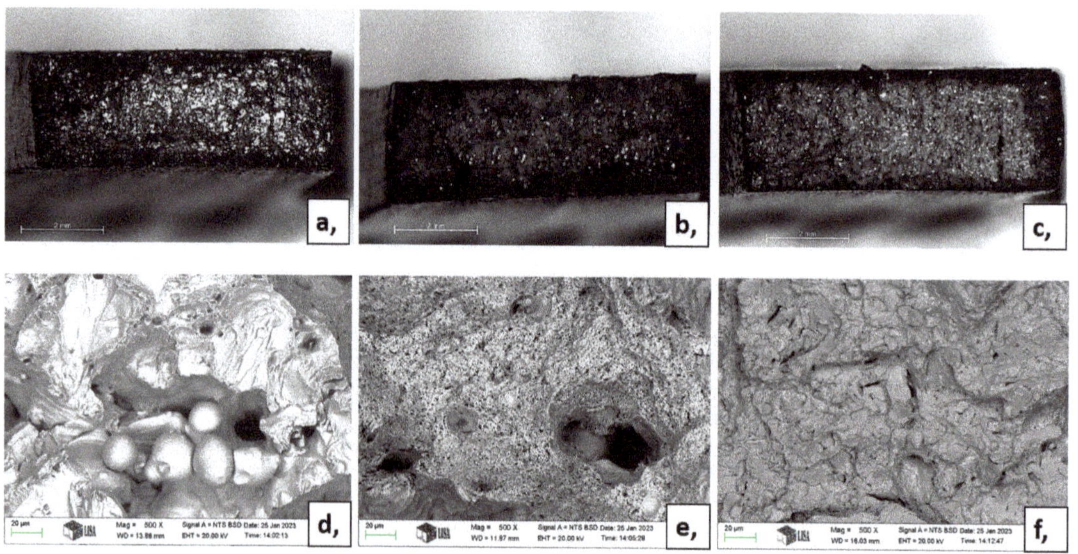

Figure 10. Fractured surfaces of (**a**,**d**) as-printed, (**b**,**e**) annealed and (**c**,**f**) H900 tensile specimens.

4. Conclusions

The present work focused on the production of LPBF samples with a desired orientation towards enhancing their properties in a specific direction. 17-4PH SS powder has been used for the production of the samples, being a widely used material for aeronautical applications, such as landing gear components wing spars and engine mounts, as well as for producing control rods, exhaust components, cockpit fasteners or even simple brackets. Such parts are predominantly loaded in one direction. By optimizing the properties of an additively manufactured 17-4PH alloy, it may be possible to produce parts with lighter and more efficient designs that can replace conventional parts made using traditional manufacturing methods. The properties of the produced specimens have been compared with those of corresponding parts from the literature made via traditional manufacturing methods and other AM processes. While the aviation industry places high standards on conventional crafting methods, their potential for increased complexity is limited. However, powder-based methods offer a significant improvement in this regard. Based on the results of the tensile tests, notable differences in characteristics are observed between parts made conventionally and those produced using our AM process. CP parts, in their as-made condition, exhibit superior tensile strength properties but lack elongation. On the other hand, LPBF-produced parts, when subjected to treatment, demonstrate competitive properties in this aspect. Moreover, the narrow standard deviation of the LPBF values indicates the stability of this manufacturing method. Our findings also suggest that additional treatments can enhance the potential applicability of these powder-based parts, particularly in terms of ultimate tensile strength (UTS) and yield strength (YS). This aligns with the requirement stated in MIL-STD-1530D, which specifies no yielding at the 100% design limit load [6] while still allowing for optimal design freedom to achieve greater complexity. Despite the printing jobs not being 100% accurate in terms of surface finish and the occasional presence of unmelted particles resulting in 2-3% porosity, the results are remarkable compared to other AM technologies, as supported by values reported in the literature. These unmelted zones were subsequently minimized by annealing and aging heat treatments. In conclusion, additively manufactured 17-4 PH parts have the potential to find application in the aviation sector, although further investigations are necessary to assess their actual applicability. Future research endeavors may focus on improving the stability of 3D printing jobs or exploring alternative heat treatment methods, considering their potential impact on the mechanical properties of AM parts. Additive manufacturing (AM) has the potential to revolutionize the way parts are designed and produced, particularly in industries like aerospace in which weight reduction is a critical factor in improving performance and fuel efficiency, which is also emphasized in US Army Directive 2019-29 [39]. By using AM to produce parts with reimagined, lighter geometries, it may be possible to reduce the overall weight of aircraft and spacecraft, resulting in significant improvements in fuel efficiency and cost savings. Moreover, the utilization of AM for part production brings additional advantages, including decreased material waste, enhanced design flexibility and expedited production times. These benefits make AM a promising alternative to conventional manufacturing methods in various industries, including aerospace. Nevertheless, it is important to acknowledge that the adoption of AM in the aerospace sector presents certain challenges that must be overcome. Quality control, certification processes and material limitations are among the key issues that need to be addressed before AM can be widely embraced in the industry.

Author Contributions: Conceptualization, Z.G. and T.M.; methodology, S.E.K.; software, S.E.K. and T.M.; validation, S.E.K. and T.M.; formal analysis, S.E.K., T.M., D.M. and E.T.; investigation, S.E.K. and T.M.; resources, S.E.K., T.M. and Z.G.; data curation, S.E.K., T.M., D.M. and E.T.; writing—original draft preparation, S.E.K.; writing—review and editing, T.M., D.M., E.T. and D.P.; visualization, S.E.K. and T.M.; supervision, Z.G. and L.V.; project administration, G.G.; funding acquisition, Z.G. All authors have read and agreed to the published version of the manuscript.

Funding: The work was carried out as part of the UMA3 project funded by the European Union's Horizon 2020 research and innovation program under grant agreement No 952463.

Data Availability Statement: The data presented in this study are available in the following paper: "Additive Manufacturing of 17-4PH Alloy: Tailoring Printing Orientation for Enhanced Aerospace Application Performance".

Acknowledgments: The experiments were carried out at the University of Miskolc, Faculty of Materials and Chemical Engineering, Institute of Physical Metallurgy, Metalforming and Nanotechnology and Institute of Foundry Engineering.

Conflicts of Interest: The authors declare no conflict of interest.

References

1. Nickels, L. AM and aerospace: An ideal combination. *Met. Powder Rep.* **2015**, *70*, 300–303. [CrossRef]
2. Berrocal, L.; Fernández, R.; González, S. Topology optimization and additive manufacturing for aerospace components. *Prog. Addit. Manuf.* **2019**, *4*, 83–95. [CrossRef]

3. Gibson, I.; Rosen, D.W.; Stucker, B. Additive Manufacturing Technologies: 3D Printing. In *Rapid Prototyping, and Direct Digital Manufacturing*, 2nd ed.; Springer: Berlin/Heidelberg, Germany, 2015.
4. Liu, R.; Wang, Z.; Sparks, T.; Liou, F.; Newkirk, J. Aerospace applications of laser additive manufacturing. In *Laser Additive Manufacturing*; Woodhead Publishing: Cambridge, UK, 2017; pp. 351–371. [CrossRef]
5. USAF Structures Bulletin EZ-SB-19-01, Durability and Damage Tolerance Certification for Additive Manufacturing of Aircraft Structural Metallic Parts, Wright Patterson Air Force Base, OH, USA, 10 June 2019. Available online: https://daytonaero.com/wp-content/uploads/EZ-SB-19-01.pdf (accessed on 27 February 2023).
6. *MIL-STD-1530D*; Department of Defense Standard Practice Aircraft Structural Integrity Program (ASIP). Military and Government Specs & Standards (Naval Publications and Form Center) (NPFC): New York, NY, USA, 2016.
7. Kundu, S.; Jones, R.; Peng, D.; Matthews, N.; Alankar, A.; Raman, S.R.K.; Huang, P. Review of Requirements for the Durability and Damage Tolerance Certification of Additively Manufactured Aircraft Structural Parts and AM Repairs. *Materials* **2020**, *13*, 1341. [CrossRef] [PubMed]
8. Zerbst, U.; Bruno, G.; Buffiere, J.Y.; Wegener, T.; Niendorf, T.; Wu, T.; Zhang, X.; Kashaev, N.; Meneghetti, G.; Hrabe, N.; et al. Damage tolerant design of additively manufactured metallic components subjected to cyclic loading: State of the art and challenges. *Prog. Mater. Sci.* **2021**, *121*, 100786. [CrossRef]
9. Khorasani, M.; Ghasemi, A.; Leary, M.; Downing, D.; Gibson, I.; Sharabian, E.G.; Veetil, J.K.; Brandt, M.; Bateman, S.; Rolfe, B. Benchmark models for conduction and keyhole modes in laser-based powder bed fusion of Inconel 718. *Opt. Laser Technol.* **2023**, *164*, 109509. [CrossRef]
10. Lavecchia, F.; Pellegrini, A.; Galantucci, L.M. Comparative study on the properties of 17-4 PH stainless steel parts made by metal fused filament fabrication process and atomic diffusion additive manufacturing. *Rapid Prototyp. J.* **2023**, *29*, 393–407. [CrossRef]
11. Gigànto, S.; Martínez-Pellitero, S.; Barreiro, J.; Leo, P.; Ángeles Castro-Sastre, M. Impact of the laser scanning strategy on the quality of 17-4PH stainless steel parts manufactured by selective laser melting. *J. Mater. Res. Technol.* **2022**, *20*, 2734–2747. [CrossRef]
12. Ponnusamy, P.; Sharma, B.; Masood, S.H.; Rahman Rashid, R.A.; Rashid, R.; Palanisamy, S.; Ruan, D. A study of tensile behavior of SLM processed 17-4 PH stainless steel. *Mater. Today Proc.* **2021**, *45*, 4531–4534. [CrossRef]
13. Liverani, E.; Lutey, A.; Ascari, A.; Fortunato, A. The effects of hot isostatic pressing (HIP) and solubilization heat treatment on the density, mechanical properties, and microstructure of austenitic stainless steel parts produced by selective laser melting (SLM). *Int. J. Adv. Manuf. Technol.* **2020**, *107*, 109–122. [CrossRef]
14. Andreacola, F.R.; Capasso, I.; Pilotti, L.; Brando, G. Influence of 3D-printing parameters on the mechanical properties of 17-4PH stainless steel produced through Selective Laser Melting. *Frat. Ed Integrità Strutt.* **2021**, *58*, 282–295. [CrossRef]
15. Raj, S.; Ghosn, L.; Lerch, B.; Hebsur, M.; Cosgriff, L.; Fedor, J. Mechanical properties of 17-4PH stainless steel foam panels. *Mater. Sci. Eng. A* **2007**, *456*, 305–316. [CrossRef]
16. Ponnusamy, P.; Masood, S.H.; Palanisamy, S.; Rahman Rashid, R.A.; Ruan, D. Characterization of 17-4PH alloy processed by selective laser melting. *Mater. Today Proc.* **2017**, *4*, 8498–8506. [CrossRef]
17. Liu, C.Y.; Tong, J.D.; Jiang, M.G.; Chen, Z.W.; Xu, G.; Liao, H.B.; Wang, P.; Wang, X.Y.; Xu, M.; Lao, C.S. Effect of scanning strategy on microstructure and mechanical properties of selective laser melted reduced activation ferritic/martensitic steel. *Mater. Sci. Eng. A* **2019**, *766*, 138364. [CrossRef]
18. Aripin, M.A.; Sajuri, Z.; Jamadon, N.H.; Baghdadi, A.H.; Syarif, J.; Mohamed, I.F.; Aziz, A.M. Effects of Build Orientations on Microstructure Evolution, Porosity Formation, and Mechanical Performance of Selective Laser Melted 17-4 PH Stainless Steel. *Metals* **2022**, *12*, 1968. [CrossRef]
19. *ASTM E 92*; Standard Test Method for Vickers Hardness of Metallic Materials. ASTM International: West Conshohocken, PA, USA, 2017.
20. *ASTM E8/E8M-22*; Standard Test Methods for Tension Testing of Metallic Materials. ASTM International: West Conshohocken, PA, USA, 2022.
21. *ASTM E23-16b*; Standard Test Methods for Notched Bar Impact Testing of Metallic Materials. ASTM International: West Conshohocken, PA, USA, 2018.
22. *ASTM A693-16(2022)*; Standard Specification for Precipitation-Hardening Stainless and Heat-Resisting Steel Plate, Sheet, and Strip. ASTM International: West Conshohocken, PA, USA, 2022.
23. Simchi, A.; Rota, A.; Imgrund, P. An investigation on the sintering behavious of 316L and 17-4ph stainless steel powders for graded composites. *Mater. Sci. Eng. A* **2006**, *424*, 282–289. [CrossRef]
24. Szewczyk-Nykiel, A.; Gadek, S.; Hebda, M.; Nykiel, M.; Pieczonka, T.; Kazior, J. Influence of Sintering Atmosphere on Densification Development of 17-4PH stainless steel powder. *Materials* **2023**, *16*, 760. [CrossRef]
25. Wu, Y.; Blaine, D.; Marx, B.; Schlaefer, C.; German, R.M. Sintering Densification and Microstructural Evolution of Injection Molting Grade 17-4 PH Stainless steel Powder. *Metall. Mater. Trans. A* **2002**, *33*, 2185–2194. [CrossRef]
26. Suri, P.; Smarslok, B.P.; German, R.M. Impact properties of sintered and wrought 17-4 PH stainless steel. *Powder Metall.* **2006**, *49*, 40–47. [CrossRef]
27. Pasebani, S.; Ghayoor, M.; Badwe, S.; Irrinki, H.; Atre, S.V. Effects of atomizing media and post processing on mechanical properties of 17-4 PH stainless steel manufactured via selective laser melting. *Addit. Manuf.* **2018**, *22*, 127–137. [CrossRef]

28. German, R.M. Thermal Processing Optimization of Injection Molded Stainless Steel Powders. *Mater. Manuf. Process.* **1997**, *12*, 713–735. [CrossRef]
29. Schönbauer, B.M.; Yanase, K.; Endo, M. The influence of various types of small defects on the fatigue limit of precipitation-hardened 17-4PH stainless steel. *Theor. Appl. Fract. Mech.* **2017**, *87*, 35–49. [CrossRef]
30. Bressan, J.D.; Daros, D.P.; Sokolwski, A.; Mesquita, R.A.; Barbosa, C.A. Influence of hardness on the wear resistance of 17-4 PHstainless steel evaluated by the pin-on-disc testing. *J. Mater. Process. Technol.* **2008**, *205*, 353–359. [CrossRef]
31. Sanjeev, K.C.; Nezhadfar, P.D.; Phillips, C.; Kennedy, M.S.; Shamsei, N.; Jackson, R.L. Tribological behavior of 17–4 PH stainless steel fabricated by traditional manufacturing and laser-based additive manufacturing methods. *Wear* **2019**, *440–441*, 203100.
32. Yadollahi, A.; Shamsaei, N.; Thompson, S.M.; Elwany, A.; Bian, L. Effects of building orientation and heat treatment on fatigue behavior of selective laser melted 17-4 PH stainless steel. *Int. J. Fatigue* **2017**, *94*, 218–235. [CrossRef]
33. Lashgari, H.R.; Adabifiroozjaei, E.; Kong, C.; Molina-Luna, L.; Li, S. Heat treatment response of additively manufactured 17-4PH stainless steel. *Mater. Charact.* **2023**, *197*, 112661. [CrossRef]
34. Murayama, M.; Hono, K.; Katayama, Y. Microstructural evolution in a 17-4 PH stainless steel after aging at 400 °C. *Metall. Mater. Trans. A* **1999**, *30*, 345–353. [CrossRef]
35. Carneiro, L.; Jalalahmadi, B.; Ashtekar, A.; Jiang, Y. Cyclic deformation and fatigue behavior of additively manufactured 17-4 PH stainless steel. *Int. J. Fatigue* **2019**, *123*, 22–30. [CrossRef]
36. Bhambroo, R.; Roychowdhury, S.; Kain, V.; Raja, V.S. Effect of reverted austenite on mechanical properties of precipitation hardenable 17-4 stainless steel. *Mater. Sci. Eng. A* **2013**, *568*, 127–133. [CrossRef]
37. Balajaddeh, B.M.; Naffakh-Moosavy, H. Pulsed Nd:YAG laser welding of 17-4 PH stainless steel: Microstructure, mechanical properties, and weldability investigation. *Opt. Laser Technol.* **2019**, *119*, 105651. [CrossRef]
38. Hu, Z.; Zhu, H.; Zhang, H.; Zeng, X. Experimental investigation on selective laser melting of 17-4PH stainless steel. *Opt. Laser Technol.* **2017**, *87*, 17–25. [CrossRef]
39. US Army Directive 2019-29, Enabling Readiness and Modernization Through Advanced Manufacturing, Secretary of The Army, Pentagon, Washington DC, 18 September 2019. Available online: https://armypubs.army.mil/epubs/DR_pubs/DR_a/pdf/web/ARN19451_AD2019-29_Web_Final.pdf (accessed on 27 February 2023).

MDPI
St. Alban-Anlage 66
4052 Basel
Switzerland
Tel. +41 61 683 77 34
Fax +41 61 302 89 18
www.mdpi.com

Aerospace Editorial Office
E-mail: aerospace@mdpi.com
www.mdpi.com/journal/aerospace

www.ingramcontent.com/pod-product-compliance
Lightning Source LLC
LaVergne TN
LVHW071619170726
843515LV00009B/2434

* 9 7 8 3 0 3 6 5 8 3 7 0 9 *